XLVI International Symposium on Multiparticle Dynamics (ISMD 2016)

EPJ Web of Conferences Volume 141 (2017)

Jeju Island, South Korea
29 August - 2 September 2016

Editors:

In-Kwon Yoo

ISBN: 978-1-5108-3940-3

Printed from e-media with permission by:

Curran Associates, Inc.
57 Morehouse Lane
Red Hook, NY 12571

Some format issues inherent in the e-media version may also appear in this print version.

Printed by Curran Associates, Inc. (2017)

For additional information, please contact EDP Sciences – Web of Conferences
at the address below.

EDP Sciences – Web of Conferences
17, Avenue du Hoggar
Parc d'Activité de Courtabœuf
BP 112
F-91944 Les Ulis Cedex A
France

Phone: +33 (0) 1 69 18 75 75
Fax: +33 (0) 1 69 28 84 91

contact-edps@webofconferences.org

Additional copies of this publication are available from:

Curran Associates, Inc.
57 Morehouse Lane
Red Hook, NY 12571 USA
Phone: 845-758-0400
Fax: 845-758-2633
Email: curran@proceedings.com
Web: www.proceedings.com

TABLE OF CONTENTS

Preface

March 10, 2017

The international Symposium on Multiparticle Dynamics (ISMD) is a series of major international physics conferences for theorists and experimentalists from three different, but closely related, communities, i.e., high energy, heavy-ion and astro-particle physics, with a common interest in reactions involving a large number of particles in the final state. The 46th International Symposium on Multiparticle Dynamics (ISMD2016)[1] was held from August 29 to September 2, 2016 at the Seogwipo KAL Hotel in Jeju Island, Republic of Korea. The symposium was hosted by a joint committee of the Korean nuclear, high-energy, and astrophysics communities. Total 99 researchers, including 39 young postdocs and students, from 17 countries were participated in ISMD2016.

During ISMD2016 total 63 plenary talks and 18 posters were presented in the following 9 subjects:

- Multi-particle correlations and fluctuations: from small to large systems
- Hadronic final state in high-p_T interactions
- Forward physics and diffraction
- Perturbative and non-perturbative features of QCD
- Collectivity in high-energy collisions: jets, flows and others
- Proton structure, small and large-x physics
- Cosmic ray and astroparticle physics
- Special Session I: Future facility
- Special Session II: Computing

For the first time in a series, the organizing committee set up the flash talk session in the first day of the symposium so that all poster presenters had a chance to speak about their original work in front of all participants. Especially, the Franco Rimondi Association in Italy generously donated 1,000 euros for the best poster award. The poster evaluation committee, chaired by Fabrizio Fabbri from INFN-Bologna, was formed. The committee members carefully evaluated all posters together with their flash talks, and finally selected Yeonju Go, Songkyo Lee and Jaebeom Park from Korea University and Myungkuk Kim from Pusan National University as co-winners. Each winner was awarded the certificate and the cash prize shared equally in the symposium banquet.

The members of Board of Elders (BoE) and the International Advisory Committee (IAC) not only provided useful guides to organize the symposium but also played a key role to select session conveners. Total 21 experimentalists and

---

[1] http://nuclear.korea.ac.kr/ismd2016

theorists served as session conveners, who worked tirelessly to invite plenary speakers and set up the best scientific program.

The enormous contributions by members of the Local Organizing Committee (LOC) were greatly acknowledged in all aspects: from financial support to publication of proceedings. Many volunteer students, especially, from Korea University took care of the operational details in the whole period of the symposium. The staff members, Ms. Eunyoung Choi in particular, from Asia Pacific Center for Theoretical Physics (APCTP) were also acknowledged for taking care of the registration.

We also wish to acknowledge generous supports from Korea University, Pusan National University, Jeju National University, Korean CMS Collaboration, Korean ALICE collaboration, APCTP, Rare Isotope Science Project (RISP), Korean Institute of Science and Technology Information (KISTI) and Global Science Data Center (GSDC).

Finally, we publish the proceeding contributions that have been subjected to peer review and accepted after reasonable revisions requested by conveners of each session.

The next ISMD will take place in the city of Tlaxcala, Mexico, in September 11-15, 2017.

Byungsik Hong (Korea Univ.) Chair

In-Kwon Yoo (Pusan Nat. Univ.) Co-chair

Committees

Board of Elders

Siegfried Bethke
Andrzej Białas
Martin Block
Wojciech Broniowski
Sergei Chekanov
Tamás Csörgő
Jorge Dias de Deus
Eddi de Wolf
Igor Dremin
Fabrizio Fabbri
Krzysztof Fiałkowski
William Gary
Paolo Giacomelli
Alberto Giovannini
Gösta Gustafson
Yogiro Hama
Kensuke Homma
László Jenkovszky
Hannes Jung
Wolfram Kittel
Stefan Kluth
Takeshi Kodama
Elena Kokoulina
Vyatcheslav Kuvshinov
Pierre van
Wesley Metzger
Atsushi Nakamura
Giulia Pancheri
Carlos Pajares
Nick van Remortel
Ladislav Sandor
Ina Sarcevic
Norbert Schmitz
Vladislav Ŝimák
Zack Sullivan
Michal Sumbera
Chung-I Tan
Thomas Trainor
Alice Valkarova
Alan R. White
Georg Wolschin
Nu Xu
Yuanfang Wu

International Advisory Committee

Peter Braun-Munzinger (GSI, Germany)
Wojciech Broniowski (Kielce University, Poland)
Kuang-Ta Chao (Peking University, China)
Sergei Chekanov (ANL, USA)
Kiwoon Choi (IBS, Korea)
Tamás Csörgő (Wigner RCP Budapest and KRF Gyöngyös, Hungary)
Igor Dremin (Lebedev Physics Institute-Moscow, Russia)
Fabrizio Fabbri (INFN-Bologna, Italy)
William Gary (University of California-Riverside, USA)
Paolo Giacomelli (INFN-Bologna, Italy)
Gösta Gustafson (Lund University, Sweden)
Tetsuo Hatsuda (RIKEN, Japan)
Sangyong Jeon (McGill University, Canada)
Stefan Kluth (MPI-Munich, Germany)
Jungil Lee (Korea University, Korea)
Su Houng Lee (Yonsei University, Korea)

Larry McLerran (BNL, USA)
Wesley Metzger (Radboud University, Netherlands)
Giulia Pancheri (INFN-Frascati, Italy)
Gunther Roland (MIT, USA)
Hiroyuki Sagawa (University of Tokyo, Japan)
Norbert Schmitz (MPI-Munich, Germany)
Zack Sullivan (Illinois Institute of Technology, USA)
Chung-I Tan (Brown University, USA)
Thomas Trainor (University of Washington, USA)
Georg Wolschin (Heidelberg University, Germany)
Nu Xu (CCNU, China and LBNL, USA)

Local Organization Committee

Suyong Choi (Korea University)
Byungsik Hong (Korea University) Chair
Yong-Joo Kim (Jeju National University)
Byul Moon (Korea University) Web-manager
Chang-Hwan Lee (Pusan National University)
Inkyu Park (University of Seoul)
Jong-Kwan Woo (Jeju National University)
In-Kwon Yoo (Pusan National University) Co-chair
Un-Ki Yang (Seoul National University)
Jin-Hee Yoon (Inha University)

Poster Evaluation Committee

Fabbri, Fabrizio (INFN-Bologna) Chair
Gosta Gustafson (Lund University, Sweden)
Wesley Metzger (Radboud University, Netherlands)
Helmut Oeschler (Heidelberg University, Germany)
Fuqiang Wang (Purdue University, USA)
Alejandro Ayala (Universidad Nacional Autonoma de Mexico, Mexico)
ShinIchi Esumi (University of Tsukuba, Japan)

Session Conveners

- Multi-particle correlations and fluctuations: from small to large systems
  Ilya Selyuzhenkov, EMMI/GSI, Germany
  Soumya Mohapatra, Columbia University, USA
  Wei Li, Rice University, USA
  Grzegorz Wilk, National Centre for Nuclear Research, Poland
  Gabriel Silveira Denicol, Federal university Fluminense (UFF), Brazil
- Hadronic final state in high-p_T interactions

Un-Ki Yang, Seoul National University, Korea
Manjit Kaur, Panjab Uni, India
Nicolas Greiner, Zurich University, Switzerland

- Forward physics and diffraction
  Christophe Royon, University of Kansas, USA
  Hans Jozef H Van Haevermaet, University Antwerpen, Belgium
  Francesco Hautmann, Oxford University, UK

- Perturbative and non-perturbative features of QCD
  Maxim Gouzevitch, Lyon, France
  Bowen Xiao, Central China Normal University, China

- Collectivity in high-energy collisions: jets, flows or other
  Yen-Jie Lee, MIT, USA
  Minjung Kweon, Inha University, Korea
  Wojciech Broniowski, Jan Kochanowski University, Poland

- Proton structure, small and large-x physics
  Matthias Grosse-Perdekamp, UIUC, USA
  Yoshitaka Hatta, Kyoto University, Japan

- Cosmic ray and astroparticle physics
  Maurizio Spurio, Universita di Bologna, Italy
  Martin Pohl, DESY, Germany
  Hang Bae Kim, Hanyang University, Korea

edp sciences

EPJ Web of Conferences

Statement of Peer review

In submitting conference proceedings to EPJ Web of Conferences, I certify to the Publisher that I adhere to the **Policy on Publishing Integrity** of the journal in order to safeguard good scientific practice in publishing.

1. All articles have been subjected to peer review administered by the proceedings editors.
2. Reviews have been conducted by expert referees, who have been requested to provide unbiased and constructive comments aimed, whenever possible, at improving the work.
3. Proceedings editors have taken all reasonable steps to ensure the quality of the materials they publish and their decision to accept or reject a paper for publication has been based only on the merits of the work and the relevance to the journal.

Title, date and place of the conference

XLVI International Symposium on Multiparticle Dynamics (ISMD 2016) in Jeju island, South Korea

Proceedings editor(s):

Prof. IN-KWON YOO

Date and editor's signature Dec. 16, 2016

17, avenue du Hoggar - PA de Courtabœuf – BP 112 - 91944 Les Ulis Cedex A (France)
Tél. : 33 (0)1 69 18 75 75 - Fax : 33(0)1 69 07 45 17 – www.edpsciences.org

EPJ Web of Conferences **141**, 01001 (2017) DOI: 10.1051/epjconf/201714101001
ISMD 2016

News on collectivity in Pb-Pb collisions from the ALICE experiment

D.J Kim[1,a] for the ALICE Collaboration

[1] *University of Jyväskylä and Helsinki Institute of Physics (HIP), Finland*

Abstract. The collective expansion of the color-deconfined fireball created in ultra-relativistic heavy-ion collisions maps the initial state of the quark-gluon plasma (QGP) to the final-state particle spectrum. The ALICE experiment has been leading important roles for completing the individual flow harmonic measurements at the highest energies to date as well as improving flow harmonic correlation techniques to understand the properties of the QGP and the full evolution of the heavy-ion collisions. In this article, a brief summary of the individual flow harmonic measurements, the details of the new observables developed in recent years from ALICE collaboration and their implications to future studies are discussed.

1 Introduction

The main emphasis of the ultra-relativistic heavy-ion collisions at the Relativistic Heavy Ion Collider (RHIC) and the Large Hadron Collider (LHC) is to study deconfined phase of the strongly interacting nuclear matter, the Quark-Gluon Plasma (QGP). This matter exhibits strong collective and anisotropic flow in the plane transverse to the beam direction, which is driven by the anisotropic pressure gradients, resulting in more particles emitted in the direction of the largest gradients. The large elliptic flow discovered at RHIC energies [1] continues to increase also at LHC energies [2, 3]. This has been predicted by calculations utilizing viscous hydrodynamics [4–9]. These calculations also demonstrated that the shear viscosity to the entropy density ratio (η/s) of QGP is close to a universal lower bound $1/4\pi$ [10] in heavy-ion collisions at RHIC and LHC energies.

The temperature dependence of the η/s has some generic features that most of the known fluids obey. For instance, one such general behavior is that the ratio typically reaches its minimum value close to the phase transition region [11]. It was shown, using kinetic theory and quantum mechanical considerations [12], that $\eta/s \sim 0.1$ would be the correct order of magnitude for the lowest possible shear viscosity to entropy ratio value found in nature. Later it was demonstrated that an exact lower bound $(\eta/s)_{\min} = 1/4\pi \approx 0.08$ can be calculated using the AdS/CFT correspondence [10]. Hydrodynamical simulations support as well the view that the QGP matter is close to that limit [8]. This in turn may have important implications for other fundamental physics goals. It is argued that such a low value might imply that thermodynamic trajectories for the expanding matter would lie close to the quantum chromodynamics (QCD) critical end point, which is another subject of intensive experimental quest [11].

[a] e-mail: djkim@jyu.fi

Anisotropic flow [13] is traditionally quantified with harmonics v_n and corresponding symmetry plane angles Ψ_n in the Fourier series decomposition of the azimuthal particle distribution in the plane transverse to the beam direction [14]:

$$E\frac{d^3N}{dp^3} = \frac{1}{2\pi}\frac{d^2N}{p_T dp_T d\eta}\left\{1 + 2\sum_{n=1}^{\infty} v_n(p_T,\eta)\cos[n(\varphi - \Psi_n)]\right\}, \quad (1)$$

where E, N, p, p_T, φ and η are the energy, particle yield, total momentum, transverse momentum, azimuthal angle and pseudorapidity of particles, respectively, and Ψ_n is the azimuthal angle of the symmetry plane of the nth-order harmonic. The nth-order flow coefficients are denoted as v_n and can be calculated as $v_n = \langle\cos[n(\varphi - \Psi_n)]\rangle$, where the brackets denote an average over all particles in all events. The anisotropic flow in heavy-ion collisions is understood as the hydrodynamic response of produced matter to spatial deformations of the initial energy density profile [15]. This profile fluctuates event-by-event due to fluctuations of the positions of the constituents inside the colliding nuclei, which in turn implies that the flow also fluctuates [16, 17]. The recognition of the importance of flow fluctuations has led to measurements of triangular flow and higher flow harmonics [18, 19] as well as the correlations between different Fourier harmonics [20, 21]. The higher order harmonics are expected to be particularly sensitive to fluctuations in the initial conditions and to the η/s [22, 23], while correlations have the potential to discriminate the two respective contributions to anisotropic flow development [20].

However, difficulties on extracting η/s in heavy-ion collisions can be attributed mostly to the fact that it strongly depends on the specific choice of the initial conditions [4, 23, 24]. The viscous effects reduce the magnitude of the elliptic flow. Furthermore, the magnitude of η/s used in hydrodynamic model calculations should be considered as an average over the temperature history of the expanding fireball as it is known that η/s of other fluids depends on temperature. In addition, part of the elliptic flow can also originate from the hadronic phase [25–27]. Therefore, knowledge of both the temperature dependence and the relative contributions from the partonic and hadronic phases should be understood better to quantify η/s of the partonic fluid.

The ALICE experiment has been leading important roles for completing the individual flow harmonic measurements at the highest energies to date as well as improving flow harmonic correlation techniques to understand the properties of the QGP and the full evolution of the heavy-ion collisions. In this article, few selected results from recently published papers from the ALICE experiment are discussed in the following sections on the emphasis of extracting η/s.

2 Results

Firstly, we show the multiparticle observables, the Symmetric 2-harmonic 4-particle Cumulants (SC), which quantify the relationship between event-by-event fluctuations of two different flow harmonics [28]. These observables are particularly robust against few-particle non-flow correlations and they provide orthogonal information to recently analyzed symmetry plane correlators. It was demonstrated that they are sensitive to the temperature dependence of η/s of the expanding medium and therefore simultaneous descriptions of different order harmonic correlations would constrain both the initial conditions and the medium properties [28, 29]. These results are discussed in Sec. 2.1. Second, pseudorapidity dependent charge particle v_2, v_3 and v_4 are measured in wide range of pseudorapidity and shown in Sec. 2.2. These results can provide access to a range of varying medium properties, even at a fixed collision energy. Thirdly, the first results of the charged hadron v_n with the highest LHC beam energy are presented in Sec. 2.3.

2.1 Event-by-event fluctuations of two different flow harmonics

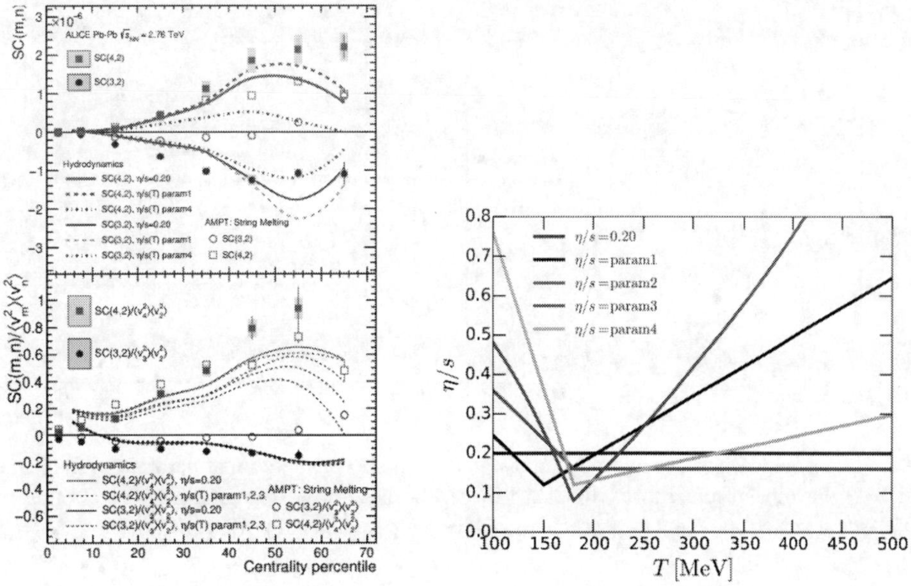

Figure 1. (Top Left) Comparison of observables SC(4,2) (red filled squares) and SC(3,2) (blue filled circles) to theoretical model from [30]. (Left Bottom) The results from top left panel have been rescaled with $\left\langle v_m^2\right\rangle\left\langle v_n^2\right\rangle$. (Right) The $\eta/s(T)$ parameterizations used in theoretical model [30].

The SC observables are defined as:

$$\begin{aligned}
\langle\langle\cos(m\varphi_1+n\varphi_2-m\varphi_3-n\varphi_4)\rangle\rangle_c &= \langle\langle\cos(m\varphi_1+n\varphi_2-m\varphi_3-n\varphi_4)\rangle\rangle \\
&\quad -\langle\langle\cos[m(\varphi_1-\varphi_2)]\rangle\rangle\langle\langle\cos[n(\varphi_1-\varphi_2)]\rangle\rangle \\
&= \left\langle v_m^2 v_n^2\right\rangle - \left\langle v_m^2\right\rangle\left\langle v_n^2\right\rangle , \qquad (2)
\end{aligned}$$

with the condition $m \neq n$ for two positive integers m and n. The complete discussion can be found in Section IV C of Ref. [31]. SC(m,n) can be normalized with the product $\left\langle v_m^2\right\rangle\left\langle v_n^2\right\rangle$ to obtain *normalized* symmetric cumulants [28, 32]. Normalized symmetric cumulants reflect only the degree of the correlation which is expected to be insensitive to the magnitudes of v_m and v_n, while SC(m,n) contains both the degree of the correlations between two different flow harmonics and individual v_n harmonics. That products in the denominator are obtained with two-particle correlations and using a pseudorapidity gap of $|\Delta\eta| > 1.0$ to suppress biases from few-particle nonflow correlations. On the other hand, in the two two-particle correlations which appear in the definition of SC(m,n) in Eq. 2 the pseudorapidity gap is not needed, since nonflow is suppressed by construction in SC observable, as the study based on HIJING model has clearly demonstrated in Ref. [28].

The measurements of SC observables have revealed that fluctuations of v_2 and v_3 are anti-correlated, while fluctuations of v_2 and v_4 are correlated in all centralities as shown in Figure. 1 (Top Left). The comparison between experimental data and the theoretical calculations [30], which incorporate both the initial conditions and system evolution, is shown in Figure. 1 (Top Left). The model captures qualitatively the centrality dependence, but not quantitatively. Most notably, there is

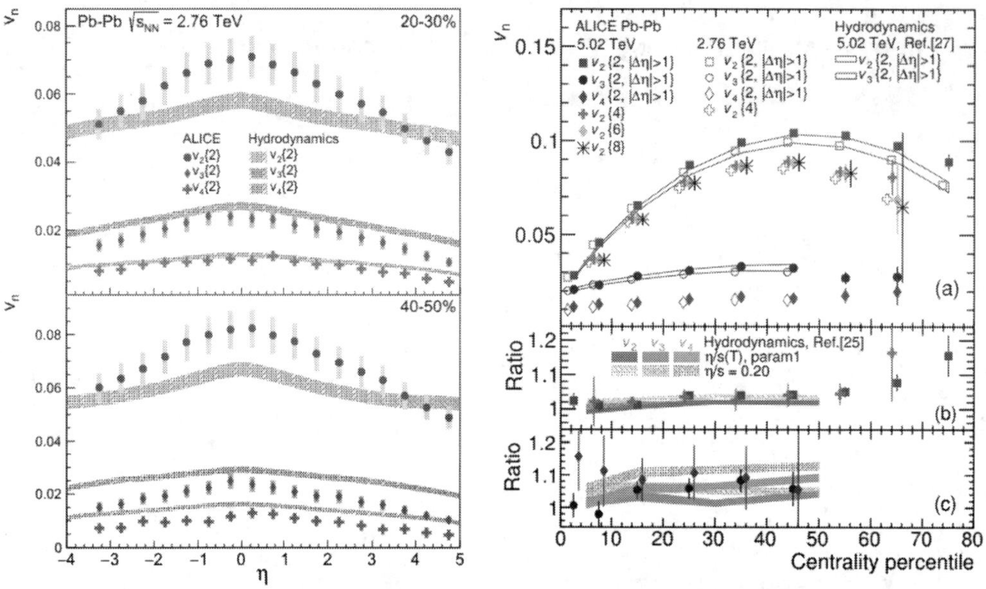

Figure 2. (Left) Comparisons to hydrodynamics predictions [33], where input parameters (temperature dependence of η/s) have been tuned to RHIC data for the Pb-Pb 20-30% (top) and 40-50% (Bottom) centralities. (Right) The first anisotropic flow measurements of charged particles in Pb–Pb collisions at $\sqrt{s_{\rm NN}}$ = 5.02 TeV.

no single centrality for which a given $\eta/s(T)$ parameterization describes simultaneously both SC(4, 2) and SC(3, 2). On the other hand, the same theoretical model captures quantitatively the centrality dependence of individual v_2, v_3 and v_4 harmonics with a precision better than 10% in central and mid-central collisions [30]. We therefore conclude that individual flow harmonics v_n and new SC(m, n) observables together provide a better handle on the initial conditions and $\eta/s(T)$ than each of them alone. This is further emphasized in Figure 1 (Bottom), where SC(3, 2) and SC(4, 2) observables were divided with the products $\left\langle v_3^2\right\rangle\left\langle v_2^2\right\rangle$ and $\left\langle v_4^2\right\rangle\left\langle v_2^2\right\rangle$, respectively. We have found that the normalized SC(4, 2) observable exhibits stronger sensitivity to different $\eta/s(T)$ parameterizations than the normalized SC(3, 2) observable, see Figure 1 (Bottom), and than the individual flow harmonics [30]. These findings indicate that the normalized SC(3, 2) observable is sensitive mainly to the initial conditions, while the normalized SC(4, 2) observable is sensitive to both the initial conditions and the system properties like η/s, which is consistent with the prediction from [20].

2.2 Moving forward to constrain the shear viscosity of QCD matter

ALICE recently published peudorapidity dependence of flow harmonics up to 4$^{\rm th}$ order harmonics [34]. At forward rapidities, the system will spend less time in the QGP phase. This implies that the viscosity from the hadronic phase would play a greater role in affecting the flow harmonics [33, 35]. Therefore, the relative decrease of the flow harmonics in different η ranges may help to disentangle the viscous effects from the hadronic phase with those from the QGP phase. The results are shown on the left panel in Fig. 2, the shape of $v_n(\eta)$ is largely independent of centrality for all order harmonic coefficients measured (v_2, v_3 and v_4). The results are compared to hydrodynamic calculations tuned to RHIC data [33]. The tuning involves finding a parameterization of the temperature dependence of η/s, so that the hydrodynamical calculations describe PHOBOS measurements of $v_2(\eta)$ [36, 37]. It is clear that the same parameterization does not describe the LHC data. For both centralities shown on the left in Figure 2, the elliptic flow coefficient v_2 is generally underestimated, while the higher order coefficients v_3 and v_4 are overestimated. These results could provide better independent constraints for the initial state fluctuations in three-dimensional space and $\eta/s(T)$ even at a fixed collision energy.

2.3 Anisotropic flow of charged particles at $\sqrt{s_{\rm NN}}$=5.02 TeV

We have presented the first anisotropic flow measurements of charged particles in Pb–Pb collisions at $\sqrt{s_{\rm NN}}$ = 5.02 from the data taken in November 2015 in Run 2 at the LHC in Figure 2 [3]. Only one low luminosity run (with trigger rate of 27 Hz) was used, being least affected by pile-up and distortions from space charge in the main tracking detector, the Time Projection Chamber (TPC). In the right panel of Figure 2 (a), the centrality dependence of v_2, v_3 and v_4 is presented from two- and multi-particle cumulants, integrated over the $p_{\rm T}$ range $0.2 < p_{\rm T} < 5.0$ GeV/c, for 2.76 and 5.02 TeV Pb–Pb collisions. To elucidate the energy evolution of v_2, v_3 and v_4, the ratios of anisotropic flow measured at 5.02 TeV to 2.76 TeV are presented in Figure 2 right panel (b) and (c). The predictions of anisotropic flow coefficients v_n from the hydrodynamic model [38] are compared to the measurements in Figure 2. (a). The predictions are compatible with the measured anisotropic flow v_n coefficients. At the same time, a different hydrodynamic calculation [39], which employs both constant η/s = 0.20 and temperature dependent η/s, can also describe the increase in anisotropic flow measurements of v_2. The increase of v_2 and v_3 from the two energies is rather moderate, while for v_4 it is more pronounced. An increase of (3.0±0.6)%, (4.3±1.4)% and (10.2±3.8)%, is obtained for elliptic, triangular and quadrangular flow, respectively, over the centrality range 0–50% in Pb–Pb collisions when going from 2.76 TeV to 5.02 TeV.

3 Summary

The ALICE experiment continues to play a critical role for completing the individual flow harmonic measurements at the highest energies to date. We are improving flow harmonic correlation techniques to understand the properties of the QGP and the full evolution of the heavy-ion collisions i.e different order flow harmonic correlations and peudorapidity dependence of flow harmonics. The different order flow harmonic correlation measurements provide strong constraints on the temperature dependence of the shear viscosity to the entropy density ratio in hydrodynamics in combination with the individual flow harmonics. Pseudorapidity dependence of $v_{2,3,4}$ results can provide access to a range of varying medium properties, even at a fixed collision energy. The individual flow harmonics with the highest LHC energy were measured and show a quantitative agreement with various models. Many analyses with Run2 Pb-Pb data with the full statistics are underway and the statistical and systematic errors will be reduced in the future to give valuable inputs to model calculations with discriminating powers on initial conditions and the transport properties of nuclear matter in heavy-ion collisions.

References

[1] K.H. Ackermann et al. (STAR), Phys. Rev. Lett. **86**, 402 (2001), nucl-ex/0009011
[2] K. Aamodt et al. (ALICE), Phys. Rev. Lett. **105**, 252302 (2010), 1011.3914
[3] J. Adam et al. (ALICE), Phys. Rev. Lett. **116**, 132302 (2016), 1602.01119
[4] P. Romatschke, U. Romatschke, Phys. Rev. Lett. **99**, 172301 (2007), 0706.1522
[5] C. Shen, U. Heinz, P. Huovinen, H. Song, Phys. Rev. **C84**, 044903 (2011), 1105.3226
[6] B. Schenke, S. Jeon, C. Gale, J. Phys. **G38**, 124169 (2011)

[7] P. Bozek, I. Wyskiel-Piekarska, Phys. Rev. **C85**, 064915 (2012), 1203.6513
[8] Phys. Rev. Lett. **110**, 012302 (2013), 1209.6330
[9] T. Hirano, P. Huovinen, Y. Nara, Phys. Rev. **C84**, 011901 (2011), 1012.3955
[10] P. Kovtun, D.T. Son, A.O. Starinets, Phys. Rev. Lett. **94**, 111601 (2005), hep-th/0405231
[11] R.A. Lacey, N.N. Ajitanand, J.M. Alexander, P. Chung, W.G. Holzmann, M. Issah, A. Taranenko, P. Danielewicz, H. Stoecker, Phys. Rev. Lett. **98**, 092301 (2007), nucl-ex/0609025
[12] P. Danielewicz, M. Gyulassy, Phys. Rev. D **31**, 53 (1985)
[13] J.Y. Ollitrault, Phys. Rev. **D46**, 229 (1992)
[14] S. Voloshin, Y. Zhang, Z. Phys. **C70**, 665 (1996), hep-ph/9407282
[15] S. Floerchinger, U.A. Wiedemann, A. Beraudo, L. Del Zanna, G. Inghirami, V. Rolando, Phys. Lett. **B735**, 305 (2014), 1312.5482
[16] M. Miller, R. Snellings (2003), nucl-ex/0312008
[17] B. Alver et al. (PHOBOS), Phys. Rev. Lett. **98**, 242302 (2007), nucl-ex/0610037
[18] B. Alver, G. Roland, Phys. Rev. **C81**, 054905 (2010), [Erratum: Phys. Rev.C82,039903(2010)], 1003.0194
[19] K. Aamodt et al. (ALICE), Phys. Rev. Lett. **107**, 032301 (2011), 1105.3865
[20] H. Niemi, G.S. Denicol, H. Holopainen, P. Huovinen, Phys. Rev. **C87**, 054901 (2013), 1212.1008
[21] G. Aad et al. (ATLAS), Phys. Rev. **C90**, 024905 (2014), 1403.0489
[22] B.H. Alver, C. Gombeaud, M. Luzum, J.Y. Ollitrault, Phys. Rev. **C82**, 034913 (2010), 1007.5469
[23] M. Luzum, J.Y. Ollitrault, Nucl. Phys. **A904-905**, 377c (2013), 1210.6010
[24] C. Shen, S.A. Bass, T. Hirano, P. Huovinen, Z. Qiu, H. Song, U. Heinz, J. Phys. **G38**, 124045 (2011), 1106.6350
[25] P. Bozek, Phys. Rev. **C85**, 034901 (2012), 1110.6742
[26] J.B. Rose, J.F. Paquet, G.S. Denicol, M. Luzum, B. Schenke, S. Jeon, C. Gale, Nucl. Phys. **A931**, 926 (2014), 1408.0024
[27] S. Ryu, J.F. Paquet, C. Shen, G.S. Denicol, B. Schenke, S. Jeon, C. Gale, Phys. Rev. Lett. **115**, 132301 (2015), 1502.01675
[28] J. Adam et al. (ALICE), Phys. Rev. Lett. **117**, 182301 (2016), 1604.07663
[29] X. Zhu, Y. Zhou, H. Xu, H. Song (2016), 1608.05305
[30] H. Niemi, K.J. Eskola, R. Paatelainen, Phys. Rev. **C93**, 024907 (2016), 1505.02677
[31] A. Bilandzic, C.H. Christensen, K. Gulbrandsen, A. Hansen, Y. Zhou, Phys. Rev. **C89**, 064904 (2014), 1312.3572
[32] G. Giacalone, L. Yan, J. Noronha-Hostler, J.Y. Ollitrault, Phys. Rev. **C94**, 014906 (2016), 1605.08303
[33] G. Denicol, A. Monnai, B. Schenke, Phys. Rev. Lett. **116**, 212301 (2016), 1512.01538
[34] J. Adam et al. (ALICE), Phys. Lett. **B762**, 376 (2016), 1605.02035
[35] E. Molnar, H. Holopainen, P. Huovinen, H. Niemi, Phys. Rev. **C90**, 044904 (2014), 1407.8152
[36] B.B. Back et al. (PHOBOS), Phys. Rev. **C72**, 051901 (2005), nucl-ex/0407012
[37] B.B. Back et al. (PHOBOS), Phys. Rev. Lett. **94**, 122303 (2005), nucl-ex/0406021
[38] J. Noronha-Hostler, M. Luzum, J.Y. Ollitrault, Phys. Rev. **C93**, 034912 (2016), 1511.06289
[39] H. Niemi, K.J. Eskola, R. Paatelainen, K. Tuominen, Phys. Rev. **C93**, 014912 (2016), 1511.04296

News on Collectivity in PbPb Collisions at CMS

Dong Ho Moon[1,a] on behalf of the CMS Collaboration

[1] *Chonnam National University, 61186, Gwangju, Republic of Korea*

Abstract. The flow anisotropies with the Fourier coefficients (n = 2, 3) for the charged particles produced in PbPb collisions at a nucleon-nucleon center-of-mass energy of 5.02 TeV is studied with the CMS detector. In order to extract the Fourier coefficients, several methods were used, such as the scalar product method or multi-particle cumulant method. The results cover both of the low-p_T region ($1 < p_T < 3$ GeV/c) associated with hydrodynamic flow phenomena and the high-p_T region where anisotropic azimuthal distributions may reflect the path-length dependence of the parton energy loss in the created medium for the seven bins of collision centrality, spanning the rang of 0-60% most-central events.

1 Introduction

A Quark-Gluon-Plasma (QGP) is a state of matter which quantum chromodynamic (QCD) predicts that it will occur at extremely high temperature and density. The experiments at the Relativistic Heavy Ion Collider (RHIC) have established the formation of QGP in ultra-relativistic nucleus-nucleus interactions [1–4]. One of strong signatures of the existence for QGP is the suppression of high-transverse-momentum (p_T) hadron yield via the nuclear modification factor, R_{AA}, which is the ratio of the final-state particle yields at a given p_T value to that expected based on a scaling of pp collision results in the absence of the medium [1–4]. The observation of dijet asymmetry in PbPb collisions at the Large Hadron Collider (LHC) furnishes further proof of jet quenching, indicating a large energy loss for partons traversing the QGP medium.

The measurement of R_{AA} alone is not sufficient to differentiate among various parton energy-loss models. In addition, it has been shown that R_{AA} is not sensitive to sub-nucleon spatial fluctuations [5]. Additional observables, such as the azimuthal anisotropy of high p_T hadrons, are needed to constrain the models [6–12]. The anisotropy can be characterized by the coefficient of the second-order Fourier harmonic, v_2, in the azimuthal angle ϕ distribution of the hadron yield relative to the reaction-plane angle Ψ_{RP}, $dN/d\phi \propto 1 + 2v_2 \cos(2(\phi - \Psi_{RP}))$, where Ψ_{RP} is defined by the short-axis direction of the "almond"-shaped interaction region created in a non-central heavy-ion collision. In order to reduce the ambiguity on the measurement, scalar product or multi-particle cumulant methods can be replaced.

Furthermore, v_3, the coefficient of the higher harmonic, is particularly sensitive to the initial-state fluctuations. Indeed, v_3 arises mainly from fluctuation in the soft sector and it remains to be seen if these fluctuations can generate triangular anisotropy at high p_T.

In this paper, a study of the azimuthal anisotropy of very high p_T particles (up to $p_T \approx 100$GeV/c) for PbPb collisions at $\sqrt{s_{NN}} = 5.02$ TeV at the LHC using the CMS detector is presented. The v_2 and

[a] e-mail: dmoon@cern.ch

v_3 coefficients are determined as a function of p_T and collision centrality in the pseudorapidity regions of $|\eta| < 1$, where $\eta = -\ln[\tan(\theta/2)]$ and θ is the polar angle relative to the counterclockwise beam direction. The first precise measurement of the v_2 and v_3 coefficients up to very high p_T region in heavy-ion collisions are represented by using scalar product method and the multi-particle cumulant analysis for several orders (4, 6 and 8) as a function of p_T.

2 Experimental Setup

2.1 Event Selection

For this analysis, 2015 PbPb data at $\sqrt{s_{NN}} = 5.02$ TeV is utilized with an integrated luminosity of 404 μb^{-1}. Events were selected by using several trigger combinations in different p_T regions and centrality bins. The minimum-bias trigger covers the low p_T region (< 14 GeV/c), while the dedicated triggers on high p_T tracks are used for high p_T region up to 100 GeV/c. The L1 trigger seed of the triggers fired based on the highest ET Regional Calorimeter Trigger (RCT) region in the barrel of the CMS detector ($|\eta| < 1.0$). At the High Level Trigger (HLT), the leading track is passing the strict quality selection criteria describes in section 2.2. The performances of trigger for tracks used in analysis are shown in Figure 1.

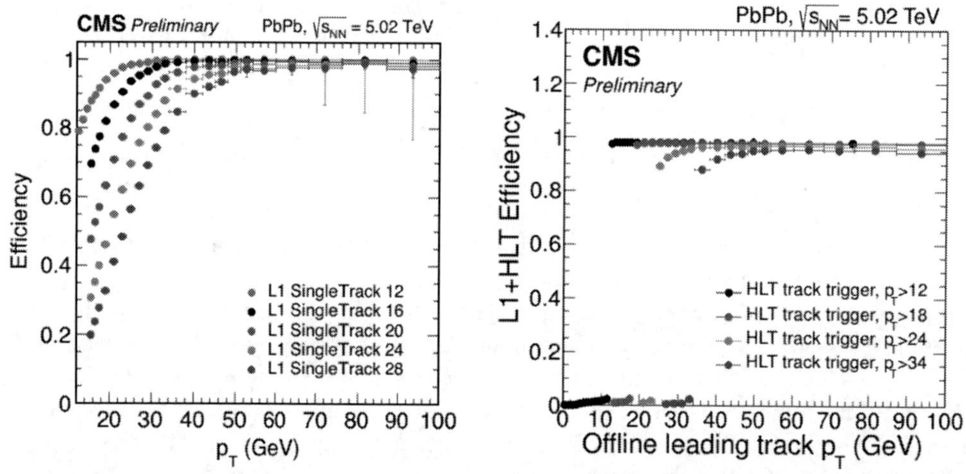

Figure 1. The efficiencies of L1 trigger (left) and High level trigger (right) as a function of p_T measured in $|\eta| <$ 1.0. L1 trigger efficiencies are estimated for p_T of 12, 16, 20, 24 and 28 GeV/c, and HLT efficiencies are done in the p_T regions of 12, 18, 24 and 34 GeV/c [13].

2.2 Track Selection

Primary-track candidates are selected by the high-purity tracks. Additional requirements are also applied to enhance the purity of primary tracks. The significance of the separation along the beam axis (z) between the track and the best vertex, $dz/\sigma(dz)$, and the significance of the impact parameter relative to the best vertex transverse to the beam, $d_T/\sigma(d_T)$, should be less than 3. The relative uncertainty of the transverse-momentum measurement, $\sigma(p_T)/p_T$, must be less than 10%. In order to reduce the fake rate of mis-reconstructed tracks and confirm high tracking efficiency, primary tracks with $|\eta| < 1.0$ and $p_T > 1.0$ GeV/c are used in the analysis. Furthermore, tracks above 20 GeV/c in the tracker are required to match with an energy deposit in the calorimeter to further decrease the fake rate.

3 Analysis Methods

3.1 Scalar product method

The "scalar-product" event-plane method is used to study azimuthal correlations with long-range of pseudorapidity. The event-plane is determined event-by-event and consists of the beam direction and the azimuthal direction of maximum particle density.

Avoiding autocorrelation effects where a given particle, or its decay products is very important to determine the event plane which we want to characterize. This is usually done by requiring a large gap in pseudorapidity between the particles used to establish the event plane and those used to determine the Fourrier coefficients.

The azimuthal dependence of the particle yield can be written in terms of an harmonic expansion with [14]

$$E\frac{d^3N}{d^3p} = \frac{1}{2\pi}\frac{d^2N}{p_T dp_T dy}\left(1 + \sum_{n=1}^{\infty} 2v_n \cos\left[n(\phi - \Psi)\right]\right) \tag{1}$$

where ϕ and E are the particle's azimuthal angle and energy, respectively. If the impact-parameter direction is known, the reference angle Ψ can be decided as the azimuthal angle of the reaction plane Ψ_R as defined by the beam and impact parameter directions.

To make the largest pseudorapidity gap possible, elliptic flow ($n = 2$) and triangular flow ($n = 3$) event planes are defined using calorimeter data, with the HF_n^- planes covering the pseudorapidity range of $-5 \leq \eta < -3$ and HF_n^+ planes covering the range $3 \leq \eta < 5$. Another event plane using tracker data with $-0.75 \leq \eta < 0.75$ was also defined and used in the three-sub-event technique for determining the resolution corrections for the HF_n^-, HF_n^+. To consider asymmetries that arise from the detector acceptance and other effects correlated with the detector conditions, a two-step process is used where the event plane is first recentered and subsequently flattened [15].

In order to measure the actual v_n coefficients, the current analysis utilizes the scalar product method introduced firstly by the STAR collaboration in a study of elliptic flow in AuAu collisions at $\sqrt{s_{NN}} =$ 130 GeV [16]. The Q-vectors are defined as

$$Q_n \equiv \sum_{i=1}^{M} e^{in\phi_i}, \tag{2}$$

where M is the number of particles in each event. Therefore, using the scalar product method, v_n harmonics can be expressed as

$$v_n\{\mathrm{SP}\} \equiv \frac{\left\langle Q_n Q_{nA}^*\right\rangle}{\sqrt{\frac{\langle Q_{nA}Q_{nB}^*\rangle\langle Q_{nA}Q_{nC}^*\rangle}{\langle Q_{nB}Q_{nC}^*\rangle}}}. \tag{3}$$

Here, the subscripts A, B, and C refer to three separate event planes reconstructed in different regions of pseudorapidity. The particles of interest are expressed by the Q_n vector and are correlated with the A event plane. The event planes B and C effectively correct for the finite resolution of the A event plane that results from finite particle multiplicities and detector effects [17].

3.2 Multi-particle correlation technique with cumulants

Another way to measure v_2 is the Q-Cumulant method [18, 19] from 4-, 6- and 8-particle correlations. This method supplies a fast, exact (no approximations), and unbiased (no interference between different harmonics) estimations for the cumulants. The cumulants are expressed in terms of the moments of the magnitude of the Q_n flow vector. We first define the m-particle (m = 2, 4, 6 or 8) correlator as

$$\begin{aligned}
\langle\langle 2\rangle\rangle &\equiv \left\langle\left\langle e^{in(\phi_1-\phi_2)}\right\rangle\right\rangle \\
\langle\langle 4\rangle\rangle &\equiv \left\langle\left\langle e^{in(\phi_1+\phi_2-\phi_3-\phi_4)}\right\rangle\right\rangle \\
\langle\langle 6\rangle\rangle &\equiv \left\langle\left\langle e^{in(\phi_1+\phi_2+\phi_3-\phi_4-\phi_5-\phi_6)}\right\rangle\right\rangle \\
\langle\langle 8\rangle\rangle &\equiv \left\langle\left\langle e^{in(\phi_1+\phi_2+\phi_3+\phi_4-\phi_5-\phi_6-\phi_7-\phi_8)}\right\rangle\right\rangle
\end{aligned} \tag{4}$$

where ϕ_i is the azimuthal angle of the i-th particle, and $\langle\langle\rangle\rangle$ indicates that the average is taken over all m-particle combinations for all events. In order to remove self-correlations, it is required that the m particles be distinctive. The unbiased estimators of the reference m-particle cumulants, c_nm, are defined as

$$\begin{aligned}
c_n\{4\} &= \langle\langle 4\rangle\rangle - 2\cdot\langle\langle 2\rangle\rangle^2 , \\
c_n\{6\} &= \langle\langle 6\rangle\rangle - 9\cdot\langle\langle 4\rangle\rangle\langle\langle 2\rangle\rangle + 12\cdot\langle\langle 2\rangle\rangle^3 , \\
c_n\{8\} &= \langle\langle 8\rangle\rangle - 16\cdot\langle\langle 6\rangle\rangle\langle\langle 2\rangle\rangle - 18\cdot\langle\langle 4\rangle\rangle^2 + 144\cdot\langle\langle 4\rangle\rangle\langle\langle 2\rangle\rangle^2 - 144\langle\langle 2\rangle\rangle^4 .
\end{aligned} \tag{5}$$

Correlating the m-particles within the reference phase space of $|\eta| < 2.4$ and p_T range of $0.3 < p_T <$ 3.0 GeV/c, we then obtain the reference flow $v_2\{m\}$, with

$$\begin{aligned}
v_4\{4\} &= \sqrt[4]{-c_n4} , \\
v_6\{6\} &= \sqrt[6]{\frac{1}{4}c_n6} , \\
v_8\{8\} &= \sqrt[8]{-\frac{1}{33}c_n8} .
\end{aligned} \tag{6}$$

With respect to the reference flow, the differential $v_2m(p_T,\eta)$ can then be expressed as

$$v_2\{4\}(p_T,\eta) = -d_2\{4\}/\sqrt[3/4]{-c_n\{4\}} , \tag{7}$$

$$v_2\{6\}(p_T,\eta) = d_2\{6\}/\sqrt[5/6]{c_n\{6\}/4^{1/6}} , \tag{8}$$

$$v_2\{8\}(p_T,\eta) = -d_2\{8\}/\sqrt[7/8]{-c_n\{8\}/33^{1/8}} , \tag{9}$$

where the differential cumulants $d_2\{m\}$ are estimated by replacing one of the m-particle cumulants in Eq. (5) with a particle from a certain particle of interest (POI) phase space in p_T or η [17].

4 Results

The v_2 and v_3 results are shown in Figure 2. Those are measured by using the scalar product (SP) method as a function of p_T ($\approx$ 100 GeV/c) in 7 centrality ranges (0–60%) of PbPb collisions at $\sqrt{s_{\rm NN}}$ = 5.02 TeV. The statistical uncertainties are represented in the bars and the systematic uncertainties are denoted in the shaded boxes.

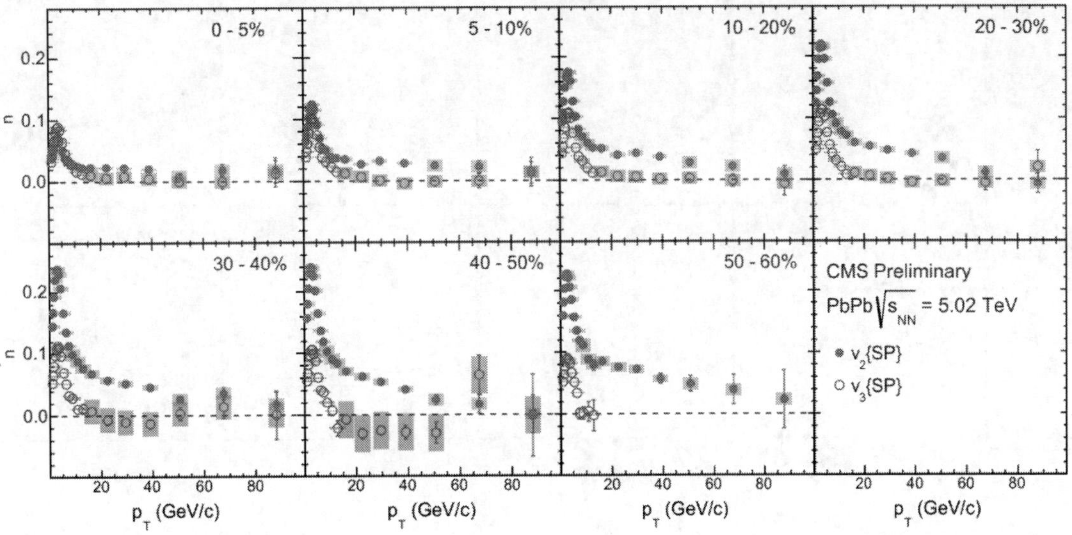

Figure 2. The v_2 and v_3 results from SP method as a function of p_T in seven centrality ranges of PbPb collisions at $\sqrt{s_{NN}} = 5.02$ TeV. Shaded boxes represent systematic uncertainties [17].

The significant positive v_2 values are observed in the most of centrality ranges up to $p_T \approx 100$ GeV/c. The v_2 value first rises and then falls off around 3 GeV/c quickly from low p_T to high p_T region. In the low p_T region, v_2 values are first increasing from most central to mid-peripheral events (30-40%), and then decreasing in the more peripheral centrality ranges. At the higher p_T region (>10 GeV/C), the v_2 values are slowly decreasing as p_T increases. The v_3 results show similar tendency as that for v_2. At low p_T, the observed v_3 values don't depend on centrality. Significant positive v_3 are found to continue up to p_T of 20 GeV/c for all centrality ranges. Above p_T= 20 GeV/c, the v_3 values are consistent with zero within given uncertainties. These results of v_2 and v_3 cover the widest p_T range ever achieved. Such a result at high p_T is sensitive to the path length dependence of hard partons in the QGP and will be expected to put more constrain on the energy loss models.

Figure 3 shows the comparison between v_2 results at $\sqrt{s_{NN}} = 5.02$ TeV and v_2 results obtained with the Event Plane (EP) at $\sqrt{s_{NN}} = 2.76$ TeV [20] and theoretical calculations from CUJET3.0 [21, 22] at 5.02 TeV. The results from SP and EP are similar up to very high p_T. The difference do not exceed 5%. The model (CUJET3.0) calculations are compatible with the results at high p_T(> 40 GeV/c) within systematic uncertainties in all over centrality ranges except for the most peripheral one where the model doesn't reproduce the data.

In order to investigate the origin of this correlation observed at very high p_T further, we obtained v_2 values from multi-particle cumulant analysis (4, 6 and 8) and compared with the SP v_2 results as shown in Figure 4. At low p_T (p_T < 3 GeV/c), the results follow the expectation from hydrodynamic models meaning $v_2\{SP\} > v_2\{4\} \approx v_2\{6\} \approx v_2\{8\}$. At high p_T(10 GeV/c), scalar product and multi-particle correlation results tend to converge to the same value.

This observation suggests strong evidence that collective anisotropic particle emission in heavy ion collisions extend to very high p_T region. Although the nature of collectivity at high p_T is likely to be related to jet quenching phenomena, the observed anisotropies at both high and low p_T are related to initial-state geometry and its event-by-event fluctuations.

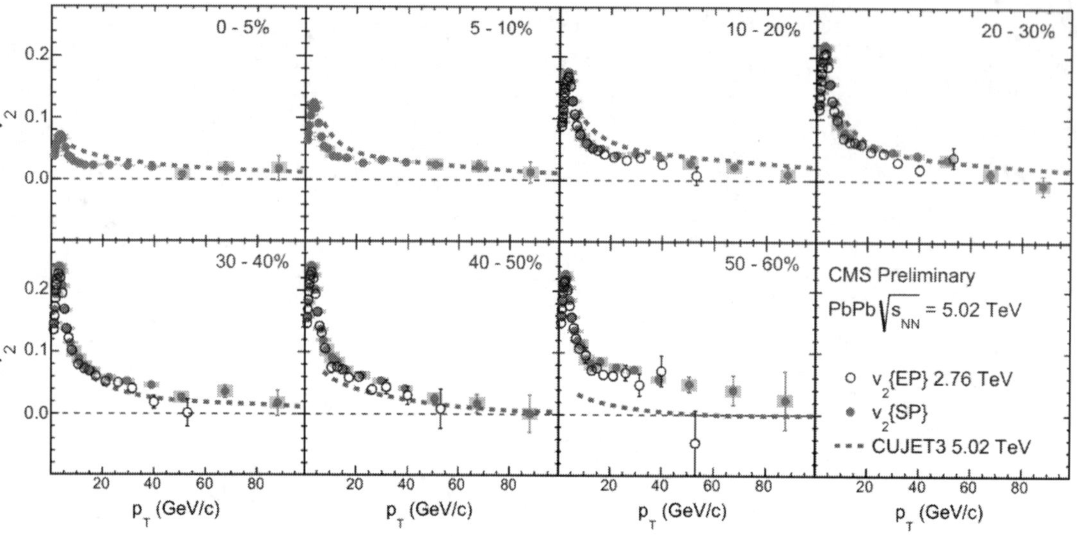

Figure 3. The v_2 results from SP and EP method as a function of p_T in seven centrality ranges of PbPb collisions at $\sqrt{s_{\mathrm{NN}}} = 5.02$ TeV and 2.76 TeV respectively. Shaded boxes represent systematic uncertainties [17] and dashed lines are predictions from CUJET3.0 model [21, 22].

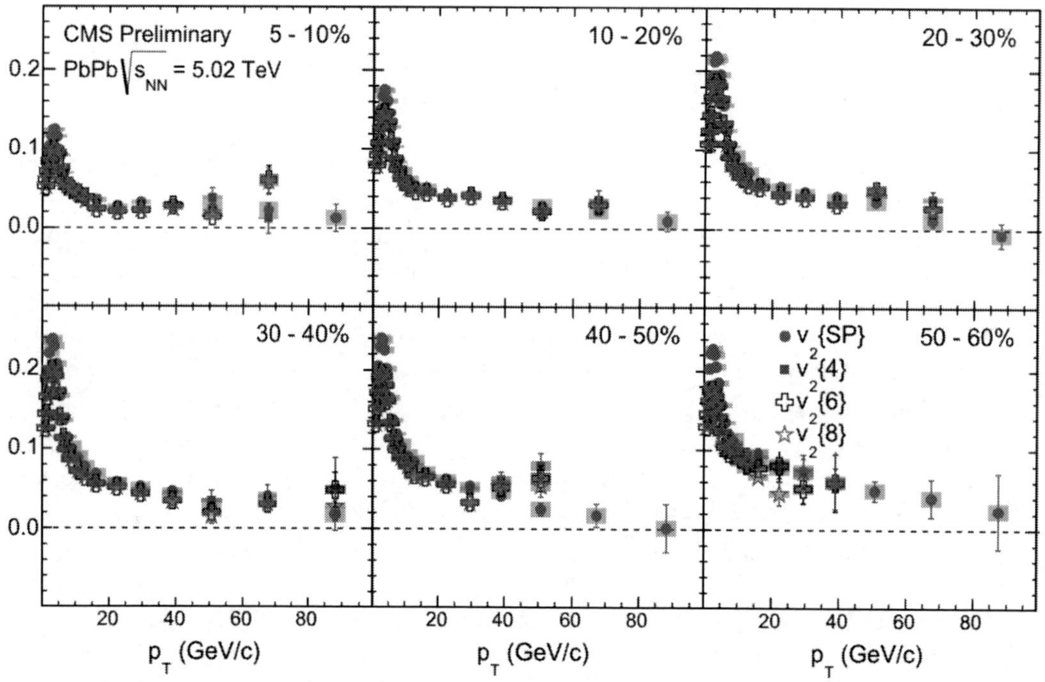

Figure 4. Comparison between the v_2 results from SP and cumulant methods as a function of p_T in six centrality ranges of PbPb collisions at $\sqrt{s_{\mathrm{NN}}} = 5.02$ TeV. Shaded boxes represent systematic uncertainties [17].

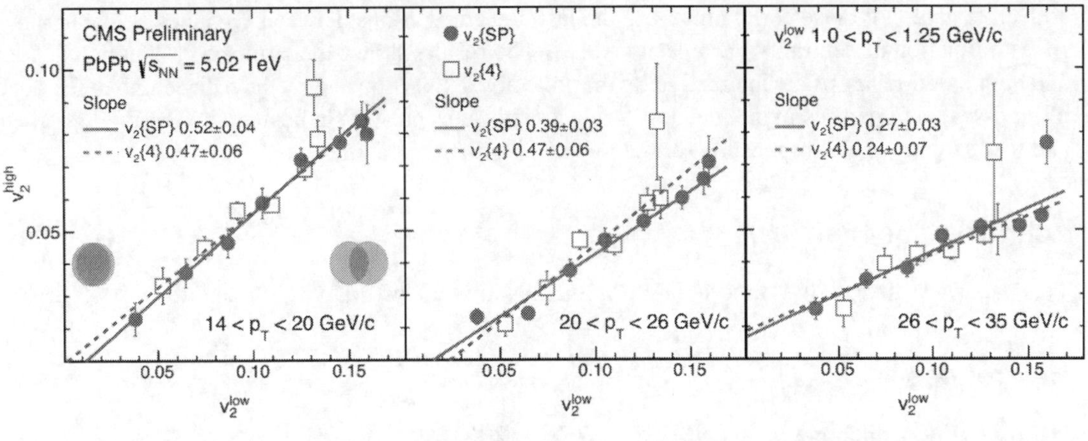

Figure 5. The v_2 values from different centrality ranges at high p_T : $12 < p_T < 14$ GeV/c, $20 < p_T < 26$ GeV/c and $26 < p_T < 35$ GeV/c vs. low p_T ($1 < p_T < 1.25$ GeV/c) obtained from SP (closed circles) and cumulant (open squares) methods in PbPb collisions at $\sqrt{s_{NN}} = 5.02$ TeV [17].

To find possible connection of azimuthal anisotropies between low-p_T (dominated by hydrodynamics) and high p_T (dominated by jet quenching) regions, the correlation between the v_2 results at low and high p_T was studied. Figure 5 shows v_2 from $12 < p_T < 14$ GeV/c, $20 < p_T < 26$ GeV/c and $26 < p_T < 35$ GeV/c as a function of v_2 from $1.0 < p_T < 1.25$ GeV/c extracted from each centrality range of PbPb collisions at $\sqrt{s_{NN}} = 5.02$ TeV. Each point represents a different centrality range starting from the most central events on the left toward peripheral events on the right. The first four points have a centrality bin width of 5% (from 0 to 20%) while, for the others, the bin width is 10%. Full circles are the results obtained from SP method and open square are extracted from 4-particle correlation analysis. A strong correlation of v_2 values between high and low p_T regions is observed over the full centrality range. While low p_T correlations can be explained by the collective expansion of the medium, high p_T correlations come from the path-length dependence of the hard-scattered partons undergo while traversing the medium. This indicates that the initial geometry and its fluctuations may be responsible for the correlations observed both at low and high p_T.

5 Summary

In summary, the azimuthal anisotropy of charged particles with respect to the event plane has been studied in PbPb collisions at $\sqrt{s_{NN}}$= 5.02 TeV using the CMS detector. The v_2 and v_3 values are measured over a wide p_T range from 1 GeV/c to 100 GeV/c in different collision centralities using scalar product method. The results cover for the first time the very high p_T region beyond 60 GeV/c. Moreover, these results provide a v_3 measurement up to a p_T region that has never before been studied (up to 100 GeV/c). The observed $v_2(p_T)$ shows a tendency of rapid increasing to a maximum at p_T of ≈ 3 GeV/c and a steep fall for all of centrality ranges. The $v_3(p_T)$ behavior is very similar to v_2 at low p_T but its value is consistent with zero over the full centrality range for $p_T > 20$ GeV/c. The observed trend in the centrality dependence of v_2 over a wide p_T range suggests a potential connection to the initial-state geometry. To further investigate the origin of the correlation at high p_T, a multi-particle correlation analysis was performed to extract v_2 from 1 GeV/c to, at maximum, 100 GeV/c. At low p_T, $v_2\{4\}$, $v_2\{6\}$ and $v_2\{8\}$ have similar values and smaller than $v_2\{SP\}$. This follows hydrodynamic expectations and is a sign of collectivity. The same behavior is observed at high p_T and reflects the collective nature of these correlations. As the last, the correlations between v_2 values at low and high p_T as a function of centrality have been studied. The results shows the strong correlations. The low and high p_T correlations are interpreted by the collective expansion and by the path-length dependence of hard-scattered partons in the medium, respectively. These results indicate that the initial geometry and its fluctuations are responsible for the low and high p_T correlations.

Acknowledgement

This paper was financially supported by Chonnam National University (Grant number: 2015-3039).

References

[1] BRAHMS Collaboration, Nucl. Phys. A **757**, 1 (2005)
[2] PHENIX Collaboration, Nucl. Phys. A **757**, 184 (2005)
[3] PHOBOS Collaboration, Nucl. Phys. A **757**, 28 (2005)
[4] STAR Collaboration, Nucl. Phys. A **757**, 102 (2005)
[5] Noronha-Hostler, Jacquelyn and Betz, Barbara and Noronha, Jorge and Gyulassy, Miklos, arXiv:1602.03788.
[6] Peigne, S. and Smilga, A. V., arXiv:0810.5702.
[7] Wicks, Simon and others, Nucl. Phys. A **784**, 426 (2007)
[8] Jia, Jiangyong and Wei, Rui, arXiv:1005.0645.
[9] Jia, Jiangyong and Horowitz, W.A. and Liao, Jinfeng, Phys. Rev. C **84**, 034904 (2011)
[10] Renk, Thorsten, Phys. Rev. C **83**, 024908 (2011)
[11] Betz, Barbara and Gyulassy, Miklos and Torrieri, Giorgio, Phys. Rev. C **84**, 024913 (2011)
[12] Betz, Barbara and Gyulassy, Miklos, arXiv:1201.0281.
[13] CMS Collaboration, https://twiki.cern.ch/twiki/bin/view/CMSPublic/HighLevelTriggerRunIIResults
[14] Poskanzer, Arthur M. and Voloshin, S. A., Phys.Rev. C **58**, 1671 (1998)
[15] Barrette, J. and others, Phys.Rev. C **56**, 3254-3264 (1997)
[16] Adler, C. and others, Phys.Rev. C **66**, 034904 (2002)
[17] CMS Collaboration, CMS PAS-HIN-15-014 (2015) http://cds.cern.ch/record/2156183?ln=en
[18] Bilandzic, Ante and Snellings, Raimond and Voloshin Sergei, Phys.Rev. C **83**, 044913 (2011)
[19] Bilandzic, Ante and Christensen, Christian Holm and Gulbrandsen, Kristjan and Hansen, Alexander and Zhou, You, Phys.Rev. C **89**, 064904 (2014)
[20] Chatrchyan, Serguei and others, Phys. Rev. Lett. **109**, 022301 (2012)
[21] Xu Jiechen and Liao Jinfeng and Miklos Gyulassy, Chinese Physics Letters **32**, 092501 (2015)
[22] Xu, Jiechen and Liao, Jinfeng and Gyulassy, Miklos, Journal of High Energy Physics **2016**, 1–50 (2016)

EPJ Web of Conferences **141**, 01003 (2017) DOI: 10.1051/epjconf/201714101003
ISMD 2016

Measurement of Azimuthal Anisotropy at RHIC-PHENIX

Maya Shimomura[1,a] for the PHENIX Collaboration

[1] *Nara Women's University, Ktauoya Nishimachi, Nara, 630-8263, Japan*

Abstract. The transverse momentum (p_T) and centrality dependence of the azimuthal anisotropy of second harmonics(v_2) are measured for charged hadron species at various collision systems and energies such as $\sqrt{s_{NN}}$ = 62.4 and 200 GeV in Cu + Cu and Au + Au collisions and at $\sqrt{s_{NN}}$ = 200 GeV in Cu + Au collisions by the PHENIX experiment at RHIC. The higher order anisotropy (v_3) are also measured for charged hadron at $\sqrt{s_{NN}}$ = 200 GeV in Cu + Cu, Au + Au and Au + Cu collisions. From these systematic study, we found that the all results are consistent with eccentricity scaling, quark number + KE_T scaling and $N_{part}^{1/3}$) scaling except at small N_{part} in Cu + Cu at 62.4 GeV. Taking these scaling (quark number, KE_T, eccentricity and $N_{part}^{1/3}$) into account, there is a universal scaling for π/K/p v_2 with different energies, collision sizes and particle species.

1 Introduction

Relativistic heavy ion collisions have been considered as a unique way to create and study the quark-gluon plasma (QGP), where the quarks and gluons are de-confined. The Relativistic Heavy Ion Collider (RHIC) at Brookhaven National Laboratory was constructed to create and study the QGP. Azimuthal anisotropy of produced particles in relativistic heavy ion collisions is a powerful probe for investigating the characteristics of the QGP. Especially the strength of the elliptic anisotropy (v_2), which is defined by the second harmonics of Fourier expansion for the azimuthal distribution of the produced particles with respect to the reaction plane, is expected to be sensitive to the early stage of heavy ion collisions. At non-central collision, the overlap region is geometrically anisotropic, like an almond shape. When the produced matter has small mean free path, interacting each other enough to reach local thermalization, it creates pressure gradient. The geometrical anisotropy transfers to the anisotropy in the momentum phase space as flow because of this pressure gradient, and v_2 indicates the strength of this elliptic flow. Thus, the measured v_2 reflects the equation of state of the dense matter such as QGP, produced in the collisions. Recently the measurement of the triangle anisotropy (v_3) has also drawn scientific attention because most of it are expected to be created by the participant fluctuations in Au + Au and the estimation of the strength of the participant fluctuations strongly depends on theoretical models at initial conditions. Therefore, v_3 is expected to be able to put some restriction on initial condition models [3]. On the other hand, as same as v_2, v_3 should also develop with the pressure gradient at QGP. The important thing here is since the produced particles randomly emit before thermalization, the geometrical anisotropy decreases with time. Therefore, to let the geometrical anisotropy make elliptic and triangular flow, the thermalization should be occurred very early stage before the geometrical anisotropy is totally gone.

[a] e-mail: mayap@bnl.gov

2 Motivation

One of the most remarkable findings at RHIC is that the strength of v_2 can be described well by hydro-dynamical models in the low transverse momentum region ($\sim$ 1 GeV/c) [1]. In the intermediate transverse momentum region (1 $\sim$ 4 GeV/c), v_2 is consistent with n_q and KE_T (= m_T - m_0) scaling, and the result supports a quark-recombination model [2]. The matter produced in the high energy heavy ion collision is expected to undergo several stages from the initial hard scattering to the final hadron emission. When the matter reaches thermalization and QGP is created, we expect hydro-dynamical behavior at quark level. Because experimentally we cannot see the QGP directly, we need a comprehensive understanding from thermalization through hadronization to freeze-out. The elliptic flow is expected to be created at QGP stage by pressure gradient, but it is important to note whenever the matter interacts with each other, there is a possibility to change v_2. [4]

For a more comprehensive understanding of azimuthal anisotropy, we have carried out systematic measurements of v_2, by measuring v_2 for identified charged hadrons in Au + Au and Cu + Cu collisions at $\sqrt{s_{NN}}$ = 200 and in Au + Au at $\sqrt{s_{NN}}$ = 62.4 GeV and also for inclusive charged hadrons in Au + Au and Cu + Cu at $\sqrt{s_{NN}}$ = 200 and 62.4 GeV. We have studied the dependence on collision energy, size and species of the produced particles comparing the $\sqrt{s_{NN}}$ = 2.76 TeV data from LHC. [9] Moreover, we measured v_2 for inclusive and identified charged hadrons in Cu + Au and v_3 for inclusive charged hadrons in Au + Au and Cu + Au at $\sqrt{s_{NN}}$ = 200. To study the detailed behavior of flow measurement, we examine the scalings to these results. We had expected that v_3 at Cu + Au is larger than that at Au + Au because Cu + Au collision provides additional triangular anisotropy at collision geometry while v_3 in Au + Au comes from initial fractuations.

3 Results

In Au+Au collisions, the values of v_2 as a function of p_T agree well at $\sqrt{s_{NN}}$ = 39, 62.4 and 200 GeV for measured centralities, 10-20, 20-30, 30-40 and 40-50%. However, the v_2 at 7.7 GeV is much lower than these. This results may indicate the energy between 7.7 and 39 is the region which switch from partonic flow to hadronic flow. We also compared the results of $\sqrt{s_{NN}}$ = 2.76 TeV data in Pb+Pb at LHC-ALICE, and it was found that v_2 at 2.76 TeV is very similar to the v_2 at 200 GeV especially at low p_T. [9]

Next, we compared different system size of collisions such as Au + Au, Cu + Cu, Pb + Pb and Cu + Au at $\sqrt{s_{NN}}$ = 200 GeV, 62.4 GeV and 2.76 TeV as a function of N_{part}. The values of v_2 agree well at $\sqrt{s_{NN}}$ = 200 GeV, 62.4 GeV and 2.76 TeV in Au + Au, but have clear differences between Cu + Cu and Au + Au, and Cu + Au results are between Cu + Cu and Au + Au. This is natural because the different nucleus collisions such as Cu + Cu, Au + Au and Cu + Au have different initial geometrical eccentricities at the same N_{part}. [10, 11] Normalizing v_2 by eccentricity, ε, (eccentricity scaling), all results follow one curve therefore, v_2 is scaled by the eccentricity at the same N_{part}. Here, we use the participant eccentricity, which includes the effect of participant fluctuations [5]. The values of v_2/ε are not a constant, therefore, v_2 can be normalized by ε at the same N_{part}, but N_{part} dependence still remains. Looking close to this dependence we empirically found that v_2/ε is proportional to $N_{part}^{1/3}$. This $N_{part}^{1/3}$ works in Au + Au, Cu + Cu collisions[10] and in Pb + Pb[9] and in Cu + Au [11] as shown in Figure 1 . Including results of $\sqrt{s_{NN}}$ = 2.76 TeV, $v_2/(\varepsilon \cdot N_{part}^{1/3})$ is independent of the collision systems except for small N_{part} in Cu+Cu at $\sqrt{s_{NN}}$ = 62.4 GeV.

One of the most famous results on RHIC v_2 measurement is that the v_2 for quark number(n_q) + KE_T scaling in Au+Au at at $\sqrt{s_{NN}}$ = 200 GeV.[2] The n_q scaling is consistent to the recombination model which assumes the quark level flow at QGP phase, and the KE_T scaling has been considered to be able

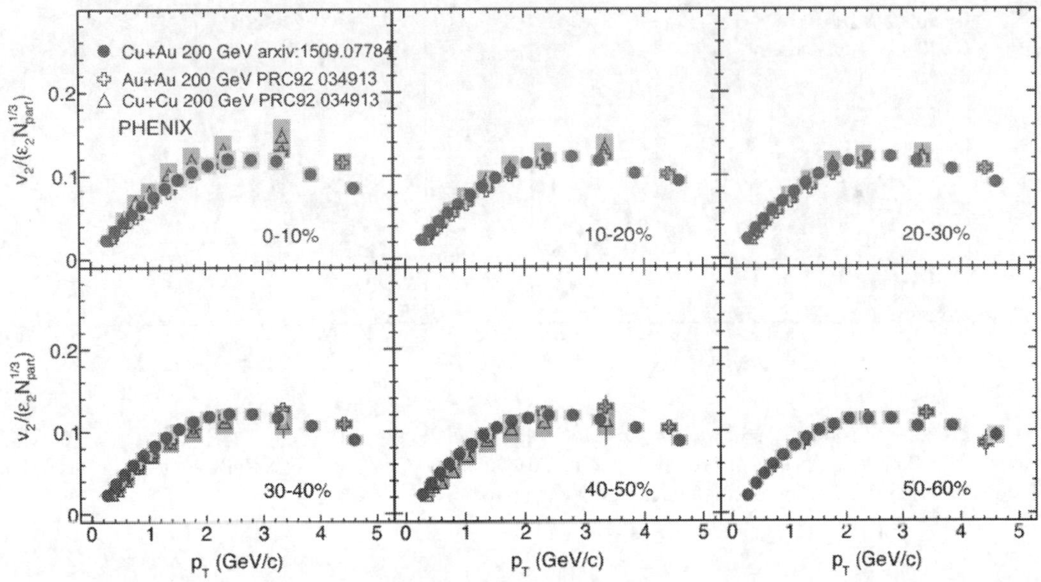

Figure 1. $v_2/(\varepsilon \cdot N_{\text{part}}^{1/3})$ vs. p_T for charged hadron in Au+Au, Cu+Cu and Cu+Au at $\sqrt{s_{NN}}$ = 200 GeV at indicated centrality bin in each panel. [9, 11, 12]

to subtract the difference of different particle v_2 at low p_T which is caused due to the radial flow effect. In Au+Au 200GeV collisions at PHENIX experiment, the large statistics and new detector allowed us to see that the both n_q and KE_T scaling works very well on various particle species including ϕ, Λ and deuteron, and even to see the break point of this scaling at KE_T = 1 GeV as shown in [8] . Above this p_T region, one can expect other mechanism is dominant to create v_2. For the systematic study, we also measured particle identified v_2 in Au+Au at $\sqrt{s_{NN}}$ = 62.4 GeV and in Cu+Cu and at $\sqrt{s_{NN}}$ = 200 GeV. It is found that v_2 in Au+Au at 62.4 GeV is also consistent with n_q + KE_T scaling. Moreover,the n_q + KE_T scaling mostly works out in Cu+Cu at $\sqrt{s_{NN}}$ = 200 GeV for central collisions. There are the small discrepancy from the KE_T scaling at peripheral collisions at low p_T and the discrepancy from the KE_T scaling depends on N_{part}. The detailed quantitative comparison are written in [10]. Comparing between π and proton, the results indicate the larger N_{part} produces more shift for proton to higher p_T based on π. This N_{part} dependence of KE_T scaling behavior for the v_2 is explained by the radial flow effect with blast wave model in [8]. Additionally, KE_T scaling does not work out at $\sqrt{s_{NN}}$ = 2.76 TeV. Proton v_2 is shifted to higher p_T compared with π more than RHIC results.[9] We also measured particle identified v_2 for π/K/p in Au+Au at $\sqrt{s_{NN}}$ = 200 GeV. Mass ordering can be seen at low p_T and baryon and meson splitting are also observed at mid p_T which seems to be consistent to n_q scaling.[12]

In addition to the fact that $v_2(p_T)$ is consistent at $\sqrt{s_{NN}}$ = 39 - 200 GeV, v_2 normalized by n_q + KE_T, eccentricity, and $N_{\text{part}}^{1/3}$ scaling follows a universal curve as shown in the right panel of Figure 2. This figure includes the 45 curves for π/K/p in Au+Au at $\sqrt{s_{NN}}$ = 200 GeV, in Au+Au at $\sqrt{s_{NN}}$ = 62.4 GeV and in Cu+Cu at $\sqrt{s_{NN}}$ = 200 GeV for the five centrality bins from 0 - 50% in 10% steps. The combined data is fit with a single 3rd-order-polynomial, producing a χ^2/NDF = 1034/490 = 2.11 (including both statistical and systematic uncertainties). [10] This is a universal scaling for

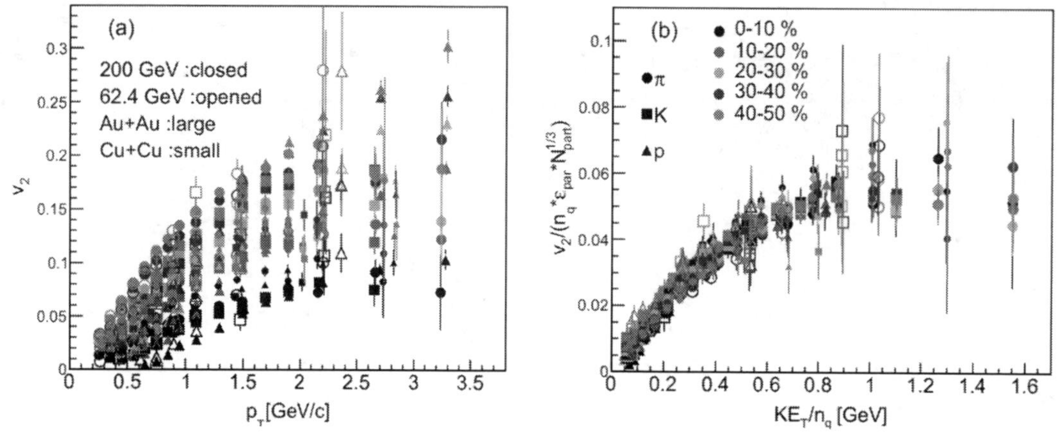

Figure 2. The left panel shows v_2 vs. p_T and the right panel shows $v_2/(\varepsilon \cdot N_{\rm part}^{1/3} \cdot n_q)$ vs. KE_T/n_q for π/K/p in Au+Au at 200 GeV, in Au+Au at 62.4 GeV and in Cu+Cu at 200 GeV for five centrality bins over 0- 50 % in 10 % steps for each system.There are 45 data sets in each panel. [10]

v_2 with different energies, collision sizes and particle species, and it indicates that v_2 is determined not only by the geometrical eccentricity but also by the size of collision. This scaling assume that differential v_2 is consistent above 39 GeV while "v_2/ε vs. (1/S)(dN/dy)" scaling plotted in [7] assumes that the higher collision energy produces higher v_2. Therefore, this $N_{\rm part}^{1/3}$ scaling works better for the differential v_2 at PHENIX results. The size dependence of v_2 can be understood as thermal freeze-out nature of produced particles based on hydrodynamical behavior, which is different from that of chemical freeze-out. [8] Moreover, this $v_n/(\varepsilon_n \cdot N_{\rm part}^{1/3} \cdot n_q)$ scaling is consistent with v_3 results in Au + Au and Cu + Au for inclusive charged hadrons.

Acknowledgements

I would like to express my great thanks to the organizers of this conference, ISMD 2016, for the opportunity to present these results.

References

[1] S. S. Adler et al., *Phys. Rev. Lett.* **91**, 182301 (2003)
[2] A. Adare et al., *Phys. Rev. Lett.* **98**, 162301 (2007)
[3] A. Adare et al., *Phys. Rev. Lett.* **107**, 252301 (2011)
[4] S. A. Voloshin, A. M. Poskanzer and R. Snellings (2008) nucl-ex/08092949
[5] B. Alver et al., *Phys. Rev. Lett.* **98**, 242302 (2007)
[6] S. Afanasiev et al., *Phys. Rev. C* **80**, 024909 (2009)
[7] S. A. Voloshin for the STAR Collaboration, *J. Phys. G* **34**, S883-886, (2007)
[8] M. Shimomura for the PHENIX Collaboration, *J. Phys.: Conf. Ser.* **270** 012041 (2011)
[9] M. Shimomura for the PHENIX Collaboration, *Pos WPCF2011***055** (2011)
[10] A. Adare et al., *Phys. Rev. C* **92**, 034913 (2015)
[11] A. Adare et al., (2015) nucl-ex/arXiv:1509.07784
[12] H. Nakagomi for the PHENIX Collaboration, *J. Phys.Conf.Ser.***736** no.1 012015 (2016)

EPJ Web of Conferences **141**, 01004 (2017) DOI: 10.1051/epjconf/201714101004
ISMD 2016

Studies of three-particle correlations and reaction-plane correlators from STAR

Prithwish Tribedy (for the STAR Collaboration)[1,a]

[1] *Physics Department, Brookhaven National Laboratory, Upton, NY 11973, USA*

Abstract. We present STAR measurements of various harmonics of three-particle correlations in $\sqrt{s_{NN}} = 200$ GeV Au+Au collisions at RHIC. The quantity $\langle\cos(m\phi_1 + n\phi_2 - (m+n)\phi_3)\rangle$ is measured for inclusive charged particles for different harmonics m and n as a function of collision centrality, transverse momentum p_T and relative pseudorapidity $\Delta\eta$. These observables provide detailed information on global event properties like correlations between event planes of different harmonics and are particularly sensitive to the expansion dynamics of the matter produced in the collisions. We compare our measurements to different viscous hydrodynamic models. We argue that these measurements probe the three dimensional structure of the initial state and provide unique ways to constrain the transport parameters involved in hydrodynamic modeling of heavy-ion collisions.

1 Introduction

By now it has been established that relativistic heavy-ion collisions produce a strongly correlated Quark Gluon Plasma (sQGP). Recent experimental studies have focused on studying the properties of such sQGP. One of the striking properties of such a phase of matter is that it exhibits nearly perfect fluidity characterized by the smallest viscosity-to-entropy-density ratio η/s amongst all known fluids in the nature. Over the past years, combined insights from both experiment through measurements of anisotropic flow coefficients v_n and from theory through viscous hydrodynamic simulations, have made precise extraction of η/s possible. However a very intrinsic characteristic about such transport parameter, i.e. its temperature dependence is not yet fully constrained by experimental measurements. A primary goal of the current measurements at RHIC is therefore to go beyond conventional measurements of flow coefficients and provide new observables that can be useful to constrain $\eta/s(T)$.

In this work we present the measurements of three particle correlations from the STAR experiment using the observable $C_{m,n,m+n} = \langle\langle\cos(m\phi_1 + n\phi_2 - (m+n)\phi_3)\rangle\rangle$, where m, n defines the harmonic coefficients and $\phi_{1,2,3}$, the azimuthal angles of three particles [1]. The inner and the outer averages are taken over all triplets and events respectively. The observable $C_{m,n,m+n}$ can be approximated as correlations of flow harmonics, v_ns, and corresponding event plane angles, Ψ_ns, as $\langle v_m v_n v_{m+n}\cos(m\Psi_m + n\Psi_n - (m+n)\Psi_{m+n})\rangle$. Theoretical studies show that such an observable can probe non-linear hydrodynamic response and therefore become more sensitive to viscosity than individual flow harmonics v_n [1–10]. Better sensitivity to viscous effects can be very useful towards

[a] e-mail: ptribedy@bnl.gov

more precise extraction of different transport parameters and possibly their temperature dependence by comparison to hydrodynamic simulations [11, 12]. Measurements of event plane and flow harmonic correlations have been performed at LHC by ATLAS collaboration [13] and recently by ALICE collaboration[14]. However measurements at a single energy is not sufficient to constrain $\eta/s\ (T)$. LHC measurements are sensitive to the η/s at higher temperatures, meanwhile full constraint on $\eta/s\ (T)$ can only be achieved with complementary measurements of $C_{m,n,m+n}$ at RHIC [11, 12, 15, 16]. In fact, measurements over the entire range of energy available under the Beam Energy Scan (BES) program at RHIC will be most preferable in this context. Measurements at RHIC have additional advantages. Since the beam rapidity is smaller one expects stronger variation of initial geometry, fluctuations, energy density, temperature, baryon density etc. over a relatively smaller window of rapidity as compared to LHC. In this context, measurements of $C_{m,n,m+n}$ on the pseudorapidity separation between particles may allow us to study the breaking of longitudinal invariance, three dimension structure of the initial state [3, 17–20] over relatively smaller widow of acceptance available for measurements at RHIC than LHC. In addition to constraining initial state and transport parameters, the charge dependence of three particle correlation can be used to search for the signals of the chiral magnetic effect (CME) [21–24]. In this work we will not study such charge dependence, however, we expect that the results presented here for inclusive charged particles will provide important baseline for the CME measurements.

2 Experiment and analysis

We analyze the data on Au+Au collisions at $\sqrt{s_{NN}}$ = 200 GeV collected by the STAR detector [25] during 2011 year running of RHIC. For the measurements of $C_{m,n,m+n}$ we use charged particles within the pseudorapidity range of $|\eta| < 1$ and transverse momentum of $p_T > 0.2$ GeV/c detected by the Time Projection Chamber (TPC), the primary tracking systems of STAR situated inside a 0.5 Tesla solenoidal magnetic field [26]. We use algebra based on Q-vectors and in order to account for imperfections in the detector acceptance we apply track-by-track weights [27, 28]. We also apply momentum dependent tracking efficiency. In such estimation, we correct for the track-merging artifacts by measuring the relative pseudo rapidity separation between any two tracks and correcting for missing pairs apparent at $\Delta\eta \approx 0$. We estimate systematic uncertainties in our measurements by analyzing datasets of different time periods, from different years, with different tracking algorithms, with different efficiency estimates, by varying z-vertex position of the collision, and by varying track selection criteria. In addition we also quantify the effects of short-range quantum and Coulomb correlations in the systematic uncertainties by studying $\Delta\eta$ dependence of $C_{m,n,m+n}$. Finally for data-model comparison we estimate the number of participant nucleons $N_{\rm part}$ using a Monte-Carlo Glauber model for different centrality intervals ($0-5\%, 5-10\%, 10-20\%, ..., 70-80\%$) used in this analysis [29, 30]. For selection of such centrality bins we use the distribution of minimum bias uncorrected multiplicity of charged particles in the pseudorapidity region $|\eta| < 0.5$ measured by the TPC.

3 Results and discussion

In this conference proceedings we present results for the correlators $C_{1,1,2}, C_{1,2,3}, C_{2,2,4}, C_{2,3,5}$. We first present differential measurements such as $\Delta\eta$ and p_T dependence of these correlators, the goal of such study is to understand how different physical scenarios effect these observables. We later on present integrated measurements i.e. the centrality dependence of $C_{m,n,m+n}$ and make comparisons to viscous hydrodynamic model calculations with different assumptions of $\eta/s(T)$.

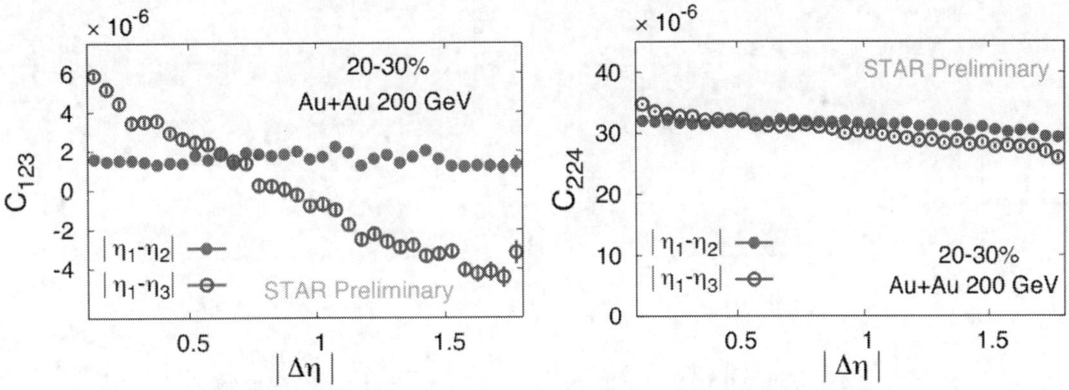

Figure 1. Relative pseudo rapidity dependence of the three particle correlator $C_{1,2,3}$ and $C_{2,2,4}$.

3.1 $\Delta\eta$ dependence

In Fig. 1 we show the $\Delta\eta$ dependence of $C_{1,2,3}$ and $C_{2,2,4}$. A similar measurement for $C_{1,1,2}$ was previously presented by STAR in Ref [22] [1]. One can clearly see in Fig. 1(left) that $C_{1,2,3}$ correlator shows a very strong dependence on $\Delta\eta_{1,3}$ (i.e. between the first and third order harmonics) but a very weak dependence on $\Delta\eta_{1,2}$ (i.e. between first and second order harmonics). We omit the curve for the variation of $C_{1,2,3}$ with $\Delta\eta_{2,3}$ for clarity which looks very similar to the curve shown for $\Delta\eta_{1,2}$. A strong dependence of $C_{1,1,2}$ correlator for $\Delta\eta_{1,2}$ (i.e. between two first order harmonics) was also observed in the previous STAR measurement in Ref [22]. In contrast, the similar measurement shown in Fig.1 (right) for the correlator $C_{2,2,4}$ with two possible combinations of $\Delta\eta$ shows a much weaker dependence compared to its absolute magnitude. These observations indicate a very specific pattern for three particle correlations. The relative rapidity dependence between either "first-first" or "first-third" harmonics show strong variations and even change of sign, whereas between second and any other harmonics the correlations show much weaker variation in relative rapidity.

Variations of $C_{m,n,m+n}$ with $\Delta\eta$ can come from hydrodynamic response to the three-dimensional structure of initial state [3, 17–20]. They can also arise from artifacts such as short-range correlations, non-flow and resonance decays [21], etc., that give rise to two-particle correlations that are correlated to an event plane (determined by the third particle) and do not vanish after averaging over many events. However, if such a variation persist up to large $\Delta\eta$, e.g. as shown in Fig.1 for $C_{1,2,3}$ vs $\Delta\eta_{1,3}$, they can not be driven by short range correlations. In a flow scenario, strong variation in $\Delta\eta$ can come from de-correlation in initial state geometry, e.g. driven by a breaking of longitudinal invariance through a forward-backward rapidity dependence of harmonic planes particularly between Ψ_1 and Ψ_3 [3]. In case of Ψ_2, one do not expect strong variation with rapidity due to geometry of collisions. In a non-flow scenario, in case of $C_{1,2,3}$ one possible source of $\Delta\eta_{1,3}$ dependence could be momentum conservation that leads to back-to-back correlations between two particles from jets that are correlated to second order event plane. We discuss such scenario in the next section.

3.2 p_T dependence

The effect of momenta conservation is expected to be dominant at higher transverse momentum and for low multiplicity events. Therefore, measurements performed in peripheral events can be a good

[1] The $\Delta\eta$ dependence for all other harmonics of $C_{m,n,m+n}$ will be presented in a future publication.

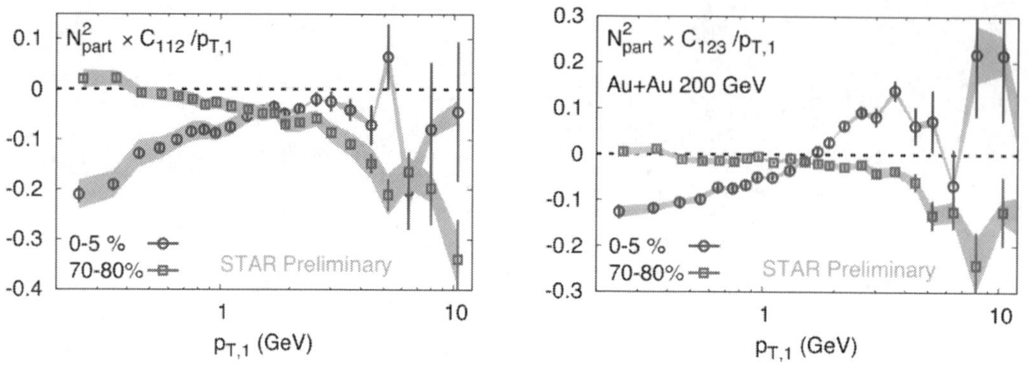

Figure 2. Transverse momentum dependence of the three particle correlator $C_{1,1,2}$ and $C_{1,2,3}$.

baseline for such studies. In the central events due to large number of particles, quenching of jet-like correlations etc., the effect of momentum conservation will not be dominant. It is therefore essential to perform this exercise in both central (e.g. 0−5%) and peripheral (e.g. 70−80%) and contrast the trend seen in data. From $\Delta\eta$ dependence of $C_{m,n,m+n}$, as discussed in previous section, we find that the correlators involving first order harmonics can be sensitive to non-flow effects such as momentum conservations from jets etc. In Fig. 2 we therefore study the variations of the correlators $C_{1,1,2}$ and $C_{1,2,3}$ with the transverse momentum p_T of the particle corresponding to the first order harmonic, i.e. for the first particle as denoted by $p_{T,1}$. In order to remove trivial increase of first order harmonic v_1 with transverse momentum and trivial dilution of correlation while going from peripheral to central events we multiply the correlator by a factor of $N^2_{\rm part}/p_{T,1}$. The results for 70−80% indicates that at high $p_{T,1}$ both $C_{1,1,2}$ and $C_{1,2,3}$ becomes negative. Such trend is consistent with a picture of momentum conservation and can be understood as follows. If a pair of back-to-back particles gets aligned along Ψ_2, they will lead to negative values for these correlators since then we have $C_{1,1,2} \approx \cos(\pi)$ and $C_{1,2,3} \approx \cos(\pm 3\pi)$. This might explain the decreasing trend for 70−80% events. However such a scenario can not explain the trend seen 0 − 5% events where one finds negative signal at small $p_{T,1}$ and nearly zero or positive signal at large $p_{T,1}$. This qualitatively different trend seen in central events can not be due to non-flow correlations from back-to-back pairs.

Clearly the differential measurements of these correlators can provide better insights of the relative contributions of different sources of correlations that can affect $C_{m,n,m+n}$. Model calculations that include full treatment of three-dimensional initial geometry, fluctuations and different other sources of correlations can improve our understanding in this context [31].

3.3 Centrality dependence

We measure the centrality dependence of $C_{m,n,m+n}$ and compare our results with three different viscous hydrodynamic model calculations. They include 1) hydrodynamic simulations by Teaney and Yan [3, 5], 2) the perturbative QCD+saturation+hydro based "EKRT" model [11] and 3) hydrodynamic simulations MUSIC [32] with IP-Glasma initial conditions [33]. In addition we also estimate the correlations from initial state geometry using Monte Carlo Glauber model by approximating $C_{m,n,m+n} = \langle \varepsilon_m \varepsilon_n \varepsilon_{m+n} \cos(m\Phi_m + n\Phi_n - (m+n)\Phi_{m+n})\rangle$, where ε_ns and Φ_ns are the initial eccentricities and the participant planes respectively. All of these models have been previously constrained by the measurements of v_n and other data on azimuthal correlations from RHIC and LHC, but they do

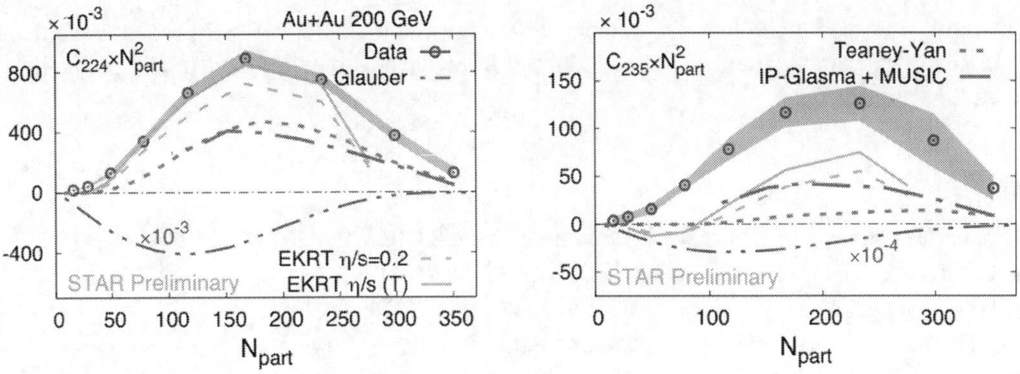

Figure 3. Centrality dependence of three particle correlator $C_{1,1,2}$ and $C_{2,2,4}$ compared to different viscous hydrodynamic model calculations.

not include longitudinal dependence in the initial state and assume boost invariance. From Fig.1 it is evident that the correlator $C_{2,2,4}$ has the least variation on $\Delta\eta$ and will provide the best opportunity for comparison to boost-invariant hydrodynamic simulations. We therefore present the centrality dependence of the correlator $C_{2,2,4}$ in Fig.3 (left). In Fig.3 (right) we also compare the centrality dependence of $C_{2,3,5}$. In order to scale out the trivial dilution of correlations due to increase of number of pairs while going from peripheral to central events we have multiplied the correlators by $N^2_{\rm part}$. The Glauber model calculations predict that purely initial state correlation of eccentricities and participant planes leads to negative values for all the correlators. Both $C_{2,2,4}$ and $C_{2,3,5}$ being positive in data indicates the dominance of non-linear hydrodynamic response of the medium to initial state geometry. This observation is consistent to the measurement at LHC by the ATLAS collaboration in Ref [13]. We however, find that although the qualitative trends predicted by different viscous hydrodynamic simulations are similar to data, some quantitative differences exist. Particularly for $C_{2,2,4}$ one can see that the current precision of the data can very well differentiate between constant and temperature dependent viscosity used in the EKRT simulations. Such comparisons would be key to constrain $\eta/s(T)$. Apart from analysis at $\sqrt{s_{NN}}$ = 200 GeV, our future studies will be focused on measurements for other lower energies under RHIC Beam Energy Scan program which will provide better constraints of $\eta/s(T)$.

4 Summary

In summary, we have presented the first measurements of three-particle correlations $C_{m,n,m+n} = \langle\langle\cos(m\phi_1 + n\phi_2 - (m+n)\phi_3)\rangle\rangle$ in $\sqrt{s_{NN}}$ = 200 GeV Au+Au collisions at RHIC. In comparison to conventional flow harmonic measurements these correlators can provide additional information such as de-correlation of event planes driven by three dimensional structure of the initial state and non-linear hydrodynamic response of the medium. When compared to viscous hydrodynamic models these measurements with the precision presented here have the potential to constrain transport parameters and their temperature dependence.

Acknowledgement

This work was supported under Department of Energy Contract No. DE-SC0012704. We thank Li Yan, Risto Paatelainen, Harri Niemi and Gabriel Denicol for providing their model predictions and helpful discussion.

References

[1] R.S. Bhalerao, J.Y. Ollitrault, S. Pal, Phys. Rev. **C88**, 024909 (2013), 1307.0980
[2] Z. Qiu, U. Heinz, Phys. Lett. **B717**, 261 (2012), 1208.1200
[3] D. Teaney, L. Yan, Phys. Rev. **C83**, 064904 (2011), 1010.1876
[4] D. Teaney, L. Yan, Phys. Rev. **C86**, 044908 (2012), 1206.1905
[5] D. Teaney, L. Yan, Phys. Rev. **C90**, 024902 (2014), 1312.3689
[6] L. Yan, J.Y. Ollitrault, Phys. Lett. **B744**, 82 (2015), 1502.02502
[7] J. Qian, U.W. Heinz, J. Liu (2016), 1602.02813
[8] J. Qian, U. Heinz (2016), 1607.01732
[9] S. McDonald, C. Shen, F. Fillion-Gourdeau, S. Jeon, C. Gale (2016), 1609.02958
[10] B. Betz et al. (2016), 1609.05171
[11] H. Niemi, K.J. Eskola, R. Paatelainen, Phys. Rev. **C93**, 024907 (2016), 1505.02677
[12] G. Denicol, A. Monnai, B. Schenke, Phys. Rev. Lett. **116**, 212301 (2016), 1512.01538
[13] G. Aad et al. (ATLAS), Phys. Rev. **C90**, 024905 (2014), 1403.0489
[14] J. Adam et al. (ALICE), Phys. Rev. Lett. **117**, 182301 (2016), 1604.07663
[15] H. Niemi, G.S. Denicol, P. Huovinen, E. Molnar, D.H. Rischke, Phys. Rev. Lett. **106**, 212302 (2011), 1101.2442
[16] C. Gale, S. Jeon, B. Schenke, P. Tribedy, R. Venugopalan, Phys. Rev. Lett. **110**, 012302 (2013), 1209.6330
[17] P. Bozek, W. Broniowski, J. Moreira, Phys. Rev. **C83**, 034911 (2011), 1011.3354
[18] J. Jia, P. Huo, Phys. Rev. **C90**, 034915 (2014), 1403.6077
[19] L.G. Pang, H. Petersen, G.Y. Qin, V. Roy, X.N. Wang, Eur. Phys. J. **A52**, 97 (2016), 1511.04131
[20] B. Schenke, S. Schlichting (2016), 1605.07158
[21] B.I. Abelev et al. (STAR), Phys. Rev. Lett. **103**, 251601 (2009), 0909.1739
[22] B.I. Abelev et al. (STAR), Phys. Rev. **C81**, 054908 (2010), 0909.1717
[23] L. Adamczyk et al. (STAR), Phys. Rev. **C88**, 064911 (2013), 1302.3802
[24] S.A. Voloshin, Phys. Rev. **C70**, 057901 (2004), hep-ph/0406311
[25] K.H. Ackermann et al. (STAR), Nucl. Instrum. Meth. **A499**, 624 (2003)
[26] M. Anderson et al., Nucl. Instrum. Meth. **A499**, 659 (2003), nucl-ex/0301015
[27] A. Bilandzic, R. Snellings, S. Voloshin, Phys. Rev. **C83**, 044913 (2011), 1010.0233
[28] A. Bilandzic, C.H. Christensen, K. Gulbrandsen, A. Hansen, Y. Zhou, Phys. Rev. **C89**, 064904 (2014), 1312.3572
[29] B.I. Abelev et al. (STAR), Phys. Rev. **C79**, 034909 (2009), 0808.2041
[30] M.L. Miller, K. Reygers, S.J. Sanders, P. Steinberg, Ann. Rev. Nucl. Part. Sci. **57**, 205 (2007), nucl-ex/0701025
[31] R.S. Longacre (2016), 1609.09848
[32] B. Schenke, S. Jeon, C. Gale, Phys. Rev. **C82**, 014903 (2010), 1004.1408
[33] B. Schenke, P. Tribedy, R. Venugopalan, Phys. Rev. Lett. **108**, 252301 (2012), 1202.6646

EPJ Web of Conferences **141**, 01005 (2017) DOI: 10.1051/epjconf/201714101005
ISMD 2016

Oscillations in counting statistics

Grzegorz Wilk[1,a] and Zbigniew Włodarczyk[2,b]

[1] *National Centre for Nuclear Research, Hoża 69, 00-681 Warsaw, Poland*
[2] *Institute of Physics, Jan Kochanowski University, Świętokrzyska 15, 25-406 Kielce, Poland*

Abstract. The very large transverse momenta and large multiplicities available in present LHC experiments on pp collisions allow a much closer look at the corresponding distributions. Some time ago we discussed a possible physical meaning of apparent log-periodic oscillations showing up in p_T distributions (suggesting that the exponent of the observed power-like behavior is complex). In this talk we concentrate on another example of oscillations, this time connected with multiplicity distributions $P(N)$. We argue that some combinations of the experimentally measured values of $P(N)$ (satisfying the recurrence relations used in the description of cascade-stochastic processes in quantum optics) exhibit distinct oscillatory behavior, not observed in the usual Negative Binomial Distributions used to fit data. These oscillations provide yet another example of oscillations seen in counting statistics in many different, apparently very disparate branches of physics further demonstrating the universality of this phenomenon.

1 Introduction

To get deeper insight into the dynamics of the multiparticle production processes one measures, in addition to single particle distributions, all kinds of correlations between produced secondaries (i.e., many particle distributions). However, we argue that the very large transverse momenta (p_T) and large multiplicities (N) available in present LHC experiments on pp collisions already allow for such insight on the level of the corresponding single particle distributions. A first example was provided in [1, 2] (and references therein) where it was shown that ratios of $data/fit$ for single particle p_T distributions show distinct, log-periodic oscillations in p_T. Because they are present in all LHC experiments, at all energies and for different colliding systems (i.e., also in *PbPb* collisions where they become rather strong for central collisions), it would be unreasonable to regard them as an experimental artifact. Taken seriously, they strongly suggest that the exponent of the observed power-like behavior of the measured p_T distributions is complex [1, 2] (the other possibility would be the existence of similar log-periodic oscillations of the scale parameter present in p_T distributions). We shall not pursue this subject here. Instead we concentrate on another example of oscillations, this time connected with the multiplicity distributions, $P(N)$. We demonstrate that some combinations of the experimentally measured values of $P(N)$, calculated from recurrence relations widely used in the description of the cascade-stochastic processes in quantum optics, exhibit distinct oscillatory behavior not observed in

[a] e-mail: grzegorz.wilk@ncbj.gov.pl
[b] e-mail: zbigniew.wlodarczyk@ujk.kielce.pl

the usual Negative Binomial Distributions used to fit data [3]. These oscillations provide therefore one more example of oscillations seen in counting statistics in many different, apparently very disparate, branches of physics, further demonstrating the universality of this phenomenon [3].

2 How to fit presently available data

Concerning $P(N)$ measured in multiparticle production processes, essentially for all collision energies studied, the most commonly used form is the two-parameter negative binomial distribution (NBD) function (see [4]) and references therein),

$$P(N;p,k) = \frac{\Gamma(N+k)}{\Gamma(k)\Gamma(N+1)} p^N (1-p)^k \qquad \text{with} \qquad p = p(m,k) = \frac{m}{m+k}, \tag{1}$$

where p is the probability of particle emission, N the observed number of particles and m and k are the two parameters of the NBD. Whereas m can be connected with the measured multiplicity, $m = \langle N \rangle$ (at least for $m = const$), the actual meaning of m and k depends on the particular dynamical picture behind the NBD form of $P(N)$. However, notwithstanding the popularity of the NBD, one observes systematic (and growing with energy) deviations of fits based on the NBD from data. The example of the recent CMS data [5] is shown in Fig. 1; the behavior of the ALICE data [6] is similar. For large multiplicities the experimental points are below the best single NBD fit. This feature is even more dramatic when one plots the ratio $R = P_{CMS}(N)/P_{NBD}(N)$, see Fig. 2 (red circles). In addition to the falling tail one can also see here some structure for smaller multiplicities. Whereas the tails of the distributions in $P(N)$ could be fitted by using weighted (incoherent) superposition of two NBD (cf., for example, [6, 7] and references therein), the structure for small multiplicities visible in Fig. 2 remained to be accounted for. To get a flat ratio R for the whole measured region of multiplicities N,

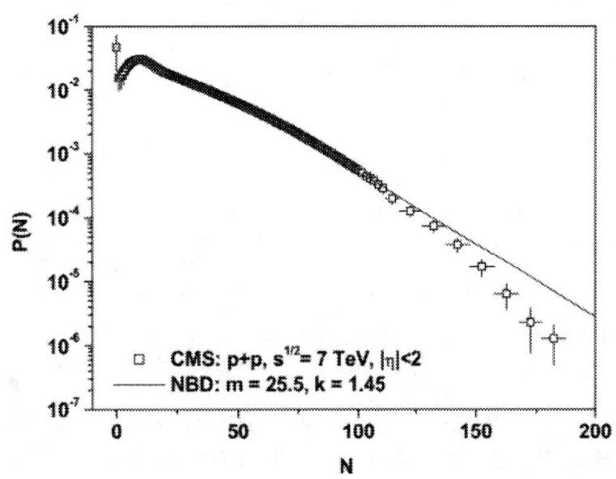

Figure 1. (Color online) Charged hadron multiplicity distributions for $|\eta| < 2$ at $\sqrt{s} = 7$ TeV, as given by the CMS experiment [5] (points), compared with the NBD for parameters $m = 25.5$ and $k = 1.45$ (solid line).

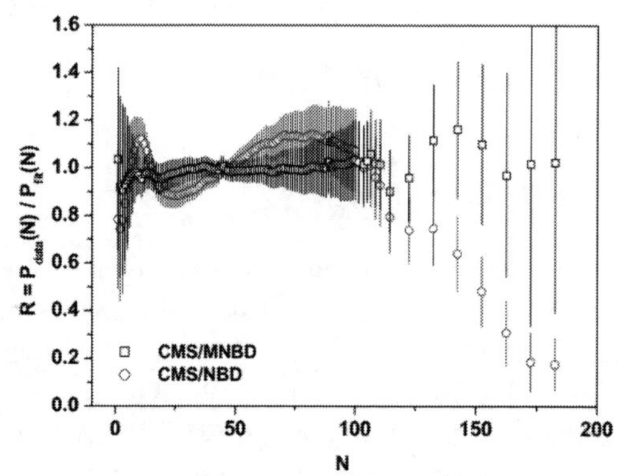

Figure 2. (Color online) Multiplicity dependence of the ratio $R = P_{CMS}(N)/P_{NBD}(N)$ for the data shown in Fig. 1 (red circles) and of the corrected ratio R (black squares) obtained from the MNBD discussed below.

cf., the black squares in Fig. 2, and still adhere to the single NBD form of $P(N)$, one has to modify it. We provide such a modification in [3] (and call it the Modified Negative Binomial Distribution - MNBD). It consists of allowing the parameter m in Eq. (1) to depend on the multiplicity N, $m = m(N)$. The desired flat ratio $R = P_{data}(N)/P_{fit}(N)$ can be obtained for

$$m = m(N) = c\exp(a|N-b|) \quad \text{corresponding to} \quad p(N) = \frac{1}{1 + \frac{c}{k}\exp(a|N-b|)} \tag{2}$$

where a, b and c are parameters. Both $m(N)$ and the probability of particle emission $p(N)$ are now non-monotonic functions of N ($p(N)$ = const in the standard NBD). This non-monoticity is located in the region of small multiplicities N [3]. Note that such a spout-like form of the proposed modification is just one of the simplest possible choices of parametrization bringing agreement with data. Whereas we cannot at this moment offer any plausible interpretation of such choice, it obviously violates the usual infinite divisibility property of the NBD, bringing it near to the predictions of QCD [8].

3 Another look at multiplicity distributions $P(N)$

To go further we note that the form of $P(N)$ can be specified by providing the relevant recurrence relation. The simplest one connects only the adjacent distributions, $P(N)$ and $P(N+1)$ (assuming that only neighboring multiplicities can influence each other):

$$(N+1)P(N+1) = g(N)P(N). \tag{3}$$

The function $g(N)$ determines the form of the $P(N)$. Its simplest nontrivial linear form, $g(N) = \alpha + \beta N$, covers, for example, the Poissonian distribution (for $\alpha = \langle N \rangle$ and $\beta = 0$), the binomial distribution (for $\alpha = \langle N \rangle k/(k - \langle N \rangle)$ and $\beta = -\alpha/k$) and the NBD (for $\alpha = \langle N \rangle k/(k + \langle N \rangle)$ and $\beta = \alpha/k$). When searching for the best $P(N)$ to fit data, $g(N)$ may be modified accordingly (for example, by introducing higher order terms [9] or by using its more involved forms [10]), in our MNBD $\alpha = k/[1 + k\exp(-a|N-b|)/c])$. However, there also exists a more general form of such a relation, widely used in all processes involving counting statistics (mainly in quantum optics and in *cascade-stochastic processes*), which connects the multiplicity $N+1$ with all smaller multiplicities [11]:

$$(N+1)P(N+1) = \langle N \rangle \sum_{j=0}^{N} C_j P(N-j). \tag{4}$$

Here $P(N)$ is defined by the coefficients C_j (used in the multiparticle phenomenology in [12, 13]). This recurrence relation provides more information than Eq. (3) and can be regarded as an expansion of $g(N)$ in series of $P(N-j)$ with coefficients C_j, which are independent of N. Dependence on N is in the $P(N-j)$ themselves and in the limits of summation. For the NBD case $g(N)$ from Eq. (3) is given by

$$g(N) = \frac{\langle N \rangle}{P(N)} \sum_{j=0}^{N} C_j P(N-j) = \frac{mk}{m+k} \frac{\Gamma(N+1)}{\Gamma(N+k)} \sum_{j=0}^{N} \frac{\Gamma(N+k-j)}{\Gamma(N+1-j)} = \frac{m}{m+k}(k+N) \tag{5}$$

(reproducing the form mentioned before). The important feature of Eq. (4) is that it can be easily inverted, i.e., knowing (for example, from the measurement process) all $P(N)$, one can use this *experimental information* to deduce the corresponding C_j by means of the following recurrence relation [3],

$$C_j = \frac{(j+1)}{\langle N \rangle} \left[\frac{P(j+1)}{P(0)} \right] - \sum_{i=0}^{j-1} C_i \left[\frac{P(j-i)}{P(0)} \right]. \tag{6}$$

In this way one obtains *additional information* on $P(N)$, namely not only their values but also how much they are influenced by the neighboring $P(N-j)$, where $j \in [1, N-1]$. Fig. 3 shows clearly that this is, indeed, the case. One observes a distinct oscillatory, exponentially damped, behavior of the C_j obtained from the CMS data [5]. This can be fitted by an exponentially damped triangular wave:

$$C_j = \frac{1}{\langle N \rangle} \left\{ a_1 \left[1 - \left| 1 - 2\left(\frac{j+\delta}{\omega} - Int\left(\frac{j+\delta}{\omega} \right) \right) \right| \right] - a_2 \right\} \cdot \exp\left(-\frac{j+\delta}{\lambda} \right). \tag{7}$$

 DOI: 10.1051/epjconf/201714101005

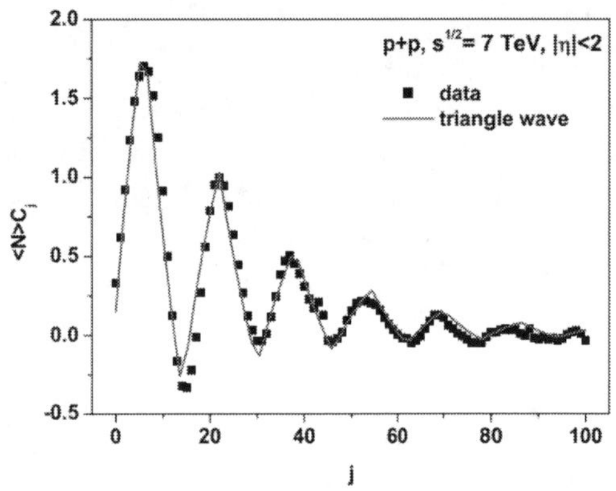

Figure 3. (Color online) coefficients C_j obtained from the pp collisions CMS data for $\sqrt{s}$ = 7 TeV and pseudorapidity window $|\eta| < 2$ [5] fitted using the parametrization given by Eq. (7) with parameters $a_1 = 3.2$, $a_2 = 0.6$, $\omega = 16$, $\delta = 1.67$ and $\lambda = 25$.

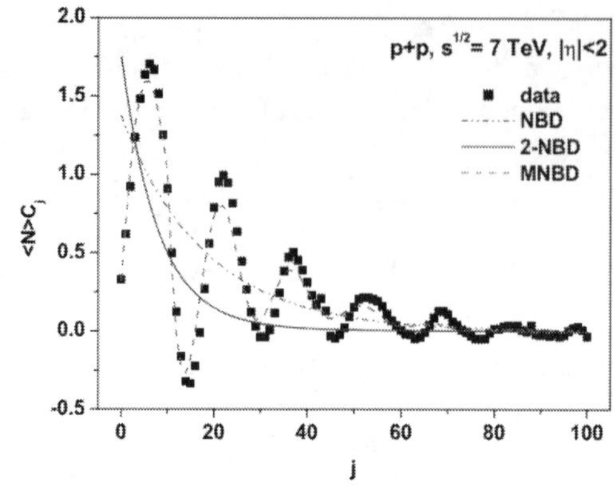

Figure 4. (Color online) comparison of the previous C_j with coefficients obtained from different fits to data on $P(N)$: using the single NBD, the 2-component NBD (2-NBD) with parameters from [7] and our MNBD fit. Only the last choice fits C_j.

(the parameter ω describes the observed periodicity period). Fig. 4 shows the same C_j compared with coefficients obtained from different fits to data on the $P(N)$, starting from the single NBD, using the 2-component NBD (2-NBD) with parameters taken from a fit to the data presented in [7] and, finally, using our MNBD fit. One can see that only our MNBD fit reproduces the experimentally obtained coefficients C_j. In [3] it is also shown that the amplitude of these oscillations decrease with the narrowing of the pseudo-rapidity window, $|\eta|$, in which the data are collected (eventually they vanish for small $|\eta|$) and that the same behavior is also observed for the ALICE data [6] (albeit with stronger oscillations in this case).

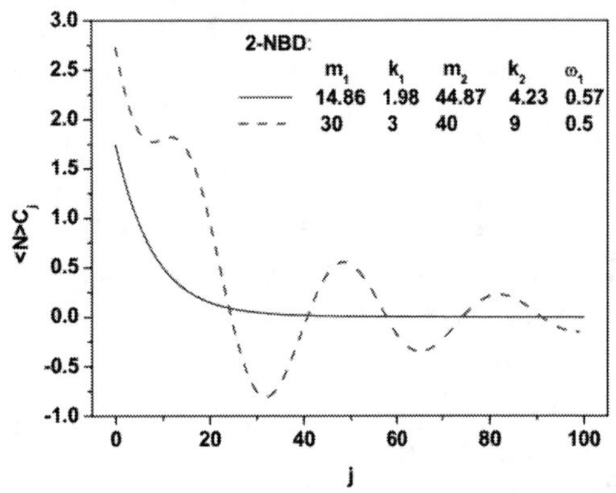

Figure 5. (Color online) Coefficients C_j from the 2-component NBD with parameters from [7] compared with C_j from the $2 - NBD$ with special choice of parameters leading to the oscillations (dashed line).

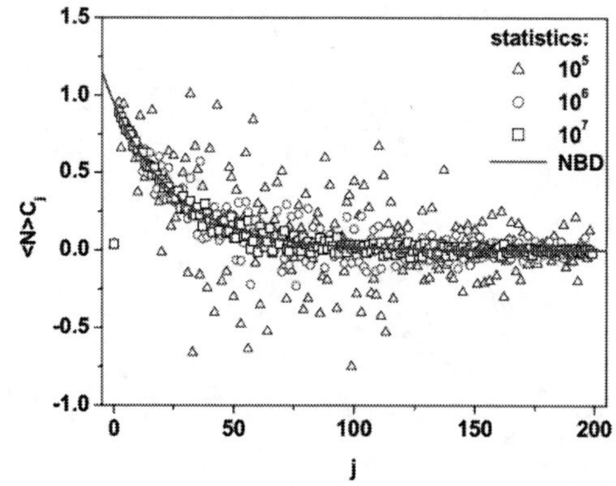

Figure 6. (Color online) Illustration of dependence on the statistics used. C_j are calculated for $P(N)$ which were constructed using different number of events sampled from the NBD.

A word of caution is in order at this point. In the analysis presented so far the measured $P(N)$ are assumed to be fully reliable. Any sensitivity of C_j to the systematic uncertainties of the measurement, to the unfolding uncertainties or, in general, to any other details of the experimental procedure used to obtain $P(N)$, which were subsequently published and which we have used above, can be checked only on the level of experiment, namely by the careful analysis of the raw data, using the proper response matrix, and so on. This exceeds our capability. This means that, in what follows, we feel free to proceed further with our reasoning. Firstly, let us observe in Fig. 4 that the coefficients C_j calculated from single or double NBD $P(N)$ used so far to fit the CMS data simply drop monotonically without any sign of fluctuations. However, when we use probabilities of particle emission $p = p(N)$ from MNBD, Eq. (2), with parameters chosen to fit the CMS data, then the coefficients C_j follow *exactly* the oscillatory behavior of the C_j obtained directly from these data, cf. Fig. 4. The observed monotonic behavior of C_j for a single NBD is expected, because in this case

$$C_j = \frac{k}{\langle N\rangle}p^{j+1} = \frac{k}{k+m}\exp(j\ln p), \tag{8}$$

and they depend only on the rank j and on the probability of particle emission p, which in this case does not depend on N and is smaller than unity. Such C_j reproduce only the monotonically falling damping exponent in Eq. (7), with $\lambda = -1/\ln p$). (It is worth noting that in the case of a binomial distribution (BD) one has $C_j = (-1)^j(k/m)[m/(k-m)]^{(j+1)}$, which oscillate very rapidly with period 2.). Also in the case of C_j obtained from the 2-NBD used to fit the CMS data [5] (with parameters given in [7]) no oscillations occur. However, as shown in Fig. 5, in this case one can find such a set of parameters that the combination of these 2 NBDs provides oscillatory behavior for the corresponding parameters C_j (although it does not necessarily fit data on $P(N)$, at least not in the case presented here, which is aimed only to demonstrate the possibility of oscillatory behavior in the case of multi-NBD compositions). As shown in [3], when we have a superposition of a number of NBD, $P(N) = \sum_i \omega_i P_{NBD}(N, p_i)$, with weights ω_i (with $\sum_i \omega_i = 1$) and emission probabilities $p_i = m_i/(m_i + k_i)$, then the corresponding coefficients are

$$\begin{aligned} C_j \;&=\; \frac{1}{P(0)}\sum_i \omega_i p_i^j (1-p_i)^{k_i+1}\,\cdot \\ &\cdot \left\{\frac{\Gamma(j+k_i+1)}{\Gamma(k_i+1)\Gamma(j+1)}\frac{m_i-\langle N\rangle}{\langle N\rangle}+1+\sum_{l=0}^{j-1}\left[1-\frac{C_l}{p_i^l(1-p_i)}\right]\frac{\Gamma(j-l+k_i)}{\Gamma(k_i)\Gamma(j-l+1)}\right\}. \end{aligned} \tag{9}$$

For $m_i < \langle N\rangle$, as well as for $C_l > p_i^l(1-p_i)$, we have negative terms which can result in nonmonotonic behavior of the coefficients C_j.

4 Summary

To summarize: our results show that a successful model of multiparticle production should describe, with the same set of parameters, both the multiplicity distributions, $P(N)$, and the corresponding coefficients, C_j. This is because these coefficients disclose interrelations between $P(N)$ and all $P(N-j)$ with $j < N$, connected with some delicate, intrinsic correlations between multiplicities, which, in turn, depend on the details of the dynamics of multiparticle production. In fact, a similar conjecture was already reached some time ago using the so called *combinants*, $C_j^{\star}$ [8, 14–18]. It turns out that our coefficients C_j are directly connected to the combinants, namely

$$C_j = \frac{(j+1)}{\langle N\rangle}C_{j+1}^{\star}. \tag{10}$$

Eq. (10) shows that the C_i are much more sensitive to oscillations than the $C^{\star}_{j}$, especially for higher values of j. Note also that our analysis is not directly connected with the wave structure observed in the data on $P(N)$ for multiplicities above $N = 25$ [6]. The coefficients C_j are completely insensitive to the $P(N > j+1)$ tail of the multiplicity distribution, while the oscillatory behavior of C_j is observed starting from the very beginning.

As shown in Fig. 6 the coefficients C_j can be measured only for events with high enough statistics. That is why they were practically not visible in previous experiments, such high statistics is available only starting from recent LHC experiments. It is also worth pointing out that, in general, $\sum_{j=1}^{\infty} C_j \geq 1$ (this is direct consequences of the fact that $\sum_{N=j}^{\infty} P(N-j) \leq 1$). The strict equality holds in the case of $P(N)$ in the form of NBD. Note that in the case of combinants $C^{\star}_{j}$ one has that for the NBD $\sum_{j=1}^{\infty} C^{\star}_{j} = -k\ln[k/(k+m)]$.

Finally, let us note the experimental drawback of the C_j. This is the necessity to know (measure) the $P(0)$, cf. Eq. (6), which serves as a normalization factor in our approach. This is not an easy (to say the least) point in any experiment because $P(0)$ depends critically on the acceptance (cf. the discussion in [3]).

Acknowledgements: We are indebted to Sandor Hegyi, Edward Grinbaum-Sarkisyan and Adam Jachołkowski for fruitful discussions and we would like to thank warmly Nicholas Keeley for reading the manuscript.

References

[1] G.Wilk and Z.Włodarczyk, Entropy **17**, 384 (2015).

[2] G.Wilk and Z.Włodarczyk, Chaos, Solitons and Fractals **81**, 487 (2015).

[3] G.Wilk and Z.Włodarczyk, *How to retrieve additional information from the multiplicity distributions*, arXiv:1601.03883.

[4] W.Kittel and E.A.De Wolf, *Soft Multihadron Dynamics*, World Scientific, Singapore, 2005.

[5] V.Khachatryan *et al.* (CMS Collaboration), J. High Energy Phys.**01**, 079 (2011).

[6] J.Adam *et al.* (ALICE Collaboration), *Charged particle multiplicities in proton-proton collisions at* $\sqrt{s} = 0.9$ *to* 8 *TeV*, arXiv:1509.07541.

[7] P.Ghosh, Phys. Rev. D **85**, 0541017 (2012).

[8] S.Hegyi, Phys. Lett. B **388**, 837 (1996).

[9] T.F.Hoang and B.Cork, Z. Phys. C **36**, 323 (1987).

[10] S.V.Chekanov and V.I.Kuvshinow, J. Phys. G **22**, 601 (1996).

[11] B.E.A.Saleh and M.K.Teich, Proc. IEEE **70**, 229 (1982).

[12] V.D.Rusov, T.N.Zelentsova, S.I.Kosenko, M.M.Ovsyanko and I.V.Sharf, Phys. Lett. B **504**, 213 (2001).

[13] V.D.Rusov and I.V.Sharf, Nucl. Phys. A **764** 460 (2006).

[14] S.K.Kauffmann and M.Gyulassy, J. Phys. A **11**, 1715 (1978).

[15] R.Vasudevan, P.R.Vittal and K.V.Parthasarathy, J. Phys. A **17** 989 (1984).

[16] A.B.Balantekin and J.E.Seger, Phys. Lett. B **266** 231 (1991).

[17] S.Hegyi, Phys. Lett. B **318**, 642 (1993).

[18] S.Hegyi, Phys. Lett. B **463**, 126 (1999).

Correlations, multiplicity distributions, and the ridge in pp and p-Pb collisions

Alice Ohlson[1,a] for the ALICE Collaboration

[1] Ruprecht-Karls-Universität Heidelberg, Heidelberg, Germany

Abstract. Measurements made by the ALICE Collaboration of single- and two-particle distributions in high-energy pp and p–Pb collisions are used to characterize the interactions in small collision systems, tune models of particle production in QCD, and serve as a baseline for heavy-ion observables. The measurements of charged-particle multiplicity density, $\langle dN_{ch}/d\eta \rangle$, and multiplicity distributions are shown in pp and p–Pb collisions, including data from the top center-of-mass energy achieved at the Large Hadron Collider (LHC), $\sqrt{s} = 13$ TeV. Two-particle angular correlations in p–Pb collisions are studied in detail to investigate long-range correlations in pseudorapidity which are reminiscent of structures previously thought unique to heavy-ion collisions.

1 Introduction

In high-energy hadronic collisions, studies of inclusive single-particle distributions are used to investigate particle production in QCD. The charged-particle multiplicity density in pp and p–Pb collisions is measured by ALICE over a range of centre-of-mass energies, including the top LHC energy of $\sqrt{s} = 13$ TeV. The multiplicity distributions are also shown, and all the experimental data is compared with Monte Carlo models. Since the produced multiplicity is dominated by soft (low-momentum) particle production, which is in the non-perturbative regime of QCD, these measurements can be used to further constrain and tune models.

Beyond single-particle inclusive measurements, two-particle correlation studies have yielded surprising results in small collision systems, showing the presence of correlations between particles over large ranges in pseudorapidity in high-multiplicity pp and p–Pb collisions. These correlations are reminiscent of features observed in heavy-ion collisions where they are commonly attributed to anisotropic flow (v_n). The transverse momentum (p_T), pseudorapidity (η), and particle species dependence of v_2 in p–Pb collisions has been measured in ALICE. In particular, in the analysis of correlations between forward muons and mid-rapidity charged hadrons it is possible to measure the v_2 for large values of pseudorapidity in both the proton-going and Pb-going directions. These observations will be used to deepen our understanding of possible collective effects in small collision systems and their implications for heavy-ion physics.

[a] e-mail: alice.ohlson@cern.ch

 DOI: 10.1051/epjconf/201714101006

2 ALICE detector

The main subsystems of the ALICE detector [1] used in the analyses reported here are: the Inner Tracking System (ITS), the Time Projection Chamber (TPC), the Forward Muon Spectrometer (FMS), and the V0 system. The ITS, used for tracking and vertex reconstruction, consists of six layers of silicon detectors; the innermost two layers comprise the Silicon Pixel Detector (SPD). Short track segments ("tracklets") can be reconstructed using only the SPD, and are used in the multiplicity density and muon-hadron correlations analyses below. Information from the ITS and TPC can also be combined to fully reconstruct charged particle tracks. Muons are detected in the FMS, which has a pseudorapidity coverage of $-4 < \eta < -2.5$. The composition of parent particles of the detected muons depends on transverse momentum: at low $p_{\rm T}$ the muons predominantly come from weak decays of pions and kaons, while at high $p_{\rm T}$ the muons are largely the result of heavy flavor decays. The V0 detectors, located at forward rapidity (the V0A at $2.8 < \eta < 5.1$ and the V0C at $-3.7 < \eta < -1.7$), are used for triggering and also to classify the overall event activity. Symmetric pseudorapidity coverage can also be achieved by utilizing only two of the four rings in each V0 detector, the innermost two rings of the V0C ($-3.7 < \eta < -2.7$) and the outermost two rings of the V0A ($2.8 < \eta < 3.9$), as was done in the muon-hadron analysis below.

3 Multiplicity density

In ALICE, the charged-particle multiplicity density, $\langle dN_{ch}/d\eta \rangle$ has been measured across a wide range in center-of-mass energy, at $\sqrt{s} = 0.9, 2.36, 2.76, 7, 8,$ and 13 TeV. The multiplicity is measured in different classes of events, including inelastic events ('INEL'), inelastic events with at least one charged particle produced within $|\eta| < 1$ ('INEL>0'), and non-single-diffractive events ('NSD'). Figure 1 shows the results for the INEL and INEL>0 classes, which demonstrate power-law scaling with $\sqrt{s}$. Results from p–Pb collisions are also shown in Fig. 1 [2].

Additionally, the charged particle multiplicity density has been measured as a function of pseudorapidity in INEL and NSD events at $\sqrt{s} = 0.9$ and 2.36 TeV, and in INEL and INEL>0 events at 13 TeV, as shown in Fig. 2. The distributions are also compared to multiple Monte Carlo (MC) models

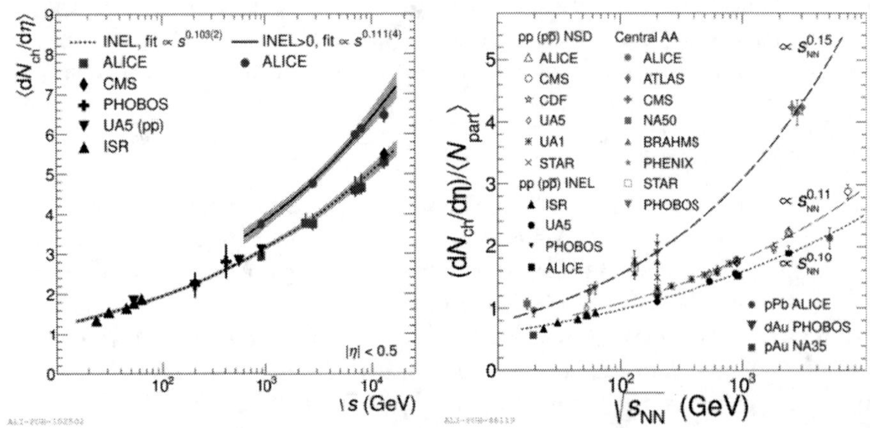

Figure 1. The mid-rapidity charged-particle multiplicity density, $\langle dN_{ch}/d\eta \rangle$, is shown as a function of center-of-mass energy for (left) pp and (right) pp, p–Pb, and Pb–Pb collisions at the LHC [2, 7].

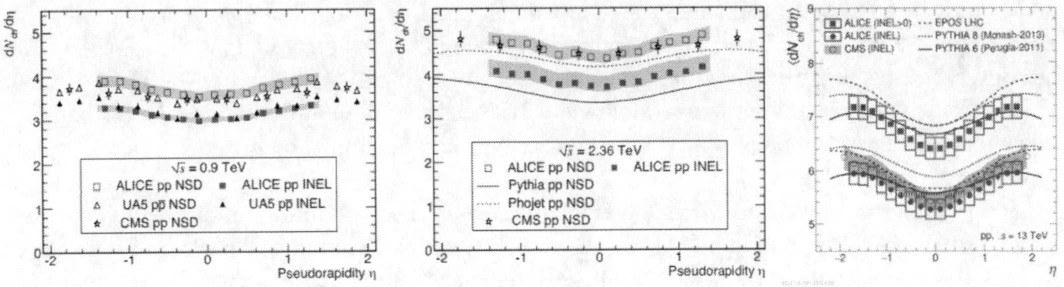

Figure 2. $\langle dN_{ch}/d\eta\rangle$ vs η in pp collisions is shown at $\sqrt{s}$ = 0.9 (left), 2.36 (center), and 13 TeV (right) [7, 8].

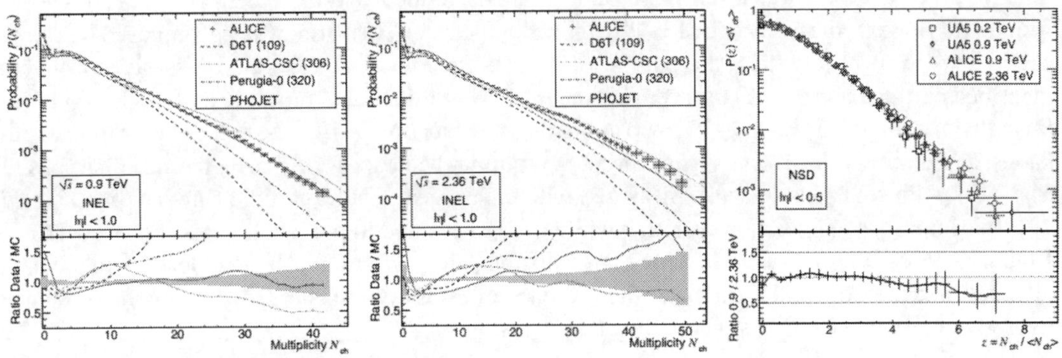

Figure 3. The multiplicity distributions are shown at $\sqrt{s}$ = 0.9 (left) and 2.36 TeV (center) with MC model comparisons. KNO scaling is also shown (right). [8]

including PYTHIA 6 [3], PYTHIA 8 [4], PHOJET [5], and EPOS LHC [6]. It can be observed in Fig. 2 that the model in best agreement with the $\sqrt{s}$ = 13 TeV data is PYTHIA 6. These results will be used for further tuning of the Monte Carlo generators.

4 Multiplicity distributions

The charged-particle multiplicity distributions, $P(N_{ch})$, were measured at $\sqrt{s}$ = 0.9 and 2.36 TeV, as shown in Fig. 3. The experimental data were compared to results from PHOJET and three PYTHIA 6 tunes (Perugia-0, ATLAS-CSC, and D6T). The best agreement with the data is achieved by PHOJET at $\sqrt{s}$ = 0.9 TeV and the ATLAS-CSC tune of PYTHIA 6 at $\sqrt{s}$ = 2.36 TeV. Furthermore, the multiplicity distributions were scaled by the mean multiplicity to obtain the distribution of $z = N_{ch}/\langle N_{ch}\rangle$, also shown in Fig. 3. The hypothesis that the distributions of $\langle N_{ch}\rangle P(z)$ are independent of center-of-mass energy is known as KNO scaling [9], and these experimental results indicate that KNO scaling holds up to approximately $z = 4$.

5 Two-particle correlations

Two-particle angular correlations, which are distributions in relative azimuthal angle ($\Delta\varphi = \varphi_{trig} - \varphi_{assoc}$) and relative pseudorapidity ($\Delta\eta = \eta_{trig} - \eta_{assoc}$) between trigger and associated particles, are used to study many aspects of the physics of heavy-ion collisions, in particular jet fragmentation and collective effects. In elementary collisions and small collision systems such as pp they show characteristic features attributed to jet production, while in heavy-ion collisions the same jet features are observed in addition to structures around $\Delta\varphi = 0$ (nearside) and $\Delta\varphi = \pi$ (awayside) extended in $\Delta\eta$. These long-range correlations, known as 'ridges,' are often attributed to hydrodynamic flow behavior in the quark-gluon plasma (QGP) and are typically quantified by the coefficients of a Fourier cosine series, v_n.

It was therefore surprising when a nearside ridge was observed in high multiplicity collisions of small systems, pp [10] and p–Pb [11]. Furthermore, it was observed that in p–Pb collisions at $\sqrt{s_{\rm NN}}$ = 5.02 TeV the nearside peak yields are mostly independent of multiplicity [12], meaning that for the same trigger and associated $p_{\rm T}$ the same jet population is selected regardless of multiplicity. This served as justification to subtract the correlations in low-multiplicity events from the high-multiplicity correlation functions in order to remove correlations due to jet and minijet fragmentation. This subtraction procedure (illustrated in Fig. 4) showed the nearside ridge more clearly and also revealed a symmetric ridge on the awayside [13, 14]. This 'double ridge' structure was decomposed into Fourier coefficients in order to extract the parameter v_2 in p–Pb collisions. The analysis was repeated with identified particles and it was observed that the v_2 shows similar mass ordering as was observed in Pb–Pb collisions. Figure 5 shows v_2 in p–Pb collisions as a function of $p_{\rm T}$ for unidentified hadrons, pions, kaons, and protons [15]. Results from CMS show similar behavior for K^0_S mesons and Λ baryons [16]. The v_2 in p–Pb collisions was also measured with the two- and multi-particle cumulant methods [17–19]. It is important to note, however, that while the v_2 measured in p–Pb collisions shows qualitatively similar features as v_2 measured in heavy ion collisions, the physical mechanism leading to a non-zero v_2 is still under theoretical debate and the presence of v_2 does *not* necessarily imply the existence of hydrodynamics or a QGP in small collision systems.

5.1 Muon-hadron correlations

In order to gain more information about potential collective effects and constrain theoretical calculations, it is important to measure the strength of the ridge to larger $\Delta\eta$ and to measure the dependence of v_2 on pseudorapidity. Both of these points are addressed in the muon-hadron analysis performed in ALICE [20], in which correlation functions between muons at forward rapidities and charged hadrons

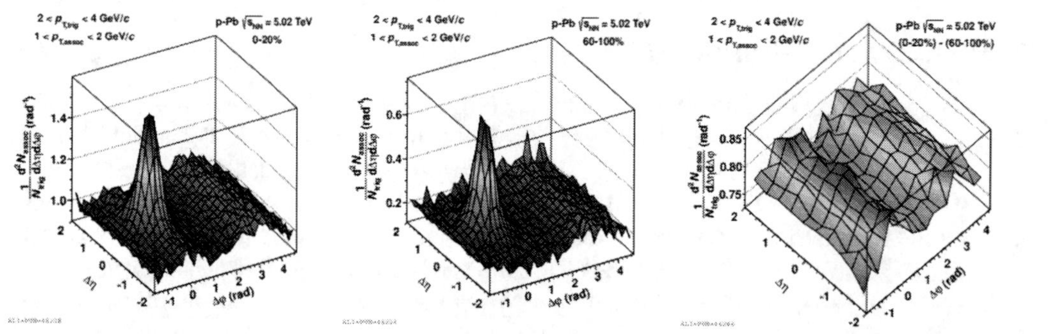

Figure 4. The two-particle correlation functions in p–Pb collisions show a nearside ridge in high-multiplicity collisions (left) while no ridge is visible in low-multiplicity collisions (center). The subtracted distribution (right) reveals a double ridge structure [13].

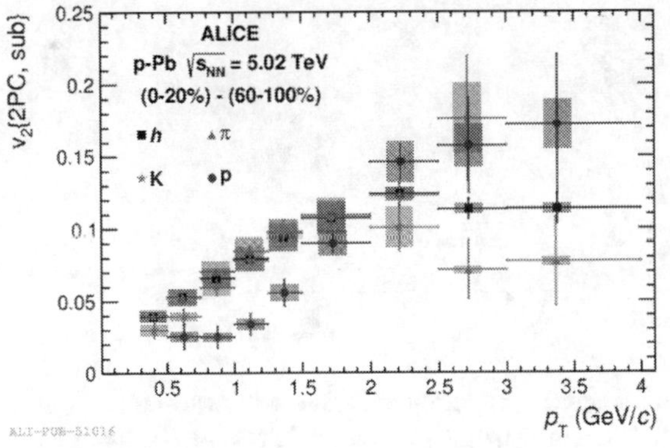

Figure 5. The v_2 of π, K, and p as well as inclusive unidentified hadrons was measured as a function of p_T in p–Pb collisions at $\sqrt{s_{NN}} = 5.02$ TeV with the subtraction method [15].

at mid-rapidity are constructed in order to investigate the long-range behavior of the double ridge structure for $-5 < \Delta\eta < -1.5$.

The correlations between muons detected in the FMS and tracklets reconstructed in the ITS were measured in high-multiplicity (the top 20% of the analyzed event sample) and in low-multiplicity (60-100%) events. As in [13], the low-multiplicity correlations are subtracted from the high-multiplicity correlation functions to remove structures associated with jet fragmentation. After subtraction, the correlation functions were projected onto $\Delta\varphi$, and then fit with a Fourier cosine series to extract v_2 for the muons detected at forward rapidities. The resulting $v_2^{\mu}\{2PC,sub\}$ values are shown in Fig. 6 for muons heading in the proton- and Pb-going directions. The data are compared with an AMPT [21] simulation in which the muon decay products are scaled to account for the efficiency of the absorber in the ALICE FMS. In Fig. 6 it is seen that while AMPT qualitatively describes the p_T-dependence at low p_T, there are significant quantitative differences in the p_T-dependence and η-dependence between data and the model. At high p_T (above $p_T \sim 2$ GeV/c), where muon production is dominated by heavy flavor decays, AMPT does not describe the data well. This could be because heavy flavor muons have a non-zero v_2, or the parent particle composition or v_2 values in data and AMPT are different. The ratio of $v_2^{\mu}\{2PC,sub\}$ in the Pb-going and p-going directions is also shown in Fig. 6 where it is observed to be independent of p_T within the statistical and systematic uncertainties. A constant fit to the data points shows that the v_2 is $(16 \pm 6)\%$ higher in the Pb-going than in the p-going direction. These results are qualitatively in agreement with model predictions. However, current theoretical calculations cannot be directly compared with experimental results, because the effects of the absorber are included in the experimental data (unfolding such effects could not be done in a model-independent way). Future model calculations should use the efficiencies provided in [20] in order to compare directly to the experimental results.

6 Conclusions

Single-particle inclusive and two-particle correlation measurements are used to characterize the pp and p–Pb collision systems. The charged-particle multiplicity density has been measured across a range of energies including the top LHC energy of $\sqrt{s} = 13$ TeV. The pseudorapidity dependence of $\langle dN_{ch}/d\eta \rangle$ has been compared with MC generators in order to further tune the models. The multiplicity distributions were also compared with models and demonstrate KNO scaling up to $z \sim 4$. In two-particle measurements, long-range correlations in pseudorapidity are observed up to $\eta \sim 4$ and $\Delta\eta \sim 5$. The presence of these correlations is reminiscent of Pb–Pb collisions where the structures are frequently attributed to hydrodynamic flow, with similar mass ordering being observed in both small and large systems. The v_2 of forward muons in the Pb-going direction is observed to be higher

 DOI: 10.1051/epjconf/201714101006

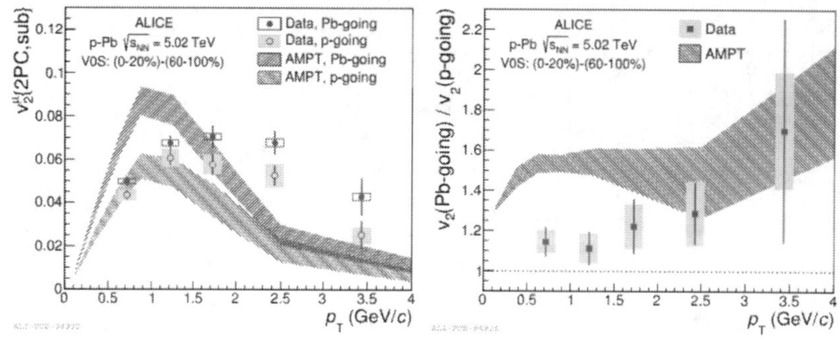

Figure 6. (left) The $v_2^{\mu}\{2PC,sub\}$ measured in the muon-hadron correlation analysis is shown in the p- and Pb-going directions and compared with results from AMPT. (right) The ratio of the $v_2^{\mu}\{2PC,sub\}$ results in the Pb- and p-going directions is compared with AMPT.

than in the p-going direction. While these features are similar to correlations observed in heavy-ion collisions, further theoretical and phenomenological investigations are needed before any inferences about collectivity in small systems can be drawn.

References

[1] K. Aamodt et al. (ALICE), JINST **3**, S08002 (2008)
[2] B. Abelev et al. (ALICE), Phys. Rev. Lett. **110**, 032301 (2013), 1210.3615
[3] T. Sjostrand, S. Mrenna, P.Z. Skands, JHEP **05**, 026 (2006), hep-ph/0603175
[4] T. Sjöstrand, S. Ask, J.R. Christiansen, R. Corke, N. Desai, P. Ilten, S. Mrenna, S. Prestel, C.O. Rasmussen, P.Z. Skands, Comput. Phys. Commun. **191**, 159 (2015), 1410.3012
[5] R. Engel, J. Ranft, S. Roesler, Phys. Rev. **D52**, 1459 (1995)
[6] T. Pierog, I. Karpenko, J.M. Katzy, E. Yatsenko, K. Werner, Phys. Rev. **C92**, 034906 (2015), 1306.0121
[7] J. Adam et al. (ALICE), Phys. Lett. **B753**, 319 (2016), 1509.08734
[8] K. Aamodt et al. (ALICE), Eur. Phys. J. **C68**, 89 (2010), 1004.3034
[9] Z. Koba, H. Nielsen, P. Olesen, Nuclear Physics B **40**, 317 (1972)
[10] V. Khachatryan et al. (CMS), JHEP **09**, 091 (2010), 1009.4122
[11] S. Chatrchyan et al. (CMS), Phys. Lett. **B718**, 795 (2013), 1210.5482
[12] B.B. Abelev et al. (ALICE), Phys. Lett. **B741**, 38 (2015), 1406.5463
[13] B. Abelev et al. (ALICE), Phys. Lett. **B719**, 29 (2013), 1212.2001
[14] G. Aad et al. (ATLAS), Phys. Rev. Lett. **110**, 182302 (2013), 1212.5198
[15] B.B. Abelev et al. (ALICE), Phys. Lett. **B726**, 164 (2013), 1307.3237
[16] V. Khachatryan et al. (CMS), Phys. Lett. **B742**, 200 (2015), 1409.3392
[17] B. Abelev et al. (ALICE), Phys. Rev. **C90**, 054901 (2014)
[18] G. Aad et al. (ATLAS), Phys. Lett. **B725**, 60 (2013), 1303.2084
[19] V. Khachatryan et al. (CMS), Phys. Rev. Lett. **115**, 012301 (2015), 1502.05382
[20] J. Adam et al. (ALICE), Phys. Lett. **B753**, 126 (2016), 1506.08032
[21] Z.W. Lin, C.M. Ko, B.A. Li, B. Zhang, S. Pal, Phys. Rev. **C72**, 064901 (2005), nucl-th/0411110

Correlations and ridge in pPb collisions in the LHCb Experiment

Mariusz Witek[1,a] on behalf of the LHCb Collaboration

[1] *Institute of Nuclear Physics PAS, Cracow, Poland*

Abstract. The LHCb experiment, besides its main programme concerning flavour physics performs also very well as a general purpose forward detector, covering the pseudo-rapidity range from 2.0 to 5.0. Exploiting the experiment's unique geometry, the LHCb collaboration is pursuing a rich programme of forward QCD measurements. In particular two-particle angular correlations are studied in proton-lead collisions at a nucleon-nucleon centre-of-mass energy of $\sqrt{s_{NN}} = 5$ TeV, with data collected by the LHCb detector at the LHC in 2013. The analysis is based on data recorded in two opposing beam configurations, in which either the direction of the proton or that of the lead remnants is analysed. This is the first measurement of a long-range correlation on the near side in proton-lead collisions in the forward region. It extends previous observations in the central region.

1 Introduction

The two-particle angular correlations are one of the tools to study the dynamics of multi-particle production in QCD and to probe collective effects arising in the dense environment of a high-energy collision. They are usually described by two-dimensional $(\Delta\eta, \Delta\phi)$-correlation functions measured in the laboratory system. The two ranges can be distinguished, the short-range ($|\Delta\eta| \lesssim 2$) and long-range ($|\Delta\eta| \gtrsim 2$) effects. On the near-side ($|\Delta\phi| \approx 0$) a short-range "jet-peak" at $\Delta\eta \approx 0$ is the dominant structure. It is caused by the fact that in the fragmentation process, the final-state particles are collimated around the initial parton. To balance the momentum, the peak is accompanied by a long-range structure on the away side ($|\Delta\phi| \approx \pi$) caused by particles that fly opposite in azimuthal angle. The near side long range correlations in the form of the so called ridge effect have been seen at the Relativistic Heavy-Ion Collider (RHIC)in p+Au and Au+Au collisions [1, 2]. Surprisingly, it turned out that in the high multiplicity class of proton-proton interactions the ridge has been also found by CMS [3]. Later, the ridge effect was reported for p+Pb collisions by Alice [4], Atlas [5] and CMS [6]. The theoretical interpretation of the mechanism responsible for the ridge in small colliding system like proton-proton is still under discussion. Various models have been proposed including colour reconnection and gluon saturation in the framework of a colour-glass condensate [7–10] or the hydrodynamic evolution of a high density partonic medium [11], multiparton interactions [12–14], jet-medium interactions [15, 16], and collective effects [17–21] induced by the formation and expansion

[a] e-mail: mariusz.witek@ifj.edu.pl

of a high-density system likely produced in these collisions. The instrumentation of the LHCb single-arm spectrometer covers a unique rapidity range $2 < \eta < 5$ in the laboratory frame [22, 23]. The measurements are based on proton-lead collisions that are dominated by single interactions (98%). The two beam configurations provided by LHC, p+Pb and Pb+p which differs by the direction of proton and ion beams can be analysed in the forward region. The simple trigger was configured to accept events with at least one reconstructed track in the vertex detector for non-empty beam bunch crossing. The events were required to have exactly one reconstructed primary vertex with at least five tracks. The tracks were selected to originate directly from interaction point based on its impact parameter with respect to primary vertex. The effect of fake tracks after track quality cuts imposed and the track reconstruction efficiency is taken into account as a weight assigned to each track. The analysis is based on a subset of the total data set recorded during the 2013 pPb running period, $0.46\,\mathrm{nb}^{-1}$ in the p+Pb configuration and $0.30\,\mathrm{nb}^{-1}$ for the Pb+p configuration. The rapidity range in the nucleon-nucleon centre-of-mass frame is $1.5 < y < 4.4$ for p+Pb and $-5.4 < y < -2.5$ for Pb+p. The results of the analysis were published in Ref. [24].

2 Analysis method

The particles are divided into three subsets according to their p_{T}: $0.15 - 1.0\,\mathrm{GeV}/c$, $1.0 - 2.0\,\mathrm{GeV}/c$ and $2.0 - 3.0\,\mathrm{GeV}/c$, and analysed separately. The analysis method follows the standard approach formulated e.g. in Ref. [25]. All particles in a given subset are taken as a trigger particle one by one. All other particles are combined with the trigger particle in a pair. Two correlation functions are constructed. The signal one $S(\Delta\eta, \Delta\phi)$ for pairs from the same event and $B(\Delta\eta, \Delta\phi)$ for pairs combined from mixed events. The correlation function reads:

$$\frac{1}{N_{\mathrm{trig}}} \frac{\mathrm{d}^2 N_{\mathrm{pair}}}{\mathrm{d}\Delta\eta\,\mathrm{d}\Delta\phi} = \frac{S(\Delta\eta, \Delta\phi)}{B(\Delta\eta, \Delta\phi)} \times B(0, 0), \tag{1}$$

where N_{pair} is the number of particle pairs found in a $(\Delta\eta, \Delta\phi)$ bin and N_{trig} is the number of trigger particles and $B(0, 0)$ is the normalization factor. By construction, the effects due to the detector occupancy, acceptance and material are accounted for by dividing the signal by the background distribution, where the latter is normalised to unity around the origin.

Since the two-particle correlations are strongly dependent on the total number of particles produced in collision, the analysis is performed for various classes of event activity. The p+Pb and Pb+p minimum bias samples each consist of about 1.1×10^8 events which are randomly selected from about 10 times larger full sample. The high-multiplicity samples contain all available events with at least 2200 hits in the vertex detector and amount to about 1.1×10^8 events in p+Pb and 1.3×10^8 events in Pb+p collisions. Five event-activity classes are defined as fractions of the hit-multiplicity distributions of the minimum-bias samples, as indicated in Fig. 1. Since collisions recorded in the Pb+p configuration reach larger hit-multiplicities compared to those in the p+Pb configuration, the relative classes are defined separately for each configuration. The 50 – 100% class contains 50% of events with the lowest event activities, followed by the 30 – 50% and 10 – 30% classes representing medium-activity events, and the 0 – 10% and 0 – 3% classes of high-activity events. The event multiplicities seen in the detector for two beam configurations are different. In order to compare the results, five common absolute activity classes (five bins labeled I – V) were defined. They are listed in Table 1. A scaling factor was introduced to achieve the same total event multiplicity in the same bin for p+Pb and Pb+p configurations.

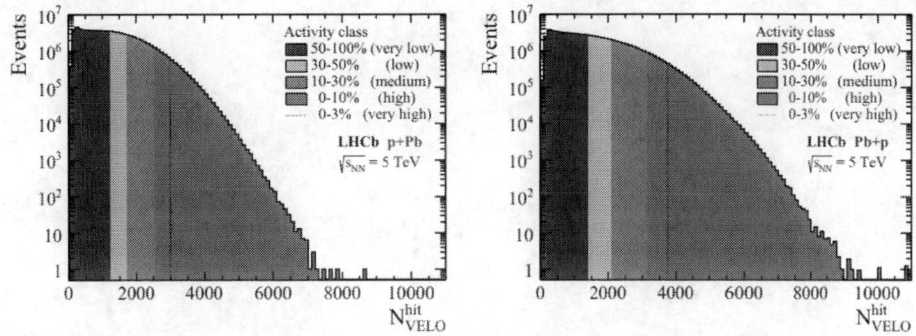

Figure 1. Hit-multiplicity distribution in the vertex detector for selected events of the minimum-bias samples in the (left) p+Pb and (right) Pb+p configurations. The activity classes are defined as fractions of the full distribution, as indicated by colours (shades). The 0 – 3% class is a sub-sample of the 0 – 10% class. The figures were taken from Ref. [24].

Table 1. Common absolute activity bins for the p+Pb and Pb+p samples. The activity of p+Pb events is scaled to match the same activity ranges of Pb+p events, as explained in the text. For illustration purposes the average number, $\langle N_{ch}\rangle_{\rm MC}$, of prompt charged particles with $p > 2\,{\rm GeV}/c$, $p_{\rm T} > 0.15\,{\rm GeV}/c$ and $2.0 < \eta < 4.9$ is listed for events simulated with the Hijing event generator.

Common absolute activity bin	$N_{\rm VELO}^{\rm hit}$-range in Pb+p scale	p+Pb $\langle N_{ch}\rangle_{\rm MC}$	Pb+p $\langle N_{ch}\rangle_{\rm MC}$
Bin I	2200 – 2400	62.8 ± 6.6	64.4
Bin II	2400 – 2600	68.4 ± 7.1	67.0
Bin III	2600 – 2800	73.7 ± 7.6	76.4
Bin IV	2800 – 3000	79.2 ± 7.9	82.4
Bin V	3000 – 3500	86.7 ± 8.2	92.9

3 Results

The correlation for particles with $1 < p_{\rm T} < 2\,{\rm GeV}/c$ is illustrated in Fig. 2 for p+Pb configuration and in Fig. 3 for Pb+p configuration. The near-side peak around $\Delta\eta = \Delta\phi = 0$ is truncated in the histograms. For low activity events only the jet-peak is observed while for highest activity class a clear ridge effect emerges. It is more pronounced in the case of Pb+p sample.

The detailed study of the evolution of the long-range correlations on the near and away sides is usually performed in one dimensional projections integrated over $\Delta\eta$:

$$Y(\Delta\phi) \equiv \frac{1}{N_{\rm trig}}\frac{{\rm d}N_{\rm pair}}{{\rm d}\Delta\phi} = \frac{1}{\Delta\eta_b - \Delta\eta_a}\int_{\Delta\eta_a}^{\Delta\eta_b}\frac{1}{N_{\rm trig}}\frac{{\rm d}^2N_{\rm pair}}{{\rm d}\Delta\eta{\rm d}\Delta\phi}{\rm d}\Delta\eta. \quad (2)$$

The jet-peak area is excluded by averaging the two-dimensional yield over the interval from $\Delta\eta_a = 2.0$ to $\Delta\eta_b = 2.9$. The non interesting correlations in the form of flat pedestals in the yield are subtracted using the zero-yield-at-minimum (ZYAM) method [26]. The subtracted one-dimensional yields for the p+Pb (full circles) and Pb+p (open circles) data samples are shown in Fig. 4 for all activity classes and $p_{\rm T}$ intervals. It can be seen that the correlation strength increases with event activity on the near side, but decreases towards higher $p_{\rm T}$ where fewer particles are found. Since more particles are emitted into the acceptance of the detector in the Pb+p compared to the p+Pb configuration, a

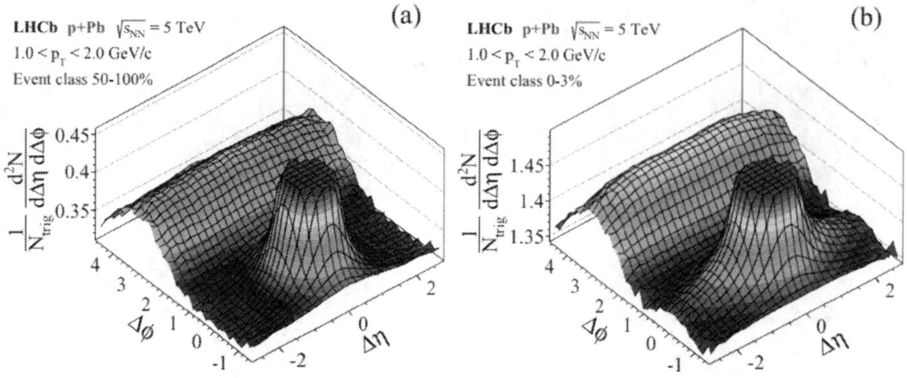

Figure 2. Two-particle correlation functions for events recorded in the p+Pb configuration, showing the (a) low and (b) high event-activity classes. The figures were taken from Ref. [24].

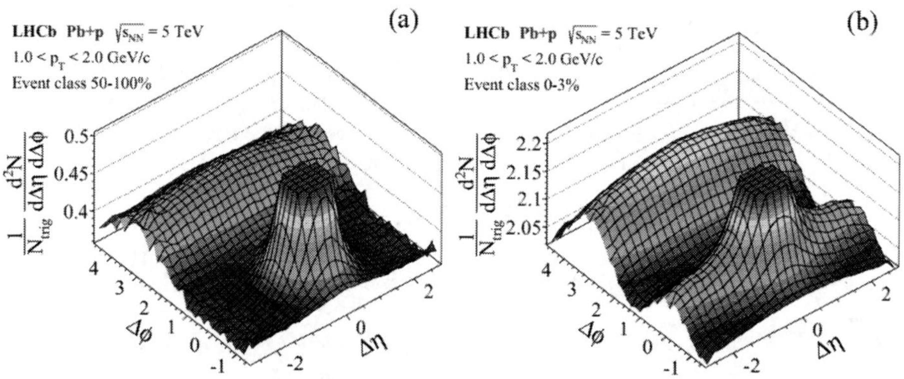

Figure 3. Two-particle correlation functions for events recorded in the Pb+p configuration, showing the (a) low and (b) high event-activity classes. The figures were taken from Ref. [24].

larger offset is observed, as indicated by the ZYAM constants. In the highest-activity class, 0 – 3%, the near-side ridge is strongly pronounced in all p_T intervals for both Pb+p and p+Pb samples. The ridge effect decreases with decreasing event activity and disappears for the activity class 30 – 50%. For p+Pb the near-side peak is significantly lower.

Similar ZYAM-subtracted yields for common bins of absolute activity are presented in Fig. 5 for $1 < p_T < 2\,\mathrm{GeV}/c$. The results for the p+Pb and Pb+p samples are compared in five event classes which probe identical activities in the range of $2.0 < \eta < 4.9$. The measured hit-multiplicities of the p+Pb sample are scaled to agree with the hit-multiplicities of the Pb+p sample. It turns out that the observed long-range correlations become compatible for p+Pb and Pb+p.

4 Summary

The samples of pPb collisions at $\sqrt{s_{NN}} = 5$ TeV collected by the LHCb experiment were used to study two-particle angular correlations for the first time in the forward region covering the pseudorapidity range $2.0 < \eta < 4.9$ in the laboratory frame. The observation of the ridge in the forward region extends

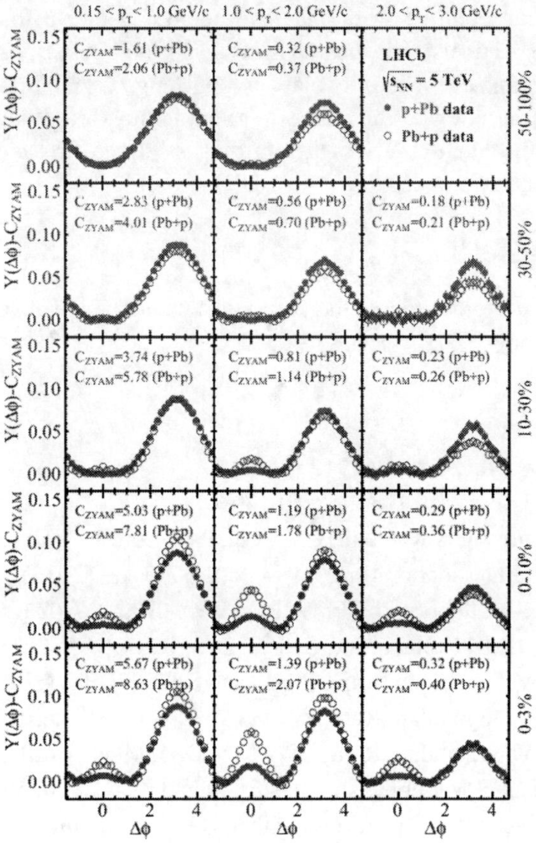

Figure 4. One-dimensional correlation yield as a function of $\Delta\phi$ obtained from the ZYAM-method by averaging over $2.0 < \Delta\eta < 2.9$. The figure was taken from Ref. [24].

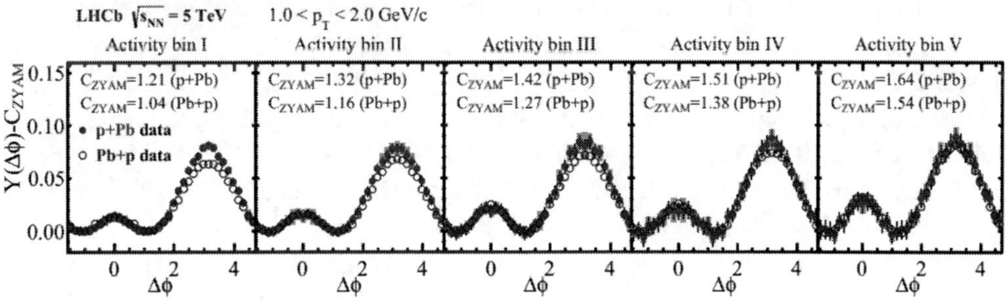

Figure 5. One-dimensional correlation yield as a function of $\Delta\phi$ obtained from the ZYAM-method by averaging the two-dimensional distribution over $2.0 < \Delta\eta < 2.9$. The figure was taken from Ref. [24].

previous LHC measurements. The analysis was performed separately for the p+Pb and Pb+p beam configurations probing the rapidities in the nucleon-nucleon centre-of-mass frame of $1.5 < y < 4.4$ and $-5.4 < y < -2.5$, respectively. The analysis was performed in various p_T ranges and for different event activity classes. In high event activity class, a long-range correlation on the near side (the ridge) is observed in both configurations. The near-side ridge is most pronounced in the range $1 < p_T < 2$ GeV/c. For identical absolute activity ranges in the p+Pb and Pb+p configurations the observed long-range correlations are compatible.

Acknowledgements

I would like to express my gratitude to the National Science Centre NCN in Poland, for financial support under the contract no. 2013/11/B/ST2/03829.

References

[1] B. Abelev et al., Phys. Rev. **C80**, 064912 (2009).
[2] B. Alver et al., Phys. Rev. Lett. **104**, 062301 (2010).
[3] Khachatryan, V., Sirunyan, A.M. et al., CMS collaboration, J. High Energ. Phys. (2010) 91.
[4] B. Abelev et al., ALICE Collaboration, Phys. Lett. **B719**, 29 (2013)
[5] G. Aad et al., ATLAS Collaboration, Phys. Rev. Lett. **110**, 182302 (2013)
[6] S. Chatrchyan et al., CMS Collaboration, Phys. Lett. **B718**, 795 (2013)
[7] Dusling, Kevin and Venugopalan, Raju, Phys.Rev. **D87**, 051502 (2013)
[8] Dusling, Kevin and Venugopalan, Raju, Phys.Rev. **D87**, 054014 (2013)
[9] Dusling, Kevin and Venugopalan, Raju, Phys.Rev. **D87**, 094034 (2013)
[10] Kovchegov, Yuri V. and Wertepny, Douglas E., Nucl.Phys. **A906**, 50-83 (2013)
[11] Bzdak, Adam and Schenke, Bjoern and Tribedy, Prithwish and Venugopalan, Raju, Phys.Rev. **C87**, 064906 (2013)
[12] Alderweireldt, Sara and Van Mechelen, Pierre, Proceedings of the 3rd International Workshop on Multiple Partonic Interactions at the LHC (MPI@LHC 2011), 33-40 (2012)
[13] Strikman, Mark, Acta Phys.Polon. **B42**, 2607-2630 (2011)
[14] Ryskin, M.G. and Martin, A.D. and Khoze, V.A., J.Phys. **G38**, 085006 (2011)
[15] Hwa, Rudolph C. and Yang, C.B., Phys.Rev. **C83**, 024911 (2011)
[16] Wong, Cheuk-Yin, Phys.Rev. **C84**, 024901 (2011)
[17] Avsar, Emil and Flensburg, Christoffer and Hatta Yoshitaka and Ollitrault, Jean-Yves and Ueda, Takahiro, Phys.Lett., **B702**,394-397 (2011)
[18] Werner, K. and Karpenko, Iu. and Pierog, T., Phys.Rev.Lett. **106**, 122004 (2011)
[19] Bozek, Piotr, Phys. Rev. **C85**, 014911 (2012)
[20] Bozek, Piotr and Broniowski, Wojciech, Phys.Lett. **B718**,1557-1561 (2013)
[21] Shuryak, Edward and Zahed, Ismail, Phys.Rev. **C88**, 044915 (2013)
[22] Alves Jr., A. A. and others, LHCb collaboration, JINST**3**, S08005 (2008)
[23] Aaij, R. and others, LHCb collaboration, Int. J. Mod. Phys. **A30**, 1530022 (2015)
[24] Aaij, R. and others, LHCb collaboration, Phys. Lett. **B762** 473-483 (2016)
[25] Abelev, Betty and others, ALICE collaboration, Phys.Lett. **B719**, 29-41 (2013)
[26] Ajitanand, N. N. and Alexander, J. M. and Chung, P. and Holzmann, W. G. and Issah, M. and Lacey, Roy A. and Shevel, A. and Taranenko, A. and Danielewicz, P., Phys. Rev. C **72**, 011902 (2005)

Femtoscopic Signature of Strong Radial Flow in High-multiplicity pp Collisions

Yuji Hirono[1,a] and Edward Shuryak[2]

[1] *Department of Physics, Brookhaven National Laboratory, Upton, New York 11973-5000*
[2] *Department of Physics and Astronomy, Stony Brook University, Stony Brook, New York 11794-3800, USA*

Abstract.
We report on the results of the pion HBT radii for high-multiplicity pp collisions calculated in relativistic hydrodynamics. By comparison with the ALICE data, we make an estimate of he magnitude of the radial flow and initial size of the system.

1 Introduction

The Hanbury-Brown-Twiss (HBT) interferometry, that originally comes from radio astronomy, is a way to measure the space-time extent of a radiation source using the influence of Bose symmetrization of the wave functions of observed particles. This technology has been extensively used in heavy-ion collision experiments, to gain information about the size and the time evolution of the created matter. The advent of LHC made it possible to trigger on high-multiplicity events. It is reported that flow anisotropies quantified by v_n show similar magnitude to those at heavy-ion collisions, suggesting the formation of a liquid-like matter. However, the interpretation of those results are under debate.

In this contribution, we look at this problem from a different angle. We use the data of HBT radii from ALICE collaboration [1], and try to extract the magnitude of the radial flow and the initial size of the system in high-multiplicity event of pp collisions. For the details of the analysis, please look at Ref. [2].

2 Method

Unlike the heavy-ion collisions, we do not have a good quantitative model to describe the initial condition for high-multiplicity pp collisions. To parametrize the solution, a particularly convenient way is to use the analytic solutions found by Gubser [3]. In those solutions, the energy density and velocity is written as

$$\epsilon(\tau, r) = \frac{\epsilon_0 (2q)^{8/3}}{\tau^{4/3}[1 + 2q^2(\tau^2 + r^2) + q^4(\tau^2 - r^2)^2]^{4/3}}, \tag{1}$$

$$v_\perp(\tau, r) = \frac{2q^2 \tau r}{1 + q^2\tau^2 + q^2 r^2}. \tag{2}$$

[a] e-mail: yhirono@bnl.gov

The space-time profiles are characterized by two parameters, q [fm^{-1}] and ϵ_0. The parameter ϵ_0 is related with the entropy per unit rapidity as

$$\epsilon_0 = f_*^{-1/3}\left(\frac{3}{16\pi}\frac{dS}{d\eta}\right)^{4/3}, \tag{3}$$

where $f_* = 11$ is the number of effective degrees of freedom in QGP [3]. We can infer the entropy per unit rapidity from the charge particle multiplicity,

$$\frac{dS}{d\eta} \simeq 7.5\frac{dN_{\rm ch}}{d\eta}. \tag{4}$$

In this way, the values of ϵ_0 are fixed from the charged particle multiplicity of the events of interest.

The other parameter q is related to the initial size of the system. As can be inferred from its dimension, a large q corresponds to a small initial size. In Fig. 1, we show two temperature profiles as functions of the radial coordinate r for $\tau = 0.6$ fm. The solution with larger q (dotted line) is more compressed and has a larger pressure gradient, which results in stronger radial flow at the freezeout.

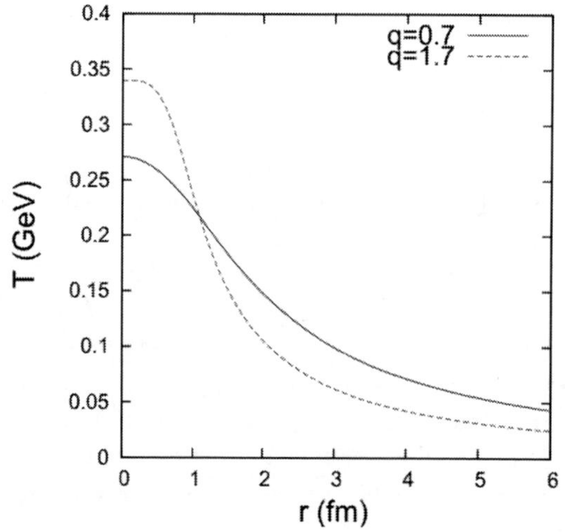

Figure 1. (Color online) Temperature profiles of the Gubser solutions for different values of the parameter q, at $\tau = 0.6$ fm. Plotted as a function of the radial coordinate r.

Although the Gubser solution can be a good description for the early stage of the time-evolution, it cannot be used at later stages close to freezeout. Near the phase transition temperature, the number of degree of freedom changes drastically, which affect the freezeout volume. The analytic Gubser solution is for a conformal EOS, and this change of degrees of freedom cannot be accommodated. Therefore, we start with the Gubser solution, and then, at certain time $\tau_0 = 0.6$ fm, we switch to the numerical evolution using a realistic EOS calculated from the lattice QCD calculations [4]. The system is evolved with the ideal relativistic hydrodynamic equations,

$$\partial_\mu T^{\mu\nu} = 0, \tag{5}$$

where $T^{\mu\nu}$ is the energy-momentum tensor. For a perfect fluid, $T^{\mu\nu}$ is expressed as

$$T^{\mu\nu} = (\epsilon + p)u^\mu u^\nu - p\eta^{\mu\nu}, \tag{6}$$

where ϵ is the energy density, p is the pressure, u^μ is the fluid four-velocity.

After solving the hydrodynamic equation, we use the standard Cooper-Frye prescription [5] to get the particle distributions,

$$p^0 \frac{d^3N}{d\eta dp_T^2} = \frac{1}{(2\pi)^3} \int \frac{p^\mu d\sigma_\mu(x)}{\exp\left[p \cdot u/T\right] \mp_{\mathrm{BF}} 1}. \tag{7}$$

The formula above is applied on an isothermal hypersurface with the freezeout temperature T_{f}. We do the Monte-Carlo sampling of pions, according to the distribution (7).

Finally, we calculate the correlations functions using the momenta and emission coordinates of pions from the sampling procedure. The two-particle correlations coming from Bose symmetrization is introduced as

$$C(k_{\mathrm{T}}, \boldsymbol{q}) = \frac{\sum_{<i,j>\in[k_{\mathrm{T}}]} \left[1 + \cos\left(q_\mu \Delta x^\mu\right)\right]}{\sum_{<i,j>\in[k_{\mathrm{T}}]} 1}, \tag{8}$$

where $\boldsymbol{k}_{\mathrm{T}} \equiv (\boldsymbol{p}_{1\mathrm{T}} + \boldsymbol{p}_{2\mathrm{T}})/2$ is the pair transverse momentum, $< i, j >\in [k_{\mathrm{T}}]$ indicates a pair of pions in a particular k_{T} bin, $q^\mu = p_1^\mu - p_2^\mu$ is four-momentum difference of a pion pair, and $\Delta x^\mu \equiv x_1^\mu - x_2^\mu$ is space-time distance of the pair. We evaluate the correlation function in the "longitudinally comoving frame", where $k_z = 0$ for each pair. We use a pseudo-rapidity cut $|\eta| < 1.0$.

The three-dimensional correlation functions are characterized by the so0-called "out-side-long" parametrization [6, 7],

$$C(k_{\mathrm{T}}, \boldsymbol{q}) = 1 + \lambda \exp\left[-R_o^2 q_o^2 - R_s^2 q_s^2 - R_\ell^2 q_\ell^2\right], \tag{9}$$

where $R_{o,s,\ell} = R_{o,s,\ell}(k_{\mathrm{T}})$ are the HBT radii, q_o is the momentum parallel to the pair transverse momentum, q_ℓ is the one parallel to the beam direction, and q_s is the one perpendicular to out and long direction. We determine the HBT radii for each k_{T} by χ^2 fitting.

3 Results

Figure 2 shows the HBT volume, $R_o R_s R_\ell$, as a function of k_{T} for several values of q. The experimental data from ALICE is also plotted. The HBT radii with $q = 1.5$ to $1.7\ \mathrm{fm}^{-1}$ reproduce the volume in the ALICE experiment pretty well.

In Fig. 3, the ratio R_o/R_s is plotted as a function of k_{T}. The ratio R_o/R_s is a decreasing function of k_{T}. For small q, R_o/R_s is almost flat. As q becomes larger, the slope becomes steeper, and R_o/R_s gets suppressed for large k_{T}. This suggests that R_o/R_s is indicative of the strength of the radial flow. The data ALICE shows even stronger suppression, which means that the initial size is very small and the expansion is strong.

The suppression of R_o/R_s is driven by correlation between emission time difference and the displacement in the out direction. For a weak flow (small q), this correlation is weak for the all region of k_{T}. As the flow gets stronger (larger q), it becomes a linear increasing function of k_{T} Since this term contributes to R_o/R_s with a negative sign, it leads to the suppression of R_o/R_s at large k_{T}. The discussion in detail can be found in Ref. [2]

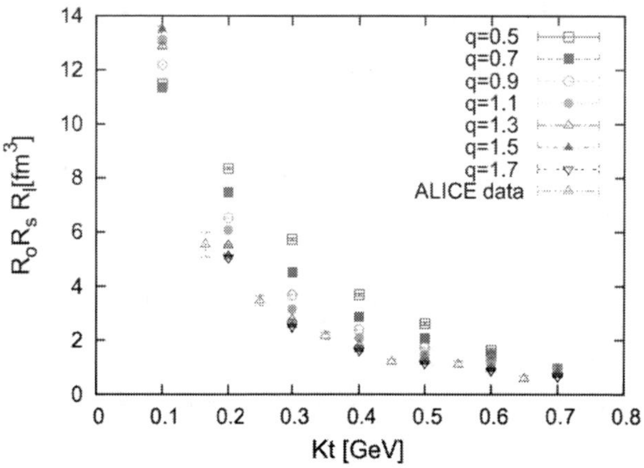

Figure 2. (Color online) HBT volume as a function of the pair transverse momentum k_T for various values of the parameter q. $dN_{ch}/d\eta = 27$, $T_f = 120$ MeV.

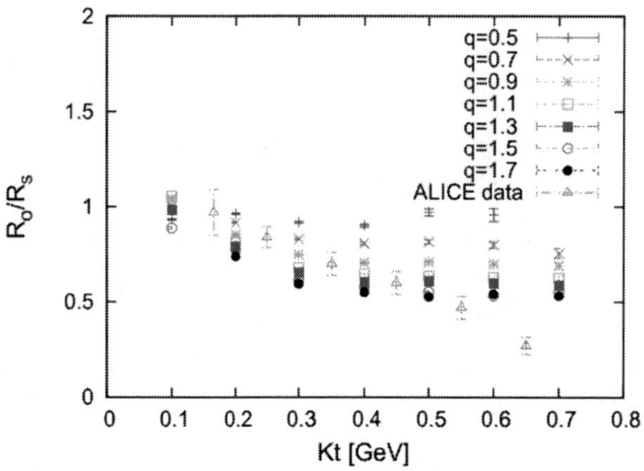

Figure 3. (Color online) The ratio R_o/R_s as a function of k_T for multiple values of q.

4 Summary

The data of HBT radii from the ALICE collaboration [1] provided a striking feature that the high multiplicity events of pp collisions at the LHC is different from others. It shows an evidence of a strong radial flow. We performed simulations of such system, basing on an ideal relativistic hydrodynamics. The early-time evolution is described by a Gubser solution, complemented by a numerical one, with a realistic EOS at later stages. Comparison of the resulting HBT radii with high multiplicity data shows the best agreement for the smallest initial fireball we study, with Gubser parameter $q = 1.5 - 1.7\ \mathrm{fm}^{-1}$. It confirms that one in fact observes the presence of collective hydrodynamical flow in an unprecedented small system, smaller than 1 fm initially.

Y.H. was supported by JSPS Research Fellowships for Young Scientists. The work of E.S. is supported in part by the U.S. Department of Energy under Contract No. DE-FG02-88ER40388.

References

[1] M. Aggarwal et al. (STAR Collaboration), Phys.Rev. **C83**, 064905 (2011), 1004.0925
[2] Y. Hirono, E. Shuryak, Phys. Rev. **C91**, 054915 (2015), 1412.0063
[3] S.S. Gubser, Phys.Rev. **D82**, 085027 (2010), 1006.0006
[4] S. Borsanyi, Z. Fodor, C. Hoelbling, S.D. Katz, S. Krieg et al., Phys.Lett. **B730**, 99 (2014), 1309.5258
[5] F. Cooper, G. Frye, Phys.Rev. **D10**, 186 (1974)
[6] S. Pratt, Phys.Rev. **D33**, 1314 (1986)
[7] G. Bertsch, M. Gong, M. Tohyama, Phys.Rev. **C37**, 1896 (1988)

Analysis of flow observables in small systems using an integrated dynamical model

Koji Kawaguchi[1,⋆], Koichi Murase[2,1,⋆⋆], and Tetsufumi Hirano[1,⋆⋆⋆]

[1] *Department of Physics, Sophia University, Tokyo 102-8554, Japan*
[2] *Department of Physics, The University of Tokyo, Tokyo 113-0033, Japan*

Abstract. We investigate collectivity in small colliding systems by using an integrated dynamical model in which Monte-Carlo initialisation of hydrodynamic fields, the ideal hydrodynamic description of the quark-gluon plasma and the kinetic description of the hadron gas are incorporated. We implement fluctuations of the multiplicity and of the longitudinal matter profile in the initial conditions which are of particular importance in small colliding systems. We also discuss the effects of hadronic rescatterings on flow observables.

1 Introduction

The perfect fluidity of the quark-gluon plasma (QGP) created in high energy nuclear collisions was discovered at Relativistic Heavy Ion Collider (RHIC) [1]. The perfect-fluid-like collective behaviour of the QGP is well described by relativistic hydrodynamics. One of the main observables to gain information of the bulk and the transport properties of the QGP is the second harmonics of azimuthal angle distributions, the so-called elliptic flow [2].

Since the system size dependence of the observables is one of the keys to get information about the property of the matter, small colliding systems attract a great deal of attention. Collective flow in small colliding systems had not been studied since the system is not supposed to reach thermal equilibrium in these systems. However, recent measurements reveal that the system exhibits collective-flow-like behaviour even in $p/d/^3$He-Au collisions at RHIC [3, 4]. All the low p_T data in small colliding systems are consistent with a hydrodynamic flow picture [5]. Thus it would be interesting to see whether flow observables really result from the presence of the QGP. Therefore we analyse flow observables in small colliding systems at the RHIC energy by employing an integrated dynamical model based on an assumption of QGP fluid formation.

2 Model

In order to describe the dynamics of high energy nuclear collisions at the RHIC energy, we employ an integrated dynamical model [6] which is combination of Monte-Carlo Glauber model (MC-Glauber)

⋆e-mail: kawaguchi@eagle.sophia.ac.jp
⋆⋆e-mail: murase@nt.phys.s.u-tokyo.ac.jp
⋆⋆⋆e-mail: hirano@sophia.ac.jp

for the initial entropy production, ideal hydrodynamics for QGP fluids in the intermediate stage and a hadronic cascade model for hadron gases in the late stage. In the initial stage, we use MC-Glauber to calculate initial entropy density distributions just after collisions of two nuclei. Space-time evolution of the QGP is described by fully (3+1) dimensional ideal hydrodynamic simulations. After that we switch from fluids to hadron gases with the Cooper-Frye formula [7] in the isothermal hypersurface at the temperature of 155 MeV. For the evolution of the hadronic gas, we use a hadron cascade code, JAM [8].

In small colliding systems, one needs to extend a model of initial conditions in hydrodynamic simulations by taking account of event-by-event fluctuations of geometry and initial entropy densities. Since an initial entropy density profile is parametrised as a smooth function in p+p collisions in the original integrated dynamical model, final multiplicity fluctuations cannot be reproduced in p+p collisions. To resolve this problem, we employ an event generator, PYTHIA [9], which naturally describes multiplicity distributions in elementary p+p collisions. For each pair of binary collisions, we run PYTHIA for one inelastic p+p collisions. For the soft particle production, it is well known that multiplicities scale with the number of participants, rather than the number of binary collisions. To account for this, we perform a rejection sampling from final particle lists in PYTHIA for pairs of all binary collisions so that the multiplicities scale with the number of participants for a given rapidity. By taking into account an idea of "rapidity triangle/trapezoid" [10–12], we parametrise the rapidity-dependent acceptance probability of the rejection sampling as

$$w(Y) = \frac{Y_{\rm b}+Y}{2Y_{\rm b}}\frac{1}{N_{\rm A}} + \frac{Y_{\rm b}-Y}{2Y_{\rm b}}\frac{1}{N_{\rm B}}, \tag{1}$$

where Y is rapidity and $Y_{\rm b}$ is beam rapidity. $N_{\rm A}$ and $N_{\rm B}$ are the number of nucleons in a nucleus A and B, respectively. We assume that the initial entropy density distribution is proportional to the number distribution of produced particles and that each produced particle is associated with a Gaussian function in the configuration space. Thus the entropy density distribution at initial proper time τ_0 in full three-dimensional space is

$$s(\tau_0, x, y, \eta_{\rm s}) = \frac{K}{\tau_0}\sum_i \frac{1}{\sqrt{2\pi\sigma_\eta^2}}\frac{1}{2\pi\sigma_\perp^2}\exp\left[-\frac{(x-x^i)^2+(y-y^i)^2}{2\sigma_\perp^2} - \frac{(\eta_{\rm s}-\eta_{\rm s}^i)^2}{2\sigma_\eta^2}\right]. \tag{2}$$

Here x and y are transverse coordinates and $\eta_{\rm s}$ is space-time rapidity. Each particle i is supposed to be produced at (x^i, y^i) in the transverse plane and $\eta_{\rm s}^i(=Y)$ in the longitudinal direction. $\sigma_\perp$ and σ_η are smearing parameters in the Gaussian function and K is an overall normalization. We set $\sigma_\perp = 0.3$ fm, $\sigma_\eta = 0.3$, $\tau_0 = 0.6$ fm/c and $K = 5.6$ and use this parameter set throughout this study.

Figure 1 shows an example of transverse ($\eta_{\rm s} = 0$) and longitudinal ($x = 0$ fm) profiles of the initial entropy density in p+Au, d+Au and ^{3}He+Au collisions. The impact parameter is chosen to be b = 0 fm. One sees differences of the numbers of separated hot spots which correspond to transverse positions of nucleons in colliding p, d and ^{3}He. In the longitudinal plane, the entropy density is distributed mainly in the negative rapidity region, *i.e.*, Au-going direction. In all cases, initial entropy density has a bumpy structure in both a transverse and longitudinal plane.

3 Results

Figure 2 shows multiplicity distributions in negative rapidity region ($-3.9 < \eta < -3.0$) in p+Au, d+Au and ^{3}He+Au collisions at $\sqrt{s_{\rm NN}}$ = 200 GeV. Since we employed, for the initial conditions,

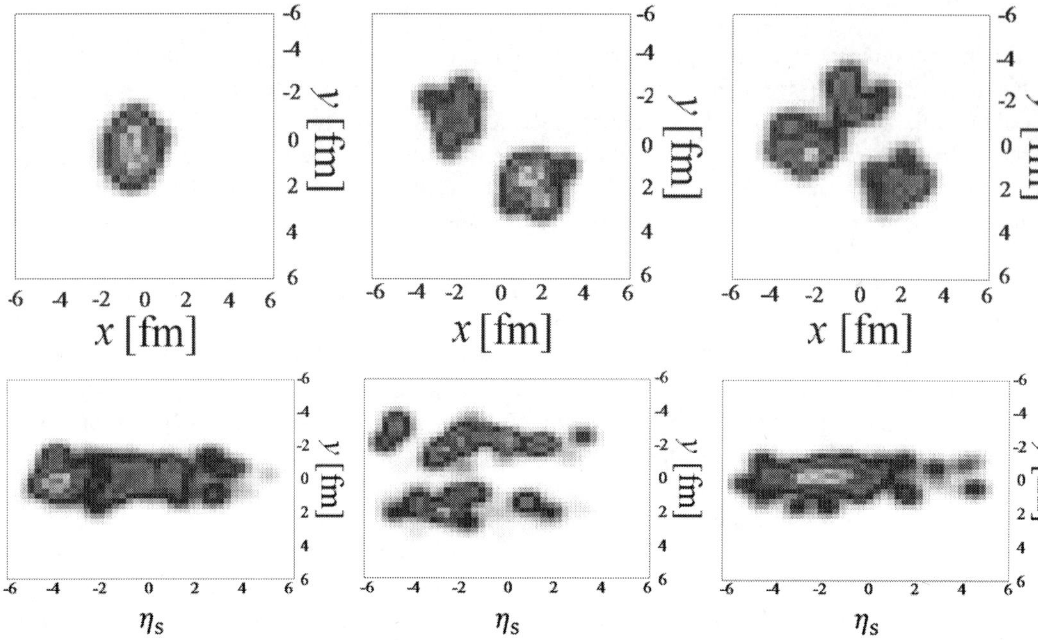

Figure 1. Examples of initial entropy density profile in the transverse plane (upper) and in the longitudinal plane (lower) in p+Au (left), d+Au (middle) and ^{3}He+Au (right) collisions at $\sqrt{s_{\rm NN}}$ = 200 GeV.

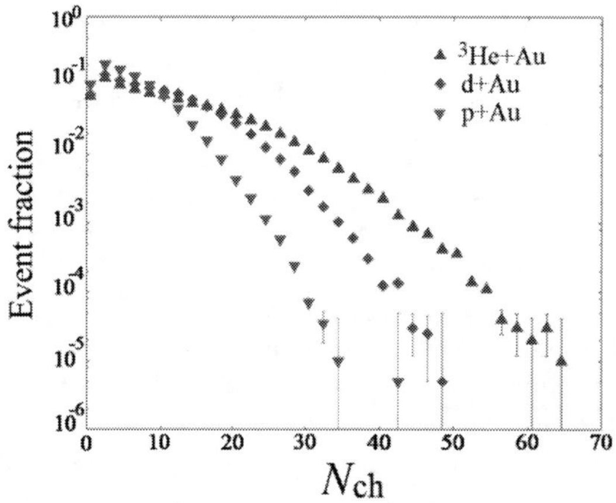

Figure 2. Multiplicity distributions of charged hadrons in $-3.9 < \eta < -3.0$ in p+Au (green), d+Au (red) and ^{3}He+Au (blue) collisions at $\sqrt{s_{\rm NN}}$ = 200 GeV.

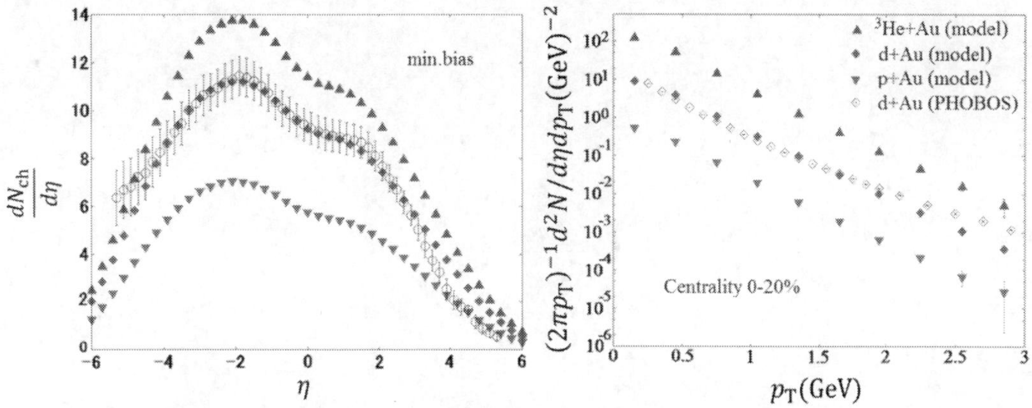

Figure 3. Pseudorapidity distributions in minimum bias (left) and transverse momentum distributions in $0.2 < \eta < 1.4$ in central 0-20% events (right). Results in p+Au (green), d+Au (red) and ^{3}He+Au (blue) collisions at $\sqrt{s_{NN}}$ = 200 GeV are compared with the PHOBOS data (black) [13, 14]. To see the plots clearly, the result in ^{3}He+Au (p+Au) collisions is multiplied (divided) by ten in the right figure.

PYTHIA in which multiplicities fluctuate in $p+p$ collisions, multiplicity distributions in small colliding systems take over these fluctuations. These multiplicity distributions are used for a later centrality cut.

Figure 3 shows the pseudorapidity distributions (left) and transverse momentum (p_T) distributions (right) of charged hadrons in p+Au, d+Au and ^{3}He+Au collisions at $\sqrt{s_{NN}}$ = 200 GeV. Pseudorapidity distributions are averaged over minimum bias events, while p_T distributions are in 0-20% centrality. The PHOBOS data [13, 14] in d+Au collisions are also shown for comparison. We reproduce the shape and yield of the pseudorapidity distribution in d+Au collisions and find the idea of initial "rapidity triangle/trapezoid" works well. The p_T slope in Fig. 3 (left) is consistent with the PHOBOS data in the low p_T region. However, our result deviates from the PHOBOS data in high p_T region since we do not have any jet components in our model.

We analyse the transverse momentum dependence of the elliptic flow parameter v_2 of charged hadrons in 0-5% central ^{3}He+Au collisions at $\sqrt{s_{NN}}$ = 200 GeV in Fig. 4. The PHENIX data are also shown for comparison. In both analyses, an event plane method is used to obtain the elliptic flow parameter v_2. We reasonably reproduce the PHENIX v_2 data [3] from the integrated dynamical model. This indicates the QGP fluid + hadronic gas picture works well also in this small colliding system.

Pseudorapidity and centrality dependences of elliptic flow parameters are shown in Fig. 5. To see how much elliptic flow is generated in the QGP fluid stage, the results with hadronic rescatterings switched off are shown [15, 16]. These elliptic flow parameters are obtained by using the two particle cumulant method. In Fig. 5 (left), the shape of elliptic flow is asymmetric with respect to pseudorapidity. Elliptic flow in the Au-going side (in the left-hand side) is larger than that in the ^{3}He-going side (in the right-hand side) due to a difference of multiplicities as shown in Fig. 3 (left). In Fig. 5 (right), top 10% results, 10-20% results and so on are shown from right to left. Our results show elliptic flow parameters increase with the multiplicity. In both figures, roughly 90% of final v_2 is generated in the QGP fluid stage. The situation is quite similar to that in Au+Au collisions at the RHIC energy. This indicates that a large fraction of v_2 is generated in the QGP fluid state even in small colliding systems.

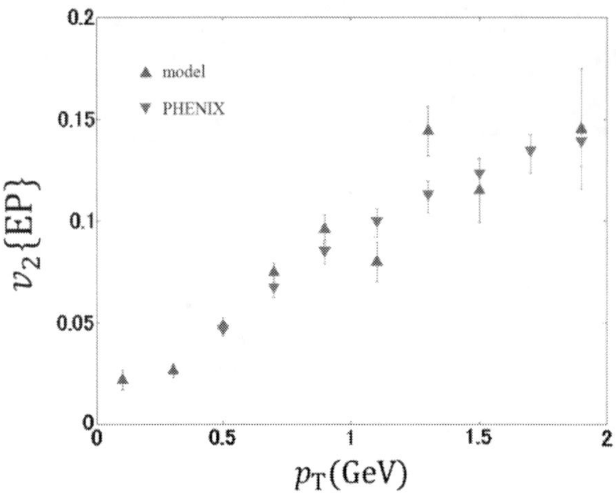

Figure 4. Preliminary result of $v_2(p_T)$ of charged hadrons at midrapidity ($|\eta| < 0.35$) in 0-5% central ^{3}He+Au collisions at $\sqrt{s_{NN}}$ = 200 GeV. Our results (blue) are compared with the PHENIX data (red) [3].

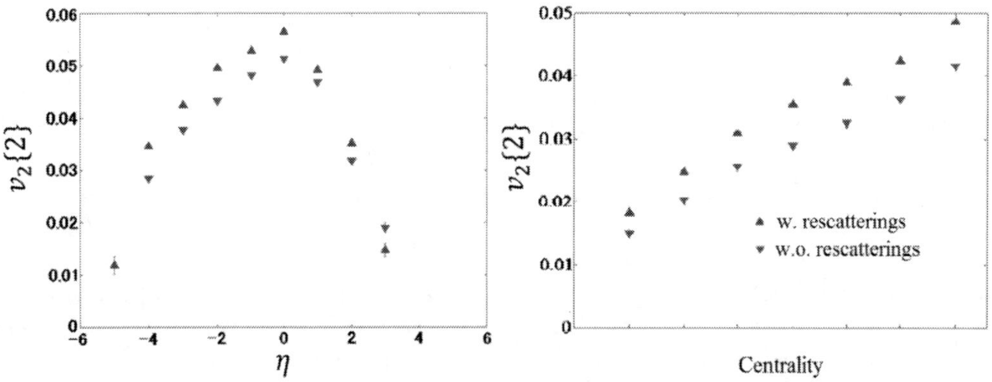

Figure 5. Pseudorapidity dependence of elliptic flow (left) and centrality dependence of elliptic flow (right) in ^{3}He+Au collisions at $\sqrt{s_{NN}}$ = 200 GeV. In the right figure, from right to left, results in 0-10%, 10-20%, 20-30%, 30-40%, 40-50%, 50-60% and 60-70% centrality are shown. Blue (Red) plots are results with (without) hadronic rescatterings.

4 Summary

We studied flow observables in small colliding systems by employing an integrated dynamical model. We extended a model of initial conditions to describe dynamics in small colliding systems by taking account of event-by-event fluctuations of geometry and entropy densities. Using the newly developed version of the model, we analysed multiplicity, pseudorapidity distributions and p_T spectra as well as p_T-/pseudorapidity-/centrality-dependence of elliptic flow in small colliding systems. From these results, we found that the observed large v_2 can be attributed to both QGP expansion and hadronic rescatterings, but a large fraction of v_2 is generated in the QGP fluid stage even in small colliding systems.

References

[1] "RHIC Scientists Serve Up Perfect Liquid", http://www.bnl.gov/newsroom/news.php?a=1303
[2] J.-Y. Ollitrault, Phys. Rev. **D46**, 229 (1992)
[3] A. Adare *et al.* (PHENIX), Phys. Rev. Lett. **444**, 212301 (2013)
[4] A. Adare *et al.* (PHENIX), Phys. Rev. Lett. **115**, 142301 (2015)
[5] A. Adare *et al.* (PHENIX), Phys. Rev. Lett. **114**, 192301 (2015)
[6] T. Hirano, P. Huovinen, K. Murase and Y. Nara, Prog. Part. Nucl. Phys. **70**, 108 (2013)
[7] F. Cooper and G. Frye, Phys. Rev. **D10**, 186 (1974)
[8] Y. Nara, N. Otuka, A. Ohnishi, K. Niita and S. Chiba, Phys. Rev. **C61**, 024901 (2000)
[9] T. Sjöstrand *et al.*, Comput. Phys. Commun. **191**, 159(2015)
[10] A. Adil and M. Gyulassy, Phys. Rev. **C72**, 034907 (2005)
[11] S.J. Brodsky, J.F. Gunion and J.H. Kuhn, Phys. Rev. Lett. **39**, 1120 (1977)
[12] T. Hirano, U.W. Heinz, D. Kharzeev, R. Lacey and Y. Nara, Phys. Lett. **B636**, 299 (2006)
[13] B.B. Back *et al.* (PHOBOS), Phys. Rev. **C72**, 031901 (2005)
[14] B.B. Back *et al.* (PHOBOS), Phys. Rev. Lett. **91**, 072302 (2003)
[15] S. Takeuchi, K. Murase, T. Hirano, P. Huovinen and Y. Nara, Phys. Rev. **C92**, 044907 (2015)
[16] T. Hirano, U.W. Heinz, D. Kharzeev, R. Lacey and Y. Nara, Phys. Rev. **C77**, 044909 (2008)

Charge dependent particle correlations motivated by chiral magnetic effect and chiral vortical effect

Jie Zhao[1,a] (for the STAR collaboration)

[1] *Department of Physics and Astronomy, Purdue University, West Lafayette, IN 47907, USA*

Abstract. In the early stage of relativistic heavy-ion collisions, a strong magnetic field can be generated. The interaction between the strong magnetic field and the hot medium would provide a way to study the nontrivial QCD topological structures, such as the chiral magnetic effect (CME) and chiral vortical effect (CVE). The three particle azimuthal correlator has been proposed to measure the separation of electric/baryonic charges with respect to the reaction plane due to the chiral effects. We report the STAR measurements of three particle azimuthal correlations of charged particles and identified particles to search for possible signatures of the CME and CVE in relativistic heavy-ion collisions of various energies and systems.

1 Introduction

Metastable domains of topological charges can form in a quark gluon plasma with approximate chiral symmetry restoration [1–5]. These topological charges can change the chirality of quarks in those local domains. This can lead to observable charge separation along the direction of the strong magnetic field produced in the early stage of relativistic heavy-ion collisions [6, 7]. This is called the chiral magnetic effect (CME). A similar effect, called the chiral vortial effect (CVE), in which the vorticity of the collision system replaces the magnetic field, can lead to baryonic charge separation [8]. Measurements of electric/baryonic charge separations from the CME and CVE can therefore provide a means to studying the nontrivial QCD topological structures. An observation of the CME/CVE-induced charge separation would confirm a fundamental property of QCD and is therefore of paramount importance.

Experimentally, the three particle correlator (Eq. 1) has been proposed [7] to measure the possible signal of CME and/or CVE.

$$\gamma \equiv \langle \cos(\phi_\alpha + \phi_\beta - 2\Psi_{\mathrm{RP}}) \rangle = \langle \cos(\phi_\alpha + \phi_\beta - 2\phi_c) \rangle / v_2 \tag{1}$$

Previous experimental results from STAR [9–12], PHENIX [13], and ALICE [14] are consistent with theoretical expectation of CME. However, those results suffer from possible physics background [7, 15]). In this paper, we will discuss new approaches to suppress the flow background, and also report results in small system d+Au collisions at STAR.

[a]e-mail: zhao656@purdue.edu

2 Experiment setup

The data used in this analysis were collected by the STAR detector [16] at Brookhaven National Laboratory from year 2003, 2004, 2007, 2010 and 2011. The main subsystems of the STAR detector used for this analysis are the Time Projection Chamber (TPC) [17] and the Time-Of-Flight detector (TOF) [18] with 2π azimuthal coverage at mid-rapidity, and three trigger subsystems, the Central Trigger Barrel (CTB), the Vertex Position Detectors (VPDs) and the Zero Degree Calorimeters (ZDCs).

Charged particle identification is performed by ionization energy loss (dE/dx) measured by the TPC and/or particle velocity measured by TOF. The correlations are reported for particles with pseudorapidity $|\eta| < 1$ and transverse momentum $p_T > 0.15$ GeV/c. We use the TPC to reconstruct the 2nd order harmonic event plane. The 1st order harmonic event plane is reconstructed by the two Zero Degree Calorimeter Shower Maximum Detectors (ZDC-SMDs) [19], which are sensitive to the directed flow of neutrons in the beam rapidity region.

3 Charge dependent correlator measurements

The three particle correlator (Eq. 1) has been proposed [7] to measure charge separation signal: Charge separation along the magnetic field ($\vec{B}$), perpendicular to Ψ_{RP} on average, would yield positive and negative values of γ for particle pairs of opposite-sign (OS) and same-sign (SS) charges, respectively. A positive $\Delta\gamma \equiv \gamma_{OS} - \gamma_{SS}$ is therefore referred to as the charge separation signal.

Figure 1 presents the OS and SS correlators as function of centrality for Au+Au collisions at $\sqrt{s_{NN}}$ = 7.7-200 GeV measured by STAR [12]. With decreasing beam energy, both OS and SS tend to rise in peripheral collisions. Similar results at 200 GeV Au+Au from PHENIX [13] and 2.76 TeV Pb+Pb from ALICE [14] were observed. This feature seems to be charge independent and can be explained by momentum conservation and elliptic flow [11]. In most cases, the difference between OS and SS is still present. The difference at low collision energies are consistent with zero within large error bars.

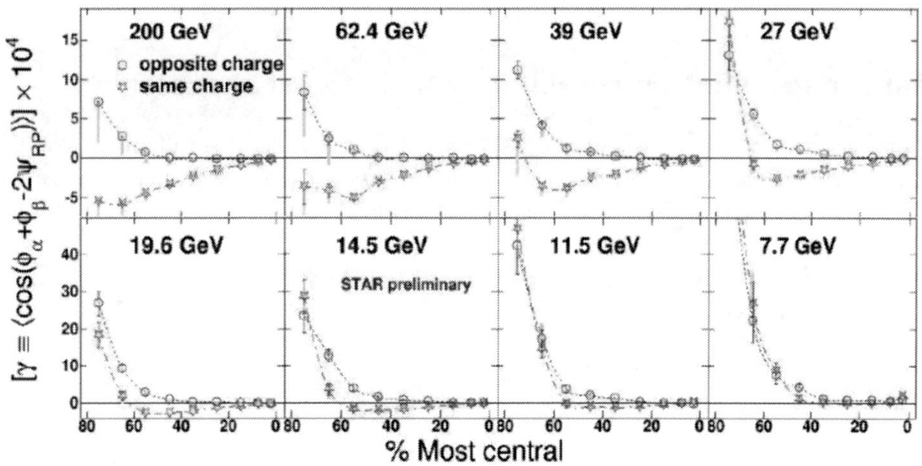

Figure 1. (Color online) Three particle correlator γ within $|\eta| < 1$ as a function of centrality for Au+Au collisions at 7.7–200 GeV. Note the different vertical scales. The filled bars (starting from the central values) represent the systematic uncertainty [12].

A hierarchal structure of correlation strength is expected from the combination of CME and CVE: p-p (CME and CVE) > p-Λ (CVE) > p-π (CME). A proton is a charged baryon, so the p-p correlation is expected to contain both the electric charge and baryonic charge separation signals; A Λ is a neutral baryon, so the p-Λ is expected to contain only the baryonic charge separation signal; A π is a charged meson, so the p-π is expected to contain only the electric charge separation signal. In order to gain insight on this physics, the identified-particle correlations are used to explore a possible hierarchal structure in the particle dependent correlations [20]. Fig. 2 shows the measured correlation of p-p, p-Λ and p-π for Au+Au at $\sqrt{s_{NN}}$ =200 GeV collisions from STAR. The results are compared to h-h correlation. The measured correlators may be contaminated by physics background (discussed below), and thus can not yet be directly compared to theoretical expectations regarding possible hierarchal structure.

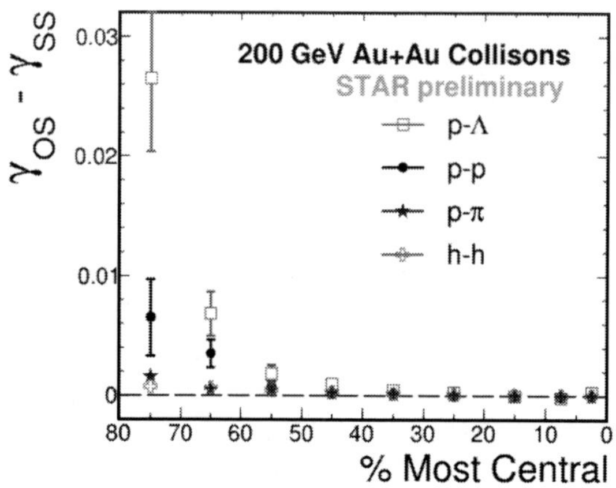

Figure 2. (Color online) Comparison of $\Delta\gamma \equiv \gamma_{OS} - \gamma_{SS}$ between p-Λ, p-p, p-π, h-h as a function of centrality for Au+Au collisions at $\sqrt{s_{NN}}$ = 200 GeV within $|\eta| < 1$.

4 Charge dependent signal with flow background suppression

While the $\Delta\gamma$ variable gives a positive signal for charge separation along the magnetic field, it also yields a positive signal for charge conservation along the reaction plane direction, for example, a positive-negative charge pair from a resonance decay. Since resonances (or generally, clusters including jets, dijets, etc.) exhibit elliptic anisotropy, there will be an overall positive $\Delta\gamma$. Those "flowing clusters" can thus mimic a CME/CVE signal [7, 10, 15]. In the case of resonance decay, the three particle correlator can be expressed as follows:

$$\langle\cos(\phi_\alpha + \phi_\beta - 2\Psi_{\rm RP})\rangle = \langle\cos(\Delta\phi + 2\phi_{res.} - 2\Psi_{\rm RP})\rangle \approx \langle\cos(\Delta\phi)\rangle \cdot v_{2,res} \qquad (2)$$

where α and β are the decay products of resonance $(res. \to \alpha + \beta)$, $\Delta\phi \equiv \phi_\alpha + \phi_\beta - 2\phi_{res.}$, and $v_{2,res} = \langle\cos 2(\phi_{res.} - \Psi_{\rm RP})\rangle$.

To suppress the flow background, STAR has exploited an event-by-event method and extracted the charge dependent signal at the observed event-by-event final-state particle anisotropy $v_{2,\rm ebye}^{\rm obs} = 0$ [21]. Since this does not necessarily guarantee zero anisotropy of the background sources (such as the resonances), residual physics background may remain [22].

In our analysis, we divide a full TPC event into two sub-events ($-1 < \eta < -0$ and $0 < \eta < 1$), and calculate the three particle correlator and $v_{2,\rm ebye}^{\rm obs}$ by using the particles from one side of the

TPC with respect to the event plane reconstructed from the other side of the TPC. Besides using the event-by-event method developed in [21] to extract signal at $v_{2,\text{ebye}}^{\text{obs}} = 0$, we also use the mixed-event method to suppress background. In the mixed-event method, we calculate the correlator and $v_{2,\text{ebye}}^{\text{obs}}$ with respect to the event plane taken from another event (i.e. we replace the Ψ_2 (Ψ_{RP}) in Eq. 2 with the event plane from another event). We subtract the mixed-event signal from the real-event signal at the same $v_{2,\text{ebye}}^{\text{obs}}$. We then average the mixed-event subtracted signal from all real events, according to the probability distribution of real events in $v_{2,\text{ebye}}^{\text{obs}}$. In other words, the subtracted mixed events are effectively weighted by the $v_{2,\text{ebye}}^{\text{obs}}$ distribution of real events. Although the signal simply averaged in mixed-events is zero, the signal averaged according to the $v_{2,\text{ebye}}^{\text{obs}}$ probability distribution of the real events (not symmetric about $v_{2,\text{ebye}}^{\text{obs}} = 0$) is no longer vanishing.

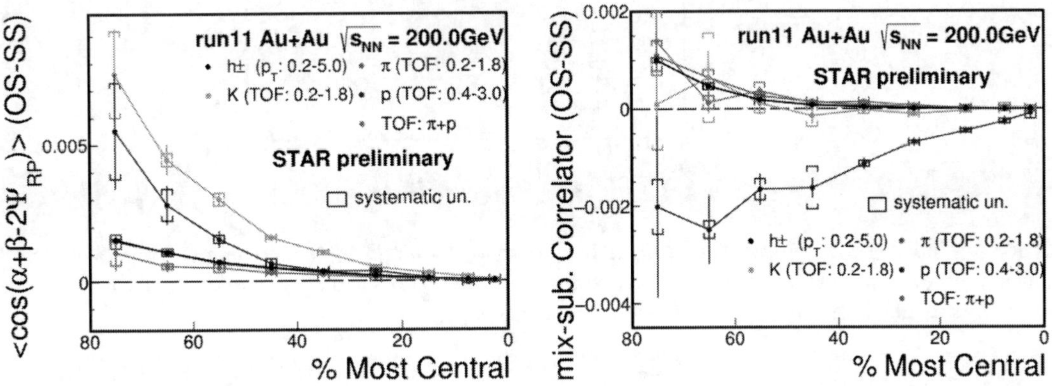

Figure 3. (Color online) Comparison of $\Delta\gamma$ (Left) and mixed-event subtracted $\Delta\gamma$ (Right) for charged particles, and identified particles (π, K, p) as a function of centrality for Au+Au collisions at $\sqrt{s_{NN}}$ = 200 GeV. The correlators and $v_{2,\text{ebye}}^{\text{obs}}$ are from one half of the TPC in η and the event plane is from the other half of the TPC. The brackets represent the systematic uncertainty.

Figure 3 left panel shows the centrality dependence of $\Delta\gamma$ for charged particles, and identified particles (π, K, p) from Au+Au collisions at $\sqrt{s_{NN}}$ = 200 GeV. Fig. 3 right panel shows the results with the mixed-event background subtraction, which is used to suppress the v_2 related background. With flow background suppression, the signals are reduced. For charged hardon and charged π the signals are reduced by ~50%, while for proton the signal becomes negative.

Figure 4 shows the collision energy dependence of the mixed-event subtracted $\Delta\gamma$ for different centralities in Au+Au collisions at $\sqrt{s_{NN}}$ form 7.7 to 200 GeV. In central collisions, charge separation signals are consistent with zero. In non-central collisions, the signals are finite, and the energy dependence is weak.

5 Charge dependent signal in small system

Recent results from CMS show that the correlator signal from p+Pb is comparable to the signal from Pb+Pb collisions at similar multiplicities [23]. We have carried out a similar control experiment using d+Au collisions. Fig. 5 shows $\Delta\gamma$ in d+Au, compared to Au+Au collisions at $\sqrt{s_{NN}}$ = 200 GeV. The data are scaled by the geometric mean of the positive and negative particle multiplicities. We observe sizeable charge dependent signal in small system d+Au collisions, also comparable to 200 GeV Au+Au collisions at similar multiplicities.

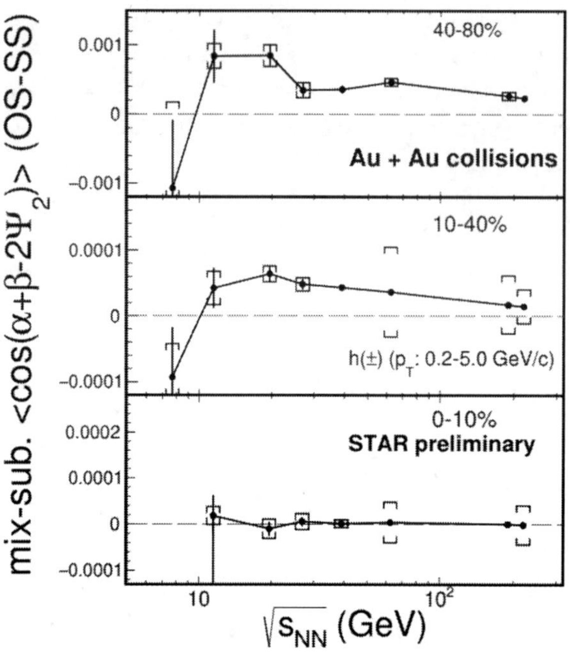

Figure 4. Collision energy dependence of the mixed-event subtracted $\Delta\gamma$ for different centralities in Au+Au collisions at $\sqrt{s_{NN}}$ = 200 GeV (Particles contributing to the correlators and $v_{2,\mathrm{ebye}}^{\mathrm{obs}}$ are from one half of the TPC in η, and the event plane is from the other half of the TPC).

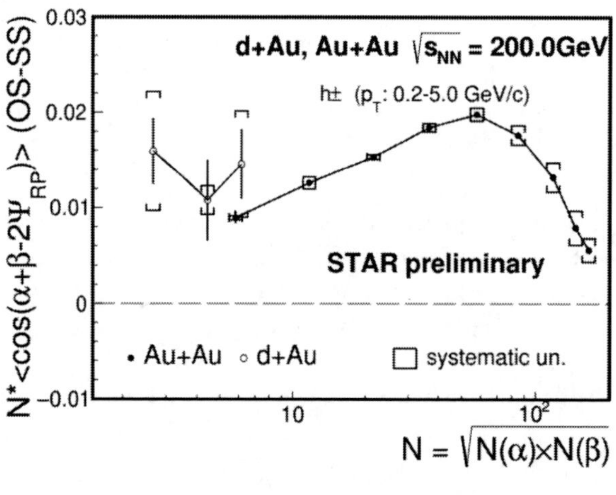

Figure 5. Multiplicity dependence of the $\Delta\gamma$ scaled by the geometric mean of the positive and negative particle multiplicities in d+Au and Au+Au collisions at $\sqrt{s_{NN}}$ = 200 GeV (Particles contributing to the correlators and $v_{2,\mathrm{ebye}}^{\mathrm{obs}}$ are from one half of the TPC in η, and the event plane is from the other half of the TPC).

6 Summary

Charge dependent three particle correlators are measured in Au+Au collisions at $\sqrt{s_{NN}}$ = 200 GeV. Identified-particle correlations are explored to search for the CME/CVE. The event-by-event anisotropy method and the mixed-event method are exploited to suppress the flow background. In Au+Au collisions at 200 GeV, with flow background suppression, proton-proton correlation signal becomes negative. In central Au+Au collisions form 7.7 to 200 GeV, with flow background suppression, charge separation signals are consistent with zero. In non-central collisions, the signals are finite, and the energy dependence is weak. In d+Au collisions at 200 GeV, we observe sizeable charge dependent signal, comparable to 200 GeV Au+Au collisions at similar multiplicities. The analysis of high statistical p+Au, d+Au data at 200 GeV are under way.

Acknowledgments

This work was partly supported by the U.S. Department of Energy (Grant No. DE-FG02-88ER40412), and the NSFC of China under Grant Nos. 11505073.

References

[1] T. D. Lee, Phys. Rev. D8, 1226 (1973).
[2] T. D. Lee and G. C. Wick, Phys. Rev. D9, 2291 (1974).
[3] P. D. Morley and I. A. Schmidt, Z. Phys. C26, 627 (1985).
[4] D. E. Kharzeev, J. Liao, S. A. Voloshin, and G. Wang, Prog. Part. Nucl. Phys. 88, 1 (2016).
[5] K. Fukushima, D. E. Kharzeev, and H. J. Warringa, Phys. Rev. D78, 074033 (2008).
[6] D. Kharzeev, R. D. Pisarski, and M. H. G. Tytgat, Phys. Rev. Lett. 81, 512 (1998); D. Kharzeev and R. D. Pisarski, Phys. Rev. D61, 111901 (2000).
[7] S. A. Voloshin, Phys. Rev. C70, 057901 (2004).
[8] D. Kharzeev, D. T. Son, Phys. Rev. Lett. 106, 062301 (2011).
[9] STAR, B. I. Abelev et al., Phys. Rev. Lett. 103, 251601 (2009).
[10] STAR, B. I. Abelev et al., Phys. Rev. C 81, 054908 (2010).
[11] STAR, L. Adamczyk et al., Phys. Rev. C 88, 064911 (2013).
[12] STAR, L. Adamczyk et al., Phys. Rev. Lett.113, 052302 (2014).
[13] N. N. Ajitanand, S. Esumi, R. A. Lacey (PHENIX Collaboration), in: Proc. of the RBRCWorkshops, vol.96, 230 (2010).
[14] ALICE, B. I. Abelev et al., Phys. Rev. Lett. 110, 012301 (2013).
[15] F. Wang, Phys.Rev. C81, 064902 (2010).
[16] K. H. Ackermann, et al., Nucl. Instr. and Meth. A 499, 624 (2003).
[17] M. Anderson, et al., Nucl. Instr. and Meth. A 499, 659 (2003).
[18] W. J. Llope, et al, doi:10.1016/j.nima.2010.07.086; STAR TOF Proposal (2004).
[19] C. Adler, H. Strobele, A. Denisov, E. Garcia, M. Murray, and S. White, Nucl. Instr. and Meth. A 461, 337 (2001); C. E. Allgower et al. (STAR Collaboration), ibid. 499, 740 (2003);
[20] F. Zhao (for STAR collaboration), Nucl. Phys. A931, 746 (2014).
[21] STAR, L. Adamczyk et al., Phys. Rev. C89, 044908 (2014).
[22] F. Wang and J. Zhao, arXiv:1608.06610[nucl-th].
[23] CMS, V. Khachatryan et al., arXiv:1610.00263 [nucl-ex].

EPJ Web of Conferences **141**, 01011 (2017) DOI: 10.1051/epjconf/201714101011
ISMD 2016

Fractal aspects of hadrons

Airton Deppman[1,a] and Eugenio Megías[2,3,b]

[1] *Instituto de Física da Universidade de São Paulo, Rua do Matão, 187 - Travessa R, São Paulo, Brazil*
[2] *Max-Planck-Institut für Physik (Werner-Heisenberg-Institut), Föhringer Ring 6, D-80805, Munich, Germany*
[3] *Departamento de Física Teórica, Universidad del País Vasco UPV/EHU, Apartado 644, 48080 Bilbao, Spain*

Abstract. The non extensive aspects of p_T distributions obtained in high energy collisions are discussed in relation to possible fractal structure in hadrons, in the sense of the thermofractal structure recently introduced. The evidences of self-similarity in both theoretical and experimental works in High Energy and in Hadron Physics are discussed, to show that the idea of fractal structure of hadrons and fireballs have being under discussion for decades. The non extensive self-consistent thermodynamics and the thermofractal structure allow one to connect non extensivity to intermittence and possibly to parton distribution functions in a single theoretical framework.

1 Introduction

In the 50's the existing experimental data on High Energy Collisions (HIC) evidenced the formation of a system in thermodynamic equilibrium. Indeed, Fermi proposed a thermodynamic model to describe hadronic collisions at high energies [1]. The main issue then was to explain how a short-lived system (half-life of a few fm/c) could reach equilibrium so fast. This system was called *fireball*. A few years later Hagedorn [2] proposed his Self-Consistent Thermodynamics (SCT) theory that was able to explain several features of hot hadronic systems. His starting point was a rather weird definition for fireballs, stating that

*A fireball is (*) a statistical equilibrium (hadronic black-body radiation) of an undetermined number of all kinds of fireballs, each o which, in turn, is considered to be (goto *)"*

The fact that the word fireball appears in the definition of fireball itself shows that this is a kind of recursive definition, not common in Physics. Self-reference is known to arise logical problems in many fields [3], but Hagedorn was able to use this definition to build the entire thermodynamics description of fireballs.

The SCT gave predictions for some quantities that could be easily accessed experimentally. It predicted the transverse momentum distribution of the particles produced in the decay of the fireball, given by

$$\left.\frac{dN}{dp_T}\right|_{y=0} = gV\frac{p_T m_T}{(2\pi)^2}\exp\left(\frac{m_T}{T}\right), \tag{1}$$

[a]e-mail: adeppman@gmail.com
[b]e-mail: emegias@mppmu.mpg.de

where g is a constant, V is the volume of the system, $m_T = \sqrt{p_T^2 + m^2}$. Another prediction was the hadron mass spectrum, given by

$$\rho(m) = \frac{\gamma}{m^{5/2}} \exp(\beta_o m)\,, \tag{2}$$

where $\beta_o = 1/T_o$ and T_o is a parameter of the theory. One interesting aspect of the theory was the prediction of a limiting temperature for the fireball, known today as Hagedorn temperature, T_H, which is numerically equal to the parameter T_o, i.e., $T_H = T_o$. The comparison between theory and experiment brought a sudden recognition of the importance of Hagedorn's theory, and its impact in HEP is notorious. Frautschi proposed a similar theory based exclusively in hadron structure, stating that *"hadrons are made of hadrons"*, again using self-reference [4]. With this so-called bootstrap model he obtained the same hadron mass spectrum formula shown in Eq. (2).

The success of SCT prompted the development of ideal gas models for hadronic systems, the Hadron Resonance Gas (HGR) models [5–12], that were able to explain many features of high energy collisions. Before that, an important paper by Dashen, Ma and Bernstein [13] had explained, through Dyson-Schwinger expansion, how a strongly interacting system could behave under some conditions as an ideal gas of resonant particles. Another important consequence of SCT came with the advent of the quark structure of hadrons, that was used by Cabibbo and Parisi [14] to propose that Hagedorn temperature was not a limiting temperature but a transition temperature between the confined quark and the deconfined quark regimes of hadronic matter. The last phase is known as Quark-Gluon Plasma and is one of the most interesting issues in nowadays Nuclear Physics. Despite its initial success and its important consequences, with the results coming from accelerators able to deliver particles at higher energies than those from existing at that time, it was found that SCT was not able to explain correctly the outcome of the new HEP data. Indeed, Hagedorn himself proposed, in substitution to this thermodynamics theory, an empirical model based on QCD [15]. Since his theory gave so many correct information about hadronic systems, the question imposes itself is: what went wrong with Hagedorn's theory?

In this work a possible answer to the question formulated above is proposed. It is shown that hadrons can present, in a very specific form, a fractal structure. Then it is shown that such fractals must be necessarily described through the Tsallis statistics [16], a generalization of the Boltzmann-Gibbs-Shannon statistics. After this, a generalization of the self-consistent theory developed by Hagedorn accommodating the generalized statistics is presented, what can be called Non Extensive Self-Consistent Thermodynamics (NESCT). Comparison of the results of NESCT and experimental data, some of its consequences and applications are discussed.

2 Thermofractals

At the same time that Hagedorn was developing his SCT theory, Mandelbrot was proposing the existence of systems with fractionary dimensions, the fractals [17]. Fractals are systems that present some special characteristics as scaling symmetry and self-similarity. Today the fractal structure is known to be present in many complex systems [18, 19]. It is interesting to note that Tsallis motivation to propose his generalized entropy was fractal systems [16].

The self-similarity is an evident feature of fireballs and hadrons according to their definitions by Hagedorn and Frautschi, respectively, as described above. However many other evidences point to the self-similar structure of hadrons: fractal dimension has been identified through the analysis of intermittence in experimental distributions obtained in HEP experiments [20–31]; a Parton Distribution Function (PDF) that describes the proton structure based on fractal properties was shown to fit experimental data rather well [32, 33]; direct evidences of self-similarity in experimental data has been observed [34–36]. In the following a system where fractal structure appears in its thermodynamical characteristics rather than in its geometrical aspects is introduced.

Thermofractals are systems presenting the following properties [37]:

1. The total energy is given by $U = F + E$, where F corresponds to the kinetic energy of N' constituent subsystems and E corresponds to the internal energy of those subsystems, which behaves as particles with an internal structure.

2. The constituent particles are thermofractals. The ratio $\langle E\rangle/\langle F\rangle$ is constant for all the subsystems. However the ratio E/F can vary according to a distribution which is self-similar (self-affine), $\tilde{P}(E)$, that is, at different levels of the subsystem hierarchy the distribution of the internal energy are equal (proportional) to those in the other levels.

3. At some level n in the hierarchy of subsystems the phase space is so narrow that one can consider $\tilde{P}(E_n)d\,E_n = \rho dE_n$, with ρ being independent of the energy E_n.

It can be shown that its thermodynamical potential [37] can be written as

$$\Omega = \int_0^{\infty} A\left[1 + \frac{\varepsilon}{NkT}\right]^{3N/2} \tilde{P}(\varepsilon)d\varepsilon\,, \tag{3}$$

where $N = N' + 2/3$. Since $\Omega = \int_0^{\infty} P(U)dU$ and $P(U)$ must be related to $\tilde{P}(\varepsilon)$ because of the second property of thermofractals, Eq. (3) must be satisfied simultaneously with the following identity:

$$P(U) = \tilde{P}(\varepsilon)\,, \tag{4}$$

expressing the self-similarity of thermofractals.

The simultaneous solution for Eqs. (3) and (4) is obtained with

$$P(\varepsilon) = A\left[1 + \frac{\varepsilon}{NkT}\right]^{-\frac{3N}{2}\frac{1}{1-\nu}}\,. \tag{5}$$

Introducing the index q by $q - 1 = (2/3N)(1 - \nu)$ and the effective temperature $\tau = 2(1 - \nu)T/3$ one finally obtains

$$P(\varepsilon) = Ae_q(-\varepsilon/k\tau)\,, \tag{6}$$

where

$$e_q(x) = \left[1 + (1 - q)x\right]^{\frac{1}{1-q}} \tag{7}$$

is the Tsallis q-exponential factor characteristic of the nonextensive statistics. It replaces the Boltzmann exponential factor by generalizing it, since for $q \to 1$, $\tilde{P}(\varepsilon)$ becomes the well-known exponential factor of Statistical Mechanics.

This result is a possible answer to the question proposed in the Introduction: what went wrong with Hagedorn's theory? Today we know that Hagedorn's fireball is a fractal, or more specifically a thermofractal. The result above shows that thermofractals must be described by Tsallis statistics, not by Boltzmann statistics as Hagedorn did. The next questions are: is it possible to generalize Hagedorn's theory with Tsallis statistics? And could the generalized theory describe experimental data? In the next section these questions will be answered.

3 Non Extensive Self-Consistent Thermodynamics

NESCT is a generalization of the SCT theory by imposing the self-consistency principle from Hagedorn in the non extensive statistics from Tsallis [38]. The basic ingredients are the two forms of partition function for fireballs proposed by Hagedorn, only this time using the Tsallis q-exponential factor in Eq. (6), that is,

$$Z_q(V_o,T) = \int_0^\infty \sigma(E)[1+(q-1)\beta E]^{-\frac{q}{(q-1)}} dE\,, \tag{8}$$

where $\sigma(E)$ is the density of state of the fireballs as a function of its energy, and

$$\ln[1+Z_q(V_o,T)] = \frac{V_o}{2\pi^2}\sum_{n=1}^{\infty}\frac{1}{n}\int_0^\infty dm \int_0^\infty dp\, p^2\rho(n;m)[1+(q-1)\beta\sqrt{p^2+m^2}]^{-\frac{nq}{(q-1)}}\,, \tag{9}$$

which is the partition function for an ideal gas of fermions and bosons with mass m. The sum in n results from the expansion of the logarithm function, and $\rho(n;m) = \rho_f - (-1)^n\rho_b$ is related to the fermionic and bosonic mass spectra. In the case $n = 1$, which will be used here, $\rho(1;m) = \rho_f + \rho_b = \rho(m)$ is the hadron mass spectrum.

Since fireballs are thermofractals and present self-similarity, the following weak-constraint must hold

$$\ln[\rho(m)] = \ln[\sigma(E)]\,. \tag{10}$$

The main task now is to find functions $\rho(m)$ and $\sigma(E)$ that satisfy the equality above and simultaneously let the two forms of partition function identical to each other. This can be achieved by choosing [38]

$$\rho(m) = \frac{\gamma}{m^{5/2}}[1+(q_o-1)\beta_o m]^{\frac{1}{q_o-1}} \qquad \text{and} \qquad \sigma(E) = bE^a[1+(q_o-1)\beta_o E]^{\frac{1}{q_o-1}}\,. \tag{11}$$

With those functions both forms of partition function result in

$$Z_q(V_o,T) \rightarrow b\Gamma(a+1)\left(\frac{1}{\beta-\beta_o}\right)^{a+1} \tag{12}$$

with $a+1 = \alpha = \gamma V_o/2\pi^2\beta^{3/2}$. From here one can see that there is a singularity at $\beta = \beta_o$, and therefore the limiting temperature is found.

Therefore it is possible to generalize Hagedorn's self-consistency principle within Tsallis statistics and as a result one gets a new hadron mass spectrum formula, given by $\rho(m)$ in Eq. (11), the limiting temperature resulting from the singularity in Eq. (12), now expressed in terms of the Tsallis temperature, but also a universal entropic index, q, which is characteristics of any hadron. It is worthy to mention that both T_o and q_o are parameters in the hadron mass spectrum, therefore they are related to the hadronic structure. As in the case of the Hagedorn's theory, here also the way to test experimentally the predictions of the theory is through the transversal momentum distribution, given by

$$\left.\frac{dN}{dp_T}\right|_{y=0} = gV\frac{p_T m_T}{(2\pi)^2}e_q\left(\frac{m_T}{T}\right)\,. \tag{13}$$

The predictions of NESCT have been compared with HEP experimental data for p_T distributions, showing a good agreement between calculation and data [42–46], resulting in $q_o = 1.14$ and $T_o = 62$ MeV. Also, the hadron mass spectrum formula has been used to describe the known hadronic states, resulting again in a very good agreement with data and leading to values of q_o and T_o very similar to those obtained with p_T analysis [44]. The extension to finite chemical potential was performed in [47], and was applied to the analysis of neutron star stability [48]. A comparison between the results from NESCT and LQCD data also has shown a good agreement between both calculations [49], in this case without any adjustable parameter. From the thermofractal structure one can obtain the fractal dimension of hadrons, what was done in Ref. [37], resulting in $D = 0.69$, a value that is close to that resulting from intermittence analysis, around $D = 0.65$ [50–54]

These analysis confirms that Hagedorn's self-consistency principle holds for hot hadronic systems even at energies as high as those found at LHC, when the appropriate statistics is used. The Tsallis statistics, on the other hand, must be used in the construction of fireballs analysis because it takes into account the complex partonic structure of hadrons. The fact that the fractal dimension obtained from the q_o and T_o values obtained in the p_T analysis or in the mass spectrum analysis shows that the hypothesis that hadrons presents a fractal structure as that described for thermofractals is a reasonable one. This is evidenced by experimental data, as explained in the introduction, and by the analysis of intermittency.

4 Conclusions

In this work it is discussed the fractal aspects of fireballs and hadrons as defined respectively by Hagedorn and Frautschi. By the introduction of the concept of thermofractal, a system presenting fractal structure in its Thermodynamic functions, it is shown that such systems must be described by Tsallis statistics rather than Boltzmann statistics.

The self-consistency principle proposed by Hagedorn is then generalized assuming Tsallis statistics, resulting in a non extensive statistics that predicts a new hadron mass spectrum formula, a limiting temperature and a universal entropic index for hadrons. This result shows that it is possible to generalize Hagedorn's theory to obtain power-law spectra for the particles produced in high energy collisions.

Several works have shown that the power-law spectra resulting from the non extensive statistics can correctly describe the general properties of particle production in HEP experiments, confirming the existence of the limiting temperature and of the universal entropic index. Also, in the present work it is discussed that the new hadron mass spectrum describes the observed hadronic states better than Hagedorn's formula. Calculations from Latice QCD are also in good agreement with the generalized thermodynamics model.

Finally, it is shown that the fractal dimension of the thermofractal obtained with values for temperature and entropic index that result from analysis of p_T distributions in high energy collisions are in agreement with those found in analyses of intermittence in experimental data from high energy collisions.

Acknowledgements

This work was supported by the Brazilian agency, CNPq, under grant 305639/2010-2, by Plan Nacional de Altas Energías Spanish MINECO grant FPA2015-64041-C2-1-P, and by the Spanish Consolider Ingenio 2010 Programme CPAN (CSD2007-00042). E.M. would like to thank the Instituto de Física of the Universidade de São Paulo, Brazil, for their hospitality and support during the completion of the initial stages of this work. The research of E.M. is supported by the European Union under a Marie Curie Intra-European fellowship (FP7-PEOPLE-2013-IEF) with project number PIEF-GA-2013-623006, and by the Universidad del País Vasco UPV/EHU, Bilbao, Spain, as a Visiting Professor.

References

[1] E. Fermi, Prog. Theor. Phys. 5 (1950) 570.
[2] R. Hagedorn, Nuovo Cimento Suppl. 3 (1965) 147.
[3] Stephen Cole Kleene, Transactions of the American Mathematical Society 53 (1943)41.
[4] S. Frautschi, Phys. Rev. D 3 (1971) 2821.
[5] R. Hagedorn, Lect. Notes Phys. 221 (1985) 53.
[6] N. O. Agasian, Phys. Lett. B519 (2001) 71.
[7] A. Tawfik, Phys. Rev. D71 (2005) 054502.
[8] E. Megias, E. Ruiz Arriola, L. Salcedo, Phys. Rev. D80 (2009) 056005.
[9] P. Huovinen, P. Petreczky, Nucl. Phys. A837 (2010) 26.
[10] S. Borsanyi, G. Endrodi, Z. Fodor, A. Jakovac, S. D. Katz, et al., JHEP 1011 (2010) 077.
[11] A. Bazavov, et al., Phys. Rev. D85 (2012) 054503.
[12] E. Ruiz Arriola, L.L. Salcedo, E. Megias, Acta Phys. Polon. B45 (2014) 2407-2454.
[13] R. Dashen, S. Ma and H. J. Bernstein, Phys. Rev. 187 (1969) 345.
[14] N. Cabibbo and G. Parisi, Phys. Lett. 59B (1975) 67.
[15] R. Hagedorn, *Multiplicities, p_T-distributions and the expected hadron → quark-gluon phase transition*, Ref.TH.3684-CERN, 1983.
[16] C. Tsallis, J. Stat. Phys. 52 (1988) 479.
[17] B. B. Mandelbrot, The Fractal Geometry of Nature, W. H. Freeman, New York 1983.
[18] T. Tél, Z. Nat. 43a (1998) 1154.
[19] M. Schroeder in Fractals, Chaos, Power Laws: Minutes from an Infinite Paradise, Dover Pub Inc. 1991 - Mineola NY.
[20] A. Bialas and R. Peschanki, Nucl. Phys. B273 (1986) 703.
[21] A. Bialas and R. Peschanki, Nucl. Phys. B308 (1988) 857.
[22] R. C. Hwa, Phys. Rev. D 41 (1990) 1456.
[23] R.C. Hwa and J. Pan, Phys. Rev. D 45 (1992) 1476.
[24] S. Hegyi, Phys. Lett. B 318 (1993) 642.
[25] I.M. Dremin and R.C. Hwa, Phys. Rev. D 49 (1994) 5805.
[26] S. Hegyi and T. Csörgö, Phys. Lett. B 296 (1992) 256.
[27] E. K. G. Sarkisyan, Phys. Lett. B 477, 1 (2000).
[28] E.K. Sarkisyan, L.K. Gelovani, G.L. Gogiberidze, and G.G. Taran, Phys. Lett. B 347 (1995) 439.
[29] W. Kittel and E. A. De Wolf, Soft Multihadron Dynamics, World Scientific, Singapore, 2005.
[30] E.A. De Wolf, I.M. Dremin and W. Kittel, Phys. Rep. 270 (1996) 1.
[31] P. Bozek, M. Ploszajczak and R. Botet, Phys. Reports 252 (1995) 101.
[32] T. Lastovicka, hep-ph/0203260.
[33] AIP Conf.Proc. 276 (1993) 454.
[34] G. Wilk and Z. Włodarczyk, Phys. Lett. B (2013) 163.
[35] M.V. Tokarev and I. Zborovsky, Theor. Math. Phys. 184 (2015) 1350.
[36] C. Beck, Physica A: Statistical Mechanics and its Applications 286 (2000) 164 - 180.
[37] A. Deppman, Phys. Rev. D 93 (2016) 054001.
[38] A. Deppman, Physica A 391 (2012) 6380.
[39] X. Yi-Long, C. Gang, W. Juan, W. Mei-Juan and C. Huan, Nuclear Physics A 920 (2013) 33.
[40] C. Gang et al., Proc. XXXI Intern. Symp. Multiparticle Dynamics, Eds. B. Yuting, Yu Meiling, W. Yuantang, World Scientific, Singapore, 2002, p. 361.

[41] C. Gang, L. Lianshou and G. Yanmin, Int. J. Mod. Phys. A 14 (1999) 3687.
[42] I. Sena and A. Deppman, Eur. Phys. J. A (2013) 49: 17.
[43] J. Cleymans and D. Worku, J. Phys. G: Nucl. Part. Phys. 39 (2012) 025006.
[44] L. Marques, E. Andrade-II and A. Deppman, Phys. Rev. D 87 (2013) 114022.
[45] M. D. Azmi and J. Cleymans, J. Phys. G: Nucl. Part. 41 (2014) 065001.
[46] M. D. Azmi and J. Cleymans, Eur. Phys. J. C 75 (2015) 430.
[47] E. Megías, D.P. Menezes and A. Deppman, Physica A 421 (2015) 15.
[48] D.P. Menezes, A. Deppman, E. Megías and L.B. Castro, Eur. Phys. J. A 51 (2015) 155.
[49] A. Deppman, J. Phys. G: Nucl. Part. Phys. 41 (2014) 055108.
[50] I. V. Ajienko et al. (EHS/NA22 Collaboration), Phys. Lett. B 222, 306 (1989).
[51] M. H. Rasool, M. A. Ahmad, and S. Ahamad, Chaos Sol- itons & Fractals 81, 197 (2015).
[52] G. Singh and P. L. Jain, Phys. Rev. C 50, 2508 (1994).
[53] C. Albajar et al., Z. Phys. C 56, 37 (1992).
[54] D. Ghosh et al., Phys. Rev. C 58, 3553 (1998).

DOI: 10.1051/epjconf/201714101012

Recent femtoscopy results from ALICE

Adam Kisiel[1,a] for the ALICE Collaboration

[1] *Faculty of Physics, Warsaw University of Technology, ul. Koszykowa 75, 00-662 Warsaw, Poland*

Abstract. ALICE, an experiment dedicated to the analysis of heavy-ion collisions is ideally suited for femtoscopic studies, with its excellent particle identification capabilities at low and intermediate momenta. It measures correlations of pions which provide crucial information on the size and the dynamics of the system. Recent results for femtoscopic radii from p+p, p–Pb and Pb–Pb collisions will be discussed with emphasis on similarities and differences between small and large systems. In particular the multiplicity and transverse momentum dependence of the radii will be discussed.

1 Introduction

Femtoscopy is a technique, which relies on the measurement of the two-particle correlations as a function of the relative momentum of the pair. In heavy-ion collisions it is most commonly performed for pairs of identical charged pions, which are the most commonly produced particles. At the Large Hadron Collider (LHC), the ALICE experiment is especially well suited for such measurement, beacause of its excellent particle identification (PID) at low relative momenta of registered particles [1]. This work discussed measurements of such correlations for p+p, p–Pb, and Pb–Pb collisions performed by ALICE.

The aim of the femtoscopic analysis is to extract the information on the size of the emitting system from the measured two-particle correlations. For pairs of identical charged pions the correlation effect arises mostly from the enhanced pair production due to the Bose-Einstein statistics. In the correlation function constructed versus the relative momentum $\vec{q} = \vec{p_1} - \vec{p_2}$ it manifests as an increased correlation at low $\vec{q}$ which has roughly a Gaussian shape (with an additional dip at lowest values of $\vec{q}$ caused by the Coulomb repulsion) [4, 5]. Its width is inversely proportional to the size of the emission region, usually called "radius". The analysis can be performed separately for each of the three independent components of $\vec{q}$, yielding three independent radii. The three directions are usually chosen according to the Bertsch-Pratt convention: "long" along the beam axis, "out" along the pair transverse momentum, and "side" perpendicular to the other two. The measured correlation function is analyzed via the Bowler-Sinyukov fitting formula:

$$C(\vec{q}) = (1 - \lambda) + K(q)\left[1 + \lambda \exp(-R_{out}^2 q_{out}^2 - R_{side}^2 q_{side}^2 - R_{long}^2 q_{long}^2)\right], \quad (1)$$

where λ characterized the strength of the correlation, R parameters are the radii, while the K function accounts for the Coulomb interaction.

[a] e-mail: kisiel@if.pw.edu.pl, Supported by the Polish National Science Centre Grant No. UMO-2014/13/B/ST2/04054

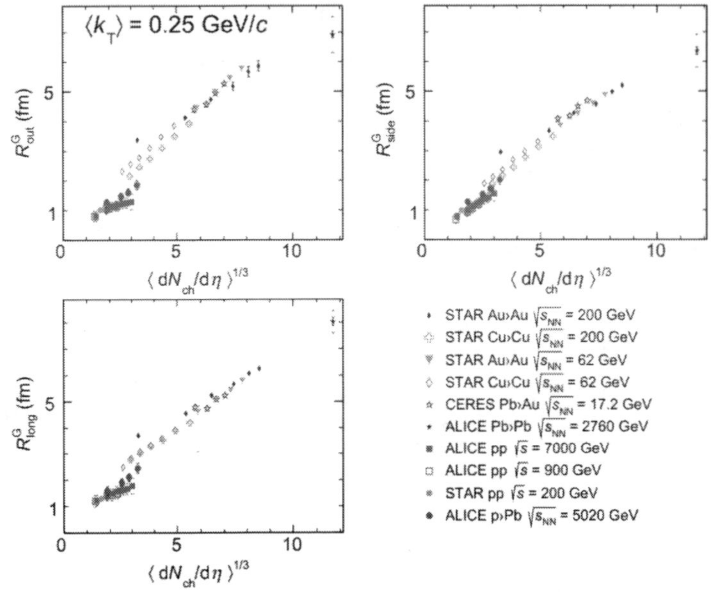

Figure 1. Comparison of femtoscopic radii (Gaussian), as a function of the measured charged-particle multiplicity, for various collision systems and energies. Figure from [5]

The value of the measured size is interesting in itself, but its dependence on event multiplicity as well as pair transverse momentum k_T is of additional significance. Firstly all experimental data collected so far have shown a linear dependence of the total femtoscopic volume of the system (obtained as a product of the radii in three dimensions) on event multiplicity. We extend this measurement to the new energy regime, and perform it, in a consistent way, in small (p+p and p–Pb) and large (Pb–Pb) systems. Such increase is universally reproduced in models of heavy-ion collisions, while its interpretation in small systems is less clear. In this work we compare these dependencies and speculate on the origin of similarities and differences between them.

The dependence of radii on k_T is interpreted in the hydrodynamic framework, which appears to work well for heavy-ion collisions in the LHC energy regime [2, 3]. In particular strong collective movement of matter (flow), which appears naturally in such models, leads to the decrease of the apparent size of the system with increasing pair momentum. Model predictions for this effect have become precise enough to speculate on the relative strenght of this decrease in different directions. In this work we compare the data in heavy-ion collisions and its model description to the effects observed in small systems and speculate on the role of collective flow there.

2 Multiplicity dependence

A comprehensive set of femtoscopic results as a function of particle multplicity is shown in Fig. 1. It presents the results for heavy-ion collision for a wide set of collision energies, for the largest possible collision system sizes (Pb and Au nuclei). These data show a linear scaling with $\langle dN_{ch}/d\eta \rangle^{(1/3)}$ in the "long" direction. In the "side" direction the scaling is also apparent, altough less precise. The scaling in the "out" direction is only approximate. In this direction the size of the system is coupled to the emission duration as well as the shape of the emission hypersurface. In particular in hydrodynamic models, it may show additional dependence on the collision energy, which would break the trivial scaling. This effect is discussed in more detail in the following section.

The heavy-ion data is also, for the first time in such detail, compared to p+p and p–Pb collisions. The p+p data is also very precisely following a liner scaling with multiplicity in all directions, but with significantly different slopes than the ones given by the heavy-ion trend. The difference may be due simply to the quite different initial stage. In particular in the region where p+p and AA data produce similar multiplicities, the initial stages are expected to be quite different - a single violent nucleon-nucleon collision in p+p versus a collection of less violent collisions in peripheral AA collisions. This interpretation however presents an interesting open question. In AA collision a growth of the system size with multiplicity is naturally explained in the Glauber model, as being simply the reflection of the growth of the nuclei overlap region, coupled to the system evolution duration growing with collision energy. However in p+p collisions the "initial stage" has identical size, regardless of final state multiplicity. Therefore, the reason why the system radius grows with multiplicity is not yet known, especially since the space-time evolution is not yet properly modelled in popular Monte-Carlo codes such as Pythia or Phojet.

An equally interesing question arises from the analysis of the p–Pb data. At low multiplicity they tend to be similar to p+p data. But the radii grow much faster with multiplicity and appear to approach the AA scaling line. This presents and intriguing possibility that the p–Pb system is a "transitional" one, and as we analyze the evolution of various observables with the increase of the final state multiplicity, we may observe a change from the "elementray" to the "complex" system.

3 Transverse momentum dependence

The dependence of radii on k_{T} for heavy-ion collisions in ALICE is shown in Fig. 2. A clear decrease of the radii with k_{T} is seen, which is roughly power-law in shape. The decrease is more steep in the "long" direction when compared to the "transvere" directions. These results are compared to the predictions of two hydrodynamics codes. Both models, even though they differ significantly on their implementation, reproduce the data very well, both qualitatively and quantitatively. Therefore, the current interpretation of data is based on hydrodynamics. The decrease is a direct consequence of collective flows. Through the mechanism known as "lengths of homogeneity" [2, 3] such flows cause the apparent size of the system to fall with k_{T}. Such dependence can then be proposed as a well calibrated and sensitive probe for collectivity in the system.

The system size in the "out" direction is comparable or smaller than in the "side" direction. This is particularly interesting, when compared to data at lower energies, where the "out" radius was usually larger. The decrease of the ratio of the two radii is a non-trivial consequence of radial flow and the change of the initial conditions for the hydrodynamic system with the increase of collision energy [2]. The modelling shows that at lower energies the system freezes out in the "outside-in" configuration. But at the LHC energies a significant change occurs. Much higher initial energy density results in the larger final system (indeed data at LHC show a size up to 30% larger than at RHIC), which also freezes out in the "inside-out" configuration. The "out" radius is sensitive to this scenario and is predicted to be smaller than "side" if such freeze-out occurs. This is precisely what is observed in ALICE data, giving additional confidence in the validity of hydrodynamic modelling of the heavy-ion collisions.

This calibrated probe can then be applied to p–Pb collisions. We show the k_{T} dependence of radii for this dataset in Fig. 3. The results are also compared to the dependence observed in p+p collisions at the highest multiplicity. The data show a clear increase of radius with final state multiplicity, as was already discussed above. The decrease of the radii with k_{T} is observed, however it is not universal. Firstly in the "long" direction the decrease is significant, quite similar in all multiplicities,

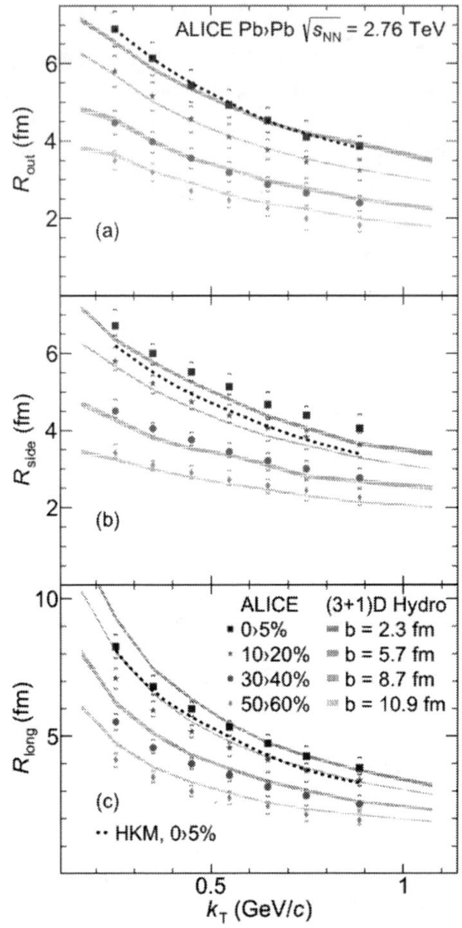

Figure 2. Comparison of femtoscopic radii, as a function of pair transverse momentum, with the calculation from the Therminator and 3+1D hydro model [2], for four centralities and with the HKM model [3] for the central data. Closed symbols are experimental data, bands and dashed lines are the model calculations. Figure from [4]

and also comparable to the one observed in p+p collisions. This resembles the situation in the AA collisions, where in this direction the dependence seemed to be universal. The "side" radius has a more complicated dependence. At small multiplicities it does not change with momentum. The slope of the dependence grows with multiplicity and is significant in the highest range. A similar evolution is also observed in p+p collisions. In the "out" direction the slope is significant at all multiplicities. The absolute values of the radii are smaller than in "side", sometimes significantly more so than in heavy-ion collisions.

Even though the decrease of radii with k_T is observed in p-Pb collisions, it is not as universal as in AA collisions. Especially the evolution of the "side" radius dependence prevents us from drawing a strong conclusion about the existence of flow in p–Pb collisions. Similar conclusions can be drawn from the comparison of the data to the hydrodynamic calculations, shown in the same Figure. The "long" radius dependence is reasonably reproduced, as is the "side" direction. Both radii show a strong preference for calculations starting with low initial size. They are however unable to fully

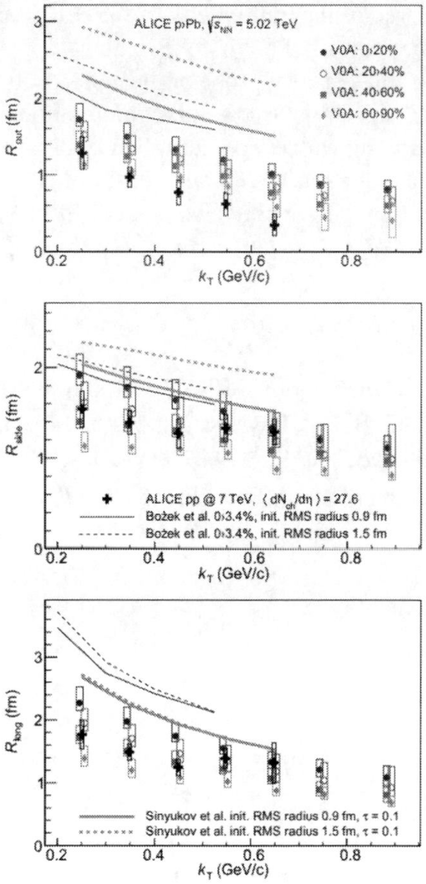

Figure 3. Comparison of femtoscopic radii (Gaussian) in p–Pb collisions at $\sqrt{s_{NN}}$ = 5.02 TeV, as a function of the measured charged-particle multiplicity, for various collision systems and energies. Figure from [5]

predict the "out" radii, showing values larger by at least 20%. The comparison is consistent with the speculation mentioned above in the discussion of muliplicity dependence. It seems that in the p–Pb collisions a transition is observed from a system at low multiplicity, which probably does not exhibit hydrodynamic collectivity, to the system at large multiplicity, which starts to resemble the collective system. Nevertheless the k_T dependence of the radii proves to be a sensitive and non-trivial observable crucial in the correct intrepretation of the physics of the elementrary and heavy-ion collisions.

4 Conclusions

We have shown a comprehensive collection of femtoscopic results from the ALICE experiment, covering the p+p, p–Pb, and Pb–Pb collisions. The measured volume increases linearly with final-state event multiplicity in all measured systems. In addition all of the three-dimensional radii separately increase linearly with cube root of the multiplicity. This dependence is natutally explained in hydrodynamic models of AA collisions. The slope of the dependence in p+p collisions is significanlty different. It suggests that other mechanisms may be responsible for this increase in the elementary collisions. The p–Pb collisions exhibit a behavior which suggests that a transition from "elementary" to "compund" system is observed as the multiplicity of the collision is increased. The pair transverse momentum dependence of the radius is Pb–Pb collisions is very well reproduced by hydrodynamic models. It is a well calibrated and sensitive probe of the collectivity of the system. In addition the results show a transition to the "outside-in" freeze-out configuration at the LHC. In the p–Pb collisions the decrease of the radii is changing with event multiplicity. It is significant at the largest multiplicity, but the hydrodynamic codes are not able to reproduce it fully. This signature again suggests that the p–Pb is a system in which a transition from "elementary" to the "collective" regime occurs as the event multiplicity is increased.

References

[1] K. Aamodt *et al.* [ALICE Collaboration], JINST **3** (2008) S08002.
[2] A. Kisiel, M. Galażyn and P. Bożek, Phys. Rev. C **90** (2014) no.6, 064914
[3] I. A. Karpenko, Y. M. Sinyukov and K. Werner, Phys. Rev. C **87** (2013) no.2, 024914
[4] J. Adam *et al.* [ALICE Collaboration], Phys. Rev. C **93** (2016) no.2, 024905
[5] J. Adam *et al.* [ALICE Collaboration], Phys. Rev. C **91** (2015) 034906

Femtoscopy in $\sqrt{s_{\mathrm{NN}}} = 5.02$ TeV p-Pb collisions with *ATLAS*

Michael Clark[1,a] on behalf of the ATLAS Collaboration

[1] *Columbia University, 116th St & Broadway, New York, NY 10027, USA*

Abstract. Bose-Einstein correlations between identified charged pions are measured for p+Pb collisions at $\sqrt{s_{\mathrm{NN}}} = 5.02$ TeV with the ATLAS detector using a total integrated luminosity of 28 nb^{-1}. Pions are identified using ionisation energy loss measured in the pixel detector. Two-particle correlation functions and the extracted source radii are presented as a function of average transverse pair momentum (k_{T}) and rapidity ($y^{\star}_{\pi\pi}$) as well as collision centrality. Pairs are selected with a rapidity $-2 < y^{\star}_{\pi\pi} < 1$ and with an average transverse momentum $0.1 < k_{\mathrm{T}} < 0.8$ GeV. The effect on the two-particle correlation function from jet fragmentation is studied, and a new method for constraining its contributions to the measured correlations is described. The measured source sizes are substantially larger in more central collisions and are observed to decrease with increasing pair k_{T}. A correlation with the local single-particle multiplicity $\mathrm{d}N_{\mathrm{ch}}/\mathrm{d}y^{\star}$ is demonstrated. The scaling of the extracted radii with the mean number of participants is also used to compare a selection of initial-geometry models. The cross term R_{ol}, which couples radial and longitudinal expansion, is measured as a function of rapidity, and a departure from zero is observed with 4.8 σ combined significance for $y^{\star}_{\pi\pi} > -1$ in the most central events.

1 Introduction

Multi-particle correlations in proton-lead (p+Pb) [1–5] and proton-proton ($p + p$) [6] collisions exhibit long-range azimuthal correlations similar to those observed in lead-lead (Pb+Pb) collisions, in which they are attributed to collective expansion of the quark-gluon plasma. Hydrodynamic models can describe the observed correlations in p+Pb [7–9], but the extent to which they are appropriate in "small" systems remains an open question.

Bose-Einstein correlations between charged pion pairs can be used to measure the spacetime extent of particle production with a technique called femtoscopy (see Reference [10] and references therein). The width of the correlations in relative momentum $\mathbf{q}$ is studied as a function of pair transverse momentum k_{T}. The measured Hanbury Brown and Twiss (HBT) radii, which are interpreted as the size of the particle-emitting region at freezeout, are expected to decrease with rising transverse momentum if the source expands [11]. Thus, HBT measurements can be used to address the questions raised by angular correlation results. While femtoscopic methods have already been applied to p+Pb systems at the LHC [12, 13], these results are obtained using a new data-driven technique to constrain the significant background contribution from jet fragmentation, and they provide the first measurements of the dependence of these source radii on the rapidity $y^{\star}_{\pi\pi}$.

[a] e-mail: michael.ryan.clark@cern.ch

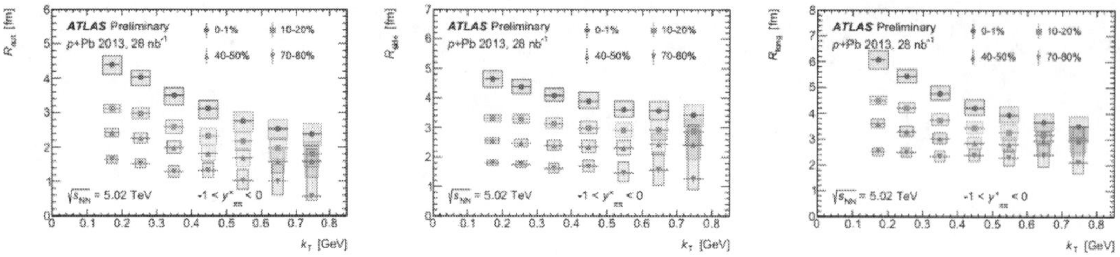

Figure 1. Three-dimensional HBT radii as a function of the average transverse momentum k_T of pion pairs [14]. The widths of the boxes vary between centrality intervals only for visual clarity.

2 Experimental Analysis

The data used in this analysis were collected by the ATLAS detector [15] during the 2013 p+Pb run at the Large Hadron Collider (LHC) with a nucleon-nucleon centre-of-momentum energy of $\sqrt{s_{NN}} =$ 5.02 TeV. The inner detector is used to reconstruct charged particles. It consists of a silicon pixel detector, a semiconductor tracker made of double-sided silicon microstrips, and a transition radiation tracker made of straw tubes. Pions are identified using a measurement of the charge deposited in the pixel detector. All three subdetectors are composed of a barrel and two symmetrically placed end-cap sections. The forward calorimeters, located in a pseudorapidity range of $3.1 < |\eta| < 4.9$, are used to estimate the centrality of each collision.

The two-particle correlation functions are taken to be of the form

$$C_{\mathbf{k}}(\mathbf{q}) = [(1-\lambda) + \lambda K(\mathbf{q}) C_{\mathrm{BE}}(\mathbf{q})]\,\Omega(\mathbf{q}), \tag{1}$$

where K is a correction factor for final-state Coulomb interactions and Ω represents non-femtoscopic background which arises due to jet fragmentation. The Bose-Einstein part of the three-dimensional correlation functions are fit to functions of the form

$$C_{\mathrm{BE}}(\mathbf{q}) = 1 + e^{-\|R\mathbf{q}\|}, \tag{2}$$

where R is a symmetric matrix in the Bertsch-Pratt coordinate system:

$$R = \begin{pmatrix} R_{\mathrm{out}} & 0 & R_{\mathrm{ol}} \\ 0 & R_{\mathrm{side}} & 0 \\ R_{\mathrm{ol}} & 0 & R_{\mathrm{long}} \end{pmatrix}. \tag{3}$$

In these coordinates, q_{out} is the projection along $\mathbf{k}_T$, q_{side} is the projection along $\hat{\mathbf{z}} \times \mathbf{k}_T$ (with the z-axis along the beamline), and q_{long} is the longitudinal component. The relative momentum of the pair is evaluated in the longitudinally comoving frame (LCMF), i.e., the frame boosted such that $k_z = 0$.

The jet fragmentation background, $\Omega(\mathbf{q})$, contributes an enhancement at low $|\mathbf{q}|$, though usually over a wider range in $|\mathbf{q}|$ and with a smaller amplitude than the Bose-Einstein correlation. This background is constrained in same-charge correlation functions by measuring the opposite-sign correlation functions, which do not contain a Bose-Einstein enhancement. Opposite-charge correlations contain contributions from resonance decays and the most prominent of these are removed by cuts on the opposite-charge pion pair mass. Even with resonances removed, due to charge conservation the fragmentation contribution is not identical between same- and opposite-charge correlations, but the shape

DOI: 10.1051/epjconf/201714101013

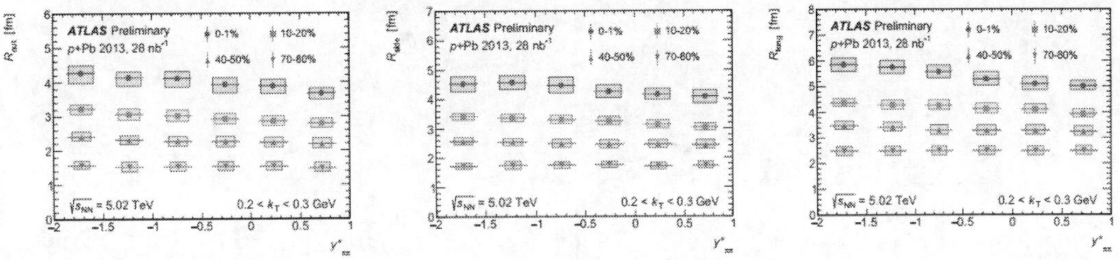

Figure 2. Three-dimensional HBT radii as a function of the rapidity $y^{\star}_{\pi\pi}$ of pion pairs [14]. The widths of the boxes vary between centrality intervals only for visual clarity.

is similar. The relationship between the two is studied in detail using Pythia 8, and a mapping from opposite- to same-charge correlation functions is derived in simulation for use with the data. For details on this and other aspects of the analysis, see Reference [14].

3 Results

Figure 1 shows the HBT radii as a function of k_T in four centrality intervals. A decrease of the radii with rising k_T, interpreted as a signature of collective expansion, is observed in central events. This trend is diminished in peripheral events.

The HBT radii are presented as a function of pair rapidity $y^{\star}_{\pi\pi}$ in Figure 2. Central collisions exhibit larger radii on the backward (Pb-going) side of the event, while peripheral events show no distinguishable variation of the radii with rapidity.

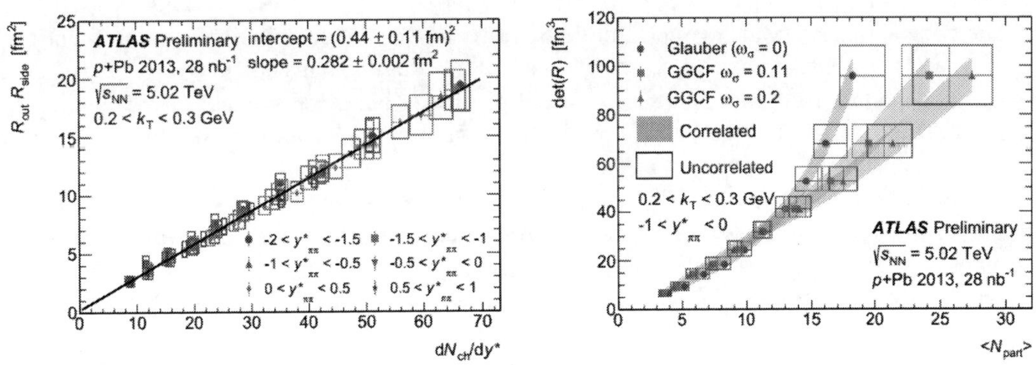

(a) (b)

Figure 3. The transverse area $R_{out}R_{side}$ is shown in (a) as a function of local multiplicity in several intervals of centrality and $y^{\star}_{\pi\pi}$ [14]. The volume element of the HBT matrix det(R) is shown in (b) as a function of N_{part} for three Glauber-Gribov colour fluctuation models.

Source radii are observed to correlate strongly with the single-particle local multiplicity $dN_{ch}/dy^{\star}$, as illustrated in Figure 3(a), which presents the transverse area $R_{out}R_{side}$ at $0.2 < k_T < 0.3$ GeV

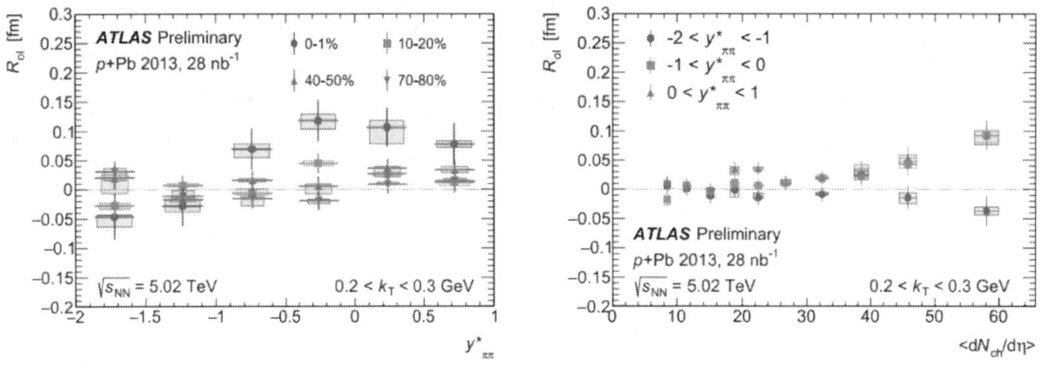

(a) (b)

Figure 4. The cross-term R_{ol}, which couples radial and longitudinal expansion, is shown as a function of rapidity (a) and average global multiplicity $\langle \mathrm{d}N_{\mathrm{ch}}/\mathrm{d}\eta \rangle$ (b) [14].

in several centrality and rapidity intervals as a function of $\mathrm{d}N_{\mathrm{ch}}/\mathrm{d}y^{\star}$. The linear behaviour with multiplicity indicates a constant areal density at low k_{T}.

Figure 3(b) compares the volume scale det(R) with the average number of nucleon participants $\langle N_{\mathrm{part}} \rangle$, both for the standard Glauber model and for two choices of the Glauber-Gribov colour fluctuation (GGCF) extension [16]. This extension of the Glauber initial-geometry model account for fluctuations in the nucleon-nucleon cross section. The volume scale increases at a rate greater than linearly for all choices shown for N_{part}. While the final-state volume indicated by det(R) should not necessarily scale linearly with the initial size N_{part}, extreme departure from a linear behaviour is difficult to describe. These results indicate that fluctuations in the size of the proton are crucial for the understanding of the initial geometry of p+Pb collisions.

The cross-term, R_{ol}, which couples to the lifetime of the source [17], is shown in Figure 4. In hydrodynamic models, a nonzero R_{ol} necessitates both longitudinal and transverse expansion. A significant departure from zero is observed in this parameter in central events, but only for rapidities $y^{\star}_{\pi\pi} \gtrsim -1$. This indicates that the particle production at mid-rapidity and in the proton-going direction is sensitive to the local z-asymmetry of the system.

These results provide a detailed description of the freeze-out geometry of proton-lead collisions. The large pseudorapidity coverage of the ATLAS detector has been utilised to probe the rapidity dependence of the asymmetric p+Pb collisions. Clear signatures of collective behaviour are observed in central collisions, but they are significantly diminished in peripheral collisions.

Acknowledgments

This work was supported by U.S. Department of Energy grant DE-FG02-86ER40281.

References

[1] ALICE Collaboration, Phys. Lett. B **719**, 29 (2013)
[2] ATLAS Collaboration, Phys. Rev. Lett. **110**, 182302 (2013)

[3] CMS Collaboration, Phys. Lett. B **724**, 213 (2013)
[4] ATLAS Collaboration, Phys. Rev. C **90**, 044906 (2014)
[5] CMS Collaboration, Phys. Rev. Lett. **115**, 012301 (2015)
[6] ATLAS Collaboration, Phys. Rev. Lett. **116**, 172301 (2016)
[7] P. Bożek, Phys. Rev. C **85**, 014911 (2012)
[8] E. Shuryak, I. Zahed, Phys. Rev. C **88**, 044915 (2013)
[9] P. Bożek, W. Broniowski, Phys. Rev. C **88**, 014903 (2013)
[10] M.A. Lisa, S. Pratt, R. Soltz, U. Wiedemann, Ann. Rev. Nucl. Part. Sci. **55**, 357 (2005)
[11] P.F. Kolb, U.W. Heinz (2003), nucl-th/0305084
[12] ALICE Collaboration, Phys. Rev. C **91**, 034906 (2015)
[13] ALICE Collaboration, Phys. Lett. B **739**, 139 (2014)
[14] ATLAS Collaboration, *ATLAS-CONF-2016-027* (2016), https://cds.cern.ch/record/2157691
[15] ATLAS Collaboration, JINST **3**, S08003 (2008)
[16] M. Alvioli, M. Strikman, Phys. Lett. B **722**, 347 (2013)
[17] S. Chapman, P. Scotto, U.W. Heinz, Phys. Rev. Lett. **74**, 4400 (1995)

EPJ Web of Conferences **141**, 02001 (2017) DOI: 10.1051/epjconf/201714102001
ISMD 2016

Measurements of the production of prompt photons, jets and vector bosons+jets in pp collisions with the ATLAS detector

Marisilvia Donadelli, for the ATLAS Collaboration[1,a]

[1] *University of Sao Paulo, Institute of Physics, 05508-090 - Sao Paulo - Brazil*

Abstract. The production of jets, photons and gauge bosons in association with jets at hadron colliders provides a stringent test of perturbative QCD and can be used to probe the gluon density function of the proton. In this letter we present recent measurements of the inclusive jet cross section in data collected at center-of-mass energies of 8 TeV and 13 TeV. The measurements have been performed differentially in various kinematic and angular observables and are compared with state-of-the-art predictions at NLO in pQCD. ATLAS has performed precise measurements of the inclusive production of isolated prompt photons in data collected at center-of-mass energies of 8 and 13 TeV, differential in both rapidity and the photon transverse momentum. The measurements are compared with NLO pQCD calculations and with several MC generators. The cross section of photons, W and Z bosons in association with one or more jets have been measured differentially in various observables at center-of-mass energies of 8 and 13 TeV and compared with state-of-the-art pQCD calculations up to NNLO QCD and NLO EWK as well as LO and NLO ME+PS generators. Dedicated phase spaces sensitive to emission of massive bosons collinear with jets or the electroweak production of vector bosons are tested.

1 Introduction

With the large data sample accumulated at the Large Hadron Collider (LHC) during the Run-1 and Run-2 periods at center-of-mass energies of 8 TeV and 13 TeV, respectively, detailed investigations of perturbative quantum chromodynamics (pQCD) and electroweak (EWK) effects were conducted by the ATLAS experiment [1]. Predictions from different Monte Carlo generators based on leading-order (LO) and next-to-leading-order (NLO) matrix elements interfaced with parton shower and hadronisation models, and fixed-order predictions at next-to-leading order (NLO) and next-to-next-to-leading order (NNLO) are compared to the observed cross sections presented differentially in various kinematic and topological observables for the production of jets, photons and vector bosons in association with jets. The Monte Carlo simulations are also used to estimate most of the contributions from background events and to unfold the data to the particle level. The unfolding procedure for all the results in the next sections follow the Bayesian Iterative method [2], as implemented in the RooUnfold package [3].

[a]e-mail: marisilvia.donadelli@cern.ch

2 Production of Jets

At high transverse momenta the scattering of partons can be calculated using pQCD and jet measurements are directly related to the scattering of quarks and gluons. The large cross sections for such processes allow for differential measurements in a wide kinematic range and stringent test of the underlying theory. Jets are reconstructed with the anti-k_t algorithm [4] with four-momentum recombination as implemented in the FastJet package [5], with radius parameter R = 0.4. The jet differential cross section results, as a function of a variety of kinematic and topological variables, are shown with the corresponding distributions presented in the analyses, which are individually unfolded to deconvolve detector effects such as inefficiencies and resolutions [6].

2.1 Four-jet production at 8 TeV

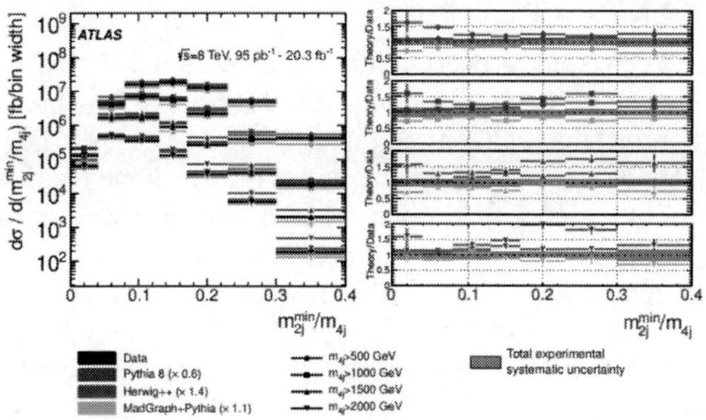

Figure 1. Unfolded four-jet differential cross section as a function of m_{2j}^{min}/m_{4j}, compared to different theoretical predictions: Pythia, Herwig++ and MadGraph+Pythia [6].

With a dataset corresponding to an integrated luminosity of 20.3 fb^{-1}, the differential cross sections for events with at least four jets were studied as a function of momentum, masses and angles [6]. The cross sections are calculated for events with at least four jets, within the rapidity range $|y| < 2.8$. Out of those four jets, the leading one must have $p_T > 100$ GeV, while the next three must have $p_T > 64$ GeV, with the condition that these four jets must be well separated from one another by $\Delta R_{min} > 0.65$. Considering that mass variables are widely used in physics searches, and that they are also sensitive to events with large separations between jets, Figure 1 shows the unfolded distribution of the minimum invariant mass of two jets relative to the invariant mass of four jets, along-

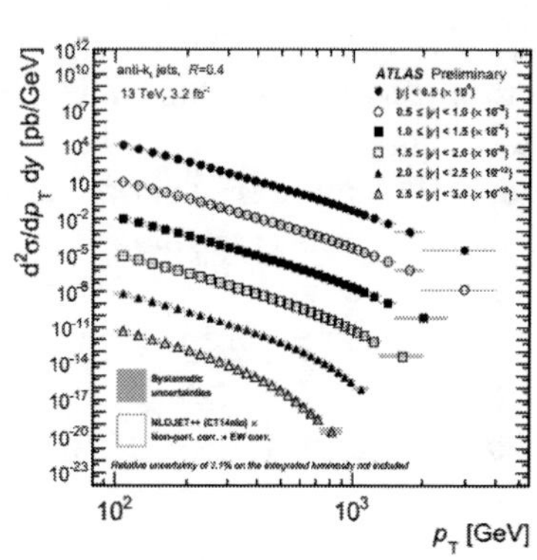

Figure 2. Inclusive-jet cross-sections as a function of p_T and $|y|$, for anti-k_t jets with R = 0.4 [7].

side with the generators used for comparison against data [6]. One can see that MadGraph+Pythia provides a very good description, with a flat ratio for all the m_{4j} cuts.

2.2 Inclusive jets at 13 TeV

Inclusive-jet cross-sections are measured using data corresponding to an integrated luminosity of 3.2 fb^{-1} [7]. The jet cross-sections are defined using jets representing those clustered from stable particles with a proper mean lifetime (τ) exceeding 10 mm, with muons and neutrinos from decaying hadrons included in the jet clustering. Figure 2 shows the inclusive-jet cross-section result for six jet rapidity intervals covering the jet $p_T^{\rm jet}$ range from 100 GeV to ~ 3.2 TeV and within a jet rapidity interval $|y| < 3$, together with the NLO pQCD predictions calculated using NLOJET++ 4.1.3 [8] with the CT14 NLO PDF set [9], to which non-perturbative and electroweak corrections are applied. The dominant systematic uncertainty for this result, which significantly extends the reach of ATLAS previous measurement [10], arises from the jet energy calibration [7].

3 Isolated prompt photons at 8 TeV

The production of prompt photons provides a probe of pQCD in a cleaner environment compared to jet production and moreover, through the dominant LO process $qg \rightarrow q\gamma$, can be used to study the gluon parton distribution function of the proton [11–15]. At LO in pQCD, two processes contribute to prompt photon production: the direct-photon process, in which the photon originates directly from the hard interaction, and the fragmentation-photon process, in which the photon is emitted in the fragmentation of a high transverse momentum parton.

With an integrated luminosity of 20.2 fb^{-1}, the differential cross section of inclusive photon production was measured as a function of E_T^{γ}, split in 4 bins of pseudorapidity covering the range $|\eta^{\gamma}| < 1.37$ and $1.56 < |\eta^{\gamma}| < 2.37$, spanning the photon transverse energy between 25 and 1500 GeV [16]. The photon candidate, identified by combining the information from the calorimeters and the inner tracker, is required to be isolated based on the amount of transverse energy inside a cone of $\Delta R = 0.4$ in the $\eta - \phi$ plane centred around the photon direction, to avoid the large contribution of photons from neutral-hadron decays. The theoretical calculations used in the analysis consist of LO Monte Carlo event generators and calculations at NLO or higher. JetPhox is used as the baseline to compare the results as it is capable of calculating the double-differential inclusive prompt photon cross section at parton level to NLO accuracy for both the direct and fragmentation photon processes. PeTeR [17] is used as a second parton-level generator to predict the differential isolated prompt photon cross section at NLO including the resummation of threshold logarithms at the next-to-next-to-next-to-leading- logarithmic (NNNLL) level. It can be seen in Figure 3 that PeTeR removes the normalisation difference seen between data and JetPhox, especially in the region $|\eta| < 1.37$. The uncertainties shown, from combining the scale, PDF and electroweak uncertainties, are about 20% lower than those from JetPhox. The PeTeR predictions match the data, within the combined measured and theoretical uncertainties, in all of the measured phase space [16].

4 Vector bosons + jets production

The relatively large production cross sections and easily identifiable decays of the boson to charged leptonic final states offer clean experimental signatures which can be precisely measured. Such processes also constitute a non-negligible background for Higgs boson studies and in searches for new phenomena. Typically in these studies, the multiplicity and kinematics of the jets are exploited to achieve a separation of the signal of interest from the Standard Model (SM) vector boson + jets process.

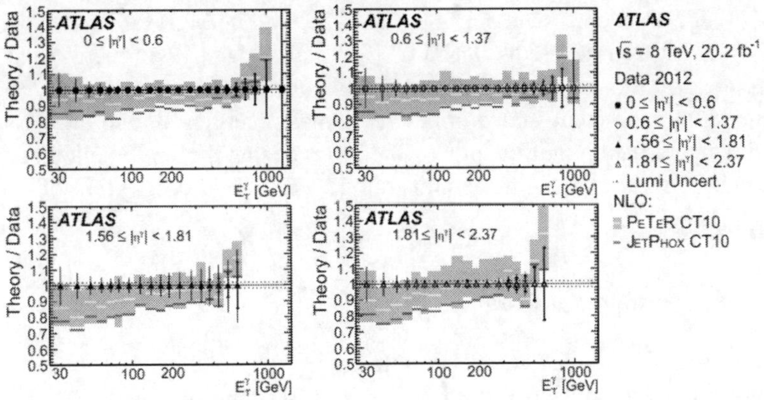

Figure 3. Ratio of theory (PeTeR and JetPhox both using the CT10 PDF) to data for the differential cross sections as a function of E_T^γ for the four η^γ regions [16].

The measurements were compared to fixed-order calculations at NLO BlackHat+Sherpa [19, 20] and at NNLO from $Z+\geq 1$ jet $N_{\rm jetti}$ and $W+\geq 1$ jet $N_{\rm jetti}$ [21–23] which use a new subtraction technique based on $N-$jettiness [24] and relies on the theoretical formalism provided in soft-collinear effective theory. The measured cross sections are also compared to predictions from the generators Alpgen v2.14 [25] interfaced to Pythia v6.426 [26] to model the parton shower (PS), denoted as Alpgen+Py6, and to Madgraph MG5_aMC NLO v2.2.2 generator [27], in two different schemes: MG5_aMC+Py8 CKKWL, using matrix elements including up to four partons at leading order, interfaced to Pythia v8.186 [28], using the CKKWL merging scheme [29] and with ME for up to two jets at NLO and with parton showers beyond this generated using the FxFx scheme [30]. The cross sections results are also compared to $W+$ j and $W+$ j j+weak shower calculation provided by Pythia v8.210 and, in this case, the W boson can either be produced by the matrix elements of the $W+$ 1 jet final state or be emitted as electroweak final-state radiation in the parton shower of a dijet event. The Sherpa + OpenLoops $W+$ j and $W+$ j j calculation incorporates NLO QCD and NLO EW corrections to both of these processes [31–35].

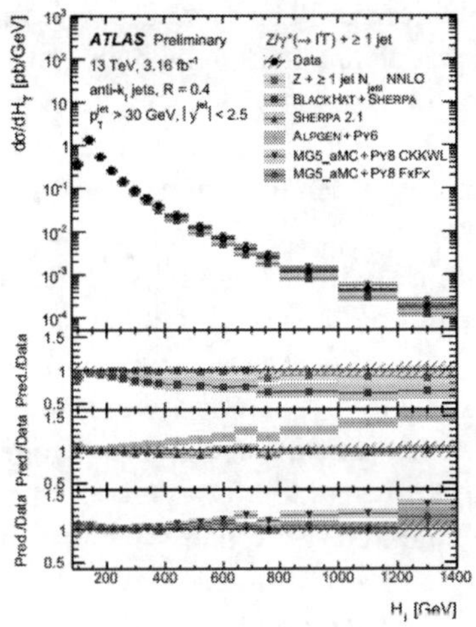

Figure 4. Measured cross section as a function of the H_T for inclusive $Z+\geq 1$ jet events [18].

4.1 Z+jets at 13 TeV

Measurements of the cross section for the production of a Z boson decaying to electrons or muons associated with up to seven jets in the jet kinematic range of transverse momentum $p_T > 30$ GeV and rapidity $|y| < 2.5$ were performed using data corresponding to an integrated luminosity of 3.16 fb^{-1}. The invariant mass of the lepton pair was selected within $71 < m_{\ell\ell} < 111$ GeV and the jet and the lepton were required to have a separation of $\Delta R(\ell, \text{jet}) > 0.4$ [18].

Figure 4 shows the measured cross sections as a function of the H_T distribution (up to 1400 GeV) in inclusive $Z+ \geq 1$ jet events. Sherpa 2.1 [36] and MG5_aMC+Py8 FxFx [30] describe well the spectrum while MG5_aMC+Py8 CKKWL [29] and Alpgen+Py6 overestimate the distribution at large values of H_T. BlackHat+Sherpa under-estimates the cross section for $H_T > 300$ GeV, as observed in similar measurements at lower centre-of-mass energies, due to the missing contributions from events with higher parton multiplicities. One can note that the agreement is recovered by adding higher orders in perturbative QCD, as demonstrated by the good description of H_T by N_{jetti} NNLO [18].

4.2 Collinear W production at 8TeV

The W boson angular distribution in events with high transverse momentum jets was measured using data corresponding to an integrated luminosity of 20.3 fb^{-1} [37]. The focus is on the contributions to $W+$ jets processes from real W emission, so that the selected events contain an isolated muon and a jet with transverse momentum $p_T > 500$ GeV. In this kinematic regime, contributions to $W+$ jets processes from real W boson emission are enhanced in the region of small angular separation between the W boson decay products and the closest jet, with the angular separation defined as the distance between the muon and the closest jet, $\Delta R(\mu, \text{jet}) = \sqrt{(\Delta\phi)^2 + (\Delta\eta)^2}$. Figure 5 shows the unfolded distribution from background-subtracted data of the angular separation between the muon and the closest jet in the signal region for events with 500 GeV < p_T (leading jet) < 600 GeV [37] along with a variety of MC generators and theoretical calculations. Alpgen+Pythia6 overestimates the total cross-section, whereas Pythia8, which is modified to explicitly include the process of W boson emission, disagrees with the measurement in the collinear region ($\Delta R(\mu, \text{jet}) < 2.4$). Agreement with the Sherpa+OpenLoops NLO QCD+EW calculation and the $W+ \geq 1$ jet N_{jetti} calculation [23] is well within the systematic and statistical uncertainties of the predictions and the measurement.

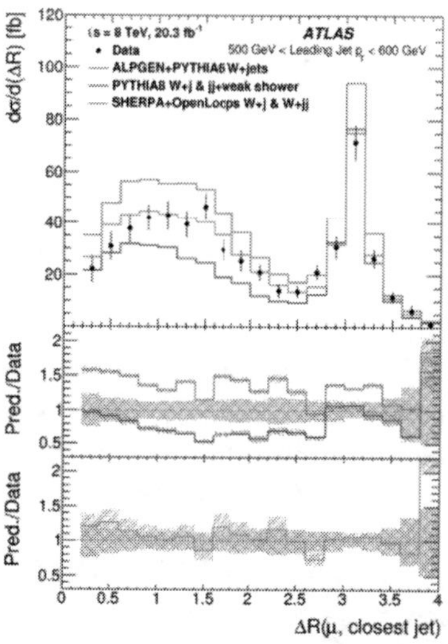

Figure 5. Unfolded distribution from background-subtracted data of the angular separation between the muon and the closest jet in the signal region along with several predictions from theory calculations for events with 500 GeV < leading jet p_T < 600 GeV [37].

5 Conclusions

Measurements of the production of jets, photons and vector bosons plus jets were presented in various kinematic and topological observables and compared to predictions from different Monte Carlo generators. MadGraph+Pythia provides the best description of mass variables for the four-jet results at 8 TeV. NLO pQCD with non-perturbative and electroweak effects corrections are compared to the inclusive jets measurement at 13 TeV. The predictions are in good agreement with data, confirming the validity of pQCD in the measured kinematic regions. The isolated prompt photons result covers ten orders of magnitude in cross section, with NLO QCD (PeTeR) with resummation+electroweak corrections describing data well within uncertainties. Vector boson+jets results provide essential input for optimisation of the Monte Carlo generators and, in particular, the collinear W measurement has implications for Monte Carlo programs that incorporate real W boson emission, a process which is only just now being probed directly at the energy of the LHC.

References

[1] G. Aad et al. (ATLAS), Journal of Instrumentation **3**, S08003 (2008)
[2] G. D'Agostini, Nucl. Instrum. Meth. **A362**, 487 (1995)
[3] T. Adye, *Unfolding algorithms and tests using RooUnfold* (2011), 1105.1160
[4] M. Cacciari et al., JHEP **04**, 063 (2008), 0802.1189
[5] M. Cacciari et al., Eur. Phys. J. **C72**, 1896 (2012), 1111.6097
[6] G. Aad et al. (ATLAS), JHEP **12**, 105 (2015), 1509.07335
[7] Tech. Rep. ATLAS-CONF-2016-092 (2016), http://cds.cern.ch/record/2209210
[8] Z. Nagy, Phys. Rev. **D68**, 094002 (2003), hep-ph/0307268
[9] S. Dulat et al., Phys. Rev. **D93**, 033006 (2016), 1506.07443
[10] G. Aad et al. (ATLAS), JHEP **02**, 153 (2015), [Erratum: JHEP09,141(2015)], 1410.8857
[11] H.L. Lai et al., Phys. Rev. **D51**, 4763 (1995), hep-ph/9410404
[12] A.D. Martin et al., Eur. Phys. J. **C4**, 463 (1998), hep-ph/9803445
[13] A.D. Martin et al., Eur. Phys. J. **C14**, 133 (2000), hep-ph/9907231
[14] P. Aurenche et al., Phys. Rev. **D73**, 094007 (2006), hep-ph/0602133
[15] D. d'Enterria, J. Rojo, Nucl. Phys. **B860**, 311 (2012), 1202.1762
[16] G. Aad et al. (ATLAS), JHEP **08**, 005 (2016), 1605.03495
[17] T. Becher, C. Lorentzen, M.D. Schwartz, Phys. Rev. **D86**, 054026 (2012), 1206.6115
[18] Tech. Rep. ATLAS-CONF-2016-046 (2016), https://cds.cern.ch/record/2206128
[19] C.F. Berger et al., Phys. Rev. **D82**, 074002 (2010), 1004.1659
[20] H. Ita et al., Phys. Rev. **D85**, 031501 (2012), 1108.2229
[21] R. Boughezal et al., Phys. Rev. Lett. **116**, 152001 (2016), 1512.01291
[22] R. Boughezal et al., Phys. Rev. **D94**, 074015 (2016), 1602.08140
[23] R. Boughezal et al. (2016), 1602.06965
[24] J. Gaunt et al., JHEP **09**, 058 (2015), 1505.04794
[25] M.L. Mangano et al., JHEP **07**, 001 (2003), hep-ph/0206293
[26] T. Sjostrand et al., JHEP **05**, 026 (2006), hep-ph/0603175
[27] J. Alwall et al., JHEP **07**, 079 (2014), 1405.0301
[28] T. Sjostrand et al., Comput. Phys. Commun. **178**, 852 (2008), 0710.3820
[29] L. Lonnblad, JHEP **05**, 046 (2002), hep-ph/0112284
[30] R. Frederix et al., JHEP **02**, 131 (2016), 1511.00847
[31] S. Kallweit et al., JHEP **04**, 021 (2016), 1511.08692
[32] F. Krauss et al., JHEP **02**, 044 (2002), hep-ph/0109036
[33] T. Gleisberg et al., Eur. Phys. J. **C53**, 501 (2008), 0709.2881
[34] F. Cascioli et al., Phys. Rev. Lett. **108**, 111601 (2012), 1111.5206
[35] A. Denner et al., PoS **LL2014**, 071 (2014), 1407.0087
[36] T. Gleisberg et al., JHEP **02**, 007 (2009), 0811.4622
[37] M. Aaboud et al. (ATLAS) (2016), 1609.07045

Search strategies for vector-like quark partners at LHC run-II

Thomas Flacke[1,2,⋆]

[1] *Center for Theoretical Physics of the Universe, Institute for Basic Science (IBS), Daejeon 34051, Korea*
[2] *Department of Physics, Korea University, Seoul 136-713, Korea*

Abstract. We present results on several new search channels and strategies for heavy vector-like quark partners or vector resonances at the LHC run-II. Run-II has sensitivity to single- and pair-produced quark partners with masses beyond 1 TeV, and for even higher masses for vector resonances. Decays of such heavy particles yield highly boosted tops, Higgses, and weak gauge bosons. At high boost, the Standard Model background of hadronic final states can be substantially suppressed when applying jet-substructure techniques. We present several case studies where the identification of hadronically decaying tops, Higgses, and/or electroweak gauge bosons allow to make new search channels competitive at run-II.

1 Motivation and Overview

Naturalness has for a long time been a guiding principle of theoretical particle physics. While the discovery of the last missing Standard Model SM particle – the Higgs – marks a great success of particle physics, it is difficult to understand why the Higgs mass is so much smaller than the Planck mass. Composite Higgs models provide a possible solution to this hierarchy problem [1, 2]. The main idea behind composite Higgs models is to realize the Higgs multilplet as a light bound state of a theory which becomes strongly coupled below a scale of a few TeV. The large hierarchy between the TeV scale and the Planck scale is realized through dimensional transmutation. If the Higgs multiplet is realized as a pseudo-Nambu-Goldstone boson (pNGB) multiplet of a global symmetry of the underlying model, a small hierarchy between the Higgs mass m_h, the electroweak scale v and the decay constant f of the pNGB can be realized at the price of a mild fine-tuning. From a low-energy perspective, the most minimal implementation of the composite Higgs paradigm is realized in models with $SO(5) \to SO(4)$ breaking [3]. Within such effective field theory descriptions, indirect bounds like deviations of Higgs couplings from the Standard Model (SM) prediction and electroweak precision tests yield constraints of $f \gtrsim 500-1000$ GeV, depending on the precise model, the top-partner masses, and the underlying UV dynamics. Flavour physics also yields important constraints, as obtaining the observed top mass requires an extended (top-)quark sector with additional sources for flavor changing neutral currents. Partial compositeness of the top can successfully address these flavor problems as recently discussed in [4].

On the other hand, composite Higgs models can be tested directly by searching for composite resonances beyond the Higgs. Composite top-partners, scalar- or vector resonances are expected to

⋆e-mail: flacke@ibs.re.kr

have masses at a scale $\sim g_{strong}f \lesssim 4\pi f$ i.e. in the multi-TeV range. Stabilizing the Higgs mass and electroweak vacuum at the SM values typically requires top-partners to be even lighter, i.e. in the TeV range. It should be noted that generically, UV embeddings of composite Higgs models require an Goldstone-boson sector with additional electroweak and colored pNGBs, as underlying models with only elementary fermions and hyper-color gauge bosons cannot realize the minimal global symmetry breaking pattern $SO(5) \to SO(4)$ [5]. The additional pNGBs are likely to be the lightest BSM states in composite Higgs model UV embeddings, and they have implications for di-boson and di-top searches (c.f. [6–9]), but in this talk we focus mainly on searches for top- and other quark-partners.

2 A sample model: the minimal composite Higgs model

In composite Higgs models, the Higgs multiplet is realized as a pNGB multiplet which is thought of as a bound state of the underlying strongly coupled theory. The vector-like quark partners are bound states of the strongly coupled theory, which are assumed to couple linearly to elementary quarks. These couplings (as well as weakly gauging the $SU(2)_L \times U(1)$ in the surviving global subgroup) explicitly break the global symmetry and induce a Higgs potential, making the Higgs a pNGB. The linear couplings between the elementary quarks and their partners (which come in multiplets of the global symmetry) induce mass-mixing between each other such that the lightest mass eigenstates (which are to be identified with the SM quarks) are a linear combination of the elementary quarks and strongly coupled states and are thus termed "partially composite". The mass-mixing also induces couplings between SM quarks, their heavy quark partners, and the electroweak bosons W, Z, h, which are relevant for the production of heavy quark partners at the LHC and the partners' decays and branching ratios. The typical strength of these couplings is the corresponding EW (or Yukawa) coupling times a mixing angle. In these proceedings we refrain from giving the full Lagrangian and interactions used for our simulations and model implementations (they can be found in Refs.[10, 11]) and instead summarize the qualitative features relevant for LHC searches:

- Quark partners come in multiplets of the global symmetry group in which the the EW group is embedded. For example a top-partner multiplet embedded into the **5** of $SO(5)$ contains one charge 5/3 partner, three charge 2/3 partners and one charge -1/3 partner.
- Quark partners are charged under $SU(3)$ and can be QCD-pair produced. This production mechanism is rather model-independent as the coupling strength is fixed, and the production cross section only depends on the mass of the quark partner.
- Quark partners can be single-produced from SM quarks and EW gauge bosons. The couplings relevant for this production mechanism are generically of electroweak strength or below, but model-dependent.
- The branching ratios of quark partners are generically model-dependent. Again taking quark partners in the **5** of $SO(5)$ as an example: the charge 5/3 state $X_{5/3}$ can only decay into W^+t due to charge conservation while the three charge 2/3 states (which we collectively denote by T) can decay into W^-b, Zt or ht, and the respective branching ratios are model- and parameter dependent.

3 Heavy quark searches at the LHC

3.1 A brief summary of existing LHC top partner searches

Both ATLAS and CMS performed numerous dedicated searches for 3rd family quark partners at the LHC run-I. The searches mostly focus on QCD pair production of charge 5/3, 2/3, and -1/3 partners and their decay into 3rd family quarks and $W/Z/h$. Searches are performed in various final states, including all-hadronic searches (using jet-substructure techniques), single- or multi-lepton final state searches, and – for charge 5/3 partner searches – also same-sign dilepton final states. The mass bounds on 3rd family partners are $m_{X53} > 940$ GeV [12] (run-II bound) , $m_T > 700-950$ GeV [13–16] (still run-I bounds), and similar bounds for m_B where the m_T and m_B bounds have a range because they depend on the T and B branching fractions into different final states.

Most ATLAS and CMS searches so far focussed on QCD pair produced top partners while for singly produced top partners bounds were obtained in a recast of other experimental searches in Ref. [17], which show that for typical couplings (in composite Higgs models), single-production channels yield at most comparable and mostly weaker bounds on the quark partner masses than the pair-production channels for LHC run-I. First ATLAS and CMS bounds for single-production (from run-II) are available only since very recently [18–22]. They constrain the single-production cross section times branching ratio for the final states th, tZ, and bW to $\sim$ 400 fb, $\sim$ 200 fb, and $\sim$ 200 fb, respectively.[1]

Bounds for vector-like quark partners of light quarks have been obtained in a recast from several experimental searches in Ref. [10]. The bounds from QCD pair production are generically weaker than for 3rd family quark partners. In particular a bound on the charge 5/3 partner is obtained at $m_{X53} > 530$ GeV. Notably, the bound on the charge 2/3 partner can be even weaker ($M_Q > 410$ GeV) if it dominantly decays into hj [23].[2] Bounds on singly-produced 1st or 2nd family quark partners can be stronger, but are model-dependent.

4 Improved search strategies for quark partners

In this section, we discuss various improved search strategies for top- and quark partners, designed for LHC run-II.

4.1 Search options for charge 2/3 top-partners

In Ref. [25], we performed a comprehensive overview on search options for singly produced charge 2/3 top partners which decay into th, tZ, or Wb. The pair-production and single-production cross section at $\sqrt{s} = 14$ TeV and for typical values of the couplings are shown in Fig. 1, left. The branching ratios into these three final states are model-dependent, such that we analyze each channel separately. For all decay channels, hadronic decays of the t, h, W, Z have larger branching ratios, but suffer from larger SM backgrounds, such that a careful analysis was necessary in order to identify the most promising set of searches. For the $T' \to tZ$ channel, the cleanest channel would be the leptonic decay of both the top and the Z, but it suffers from a low branching ratio. The most promising channels are the $t_{\rm had}Z_{ll}$ and $t_{\rm had}Z_{\rm inv}$ channel, where the first one is more standard. The $Z_{\rm inv}$ channel benefits from a lager branching ratio but has more SM background. We use jet-substructure techniques (in particular the template overlap method) for top-tagging, and a forward jet tag for both channels, a standard Z_{ll} reconstruction for the leptonic decay and a hard MET cut for the invisible decay. In our analysis we show that both channels have comparable prospects for discovery if $M_{T'} \lesssim 1$ TeV, but for larger $M_{T'}$, the prospects for the $Z_{\rm inv}$ become more favorable, as a hard MET cut reduces the SM background very effectively. For details, c.f. [25, 26].

We also performed analyses for various th and Wb decay channels and find the best performance for the $t_{\rm had}h_{bb}$ and $W_{\rm lep}b$ channels. Fig. 2 shows a projection of the production cross sections of T'

[1] The bound on the cross section depends on the mass of the resonance, and the values are only given for orientation. For more precise information see Refs. [18–22].

[2] The sensitivity on this di-Higgs, di-jet channel can be substantially improved at LHC run-II [24].

DOI: 10.1051/epjconf/201714102002

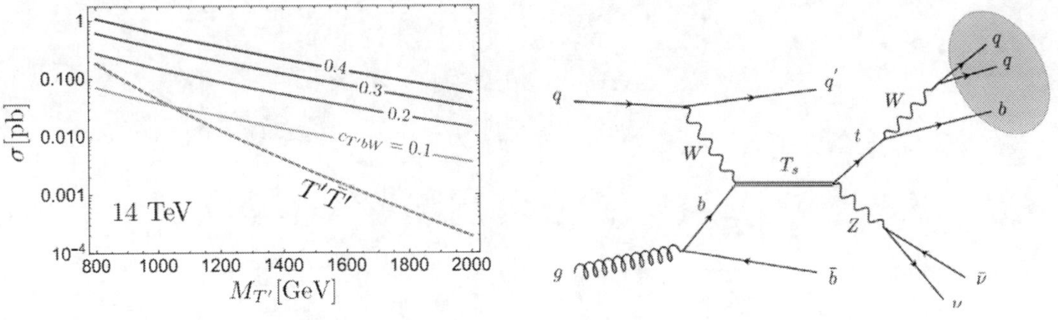

Figure 1. Left: Single- and pair-production cross section of top partners at $\sqrt{s}$ = 14 TeV. Right: Single-production of a top partner with subsequent decay to $t_{\text{had}}Z_{\text{inv}}$.

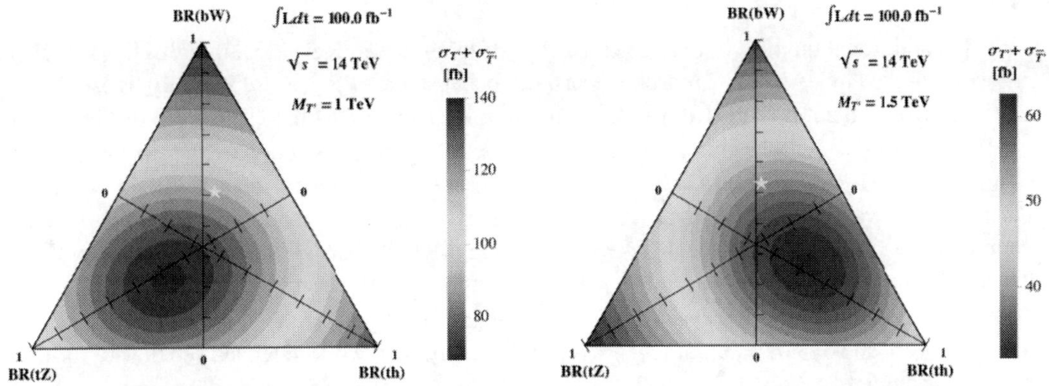

Figure 2. Expected discovery reach for a T' with mass of 1 TeV (left) and 1.5 TeV (right) in terms of T' production cross section for the LHC at 14 TeV with 100 fb^{-1} of data.

at which a T' with a mass of 1 TeV (left) or 1.5 TeV (right) can be discovered with an integrated luminosity of 100 fb^{-1}. The analogous figures for exclusion reach can be found in Ref.[25]. For an analysis of the $X_{5/3}$ single-production channel with semi-leptonic decay of tW c.f. Ref. [11].

4.2 Di-Higgs signatures from composite quark partners

As another example, we studied pair production of a quark partner U_h which decays into uh.[3] For definiteness we chose the vector-like quark to be the partner of the up-quark. Fig. 3 (left) shows the QCD pair production cross section of the up-partner, while the topology of the event is displayed in Fig. 3 (left). As can be seen, this decay can lead to a substantial cross section for a di-higgs final state

[3]For the purpose of this study, we assume a uh branching ratio of 100%. C.f. Ref.[24] for a model which yields such a dominant qh decay.

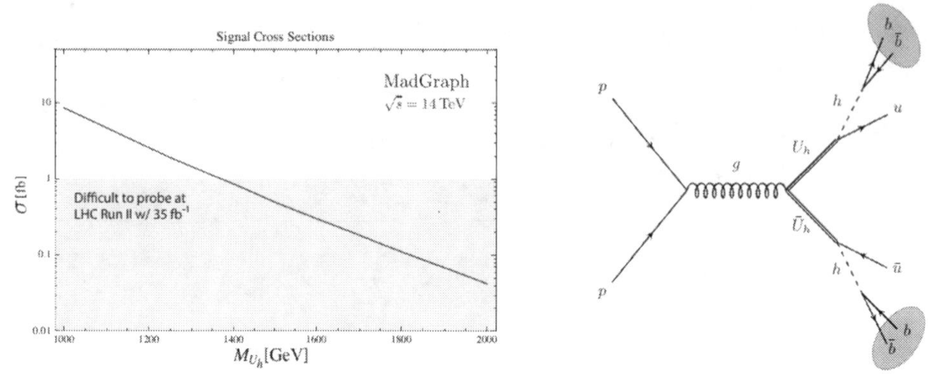

Figure 3. Left: QCD pair-production of quark partners at $\sqrt{s}$ = 14 TeV. Right: parameter-independent topology of di-Higgs - di-quark production in models with composite light quark partners.

with two additional quarks. Nevertheless, as has been shown in Ref. [23], this di-Higgs final state is not well covered by early LHC searches, partially because the Higgses are strongly boosted, but early LHC searches did not provide information on the p_T distribution of the Higgses in di-higgs searches.

We designed a search strategy in order to maximize the sensitivity to the $qqh_{bb}h_{bb}$ final state (with boosted Higgses). Again, we use jet-substructure techniques to identify the boosted Higgses. Furthermore, we use that the invariant mass of the two qh pairs are identical (if paired correctly) which can be checked on an event-by-event basis, and which also yields a good mass resolution for the determination of M_{U_h}. We find that with a luminosity of 35 fb^{-1}, a light quark partner with 100% BR into qh can be discovered at a mass of above 1 TeV, with an expected $S/\sqrt{B} = 7.5$ at $M_{U_h} = 1$ TeV and $S/\sqrt{B} = 3.8$ at $M_{U_h} = 1.2$ TeV, implying a well increased sensitivity for this most "elusive" channel.

The above analysis has been performed at leading order, only, and due to the large number of quarks in the all-hadronic final state, an NLO analysis of the production cross section and kinematical distributions are desirable in order to obtain more reliable results. A joint group effort in this direction has been started in Les Houches 2015 [27], followed by first published results [28].

5 Outlook on searches for vector resonances with decays to top partners

In the last section, we discussed production and decay of quark- or top-partners in absence of other composite resonances. However, composite Higgs models are expected to also contain vector resonances. If they are present, but heavier than the top-partners, two interesting effects occur. The interaction between the vector- and top partners arises from the strong sector, and is expected to dominate over the couplings of vector resonances to the SM particles. Thus, the vector resonances dominantly decay to top and top-partner (if $m_\rho < 2m_{T'}$) or into two top partners (if $m_\rho > 2m_{T'}$). This on the one hand reduces the bounds on vector resonance searches in SM decay channels, and on the other hand, it provides an additional, resonant, production mode for top-partner pairs or a top-top-partner final state, which lead to 3- or 4-body final states from the vector-resonance decays. Such channels provide interesting signatures for future collider searches of vector resonance and top-partner searches.

6 Conclusions

Motivated by composite Higgs models, we presented several new search strategies for vector-like top-partners and other quark-partners for LHC run II. We found that jet-substructure techniques strongly increase the sensitivity in these searches, because the discrimination power to background for the decay products from heavy vector-like quarks (in particular tops, W, Z bosons and Higgses) can be substantially increased as the decay products are strongly boosted. We also pointed out that the interplay of vector resonances and vector-like quarks can alter the interpretation of existing searches and promises interesting signatures to be explored in run-II.

Note added: After this talk was given, we published a first work on identifying Z' resonances in decays to tT' in the $\bar{t}tZ$ final state, which quantifies our statements on the interplay of vector resonances and top=partners [29].

Acknowledgements

I would like to thank the ISMD organizers for setting up this extraordinary conference. My work was supported by the Basic Science Research Program through the National Research Foundation of Korea (NRF) funded by the ministry of Education, Science and Technology (No. 2013R1A1A1062597), by the France-Korea Particle Physics Lab (FKPPL), and by IBS under the project code, IBS-R018-D1.

References

[1] D.B. Kaplan, H. Georgi, Phys. Lett. **B136**, 183 (1984)
[2] D.B. Kaplan, H. Georgi, S. Dimopoulos, Phys. Lett. **B136**, 187 (1984)
[3] K. Agashe, R. Contino, A. Pomarol, Nucl. Phys. **B719**, 165 (2005), hep-ph/0412089
[4] G. Cacciapaglia, H. Cai, T. Flacke, S.J. Lee, A. Parolini, H. Serôdio, JHEP **06**, 085 (2015), 1501.03818
[5] G. Ferretti, D. Karateev, JHEP **03**, 077 (2014), 1312.5330
[6] H. Cai, T. Flacke, M. Lespinasse (2015), 1512.04508
[7] A. Belyaev, G. Cacciapaglia, H. Cai, T. Flacke, A. Parolini, H. Serôdio, Phys. Rev. **D94**, 015004 (2016), 1512.07242
[8] G. Cacciapaglia, H. Cai, A. Deandrea, T. Flacke, S.J. Lee, A. Parolini, JHEP **11**, 201 (2015), 1507.02283
[9] A. Belyaev, G. Cacciapaglia, H. Cai, G. Ferretti, T. Flacke, A. Parolini, H. Serôdio (2016), 1610.06591
[10] C. Delaunay, T. Flacke, J. Gonzalez-Fraile, S.J. Lee, G. Panico, G. Perez, JHEP **02**, 055 (2014), 1311.2072
[11] M. Backović, T. Flacke, S.J. Lee, G. Perez, JHEP **09**, 022 (2015), 1409.0409
[12] CMS (2015), CMS-PAS-B2G-15-006
[13] G. Aad et al. (ATLAS), JHEP **10**, 150 (2015), 1504.04605
[14] G. Aad et al. (ATLAS), JHEP **11**, 104 (2014), 1409.5500
[15] G. Aad et al. (ATLAS), JHEP **08**, 105 (2015), 1505.04306
[16] V. Khachatryan et al. (CMS), Phys. Rev. **D93**, 012003 (2016), 1509.04177
[17] O. Matsedonskyi, G. Panico, A. Wulzer, JHEP **12**, 097 (2014), 1409.0100
[18] CMS (2015), CMS-PAS-B2G-15-008

[19] CMS (2016), CMS-PAS-B2G-16-001
[20] CMS (2016), CMS-PAS-B2G-16-005
[21] CMS (2016), CMS-PAS-B2G-16-006
[22] ATLAS (2016), ATLAS-CONF-2016-072
[23] T. Flacke, J.H. Kim, S.J. Lee, S.H. Lim, JHEP **05**, 123 (2014), 1312.5316
[24] M. Backović, T. Flacke, J.H. Kim, S.J. Lee, JHEP **04**, 082 (2015), 1410.8131
[25] M. Backović, T. Flacke, J.H. Kim, S.J. Lee, JHEP **04**, 014 (2016), 1507.06568
[26] M. Backović, T. Flacke, J.H. Kim, S.J. Lee, Phys. Rev. **D92**, 011701 (2015), 1501.07456
[27] G. Brooijmans et al., *Les Houches 2015: Physics at TeV colliders - new physics working group report*, in *9th Les Houches Workshop on Physics at TeV Colliders (PhysTeV 2015) Les Houches, France, June 1-19, 2015* (2016), 1605.02684
[28] B. Fuks, H.S. Shao (2016), 1610.04622
[29] M. Backović, T. Flacke, B. Jain, S.J. Lee (2016), 1610.08810

EPJ Web of Conferences **141**, 02003 (2017) DOI: 10.1051/epjconf/201714102003
ISMD 2016

Measurements of jet rates with the anti-k_t and SISCone algorithms at LEP with the OPAL detector

Stefan Kluth[1,a] and Andrii Verbytskyi[1]

[1] *MPI für Physik, Föhringer Ring 6, 80805 München, Germany*

Abstract. We study jet production in e^+e^- annihilation to hadrons with data recorded by the OPAL experiment at LEP at centre-of-mass energies between 90 GeV and 207 GeV. The jet production rates were measured for the first time with the anti-k_t and SISCone jet clustering algorithms. We compare the data with predictions by modern Monte Carlo event generators.

1 Introduction

Hadronic final states in high energy particle collisions are characterised by the so-called jets, which describe the phenomenon that final state particles appear in collimated groups. These jets are quantified by defining jet clustering algorithms, which assign particles to groups in order to identify the jet structure of the events. Local parton hadron duality [1, 2] establishes the close correspondence between the hadronic final state jets and the partons as objects of the theory of strong interactions, Quantum Chromo Dynamics (QCD). The production of hadronic final states is described by QCD together with a model for the transition from partons to hadrons.

At the LHC the understanding of the properties of jets and their connection to the partons produced in the hard interactions is essential for the success of its physics program. Before LHC jets were studied in electron-proton (e^+p) collisions by the HERA experiments [3] and in electron positron (e^+e^-) collisions at LEP and earlier colliders [4, 5], and in $p\bar{p}$ collisions, see e.g. [6].

The introduction of new jet algorithms for LHC [7, 8] raises the question how these algorithms perform in e^+e^- or $e^\pm p$ collisions in comparison to established algorithms. A cone jet algorithm for e^+e^- was introduced and studied by OPAL previously [9]. Two studies by ZEUS [10, 11] compared the k_t, anti-k_t and SISCone algorithms in $e^\pm p$ DIS and photoproduction.

We report here on studies of the anti-k_t and SISCone algorithms with e^+e^- data from the OPAL experiment at LEP. In e^+e^- collisions a color singlet is produced in the decay of the virtual Z^0 boson or photon to quarks thus providing a very clean environment to study jet physics. The anti-k_t and SISCone algorithms have not yet been tested in e^+e^- collisions.

2 LEP and OPAL

The large electron-positron collider LEP operated at CERN from 1989 to 2000 at centre-of-mass energies from around 91.2 GeV to a maximum of 209 GeV. The OPAL[1] experiment was one of the

[a] e-mail: skluth@mpp.mpg.de
[1] Omni-Purpose Apparatus at LEP

four large detectors at LEP. The OPAL detector [12] had almost complete coverage of the solid angle around the collision point, good momentum (energy) resolution for charged (neutral) particles.

We use data from OPAL recorded during the so-called LEP 2 phase, where LEP was operated at the highest possible energies. A sample in total of about 400,000 hadronic Z^0 decays was recorded to calibrate the detector together with the high energy samples of a few hundred to several thousand high energy hadronic events, for details see [13].

The analysis uses large samples of simulated signal events produced on the Z^0 peak with JETSET 7.4 [14] or HERWIG 6.2 [15] and at higher energies with the KK2f generator [16] coupled with PYTHIA 6 [17] or HERWIG 6 for parton shower and hadronisation. Background events from production of W-boson pairs are simulated with grc4f 2.1 [18] and KORALW [19]. Other backgrounds from two-photon interactions and τ-pair production are negligible after the event selection.

3 Event selection and jet reconstruction

In order to select hadronic final states produced at high energies above the Z^0 peak the event selection has to suppress so-called radiative return events as well as production of W-boson pairs for $\sqrt{s} >$ 161 GeV. Radiative return is a process where radiation of a high energy photon from the initial state (ISR) causes the production of a hadronic final state at reduced effective centre-of-mass (cm) energy $\sqrt{s'}$. At energies above the Z^0 peak the cross section for this process is enhanced when the remaining effective cm energy $\sqrt{s'} \approx m_Z$.

Briefly, in OPAL an algorithm searches for isolated high energy photons in the detector and reconstructs up to four jets in the hadronic final state. Then kinematic fits are performed for the energy of zero, one or two photons emitted parallel to the beam and constrained to the nominal cm energy. The fit with the lowest number of photons and an acceptable χ^2 is chosen and the effective cm energy $\sqrt{s'}$ is calculated. High energy hadronic events are required to have $\sqrt{s'}$ within 10 GeV of the nominal value. Any remaining background is accounted for in the experimental correction procedure.

The production of W-boson pairs decaying into hadronic final states is identified and rejected with the well-developed event selection algorithms used by OPAL analyses of W-boson production and properties [20]. Any remaining background is subtracted using simulated samples of W-boson production.

The number of selected events at energies above m_Z ranges from less than 300 at $\sqrt{s} = 130$ and 136 GeV to more than 3000 at $\sqrt{s} = 189$ GeV. At the highest energy of $\sqrt{s} = 207$ GeV about 1700 events were selected.

Jets are reconstructed from all hits in the electromagentic calorimeter and all tracks found in tracking chambers passing basic quality criteria. A so-called matching algorithm identifies hits in the calorimeter connected with tracks and subtracts the estimated energy contributed by the charged particle from the reconstructed energy in the calorimeter [21].

The anti-k_t algorithm [8] in its variant for e^+e^- collisions defines as distance measure between two particles i and j with energies E_i and E_j

$$d_{ij} = \min(E_i^{-2}, E_j^{-2})(1 - \cos\theta_{ij})/(1 - \cos R); \quad d_{iB} = E_i^{-2} \tag{1}$$

where $\cos\theta_{ij}$ is the angle between the two particles, R is a parameter and d_{iB} is the analogue of the distance between a particle i and the beam direction

If d_{ij} is the smallest of all phase space distances the two particles are combined into a pseudo-particle by adding their 4-vectors and removed from the list while the pseudo-particle is added to the list. If d_{iB} is the smallest distance particle i is removed from the list. This procedure continues until the list of (pseudo-) particles is exhausted. The anti-k_t algorithm clusters particles beginning from the most energetic ones usually at the centre of jets and effectively excludes particles at an angle larger than R from the jet axis. The result are jets with a well defined cone-like shape with opening half-angle R.

In the SISCone algorithm [7] all final state particles are used as seeds to center a cone of opening half-angle R. For all particles in the cone the sum of their momenta is calculated and the polar and azimuth angles θ and ϕ of the momentum sum vector are compared with the angles θ_C and ϕ_C of the cone axis. The cone directions are changed until the directions of the momentum sum vector and the cone match. The resulting stable cones are merged if their shared momentum sum is larger than a predefined threshold. For the presentation of the SISCone algorithm results the transformation $y = 1 - \cos R$ is used.

For both algorithms a cut on the minimum jet energy $E_{cut}/E_{vis.} > \epsilon = 0.077$ is applied when R is varied or a fixed value of $R = 0.36$ is chosen when ϵ is varied.

4 Experimental correction procedure

The experimental correction is performed using a novel procedure [22]. The simulated events from a generator like PYTHIA 6 (Py6) or HERWIG 6 (He6) at "particle level", i.e. after hard interaction, parton shower, hadronisation and decays of unstable particles[2], are passed through a full simulation of the OPAL experiment [23]. The hard interaction in the simulation is required to have effective cm energy $\sqrt{s'}$ within 1 GeV of the nominal value in order to suppress ISR. In this way the experimental correction also accounts for ISR effects.

The simulated detector response data are then reconstructed with the same software as the real data to obtain simulated events at "detector level". Events are labeled by the numbers of reconstructed jets at the values of y or ϵ of a given distribution. For example, with four increasing values of y an event could have the sequence $\{4322\}$, i.e. at y_1 4 jets, at y_2 3 jets and at y_3 and y_4 2 jets are found. The ratio of the numbers of events for each label at particle and detector level defines the correction factor. The advantage over the established method of comparing numbers of reconstructed jets directly is that with labeling events are essentially sorted by similarity in topology and the corresponding correction factors are applied to similar events.

Background is subtracted for each label using the predicted contribution from samples of simulated background events. Background is dominated by W-boson pair production with hadronic decays of both Ws. Therefore labels containing events with 4-jet topology for larger values of y or ϵ receive most of the background.

After subtraction of background and application of experimental correction factors the jet fractions are calculated by summing over appropriate labels. For example the 3-jet fraction for y_2 is

$$R_3(y_2) = 1/N \sum_{\{ijkl\},j=3} n_{ijkl} \tag{2}$$

where N is the total number of events and n_{ijkl} is the number of events with label $\{ijkl\}$. The statistical covariances between y or ϵ points can be computed directly based on equ. (2).

Figure 1 shows examples of comparisons of corrected data to simulation produced by the OPAL collaboration as explained in section 2. The simulations provide an excellent description of the data at $\sqrt{s} = 91$ GeV where they were tuned and a reasonable description of the 207 GeV data. We conclude that the experimental corrections derived from these simulations are reliable. Differences between the experimental corrections derived from PYTHIA 6 or HERWIG 6 are included in the experimental systematic uncertainties.

[2] All particles with lifetimes greater than 300 ps are treated as stable.

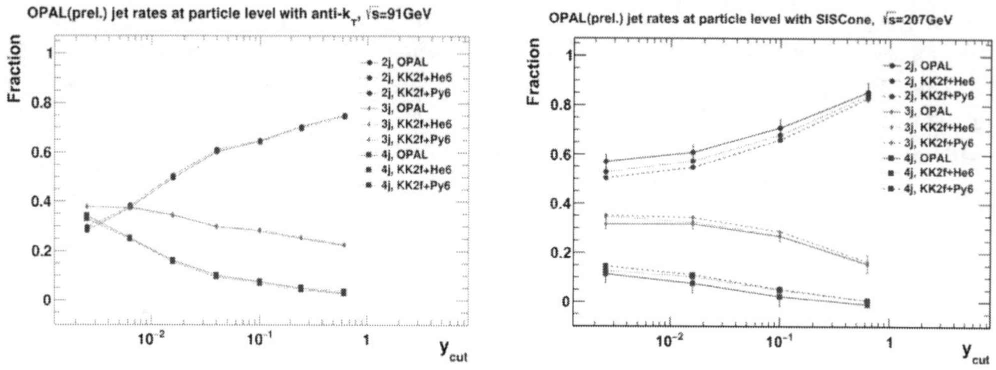

Figure 1. (left) Jet rates at particle level with anti-k$_t$ at $\sqrt{s}$ = 91 GeV compared with old simulation tuned by OPAL. The first y point corresponds to R = 0.07 and the last to R = 1.19. (right) The same for SISCone at $\sqrt{s}$ = 207 GeV.

5 Results

We present measurements of the anti-k$_t$ and SISCone jetrates corrected to the particle level at different LEP energy points and compare with predictions by the modern Monte Carlo generator programs Pythia8 (Py8) and Herwig++ (H++). In recent years significant progress in understanding and implementation of event generators has been made and we have taken Pythia8 and Herwig++ as representative examples [24]. The generator programs were used with their default parameter settings. Figure 2 shows results for anti-k$_t$ at $\sqrt{s}$ = 91 and 207 GeV as a function of y and for SISCone at $\sqrt{s}$ = 91 and 207 GeV as function of $\epsilon = E_{cut}/E_{vis.}$.

We observe a good description of the jet rate data at all energy points by the simulation predictions. The description of the high energy data at 207 GeV is somewhat improved compared to the old OPAL programs shown in figure 1. This might be a result of improved treatment of perturbative QCD in the new simulations responsible for predicting effects evolving with the cm energy $\sqrt{s}$. We note that the general behaviour of the anti-k$_t$ and SISCone algorithms is very similar for the same values of y and ϵ and thus their behaviour is dominated by their parameter settings and not by algorithm details.

In figure 3 we study predictions for hadronisation corrections for the jet rates measured with the new algorithms. The corrections are defined as the ratio of predictions at the parton and the particle level from the Monte Carlo generators. Parton level is defined as the final state consisting of quarks and gluons of simulated events after hard interaction and parton shower before the hadronisation model is applied. The displays in figure 3 show the hadronisation corrections calculated at several points of ϵ for a fixed value of opening half-angle R = 0.36 for the SISCone (left) and anti-k$_t$ (right) algorithms.

The corrections for SISCone from KK2f+Py6 (old OPAL tuned MC) and Pythia8 agree well while Herwig++ tends to predict smaller corrections. For the SISCone 3-jet rate the corrections are about 10% with a trend to decrease with decreasing ϵ. For the SISCone 2-jet rate the corrections are about 10-20% while the SISCone 4-jet rate has corrections ranging from 90% to 10-20% at small ϵ. The corrections for the 2-jet rate and the other jet rates are below and above one, respectively. This effect can be explained with the well-known general broadening of jets due to hadronisation reducing the 2-jet and increasing the >2 jet rates.

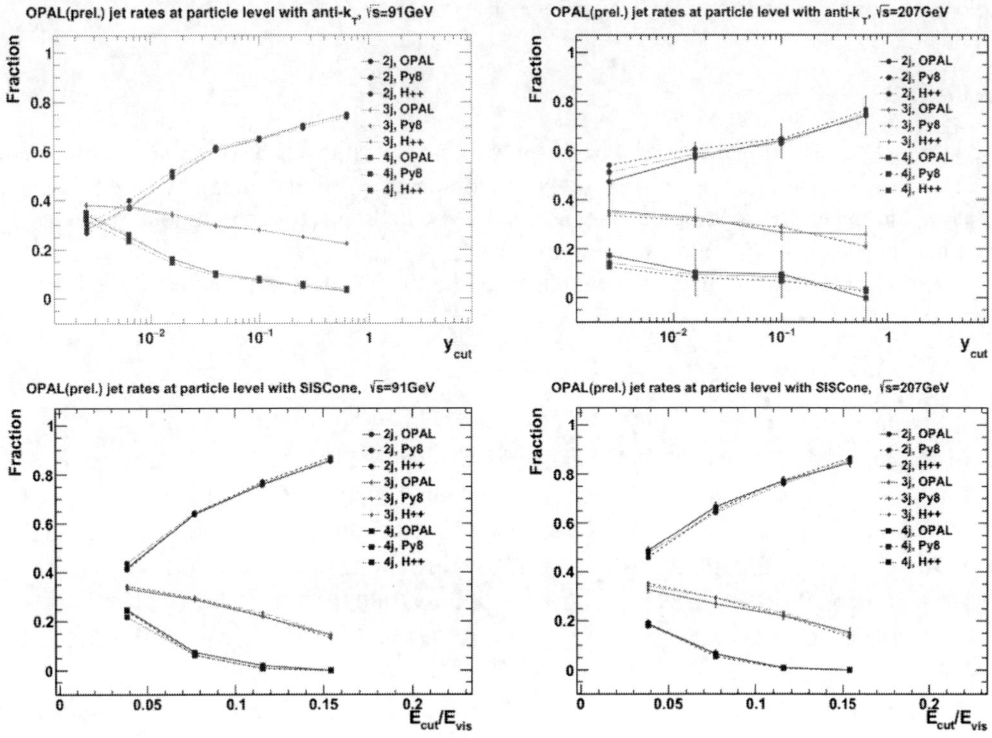

Figure 2. (upper row) Jet rates at particle level with anti-k_t at $\sqrt{s} = 91$ and 207 GeV compared with new Monte Carlo predictions as indicated. (lower row) The same for SISCone as a function of ϵ for $R = 0.36$.

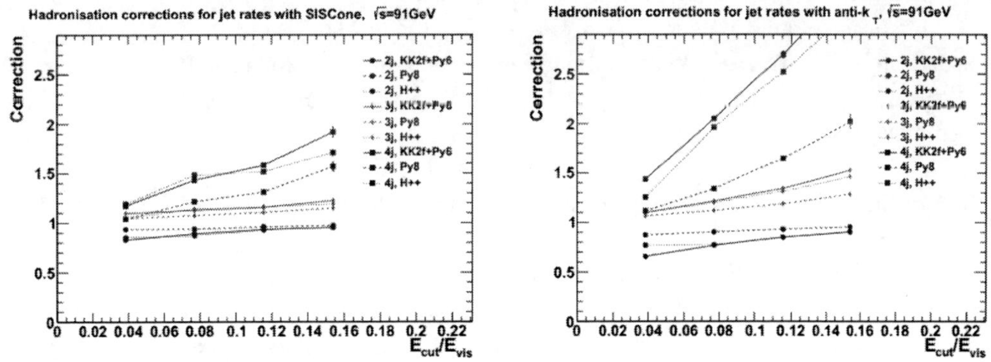

Figure 3. (left) Hadronisation corrections predicted by simulations as indicated for SISCone at $\sqrt{s} = 91$ as a function of ϵ for $R = 0.36$. (right) The same for anti-k_t.

For the anti-k_t algorithm we find a similar pattern of the corrections compared with SISCone, but the values of the correction factors are generally larger. For the Durham algorithm with additional requirement on the jet energies (not shown) we find hadronisation corrections essentially flat in ϵ and significantly smaller w.r.t. both SISCone and anti-k_t.

6 Summary

We have shown results for jet rates reconstructed with the algorithms SISCone and anti-k_t in hadronic final states in e^+e^- annihilation. The jet rates are compared with predictions by Monte Carlo event generators and generally good agreement is found. The hadronisation corrections predicted by the generators show interesting behaviour as a function of $\epsilon = E_{cut}/E_{vis.}$.

The speaker would like to thank the organisers of the ISMD 2016 for a interesting and stimulating conference.

References

[1] Y.I. Azimov, Y.L. Dokshitzer, V.A. Khoze, S.I. Troian, Z. Phys. **C27**, 65 (1985)
[2] Y.I. Azimov, Y.L. Dokshitzer, V.A. Khoze, S.I. Troian, Z. Phys. **C31**, 213 (1986)
[3] T. Schörner-Sadenius, Eur. Phys. J. **C72**, 2060 (2012), erratum: Eur. Phys. J.C72, 2133 (2012)
[4] A. Ali, G. Kramer, Eur. Phys. J. **H36**, 245 (2011), 1012.2288
[5] S. Kluth, Rept. Prog. Phys. **69**, 1771 (2006), hep-ex/0603011
[6] C. Mesropian, D. Bandurin, Int. J. Mod. Phys. **A30**, 1541002 (2015), arXiv:1409.5639
[7] G.P. Salam, G. Soyez, J. High Energy Phys. **05**, 086 (2007), arxiv:0704.0292
[8] M. Cacciari, G.P. Salam, G. Soyez, J. High Energy Phys. **04**, 063 (2008), arxiv:0802.1189
[9] R. Akers et al. (OPAL), Z. Phys. **C63**, 197 (1994)
[10] H. Abramowicz et al. (ZEUS), Phys. Lett. **B691**, 127 (2010), 1003.2923
[11] H. Abramowicz et al. (ZEUS), Nucl. Phys. **B864**, 1 (2012), arXiv:1205.6153
[12] K. Ahmet et al. (OPAL), Nucl. Instrum. Methods **A305**, 275 (1991)
[13] G. Abbiendi et al. (OPAL), Eur. Phys. J. **C40**, 287 (2005), hep-ex/0503051
[14] T. Sjöstrand, Comput. Phys. Commun. **82**, 74 (1994)
[15] G. Marchesini et al., Comput. Phys. Commun. **67**, 465 (1992)
[16] S. Jadach, B.F.L. Ward, Z. Was, Comput. Phys. Commun. **130**, 260 (2000)
[17] T. Sjöstrand, L. Lönnblad, S. Mrenna, *Pythia 6.2: Physics and manual*, LU-TP-01-21 (2001), hep-ph/0108264
[18] J. Fujimoto et al., Comput. Phys. Commun. pp. 128–156 (1997)
[19] S. Jadach, W. Placzek, M. Skrzypek, B.F.L. Ward, Z. Was, Comput. Phys. Commun. **119**, 272 (1999)
[20] G. Abbiendi et al. (OPAL), Phys. Lett. **B493**, 249 (2000)
[21] G. Abbiendi et al. (OPAL), Eur. Phys. J. **C45**, 547 (2006), hep-ex/0507047
[22] A. Verbytskyi, *Studies of correlations between measurements of jet observables* (2016), arXiv:1609.06898
[23] J. Allison et al., Nucl. Instrum. Methods **A**, 47 (1992)
[24] T. Sjöstrand, *Status and developments of event generators*, in *4th Large Hadron Collider Physics Conference (LHCP 2016) Lund, Sweden, June 13-18, 2016* (2016), arXiv:1608.06425

Low energy exclusive $e^+e^- \to$ hadrons cross sections and $g-2$ of the muon

J. William Gary[1,a]

[1] *Department of Physics and Astronomy, University of California, Riverside, CA 92521, USA*

Abstract. The 3.5 standard deviation discrepancy between the standard model prediction for the muon anomalous magnetic moment $g-2$ and the corresponding experimental measurement is one of the most intriguing potential signals for physics beyond the standard model. The largest uncertainty in the standard model prediction for the muon $g-2$ arises from uncertainties in the measured low-energy $e^+e^- \to$ hadrons cross sections. New cross section results from the Babar experiment at SLAC for $e^+e^- \to \pi^+\pi^-\pi^0\pi^0$ and for $e^+e^- \to KK\pi\pi$ channels are presented that significantly reduce some of the uncertainties in the standard model prediction for the muon $g-2$.

1 Introduction

The gyromagnetic ratio g specifies the relationship between an angular momentum and the corresponding magnetic moment. The Dirac result for the gyromagnetic ratio g_ℓ of an elementary charged spin-1/2 lepton ℓ is $g_\ell = 2$ exactly. The Dirac result corresponds to the tree level diagram shown in the leftmost plot of Fig. 1. Radiative corrections alter the prediction to $g_\ell = 2(1 + a_\ell)$, introducing sensitivity to new physics through loops. The term a_ℓ is known as the anomalous moment, expressed as $a_\ell = (g_\ell - 2)/2$ and simplified in common parlance to "$g-2$". The muon anomaly a_μ is much more sensitive to virtual heavy particle production in loops than the electron anomaly a_e because the relative virtual terms scale like $(m_\mu/m_e)^2 \approx 43,000$, with m_ℓ the lepton mass.

In the standard model (SM), the muon anomaly can be expressed as $a_\mu = a_\mu^{QED} + a_\mu^{EW} + a_\mu^{had}$, receiving contributions from electromagnetic, weak, and hadronic loop corrections. The purely electromagnetic term a_μ^{QED} has been calculated [2] to high order in perturbation theory. The electroweak term a_μ^{EW} [3] includes contributions from the recently discovered Higgs boson. The hadronic term a_μ^{had} is itself divided into three parts, illustrated by the three rightmost plots of Fig. 1: the leading-order hadronic vacuum polarization term $a_\mu^{had,LO-VP}$, the higher-order hadronic vacuum polarization terms $a_\mu^{had,HO-VP}$, and the so-called light-by-light scattering term $a_\mu^{had,LbLs}$. A summary of the individual contributions to the SM result for a_μ is given in Table 1. Summing these contributions yields a total SM prediction of $a_\mu^{SM} = 116\,591\,802 \pm 49 \times 10^{-11}$ [4], which lies $287 \pm 80 \times 10^{-11}$ below the average measured result, $a_\mu^{data} = 116\,592\,091 \pm 63 \times 10^{-11}$ [5, 6], corresponding to a discrepancy of about 3.5 standard deviations. From Table 1 it is seen that the uncertainty in the SM prediction is dominated by the uncertainty in the leading-order vacuum polarization term $a_\mu^{had,LO-VP}$.

[a] e-mail: bill.gary@ucr.edu

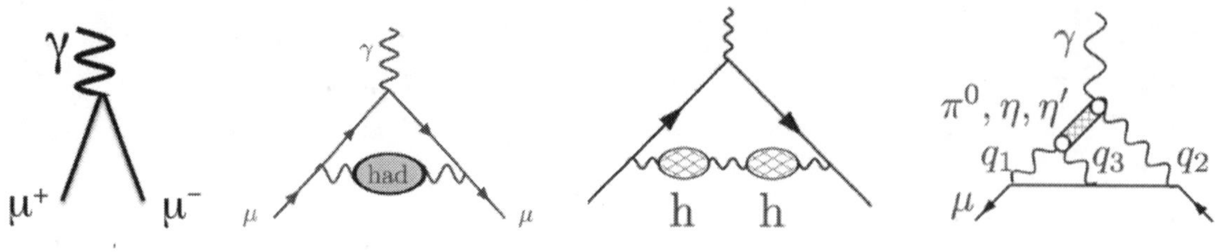

Figure 1. (left) Tree-level vertex diagram for a muon coupling to a photon and (left-center) the diagram corresponding to the leading-order hadronic correction. Examples of diagrams for the higher-order hadronic corrections and for light-by-light scattering are shown in the right-center and right plots, respectively. The three leftmost plots are taken from Ref. [1].

Table 1. SM results for the different contributions to the muon anomaly a_μ, taken from Ref. [4].

a_μ term	SM prediction ($\times 10^{-11}$)
a_μ^{QED}	116584718.951 ± 0.080
a_μ^{EW}	153.6 ± 1.0
$a_\mu^{had,LO-VP}$	6923 ± 42
$a_\mu^{had,HO-VP}$	-98.4 ± 0.7
$a_\mu^{had,LbLs}$	105 ± 26

The energy scale is too low for $a_\mu^{had,LO-VP}$ to be calculated perturbatively, and lattice calculations [7] are not yet sufficiently precise. Instead, the most precise result for $a_\mu^{had,LO-VP}$ is currently obtained from low-energy $e^+e^- \to hadrons$ data, using the optical theorem and the following dispersion integral (see Ref. [6]):

$$a_\mu^{had,LO-VP} = \left(\frac{\alpha}{3\pi}\right)^2 \int_{m_\pi^2}^{\infty} \frac{K(s)R_{had}}{s} ds, \tag{1}$$

where $K(s)$ is a kinematic factor and R_{had} is the $e^+e^- \to hadrons$ cross section normalized to the Born-level $e^+e^- \to \mu^+\mu^-$ cross section. Because of the $1/s$ factor, low-energy contributions dominate the dispersion integral. Therefore precise measurements of R_{had} are needed at low values of center-of-mass energy $\sqrt{s}$. Below $\sqrt{s} = 2$ GeV, the sum of exclusive hadronic channels is used because the background from $e^+e^- \to e^+e^-$ and $e^+e^- \to \mu^+\mu^-$ events is difficult to distinguish from inclusive multihadronic events at such low energies. Also, the inclusive detection efficiency is difficult to determine with precision at very low $\sqrt{s}$. For energies above 2 GeV, inclusive $e^+e^- \to hadrons$ data and perturbative calculations can be used.

Figure 2 shows measurements of exclusive hadronic cross sections from the Babar Collaboration at SLAC. It is seen that the $e^+e^- \to \pi^+\pi^-$ channel dominates the low-energy hadronic cross section. It contributes about 75% of the area to the dispersion integral. The $e^+e^- \to \pi^+\pi^-$ cross section has been measured by the Babar [8], KLOE [9, 10], and BES-III [11] experiments, each claiming around 1% precision in their results. However, the results of the three experiments differ in some kinematic regions by more than the claimed 1% uncertainties. Future results, expected from the Babar, BES-III, and CMD3 experiments, may clarify the $e^+e^- \to \pi^+\pi^-$ results.

Table 2 lists the channels whose uncertainties in the cross section contribute the most to the uncertainty $\Delta a_\mu^{had,LO-VP}$ in $a_\mu^{had,LO-VP}$. The largest contribution to $\Delta a_\mu^{had,LO-VP}$ comes from the $\pi^+\pi^-$ channel, followed by the $\pi^+\pi^-\pi^0$ and $\pi^+\pi^-\pi^0\pi^0$ channels. Channels with two kaons are also seen

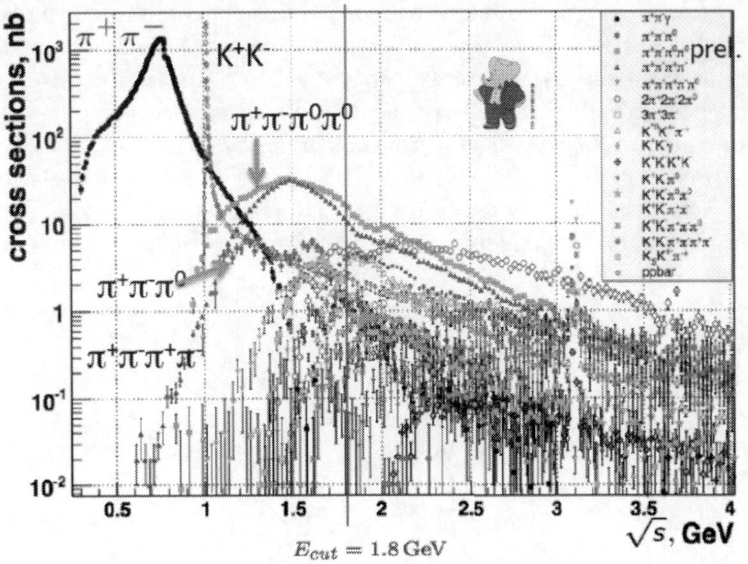

Figure 2. Summary of Babar measurements of low-energy exclusive hadronic cross sections.

to be important. Babar has recently measured the cross sections for $e^+e^- \to K^+K^-$ [12, 13] and $e^+e^- \to K_L K_S$ [14]. These new results will significantly reduce the respective contributions to $\Delta a_\mu^{had,LO-VP}$.

In the following, new Babar measurements of the $e^+e^- \to \pi^+\pi^-\pi^0\pi^0$, $K_S K^+\pi^-\pi^0$, $K_S K^+\pi^-\eta$, $K_S K_L \pi^0$, $K_S K_L \eta$, and $K_S K_L \pi^0\pi^0$ cross sections are presented, and the implications for $\Delta a_\mu^{had,LO-VP}$ discussed. These results are preliminary. The data are collected with the initial-state radiation (ISR) method. In the ISR method, an effective reduced hadronic energy $\sqrt{s'}$ is obtained by selecting events with a hard ISR photon radiated from either the incoming electron or incoming positron in fixed-energy e^+e^- collisions. In the Babar studies, the ISR photon candidate is taken to be the highest energy photon in an event and must have a center-of-mass energy larger than 3 GeV.

Table 2. Exclusive $e^+e^- \to hadrons$ channels contributing the largest uncertainty $\Delta a_\mu^{had,LO-VP}$ to the SM prediction of the muon anomaly term $a_\mu^{had,LO-VP}$ (taken from Ref. [4]).

Channel	$\pi^+\pi^-$	$\pi^+\pi^-\pi^0$	$\pi^+\pi^-\pi^0\pi^0$	K^+K^-	$\pi^+\pi^-\pi^+\pi^-$	$KK\pi\pi$	$K_S K_L$
$\Delta a_\mu^{had,LO-VP}$ ($\times 10^{-11}$)	28	15	12	7	5	4	4

2 The $e^+e^- \to \pi^+\pi^-\pi^0\pi^0$ cross section

As illustrated in Table 2, the $e^+e^- \to \pi^+\pi^-\pi^0\pi^0$ process represents one of the least known cross sections important for the muon anomaly. In the Babar analysis, events with exactly two oppositely charged tracks are selected. The two tracks must have momentum $p > 100$ MeV and be consistent with originating at the primary event vertex. Besides the ISR photon candidate, there must be at least four other photon candidates in the event, each with energy $E > 50$ MeV. A kinematic fit is performed to the signal hypothesis, constraining two 2γ combinations to the π^0 mass. The overall combination of four photons that yields the smallest χ^2 value is selected, requiring $\chi^2 < 30$.

Background from the ISR processes $e^+e^- \to \pi^+\pi^-\pi^0\gamma$, $e^+e^- \to \pi^+\pi^-\pi^0\eta\gamma$, and $e^+e^- \to \pi^+\pi^-\eta\eta\gamma$ is suppressed by rejecting events with kinematic fits consistent with those hypotheses. Background from ISR $e^+e^- \to \pi^+\pi^-\pi^0\pi^0\pi^0\gamma$ events is evaluated by measuring the cross section for these events using an analogous technique to that described above for $e^+e^- \to \pi^+\pi^-\pi^0\pi^0\gamma$ events, and then adjusting the simulation accordingly to reliably predict the rate of these events in the $e^+e^- \to \pi^+\pi^-\pi^0\pi^0\gamma$ sample. The largest non-ISR background is from $q\bar{q}$ continuum events arising from the misidentification of a photon from π^0 decay as an ISR photon. This background is subtracted using simulation that is normalized to data based on the π^0 peak from $\gamma_{ISR}\gamma$ combinations, where γ_{ISR} is the ISR photon candidate, with γ any other photon in the event not assigned to a signal π^0 candidate.

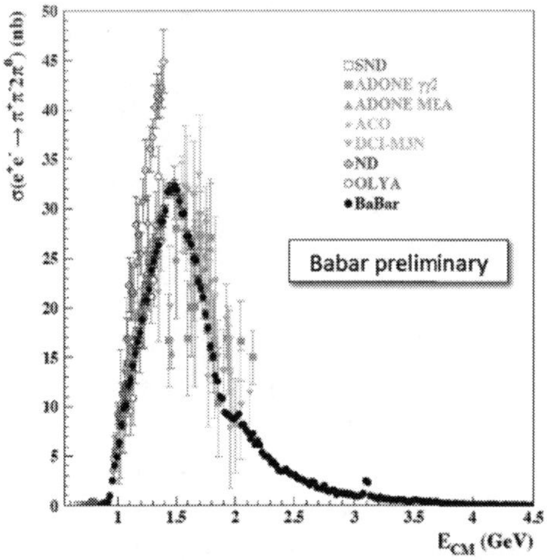

Figure 3. Babar measurement of the $e^+e^- \to \pi^+\pi^-\pi^0\pi^0$ cross section in comparison with published results from previous experiments.

The resulting measurement of the $e^+e^- \to \pi^+\pi^-\pi^0\pi^0$ cross section is shown in Fig. 3. The Babar results are far more precise than previously published results, and cover a wider energy range. The contribution to the muon anomaly for $1.02 < \sqrt{s} < 1.8$ GeV is measured with 3.4% precision, compared to the previous precision of 6.7% based on preliminary Babar data from 2007.

3 First measurements of the $e^+e^- \to K_S K^+\pi^-\pi^0$ and $K_S K^+\pi^-\eta$ cross sections

A similar analysis is performed for the $e^+e^- \to K_S K^+\pi^-\pi^0$ and $K_S K^+\pi^-\eta$ cross sections. Besides the ISR photon candidate, at least one $K_S \to \pi^+\pi^-$ candidate consistent with the interaction point, and at least two additional photons with a diphoton mass consistent with the π^0 or η mass, are required. Two oppositely charged tracks must be present, one identified as a pion and the other as a kaon. Background from non-ISR $q\bar{q}$ events is evaluated from simulation normalized to data using the $\gamma_{ISR}\gamma$ peak at the π^0 mass. Background from ISR processes is evaluated from data sidebands.

The resulting cross section measurements are shown in Fig. 4. These are the first measurements of these cross sections. Very clear J/ψ peaks are visible for both channels The $J/\psi \to K_S K^+\pi^-\pi^0$ branching fraction is measured to be $(5.7 \pm 0.3 \pm 0.4) \times 10^{-3}$, where the first uncertainty is statistical and the second systematic. This is the first observation of this decay channel. The $J/\psi \to K_S K^+\pi^-\eta$

DOI: 10.1051/epjconf/201714102004

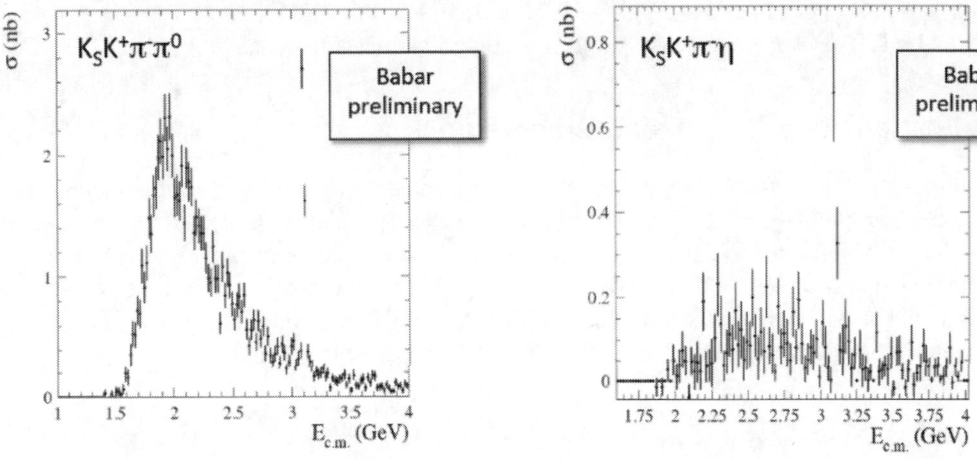

Figure 4. Babar measurements of the (left) $e^+e^- \to K_S K^+\pi^-\pi^0$ and (right) $K_S K^+\pi^-\eta$ cross sections.

branching fraction is measured to be $(1.30 \pm 0.25 \pm 0.07) \times 10^{-3}$, around 2 standard deviations below a previous result from the BES Collaboration [15].

4 First measurements of the $e^+e^- \to K_SK_L\pi^0$, $K_SK_L\eta$, and $K_SK_L\pi^0\pi^0$ cross sections

As in the analysis described in the previous section, at least one $K_S \to \pi^+\pi^-$ candidate and at least two additional photons consistent with π^0 or η decay are required. A K_L candidate is identified as an isolated calorimeter cluster with energy larger than 0.2 GeV. The detection efficiency for K_L particles is determined from data using $e^+e^- \to \phi\gamma \to K_S K_L \gamma$ events and a tag-and-probe method. Background is suppressed, with the residual background subtracted, using methods similar to those described in the previous sections.

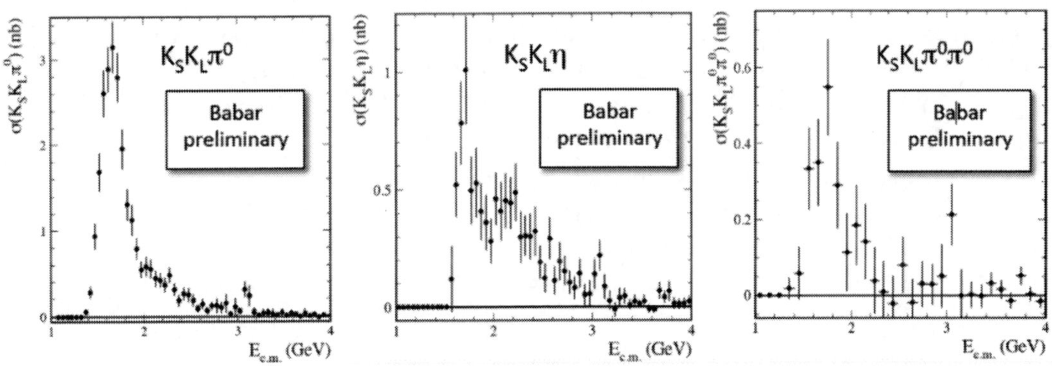

Figure 5. Babar measurements of the (left) $e^+e^- \to K_S K_L\pi^0$, (center) $K_S K_L\eta$, and (right) $K_S K_L\pi^0\pi^0$ cross sections.

The cross section results are shown in Fig. 5. All of these are first measurements. The corresponding J/ψ branching fractions, $(2.06 \pm 0.24 \pm 0.10) \times 10^{-3}$, $(1.45 \pm 0.32 \pm 0.08) \times 10^{-3}$, and $(1.86 \pm 0.43 \pm 0.10) \times 10^{-3}$, respectively, also represent first measurements.

5 Implications of $KK\pi\pi$ measurements on a_μ

The results for $e^+e^- \to K_S K^+\pi^-\pi^0$ and $K_S K_L \pi^0\pi^0$ presented above can be combined with previous Babar results for $e^+e^- \to K^+K^-\pi^+\pi^-$, $K^+K^-\pi^0\pi^0$, $K_S K_S \pi^+\pi^-$, and $K_S K_L \pi^+\pi^-$ [14, 16], which all include studies of the intermediate resonant states, to calculate the contribution of $KK\pi\pi$ channels to the muon anomaly a_μ. Heretofore, the contributions of these previously unmeasured channels had been estimated using isospin relations. The contribution of $KK\pi\pi$ channels to a_μ for $\sqrt{s} < 1.8$ GeV is determined to 6% precision, compared to 30% precision available previously.

6 Summary

Precise low-energy e^+e^- hadronic cross section data are needed to obtain an accurate standard model prediction for the part of the muon magnetic moment anomaly associated with leading-order vacuum polarization terms. New results from the Babar experiment at SLAC reduce the respective uncertainty in this term from 6.7% to 3.4% for the $\pi^+\pi^-\pi^0\pi^0$ channel and from 30% to 6% for $KK\pi\pi$ channels. Besides the implications for the muon anomaly, the Babar ISR program has provided new tests of QCD, such as in measurements of koan form factors, a wealth of information about low-mass resonances, first observations of cross sections, and first observations of J/ψ and $\psi(2S)$ branching fractions.

References

[1] F. Jegerlehner and A. Nyffeler, Phys. Rept. **477**, 1 (2009).
[2] T. Aoyama, M. Hayakawa, T, Kinoshita, and M. Nio, Phys. Rev. Lett. **109**, 111808 (2012).
[3] C. Gnendiger, D. Stöckinger, and H. Stöckinger-Kim, Phys. Rev. D **88**, 053005 (2013).
[4] M. Davier, A. Hoecker, B, Malaescu, and Z. Zhang, Z. Phys. C **71**, 1515 (2011).
[5] G. W. Bennett et al. (E821 Collaboration), Phys. Rev. D **73**, 072003 (2006).
[6] K. A. Olive et al. (Particle Data Group), Chin. Phys. C **38**, 090001 (2014).
[7] M. Della Morte, B. Jäger, A. Jüttner, and H. Wittig, JHEP **03**, 055 (2012).
[8] J. P. Lees et al. (Babar Collaboration), Phys. Rev. D **86**, 032013 (2012).
[9] F. Ambrosino et al. (KLOE Collaboration), Phys. Lett. B **700**, 102 (2011).
[10] D. Babusci et al. (KLOE Collaboration), Phys. Lett. B **720**, 336 (2013).
[11] M. Ablikim et al. (BESIII Collaboration), Phys. Lett. B **753**, 629 (2016).
[12] J. P. Lees et al. (Babar Collaboration), Phys. Rev. D **88**, 032013 (2013).
[13] J. P. Lees et al. (Babar Collaboration), Phys. Rev. D **92**, 072008 (2015).
[14] J. P. Lees et al. (Babar Collaboration), Phys. Rev. D **89**, 092002 (2014).
[15] M. Ablikim et al. (BES Collaboration), Phys. Rev. D **77**, 032005 (2008).
[16] J. P. Lees et al. (Babar Collaboration), Phys. Rev. D **86**, 012008 (2012).

Search for Muonic Dark Forces at BABAR

Romulus Godang[1,a]
On Behalf of the BABAR Collaboration

[1] *Department of Physics*
University of South Alabama
411 University Boulevard, North
Mobile, AL 36688

Abstract. Many models of physics beyond Standard Model predict the existence of light Higgs states, dark photons, and new gauge bosons mediating interactions between dark sectors and the Standard Model. Using a full data sample collected with the BABAR detector at the PEP-II e^+e^- collider, we report searches for a light non-Standard Model Higgs boson, dark photon, and a new muonic dark force mediated by a gauge boson (Z') coupling only to the second and third lepton families. Our results significantly improve upon the current bounds and further constrain the remaining region of the allowed parameter space.

1 Introduction

Many astrophysical and cosmological observations indicate that a fraction of energy density in the universe is due to non-baryonic matter. The microscopic nature of dark matter is currently unknown. Models of physics beyond Standard Model predict the existence of a new non-Abelian gauge group Higgs with gauge boson masses below 10 GeV [1]. The WIMP hypothesis suggested that dark matter is assumed to consist of stable particle with low masses. Such new gauge bosons can typically interact with other Standard Model elementary particles. The new gauge boson, Z', can couple to the Standard Model field. SM fields can be directly charged under new gauge boson or the Z' may couple with the SM hypercharge boson [2].

Based on the $L_\mu - L_\tau$ model [3] one of the most promising candidates based on gauging the existing approximate global symmetries of the Standard Model (SM) is the gauge group associated with the difference between muon and tau-lepton number. The gauge $L_\mu - L_\tau$ model portal to the Z' has all features of being coupled only to the leptons of the second and third generation. The $L_\mu - L_\tau$ model has been studied in the neutrino mass model. It has also been studied in the (g-2) current discrepancy. Consequently it was found that the tentative explanation of the (g-2) anomaly in models with large electron coupling is now excluded at least for multi-GeV and heavier Z' bosons. The model explains the possible production of Z' gauge boson production via $e^+e^- \rightarrow\rightarrow \mu^+\mu^-Z'$, $Z' \rightarrow \mu^+\mu^-$ that primarily comes from the radiation of heavy-flavor leptons and is shown in the following Fig. 1.

[a] e-mail: godang@southalabama.edu

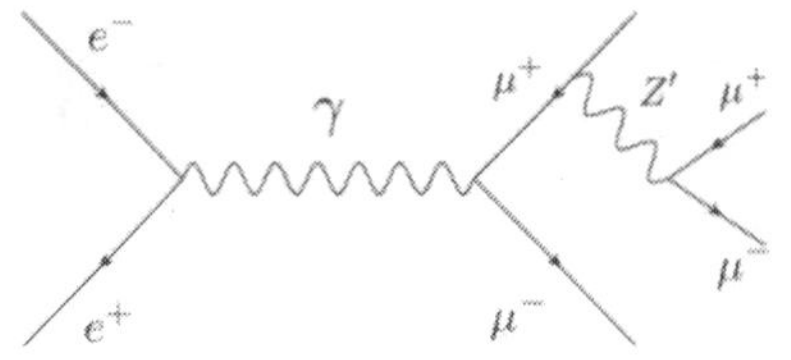

Figure 1. Feynman Diagram for production of gauge boson Z' based on the $L_\mu - L_\tau$ model at an e^+e^- collider.

2 Data and Event Selection

We used the data collected by the BABAR detector5 [4] with the total luminosity of 514 fb^{-1}. Most of the data were taken at the $\Upsilon(4S)$ resonance plus including about 28 fb^{-1} data at $\Upsilon(3S)$ and 14 fb^{-1} data at $\Upsilon(2S)$ and 48 fb^{-1} data at the off-resonance. The $\Upsilon(4S)$ resonance decays to a pair of $B\bar{B}$ [5]. We used about 5% of the data set to validate and optimize the analysis method. The rest of the data was only examined after finish finalizing the analysis method. For the background study we generated signal Monte Carlo (MC) samples.

Signal MC events are generated using MadGraph 5 [6], which calculates matrix elements for the sample. The MC then were showered using Pythia 6 [7] for about 30 different Z' mass hypotheses. The main background comes from the QED processes. We generate the direct processes of $e^+e^- \to \mu^+\mu^-\mu^+\mu^-$ using Diag36 [8], which includes the full set of the lowest order diagrams. The Diag36 does not include initial state radiation (ISR) samples. The events of the process of $e^+e^- \to e^+e^-(\gamma)$ are generated using BHWIDE [9] and the MC events of $e^+e^- \to \mu^+\mu^-(\gamma)$ and $e^+e^- \to \tau^+\tau^-(\gamma)$ are generated using KK [10]. The off-resonance data samples, $e^+e^- \to \bar{q}q$ (q = u, d, s, c), are simulated using EvtGen [11]. The events processes of $e^+e^- \to \psi(2S)\gamma$ then $\psi(2S) \to \pi^+\pi^- J/\psi$ and $J/\psi \to \mu^+\mu^-$ were generated using a structure function technique [12, 13]. Finally the detector acceptance and reconstruction efficiency are determined using MC simulation based on GEANT4 [14].

3 Z' Measurement

We select events containing exactly two pairs of oppositely charged tracks, consistent with the topology of the process: $e^+e^- \to \mu^+\mu^- Z'$ and $Z' \to \mu^+\mu^-$ final state. The muons are identified by particle identification algorithms for each track. We require the sum of energies of the electromagnetic clusters that are not associated to any track must be less than 200 MeV. We finally reject events that come from the $\Upsilon(3S)$ and $\Upsilon(2S)$, where $\Upsilon(2S, 3S) \to \pi^+\pi^-\Upsilon(1S)$, $\Upsilon(1S) \to \mu^+\mu^-$ decays if the dimuon combination is within 100 MeV of the $\Upsilon(1S)$ where pions are misidentified as muons.

The distribution of the four-muon invariant mass after all selections is shown in Fig. 2 (left). At the low mass of the four-muon invariant mass, $m(4\mu) < 9$ GeV, is well reproduced by the Monte Carlo simulation including direct decays of $e^+e^- \to \mu^+\mu^-\mu^+\mu^-$, however, the Monte Carlo simulation overestimates the full energy peak by ~ 30% and fails to reproduce the radiactive tail. The overestimate simulation is expected because the Diag36 simulation does not simulate the initial state radiation (ISR) events. To estimate the potential ISR emission we select $e^+e^- \to \mu^+\mu^-\mu^+\mu^-$ events by requiring a four-muon invariant mass distribution within 500 MeV of the nominal center-of-mass energy. We also require the tracks to originate from the interaction point to within its uncertainty and constraining the center-of-mass energy of the system to be within the beam energy spread. The four-muon invariant mass after allowing the potential ISR emission is now fit as shown in Fig. 2 (right).

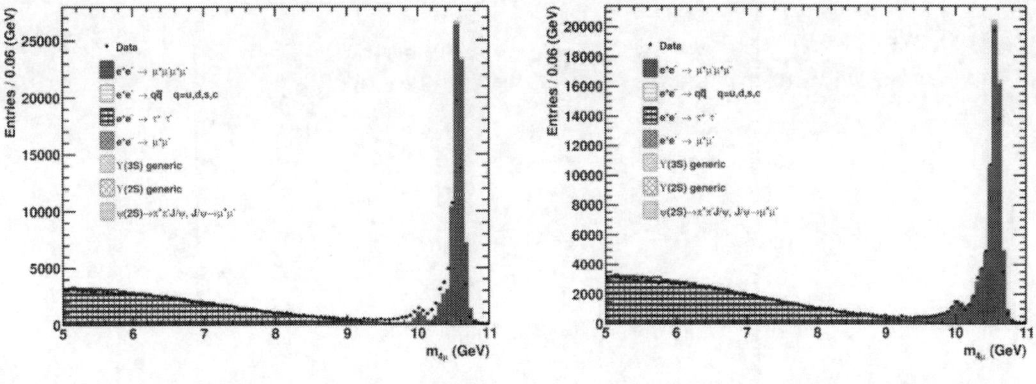

Figure 2. (Left) The four-muon invariant mass distribution with the Monte Carlo predictions of various processes including direct decay of $e^+e^- \to \mu^+\mu^-\mu^+\mu^-$ normalized to the data luminosity. (Right) The four-muon invariant mass distribution with the Monte Carlo predictions of various processes by allowing the initial state radiation emission.

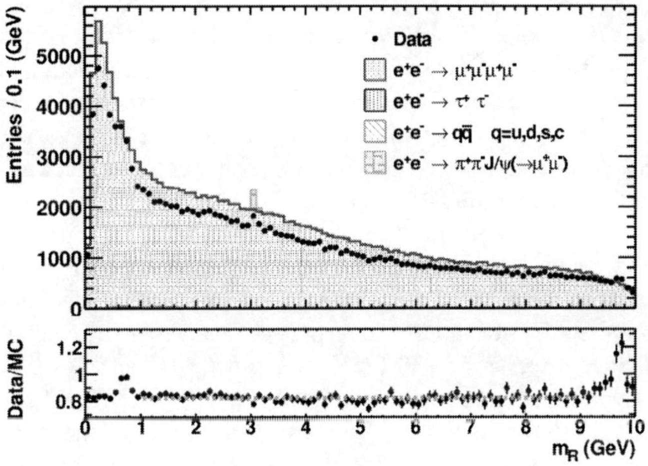

Figure 3. The distribution of the reduced dimuon mass, $m_R = \sqrt{m^2_{\mu^+\mu^-} - 4m^2_\mu}$, for the optimization sample after applying all selections, together with the Monte Carlo predictions of various processes with normalized to the data luminosity. The fit ratio between the reconstructed and simulated events is shown as a blue dashed line.

We also show the distribution of the reduced dimuon mass. The reduced dimuon mass is calculated using the following equation $m_R = \sqrt{m^2_{\mu^+\mu^-} - 4m^2_\mu}$ in linear scale as shown in Fig. 3 and in log scale as shown in Fig. 4. The most dominant samples is coming from the direct decay of $e^+e^- \to \mu^+\mu^-\mu^+\mu^-$ process. The contribution from the decay of $\Upsilon(2S) \to \pi^+\pi^- J/\psi$, $J/\psi \to \mu^+\mu^-$ as shown around 3 GeV. The reduced dimoun mass distribution has a better behavior near threshold and it is also easier to model compare to the dimuon mass distribution. The signal efficiency at low masses is about 35% and it rises to about 50% around higher mass of the reduced dimuon mass. We exclude the J/ψ region when calculating the correction factors by fitting the simulated and reconstructed reduced dimuon

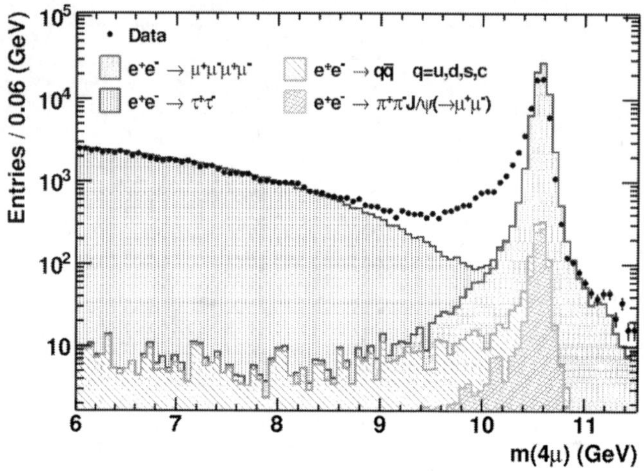

Figure 4. The distribution of the reduced dimuon mass, $m_R = \sqrt{m^2_{\mu^+\mu^-} - 4m^2_\mu}$ in log scale together with the Monte Carlo predictions of various processes.

masses in the range of $1 < m_R < 9$ GeV. We obtain a correction factor of 0.82 as shown in Fig. 3. It includes the ISR emission and differences in the trigger efficiency, charged particle identification, and track and photon reconstruction efficiencies. We assign a systematic uncertainty of 5% to cover the small variations between the uncertainties on the $e^+e^- \to \mu^+\mu^-\mu^+\mu^-$ and data taking period. We calculate the ISR contribution based on the quasi real electron approximation [15]. We assess the sidebands of the four dimuon mass distribution in the range of 5.0 - 8.0 GeV. In this region the process of $e^+e^- \to \tau^+\tau^-(\gamma)$ is dominant. The correction factors are in agreement with the correction factors obtain from the reduce dimuon mass spectrum. The signal yield is extracted bu a series of unbinned likelihood fits to the reduced dimuon mass spectrum within the range of $0.212 < m_R < 10$ GeV and $0.212 < m_R < 9$ GeV for the $\Upsilon(4S)$ resonance data and $\Upsilon(2S)$ and $\Upsilon(3S)$ resonances data, respectively. We exclude a region of ±30 MeV around the nominal known J/ψ mass. We probe a total of 2219 mass hypothesis. The cross section of $e^+e^- \to \mu^+\mu^-Z'$, $Z' \to \mu^+\mu^-$ is extracted as a function of Z' mass as shown in Fig. 5. The gray band indicates the excluded region. We find the largest local significance is 4.3σ around Z' mass of 8.2 GeV that is corresponding to the global significance of 1.6σ and it is consistent with the zero-hypothesis. We also derive 90% confidence level (CL) Bayesian upper limit on the cross section of $e^+e^- \to \mu^+\mu^-Z'$, $Z' \to \mu^+\mu^-$ as shown in Fig. 6.

We consider all uncertainties to be uncorrelated except for the uncertainties of the luminosity and efficiency. We finally extract the corresponding 90% CL on the coupling parameter g' by assuming the equal magnitude vector couplings muons, taus and the corresponding neutrinos together with the existing limits from Borexino and neutrino experiments as shown in Fig. 7. We set down to 7×10^{-4} near the dimuon threshold.

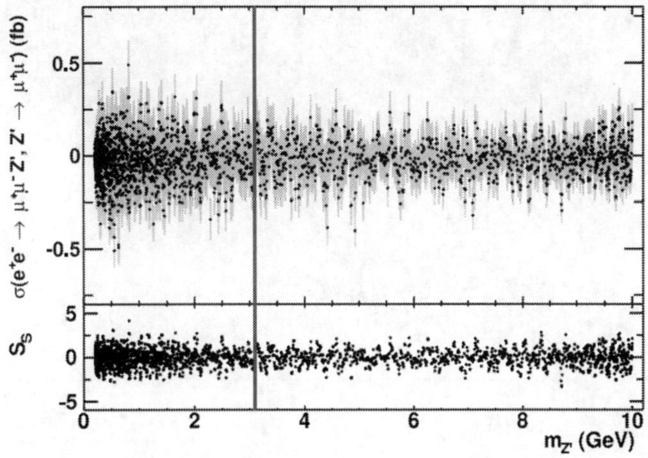

Figure 5. The measurement of $e^+e^- \to \mu^+\mu^- Z'$, $Z' \to \mu^+\mu^-$ cross section with its statistical significance as a function of the Z' mass. The excluded region is indicated by the gray band.

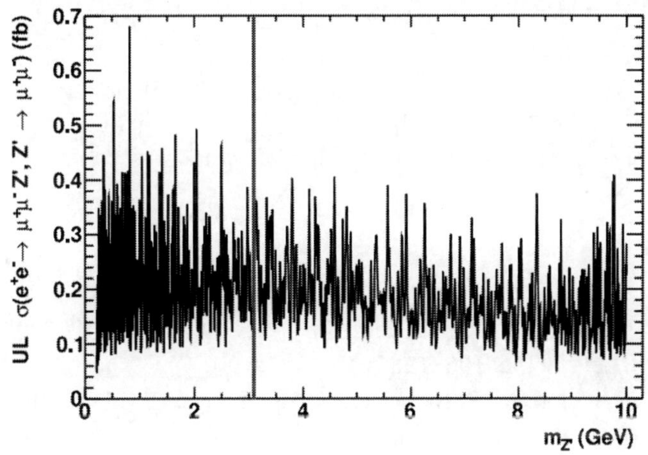

Figure 6. The limit on the cross section $\sigma(e^+e^- \to \mu^+\mu^- Z', Z' \to \mu^+\mu^-)$ as a function of the Z' mass. The excluded region is indicated by the gray band.

4 Conclusion

In conclusion, we have performed the first direct measurement of Z' production from the decay of $e^+e^- \to \mu^+\mu^- Z'$, $Z' \to \mu^+\mu^-$ an e^+e^- collider at BABAR. No significant signal is observed for Z' masses in the range of 0.212 - 10 GeV. We set limits on the coupling parameters g' down to 7×10^{-4}. We set a strongest bounds for many parameter space below 3 GeV. We exclude most of the remaining parameter space preferred by the discrepancy between the calculated and measured anomalous magnetic moment of the muon above the dimuon threshold [16].

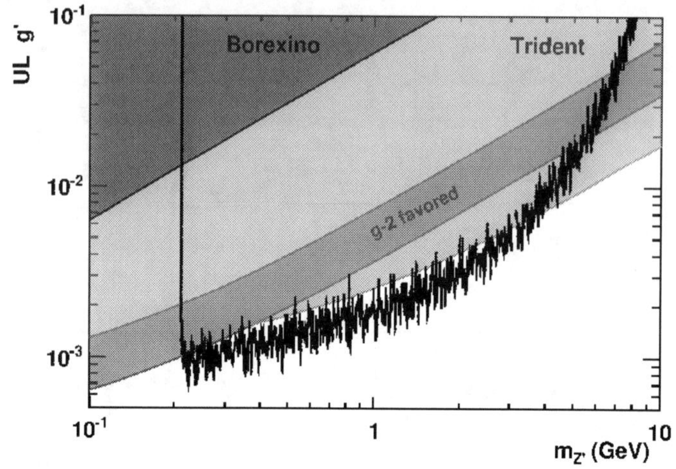

Figure 7. Upper limit on the new gauge coupling g' as a function of the mass of Z' together with the existing limits from Borexino and neutrino experiments.

ACKNOWLEDGMENTS

The author would like to thank B. Echenard for the helpful conversations. The author also thanks the organizers of the XLVI International Symposium on Multiparticle Dynamics, Jeju Island, South Korea. The support from the BABAR Collaboration, the University of South Alabama, and the University of Mississippi is gratefully acknowledged. This work was partially supported by the U.S. Department of Energy.

References

[1] A. Birkedal, K. Matchev, and M. Perelstein, Phys. Rev. D **70**, 077701 (2004)
[2] B. Holdom, Phys. Lett. B **166**, 196 (1986)
[3] W. Altmannshofer, S. Gori, M. Pospelov, and I. Yavin Phys. Rev. D **89**, 095033 (2014)
[4] B. Aubert *et al.,* (BABAR Collaboration), Nucl. Instrum. Meth. A **729**, 1 (2013)
[5] B. Aubert *et al.,* (BABAR Collaboration), Phys. Rev. Lett. **95**, 042001 (2005)
[6] J. Alwall *et al.,* JHEP **1407**, 079 (2014)
[7] T. Sjostrand, S. Mrenna, and P. Z. Skands, JHEP **0605**, 026 (2006)
[8] F. A. Berends, P. H. Daverveld and R. Kleiss, Nucl. Phys. B **253**, 441 (1985)
[9] S. Jadach, W. Placzek, and B. F. L. Ward, Phys. Lett. B **390**, 298 (1997)
[10] S. Jadach, B. F. L. Ward, and Z. Was, Phys. Rev. D **63**, 113009 (2001)
[11] D. J. Lange, Nucl. Instrum. Meth. A **462**, 152 (2001)
[12] A. B. Arbuzov *et al.,* J. High Energy Phys. **9710**, 001 (1997)
[13] M. Caffo, H. Czyz, and E. Remiddi, Nuovo Cim. **A110** 515 (1997); Phys. Lett. **B327**, 369 (1994)
[14] S. Agostinelli *et al.,* (GEANT4 Collaboration), Nucl. Instrum. Meth. A **506**, 250 (2003)
[15] M. Benayoun, S. I. Eidelman, V. N. Ivanchenko and Z. K. Silagadze, Mod. Phys. Lett. A **14**, 2605 (1999)
[16] J. P. Lees *et al.,* (BABAR Collaboration), Phys. Rev. D **94**, 011102 (2016)

Top-Quark p_T-Spectra at LHC and Flavor Independence of z-Scaling

Mikhail Tokarev[1,a], and Imrich Zborovsky[2,b]

[1] *Joint Institute for Nuclear Research, 141980 Dubna, Russia*
[2] *Nuclear Physics Institute, Academy of Sciences of the Czech Republic 250 68 Řež, Czech Republic*

Abstract. New results of analysis of *top*-quark differential cross sections obtained by the CMS and ATLAS Collaborations in pp collisions at the LHC in the framework of z-scaling approach are presented. The spectra are measured over a wide range of collision energy $\sqrt{s} = 7, 8, 13$ TeV and transverse momentum $p_T = 30 - 1000$ GeV/c of *top*-quark using leptonic and jet decay modes. The LHC data on transverse momentum of the *top*-quark spectra are compared with the Tevatron data obtained by the DØ Collaboration in $p\bar{p}$ collisions at $\sqrt{s} = 1.96$ GeV. Flavour independence of the scaling function $\psi(z)$ in pp and $p\bar{p}$ interactions over a wide collision energy range $\sqrt{s} = 19 - 13000$ GeV is verified. This property of $\psi(z)$ was found for different hadrons from π-meson up to Υ particle. The flavour independence of $\psi(z)$ is used as indication on self-similarity of *top*-quark production. A tendency to saturation of $\psi(z)$ at low z for *top*-quark is demonstrated. We anticipate that data on low- and high-p_T inclusive spectra of *top*-quark production at LHC energies could be of interest for verification of the self-similarity over a wide range of masses and different flavour content of produced particles.

1 Introduction

The *top*-quark is the heaviest known elementary particle. It was discovered at the Tevatron $p\bar{p}$ collider in 1995 by the CDF and DØ Collaborations [1, 2] at a mass of around 170 GeV. The first measurements of the differential cross section as a function of the transverse momentum of the *top*-quark were presented by the DØ Collaboration in [3]. It is expected that *top* physics is extremely important to search for new and study of known symmetries in high-p_T region. One of them is self-similarity related to scale transformations in the space of momentum fractions. The symmetry related to the ideas of self-similarity of hadron interactions at a constituent level is manifested by the z-scaling [4, 5]. The scaling was used for analysis of inclusive spectra obtained at the accelerators U70, S$\bar{p}$pS, SPS, ISR, Tevatron and RHIC [6, 7]. The scaling is treated as manifestation of self-similarity of the structure of the colliding objects (hadrons or nuclei), the interaction mechanism of their constituents, and the process of fragmentation into real hadrons. In the framework of this approach, the transverse momentum spectra of inclusive particles are described in therms of the scaling function $\psi(z)$ in dependence

[a] e-mail: tokarev@jinr.ru
[b] e-mail: zborovsky@ujf.cas.cz

on self-similarity parameter z. Both are expressed via experimentally measurable quantities (inclusive cross section, multiplicity density, momenta and masses of the colliding and produced particles). The shape of the scaling function was found to be independent of the collision energy, multiplicity density, detection angle and hadron type including production of particles with heavy flavor content. The power behavior $\psi(z) \sim z^{-\beta}$ was established in the high-z (high-p_T) range. At low z (low p_T), a saturation of the scaling function was found [7, 8]. The z-scaling reflects self-similarity of hadron structure, interaction of hadron constituents and hadronization process. The energy, angular, and multiplicity independence of the scaling function $\psi(z)$ gives strong constraints on the values of the model parameters c, δ, and ϵ entering in the definition of z. The parameter c is interpreted as a "specific heat" of the produced medium associated with the production of an inclusive particle. It controls the behavior of $\psi(z)$ at low z. The structure of the colliding protons at high momenta is characterized by the parameter δ interpreted as a fractal dimension. The fragmentation process is described in terms of the fragmentation dimension ϵ_F. Universality of the shape of the scaling function is given by its flavour independence. It means that spectra of particles with different flavour content can be described by the same function $\psi(z)$ with values of z and ψ rescaled by a scale factor α_F.

In the present paper we analyze the data on transverse momentum spectra of *top*-quark production in pp collisions at the energy $\sqrt{s} = 7, 8$ and 13 TeV in the middle rapidity range obtained by the CMS [9–11] and ATLAS [12, 13] Collaborations at the LHC in the framework of z-scaling approach. The results of analysis of the *top*-quark spectra obtained in pp collisions at the LHC are compared with similar spectra [14] measured by the DØ Collaboration in $\bar{p}p$ collisions at the Tevatron energy $\sqrt{s} = 1.96$ TeV and with the reference curve for π-mesons. We verify the flavor independence of $\psi(z)$ for *top*-quark production in pp collisions in the new kinematic range. A tendency to saturation of $\psi(z)$ for the process at low z is demonstrated. A power-law behavior of $\psi(z)$ at high z is also observed. The measurements of high-p_T spectra of the *top*-quark production at highest LHC energy is of interest for verification of the self-similarity of particle production, understanding flavor origin and search for new physics symmetries with *top*-quark probe.

2 z-Scaling

Here we follow the basic ideas of the z-scaling concept [6, 7]. At sufficiently high energies, a collision of hadrons or nuclei is considered as an ensemble of individual interactions of their constituents (partons, quarks, gluons). Structures of the colliding objects are characterized by parameters δ_1 and δ_2. Interacting constituents carry the fractions x_1, x_2 of the momenta P_1, P_2 of the incoming hadrons (or nuclei). The inclusive particle with mass m carries the fraction y_a of momentum of the produced object in constituent interaction. The object moves in the observed direction and its fragmentation is characterized by parameter ϵ_a. The fragmentation in the recoil direction is described by parameter ϵ_b and the momentum fraction y_b. Multiple interactions of the constituents are considered to be similar. This property reflects self-similarity of hadron interactions at the constituent level.

2.1 Momentum fractions x_1, x_2, y_a, and y_b

The elementary sub-process is considered as a binary collision of the constituents with masses x_1M_1 and x_2M_2 resulting in the scattered and recoil objects with masses m/y_a and $x_1M_1 + x_2M_2 + \overline{m}/y_b$ in the final state. The produced secondary objects transform into real particles after the constituent collisions. The registered particle with mass m and 4-momentum p is produced with its hadron counterpart with mass $\overline{m}$ carrying the momentum fractions y_b of the produced recoil. The momentum conservation law in the constituent sub-process is subject to the condition

$$(x_1P_1 + x_2P_2 - p/y_a)^2 = M_X^2, \tag{1}$$

with the recoil mass $M_X = x_1 M_1 + x_2 M_2 + \overline{m}/y_b$. The production of associated particle with mass $\overline{m}$ ensures conservation of the additive quantum numbers. The conservation law expresses locality of the hadron interactions at a constituent level. It represents a kinematic constraint on the momentum fractions x_1, x_2, y_a, and y_b which determine the underlying elementary sub-process. Structure of the colliding objects and fragmentation of the systems formed in the scattered and recoil directions are characterized by the parameters δ_1, δ_2, and ϵ_a, ϵ_b, respectively. The parameters are connected with the corresponding momentum fractions by the function

$$\Omega = (1 - x_1)^{\delta_1}(1 - x_2)^{\delta_2}(1 - y_a)^{\epsilon_a}(1 - y_b)^{\epsilon_b}. \quad (2)$$

The quantity Ω is proportional to the relative number of all such constituent configurations in the inclusive reaction which contain the configuration defined by the fractions x_1, x_2, y_a, and y_b. The Ω plays the role of a relative volume which occupy these configurations in the space of the momentum fractions. The parameters δ_1, δ_2, and ϵ_a, ϵ_b are interpreted as fractal dimensions in the relevant parts of the space which correspond to the colliding objects and fragmentation processes, respectively. For given values of δ_1, δ_2, and ϵ_a, ϵ_b, the fractions x_1, x_2, y_a, and y_b are determined in a way to maximize the function Ω, simultaneously fulfilling the condition (1). In the case of pp ($p\bar{p}$) interactions we have $M_1 = M_2 = m_N$ and set $\delta_1 = \delta_2 \equiv \delta$. We assume that the fragmentation of the objects moving in the scattered and recoil directions can be described by the same parameter $\epsilon_a = \epsilon_b \equiv \epsilon_F$ which depends on the type of the inclusive particle. The relation $m = \overline{m}$ is used for each particle species. The parameters δ, and ϵ_F are determined in accordance with self-similarity requirements and experimental data. They were found to have constant values in pp and $p\bar{p}$ collisions at high energies.

2.2 Scaling function $\psi(z)$ and self-similarity parameter z

The self-similarity of hadron interactions reflects the property that hadron constituents and their interactions are similar. The self-similarity variable z is defined as follows

$$z = z_0 \Omega^{-1}, \quad (3)$$

where $z_0 = \sqrt{s_\perp}/(dN_{ch}/d\eta|_0)^c m_N$ and Ω is maximal value of (2) with the condition (1). For a given inclusive reaction, the quantity z is proportional to the transverse kinetic energy $\sqrt{s_\perp}$ of the constituent sub-process consumed on the production of the inclusive particle and its counterpart with masses m and $\overline{m}$, respectively. The quantity $dN_{ch}/d\eta|_0$ is the corresponding multiplicity density of charged particles in the central interaction region at the pseudo-rapidity $\eta = 0$. The parameter c characterizes properties of the produced medium. It is interpreted as a "specific heat" of the medium in the state with the respective value of $dN_{ch}/d\eta|_0$. The constant m_N is the nucleon mass.

The scaling function $\psi(z)$ is expressed in terms of the experimentally measured inclusive cross section $Ed^3\sigma/dp^3$, multiplicity density $dN/d\eta$, and the total inelastic cross section σ_{in} as follows [6]

$$\psi(z) = \frac{\pi}{(dN/d\eta)\,\sigma_{\text{in}}} J^{-1} E\frac{d^3\sigma}{dp^3}, \quad (4)$$

where s is the square of the center-of-mass energy and J is the Jacobian for the transformation from $\{p_T^2, y\}$ to $\{z, \eta\}$. The multiplicity density $dN/d\eta$ in (4) depends on the center-of-mass energy, centrality, and on the production angles at which the inclusive spectra were measured. The scaling function is normalized as follows

$$\int_0^\infty \psi(z)dz = 1. \quad (5)$$

It allows us to interpret $\psi(z)$ as a probability density of the production of an inclusive particle with the corresponding value of the variable z.

DOI: 10.1051/epjconf/201714102006

3 Flavor independence of $\psi(z)$

Universality of z-scaling is given by its flavour independence. It means that spectra of particles with the different flavour content can be described by an universal scaling function $\psi(z)$ in z-presentation [6, 7]. Our previous analysis was based on the observation that simultaneous energy, angular, and multiplicity independence of the z-scaling for negative pions, kaons, and anti-protons produced in proton-proton collisions gives the same shape of the scaling function. We exploit the scaling transformation

$$z \to \alpha_F z, \quad \psi \to \alpha_F^{-1}\psi \tag{6}$$

for comparison of the shape of the scaling function $\psi(z)$ for different hadron species. The parameter α_F is a scale dependent quantity. The transformation does not change the shape of $\psi(z)$. It preserves the normalization equation (5) and does not destroy the energy, angular, and multiplicity independence of the z-presentation of particle spectra.

Figure 1(a) shows spectra of negative pions, kaons, antiprotons, and Λ's produced in pp collisions over the range $\sqrt{s} = 19 - 200$ GeV and $\theta_{cms} = 3^0 - 90^0$ in z-presentation. The symbols correspond to the data on differential cross sections measured in the central and fragmentation regions. The solid lines represent the same curve shifted by multiplicative factors for reasons of clarity. One can see that the distributions of different hadrons are sufficiently well described by a single curve over a wide z-range. The function $\psi(z)$ changes more than ten orders of magnitude. The indicated values of the parameter ϵ_F are consistent with the energy, angular, and multiplicity independence of z-presentation of particle spectra. The scale factors α_F are constants which allow us to describe the scaling function for different hadron species by a single curve.

Figure 1(b) shows z-presentation of the transverse momentum spectra of strange mesons and baryons measured by the STAR and PHENIX Collaborations in pp collisions at $\sqrt{s} = 200$ GeV in the central rapidity region at the RHIC. The symbols representing data on differential cross sections include baryons which consist of one, two and three strange valence quarks. The multiplicative factors $10^0, 10^{-1}$ and 10^{-2} are used to show the data z-presentation separately for mesons, single-strange $(\Lambda, \Lambda^*, \Sigma^*)$ and multi-strange (Ξ^-, Ω) baryons, respectively. The pion distributions obtained

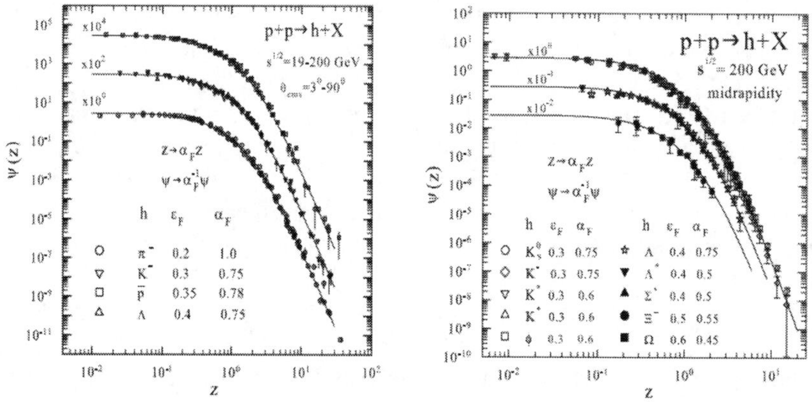

a) b)

Figure 1. Inclusive spectra of hadrons produced in pp collisions in the z-presentation. The symbols denote the experimental data obtained in the experiments performed at CERN, FNAL, and BNL. The plots are taken from Ref. [7, 15]. The solid line is a reference curve corresponding to π^--meson production in pp collisions.

 DOI: 10.1051/epjconf/201714102006

by the STAR Collaboration are used for comparison as reference points. The shape of $\psi(z)$ for all these particles is described by the same curve (solid line) which is depicted in Fig. 1(a). Based on the obtained results we conclude that the RHIC data confirm the flavor independence of the z-scaling in pp collisions including particles with very small p_T. Next we show that the flavor independence of z-presentation of hadron spectra is valid for the *top*-quark production as well.

4 Self-similarity of *top*-quark production

The first analysis of *top*-quark spectra [3] in the framework of the z-scaling approach was performed in [16]. In the present paper we analyze spectra of *top*-quark production obtained at the LHC and compare it with data at lower energy from the Tevatron. The differential cross sections were measured in dependence on the *top*-quark transverse momentum p_T at middle rapidity.

Figure 2 shows the z-presentation of the spectra of *top*-quark production obtained in pp collisions at the LHC energies $\sqrt{s} = 7, 8$ and 13 TeV in the central rapidity region. The data were obtained by the CMS [9–11] and ATLAS [12, 13] Collaborations. The measurements of the inclusive production cross section were performed in the dilepton and jet channels. The data include measurements over a wide range of the transverse momentum $30 < p_T < 1000$ GeV. The scaling function $\psi(z)$ for the *top*-quark distributions was calculated according to Eq. (4). Data on π^--meson spectra shown by the solid line serves as a reference data. The values of the fractal dimension $\delta = 0.5$ and the parameter $c = 0.25$ are the same as used in our previous analyses [6, 7, 17]. We have set $\epsilon_{top} = 0$ in the case of the *top*-quark, as no or negligible energy loss is assumed in the elementary $t\bar{t}$ production process. This choice corresponds to $y_a = y_b = 1$ in the whole p_T range. The scale parameter α_F in the transformation (6) is found to be $\alpha_{top} \simeq 0.0045$. No additional parameters were used.

The new data [14] on spectra of the *top*-quark production in $p\bar{p}$ collisions obtained by the DØ Collaboration at the Tevatron energy $\sqrt{s} = 1.96$ GeV are shown for comparison with the data measured at the LHC. As seen from Fig. 2, the scaling function demonstrates the energy independence

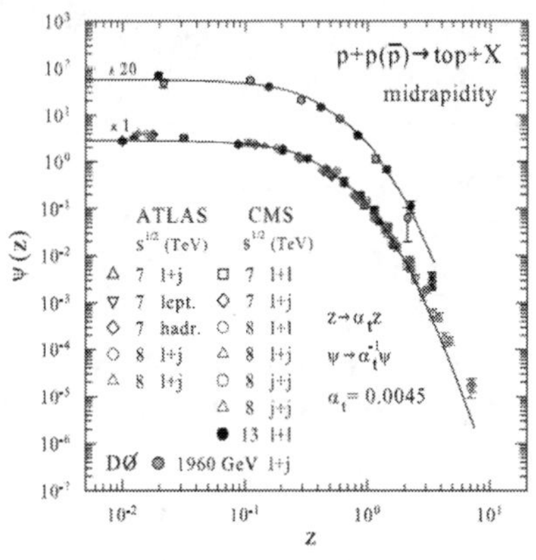

Figure 2. The scaling function $\psi(z)$ of the *top*-quark production in pp and $p\bar{p}$ collisions at the LHC energies $\sqrt{s} = 7, 8, 13$ TeV and the Tevatron energy $\sqrt{s} = 1.96$ TeV. The symbols denote the experimental data obtained by the CMS, ATLAS and DØ Collaborations are taken from [9–11],[12, 13] and [14], respectively. The solid line is a reference curve corresponding to π^--meson production in pp collisions.

over a wide range of the self-similarity parameter $z = 0.001 - 8$. The function $\psi(z)$ changes more than five orders of magnitude in this region.

As seen from Fig. 2, the z-presentation of the *top*-quark transverse momentum distribution follows the shape of the z-scaling in pp ($p\bar{p}$) collisions for other particles sufficiently well. Though the *top* spectrum is in the limited kinematic region, we would like to stress that it is compared with existing analyses of p_T-distributions of inclusive cross sections $Ed^3\sigma/dp^3$ for other hadrons which reveal strong dependence on the energy, angle, multiplicity, and type of the produced particles. Based on the above comparison we conclude that the new LHC and Tevatron data on inclusive spectra of the *top*-quark production support the flavor independence of the scaling function $\psi(z)$ over the interval of $z = 0.001 - 8$. This result gives us indication on self-similarity of *top*-quark production in pp and $p\bar{p}$ interactions up to the *top*-quark transverse momentum $p_T < 1000$ GeV and for a wide range of the collision energies $\sqrt{s}$.

Acknowledgments

The investigations have been supported by the RVO61389005 institutional support and by the grant LG 15052 of the Ministry of Education of the Czech Republic.

References

[1] F.Abe et al. (CDF Collaboration), Phys. Rev. Lett. **74**, 2626 (1995).
[2] S.Abachi et al. (D∅ Collaboration), Phys. Rev. Lett. **74**, 2632 (1995).
[3] V.M.Abazov et al. (D∅ Collaboration), Phys. Lett. **B693**, 515 (2010).
[4] I.Zborovský, Yu.Panebratsev, M.Tokarev, and G.Škoro, Phys. Rev. D **D54**, 5548 (1996).
[5] M.Tokarev, I.Zborovský, Yu.Panebratsev, and G.Škoro, Int. J. Mod. Phys. **A16**, 1281 (2001).
[6] I. Zborovský and M.V. Tokarev, Phys. Rev. **D75**, 094008 (2007).
[7] I. Zborovský and M.V. Tokarev, Int. J. Mod. Phys. **A24**, 1417 (2009).
[8] M.V.Tokarev, I.Zborovský, Nucl. Phys. B (Proc. Suppl.) **219-220**, 301 (2011).
[9] S. Chatrchyan et al. (CMS Collaboration), Eur. Phys. J. **C73**, 2339 (2013)
[10] V. Khachatryan et al. (CMS Collaboration), Eur. Phys. J. **C75**, 542 (2015);
V. Khachatryan et al. (CMS Collaboration), Eur. Phys. J. **C76**, 128 (2016);
V. Khachatryan et al. (CMS Collaboration), arXiv:1505.04480 (2015);
V. Khachatryan et al. (CMS Collaboration), arXiv:1605.00116 (2016).
[11] CMS Collaboration, CMS PAS TOP-16-011.
[12] G. Aad et al. (ATLAS Collaboration), Phys. Rev. **D90**, 072004 (2014);
G. Aad et al. (ATLAS Collaboration), J. High Energ. Phys. **06**, 100 (2015).
[13] G. Aad et al. (ATLAS Collaboration), Phys. Rev. **D93**, 032009 (2016);
G. Aad et al. (ATLAS Collaboration), arXiv:1511.04716 (2015).
[14] V.M. Abazov et al. (DØ Collaboration), Phys. Rev. **D90**, 092006 (2014).
[15] M.V. Tokarev and I. Zborovský, "Search for new symmetries:Self-similarity of strangeness production in pp collisions", In: Proc. of Int. Conf. "Hadron Structure and QCD" (HSQCD'2016), Gatchina, Russia, June 27 - July 1, 2016.
[16] M.V. Tokarev and I. Zborovský, J. Mod. Phys. **3**, 815 (2012).
[17] M.V.Tokarev, I.Zborovský, J. Phys. G: Nucl. Part. Phys. **37**, 085008 (2010).

EPJ Web of Conferences **141**, 02007 (2017) DOI: 10.1051/epjconf/201714102007
ISMD 2016

Thermal photons from gluon fusion with magnetic fields

Alejandro Ayala[1,2,a], Jorge David Castaño-Yepes[1,b], C. A. Dominguez[2,c], and L. A. Hernández[2,d]

[1] *Instituto de Ciencias Nucleares, Universidad Nacional Autónoma de México, Apartado Postal 70-543, México Distrito Federal 04510, Mexico.*
[2] *Centre for Theoretical and Mathematical Physics, and Department of Physics, University of Cape Town, Rondebosch 7700, South Africa*

Abstract. We compute the production of thermal photons in relativistic heavy-ion collisions by gluon fusion in the presence of an intense magnetic field, and during the early stages of the reaction. This photon yield is an excess over calculations that do not consider magnetic field effects. We add this excess to recent hydrodynamic calculations that are close to describing the experimental transverse momentum distribution in RHIC and LHC. We then show that with reasonable values for the temperature, magnetic field strength, and strong coupling constant, our results provide a very good description of such excess. These results support the idea that the origin of at least some of the photon excess observed in heavy-ion experiments may arise from magnetic field induced processes.

The results from heavy-ion experiments carried out at the BNL Relativistic Heavy-Ion Collider (RHIC) and at the CERN Large Hadron Collider (LHC), show that a state of matter is formed where quarks and gluons are not confined to individual nucleons [1, 2]. Non-central collisions produce magnetic fields with an intensity that at the beginning of the reaction is estimated to be as high as several times the mass of the pion squared [3–6], though their intensity fades out fast with time. At early times, it is also when the largest temperatures are achieved and when the soft dynamics is dominated by gluons.

A magnetic field makes it possible to produce photons from processes otherwise not allowed. Recent calculations [7–10] invoking the presence of these magnetic fields have been implemented to try to explain the experimentally measured excess [11] of thermal photons over models that describe well other low momentum observables.

A magnetic field naturally produces an asymmetry in the emission of electromagnetic radiation. Therefore, magnetic fields can be a source of not only an excess in the photon yield, but also of the puzzling large strength of the coefficient v_2 in the Fourier expansion of the azimuthal distribution. The latter has been found to be as large as that of pions [12]. Although some recently improved hydrodynamic [13, 14] and transport [15] calculations obtain a better agreement with ALICE and PHENIX measurements of low transverse momentum photons, this agreement is not yet complete [16]. Therefore, it remains important to quantify the fraction of the yield, and of the asymmetry arising from magnetic field effects, to better characterize the initial stages of heavy-ion reactions.

[a] e-mail: ayala@nucleares.unam.mx
[b] e-mail: jorge.castano@correo.nucleares.unam.mx
[c] e-mail: cesareo.dominguez@uct.ac.za
[d] e-mail: HRNLUI001@myuct.ac.za

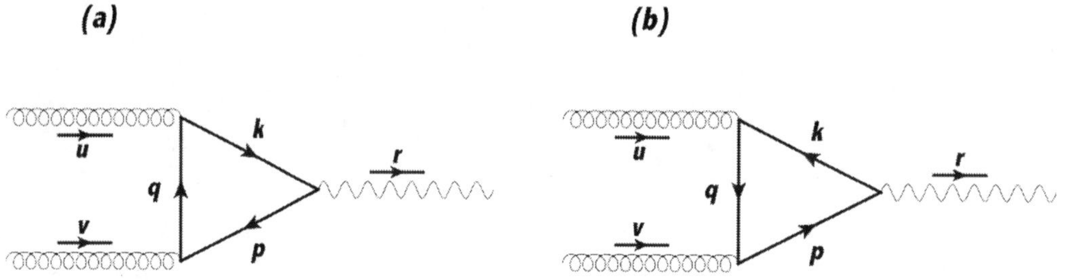

Figure 1. Feynman diagrams for the amplitude for photon production from gluon fusion. The thick lines in the loop represent the quark propagators in the presence of the magnetic field.

It is then natural to explore a mechanism where collisions of these gluons induce the emission of photons. In this work we outline the computation of the production of thermal photons from the perturbative fusion of gluons early in the collision. More details are provided in Ref. [17].

The amplitude for the process is depicted by the Feynman diagrams in Fig. 1, which also defines the kinematical variables. The thick loop lines represent the quark propagator in the presence of the magnetic field. In the absence of this field, the diagrams cancel each other. It is the presence of the field which makes it possible that both diagrams contribute with the same relative sign.

The fermion propagator in coordinate space cannot longer be written as a simple Fourier transform of a momentum propagator but instead it is written as [18]

$$S(x,x') = \Phi(x,x') \int \frac{d^4p}{(2\pi)^4} e^{-ip\cdot(x-x')} S(p)\,, \tag{1}$$

where

$$\Phi(x,x') = \exp\left\{ iq_f \int_{x'}^{x} d\xi^\mu \left[A_\mu + \frac{1}{2} F_{\mu\nu}(\xi - x')^\nu \right] \right\}, \tag{2}$$

is called the *phase factor*, and q_f is the absolute value of the quark charge. We consider the contribution of two light flavors, thus $q_u = 2|e|/3$ and $q_d = |e|/3$. The propagator in momentum-space, $S(p)$, is given by

$$\begin{aligned} iS(p) &= \int_0^\infty \frac{ds}{\cos(q_f Bs)} e^{is\left(p_\parallel^2 - p_\perp^2 \frac{\tan(q_f Bs)}{q_f Bs} - m_f^2 + i\epsilon\right)} \\ &\times \left[\left(\cos(q_f Bs) + \gamma_1\gamma_2 \sin(q_f Bs) \right) (m_f + \not{p}_\parallel) - \frac{\not{p}_\perp}{\cos(q_f Bs)} \right], \end{aligned} \tag{3}$$

where m_f is the quark mass. We have chosen the homogeneous magnetic field to point in the $\hat{z}$ direction, namely $\boldsymbol{B} = B\hat{z}$. This configuration can be obtained from an external vector potential which we choose in the so called *symmetric gauge* $A^\mu = \frac{B}{2}(0, -y, x, 0)$. We have also defined $p_\perp^\mu \equiv (0, p_1, p_2, 0)$, $p_\parallel^\mu \equiv (p_0, 0, 0, p_3)$, $p_\perp^2 \equiv p_1^2 + p_2^2$ and $p_\parallel^2 \equiv p_0^2 - p_3^2$.

Since the two Feynman diagrams of Fig. 1 give the same contribution, we concentrate on the computation of the amplitude depicted in Fig. 1a which becomes

$$\begin{aligned} \mathcal{M}^{(a)} &= -\int d^4x \int d^4y \int d^4z \int \frac{d^4p}{(2\pi)^4} \int \frac{d^4q}{(2\pi)^4} \int \frac{d^4k}{(2\pi)^4} e^{-ip\cdot(y-x)} e^{-iq\cdot(z-y)} e^{-ik\cdot(x-z)} e^{-iu\cdot z} e^{-iv\cdot y} e^{ir\cdot x} \\ &\times \mathrm{Tr}\left[iq_f\gamma_\mu iS(k) ig\gamma_\alpha t^c iS(q) ig\gamma_\nu t^d iS(p) \right] \epsilon^{*\mu}(\lambda_r)\Phi(x,y)\Phi(y,z)\Phi(z,x)\epsilon^\alpha(\lambda_u)\epsilon^\nu(\lambda_v). \end{aligned} \tag{4}$$

The product of phase factors can be written as

$$\Phi(x,y)\Phi(y,z)\Phi(z,x) = e^{i\frac{q_f B}{2}\epsilon_{ij}(z-x)_i(x-y)_j}, \tag{5}$$

where we used the explicit form of A^μ which gives $F_{12} = -F_{21} = -B$, with the rest of the components of $F_{\mu\nu}$ vanishing, and ϵ_{ij} being the Levi-Civita symbol.

We use the fact that when the magnetic field is very intense, as compared to the other energy (squared) scales involved, the quark dynamics is dominated by the lowest Landau level (LLL). For the case of quarks that have not yet thermalized, this means that the magnetic field is taken to satisfy $eB \gg m_f^2$. For the LLL, the propagator in Eq. (3) can explicitly be written as

$$iS(p) = 2ie^{-\frac{p_\perp^2}{q_f B}} \frac{(\not{p}_\parallel + m_f)}{p_\parallel^2 - m_f^2} O_\parallel^\pm. \tag{6}$$

The operator $O_\parallel^\pm = [1 \pm i\gamma_1\gamma_2 sign(eB)]/2$ projects onto the longitudinal space. The positive (negative) sign corresponds to the case when the product eB is positive (negative) and is to be used when the loop is made out of positively (negatively) charged quarks. Therefore the matrix element can be factorized into a product of transverse and longitudinal pieces, namely

$$\mathcal{M}^{(a)} = (2\pi)^4\delta^4(r-v-u)\mathcal{M}_\perp^{(a)}\mathcal{M}_\parallel^{(a)} \tag{7}$$

$$\begin{aligned}\mathcal{M}_\perp^{(a)} &= \left(\frac{4\pi}{q_f B}\right)^2 \int \frac{d^2p_\perp}{(2\pi)^2} \int \frac{d^2q_\perp}{(2\pi)^2} \int \frac{d^2k_\perp}{(2\pi)^2} e^{-\frac{k_\perp^2}{q_f B}} e^{-\frac{q_\perp^2}{q_f B}} e^{-\frac{p_\perp^2}{q_f B}} \prod_{i,j=1,2} e^{i\frac{2}{q_f B}\epsilon_{ij}(q-k+u)_i(q-p-v)_j} \\ &= \left(\frac{q_f B}{12\pi}\right) e^{-\frac{(u+v)_\perp^2}{3q_f B}},\end{aligned} \tag{8}$$

$$\begin{aligned}\mathcal{M}_\parallel^{(a)} &= -8\left(\frac{q_f g^2 \delta^{cd}}{2}\right) \int \frac{d^2p_\parallel}{(2\pi)^2} \int \frac{d^2q_\parallel}{(2\pi)^2} \int \frac{d^2k_\parallel}{(2\pi)^2} (2\pi)^4 \delta^2\left[(q-k+u)_\parallel\right] \delta^2\left[(q-p-v)_\parallel\right] \\ &\times \frac{\mathrm{Tr}\left[\gamma_\mu \not{k}_\parallel O_\parallel \gamma_\alpha \not{q}_\parallel O_\parallel \gamma_\nu \not{p}_\parallel O_\parallel\right]}{k_\parallel^2 q_\parallel^2 p_\parallel^2} \epsilon^\alpha(\lambda_u)\epsilon^\nu(\lambda_v)\epsilon^{*\mu}(\lambda_r).\end{aligned} \tag{9}$$

Since at the early stages of the collision gluons are far more abundant than quarks, we compute Eq. (9) under the assumption that quarks do not yet thermalize. Accordingly, we set $m_f = 0$ since in the absence of thermal corrections, the light-quark vacuum masses are negligible. The trace in Eq. (9) contains the product of up to twelve gamma matrices. The resulting expression is long and involved. It is however easy to show that upon squaring and summing over polarizations, only a small piece survives so that the trace can be expressed as

$$\mathrm{Tr}\left[\gamma_\mu \not{k}_\parallel O_\parallel^\pm \gamma_\alpha \not{q}_\parallel O_\parallel^\pm \gamma_\nu \not{p}_\parallel O_\parallel^\pm\right] \to k_{\parallel\nu}(p_{\parallel\mu}q_{\parallel\alpha} - p_{\parallel\alpha}q_{\parallel\mu}) + k_{\parallel\mu}(p_{\parallel\nu}q_{\parallel\alpha} + p_{\parallel\alpha}q_{\parallel\nu}) + k_{\parallel\alpha}(p_{\parallel\nu}q_{\parallel\mu} - p_{\parallel\mu}q_{\parallel\nu}), \tag{10}$$

where the arrow indicates this to be the only contributing portion. Two of the integrations in Eq. (9) become straightforward using the delta-function restrictions. We choose those two as the integrals over $k_\parallel$ and $q_\parallel$.

In order to make tractable the calculation of the longitudinal piece of the photon emission rate, Eq. (9), use is made of some simplifying assumptions, as is described in Ref. [17].

The final expression for the matrix element [Eqs. (7), (8) and (9)] squared, after adding the contribution from the Feynman diagram in Fig. 1b, summed over polarizations, becomes

$$\sum_{\mathrm{pol},f} |\mathcal{M}|^2 = \left(\frac{6256}{2187}\right)\left(\frac{\mathcal{V}\mathcal{T}}{m_g^2}\right)\delta^4(r-v-u)\sum_f \left(q_f g^2\right)^2 \left(q_f B\right)^2 e^{-2\frac{r_\perp^2}{3q_f B}}, \tag{11}$$

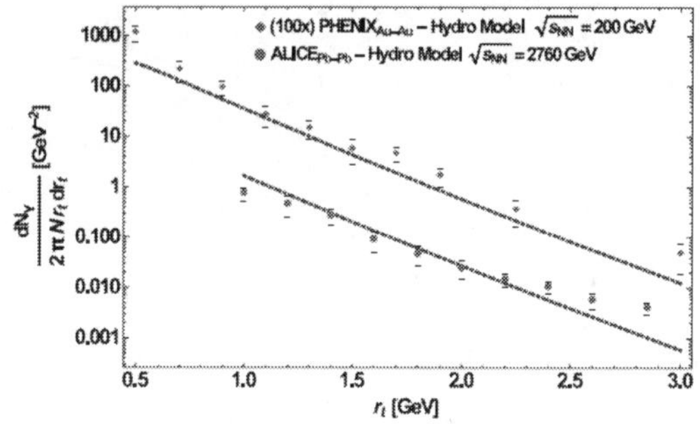

Figure 2. Experimental excess photon yield with respect to the hydro calculation of Ref. [13] compared to our calculation for the centrality class $0-20\%$. The upper set corresponds to PHENIX data (multiplied by 100) and the lower set to ALICE data.

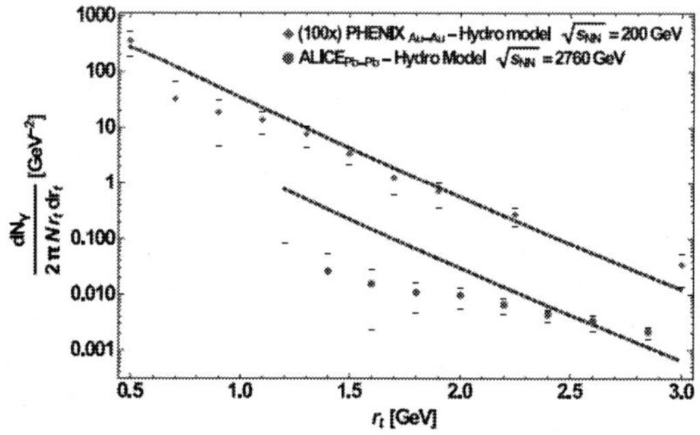

Figure 3. Experimental excess photon yield with respect to the hydro calculation of Ref. [13] compared to our calculation for the centrality class $20-40\%$. The upper set corresponds to PHENIX data (multiplied by 100) and the lower set to ALICE data.

where $\mathcal{VT}$ represents the space-time volume of the reaction coming from squaring $(2\pi)^4\delta(r-v-u)$ and we have included the sum over the two light flavors $f = u, d$. The odd-looking factor $6256/2187 \sim 2.86$ is obtained from the longitudinal piece of the matrix element squared after collecting the coefficients of the contraction of the polynomial in the components of $u_\parallel$ and $v_\parallel$. Notice that after the approximations made to compute Eq. (9), the dependence on $r_\perp$ of Eq. (11) comes exclusively from the transverse piece of the matrix element, Eq. (8).

The invariant photon yield is obtained by integrating over the corresponding phase space weighed with the thermal distribution, namely

$$\frac{r_0 dN}{d^3r} = \frac{1}{2(2\pi)^3}\int \frac{d^3u}{2u_0(2\pi)^3}\int \frac{d^3v}{2v_0(2\pi)^3}\sum_{\text{pol},f}|\mathcal{M}|^2\, n(u_0)n(v_0), \tag{12}$$

where $n(E) = 1/[\exp\sqrt{E^2+m_g^2}/T - 1]$ is the Bose-Einstein distribution and T is the temperature.

The number of photons per unit momentum transverse to the beam axis, integrated over the full azimuthal angle, and around mid-rapidity, is explicitly given by

$$\frac{dN}{dr_t} = C\left[\left(\frac{1}{3}\right)^4 e^{-2\frac{v_0^2 r_t^2}{eB}} + \left(\frac{2}{3}\right)^4 e^{-\frac{v_0^2 r_t^2}{eB}}\right] I\left(\frac{r_t}{T};\sqrt{\frac{2}{3}g}\right), \tag{13}$$

where we have evaluated the distribution at $y = y_0 = 0.5$, given that the rapidity interval $\Delta y \sim 1$ is centered around mid-rapidity. We have also defined

$$C = \frac{g^4 T e^4 B^2}{8(2\pi)^7}\left(\frac{6256}{2187}\right)\left(\frac{\mathcal{V}\mathcal{T}}{m_g^2}\right), \tag{14}$$

and

$$I(z;\lambda) \equiv \int_0^z \frac{dx x^2 n\left(\sqrt{(z+x)^2-(2x)^2}\right)n(x)}{\left(\sqrt{x^2+\lambda^2}\right)\left(\sqrt{(z+x)^2-(2x)^2+\lambda^2}\right)}. \tag{15}$$

We have also used that in QCD with two flavors, $m_g^2 = (2/3)g^2T^2$. The yield given by Eq. (13) can be properly called *thermal* because its computation assumes that the gluons are thermally distributed in phase space.

In order to be able to compare to ether data or models that do not consider photon emission induced by magnetic fields, we compute the yield normalized to the number of events. This is given by

$$\frac{1}{2\pi N r_t}\frac{dN}{dr_t} = \frac{\left[\left(\frac{1}{3}\right)^4 e^{-2\frac{y_0^2 r_t^2}{eB}} + \left(\frac{2}{3}\right)^4 e^{-\frac{y_0^2 r_t^2}{eB}}\right]\frac{I(r_t/T)}{2\pi}}{\sqrt{3\pi eB/2}\left[\left(\frac{2}{3}\right)^{9/2} + \left(\frac{1}{3}\right)^{9/2}\right]\int_0^\infty dr_t I(r_t/T)}, \tag{16}$$

where the factor $\sqrt{3\pi eB/2}$ in the denominator comes from the integration over rapidity. Notice that, despite having approximated $r_\perp \simeq 0$ for the calculation of the longitudinal matrix element, the normalization N is computed extending the integration range over r_t up to $+\infty$. This is a valid approximation, given that the main behavior of the rate is dictated by the exponential fall off in this variable coming from the transverse part of the matrix element squared. The normalized distribution is independent of the space-time region of the reaction. The impact parameter dependence is due to the dependence on the field intensity. The photon transverse yield given by Eq. (16) is an *excess yield* that should be added to calculations that do not consider magnetic field effects for photon emission. In order to compare with experimental data we first proceed to use appropriate values for the temperature, the coupling g, and the magnetic field strength. We take $g = 1$ (the results turn out to be only marginally sensitive to the value of g). The variation of the field intensity with time and impact parameter for RHIC and LHC energies is taken from Ref. [19]. We chose one of the largest values of eB which for RHIC, $\sqrt{s_{NN}} = 200$ GeV, correspond to $0.5 \times 10^4 < eB/(\text{MeV})^2 < 10^5$ and for LHC, $\sqrt{s_{NN}} = 2.76$ TeV, $eB/(\text{MeV})^2 \simeq 10^4$, with small variations coming from a slight dependence on the impact parameter. To relate the centrality class of the collision with the impact parameter we follow the geometrical model of Ref. [20]. Figures 2 and 3 show our results compared to the experimental excess photon yield with respect to one recent hydrodynamical calculation [13]. The latter has been shown to approach the description of the experimental data within the lowest part of the uncertainties. Figure 2 (Fig. 3) shows a comparison with the centrality class case 0 – 20% (20 – 40%). In each graph the upper set corresponds to PHENIX (multiplied by 100) and the lower to ALICE data. Notice that even with the ballpark choices of the parameters involved, our calculation provides a very good description of the excess photons. For the case of ALICE 20 – 40% our calculation overshoots the data. This can be due to the fact that the hydrodynamic calculation [13] which we use as the reference to compute the excess also overshoots the data, for this centrality class at low p_t.

In conclusion we have computed the matrix element squared for photon production from th gluon fusion process in the presence of a magnetic field in the context of heavy-ion collisions. Assuming that at very early times, the gluon population is thermal, we have used this matrix element to compute the thermal photon spectrum. For this purpose we have used a simple scenario using conservative values for the temperature and strength of the magnetic field. We have compared this photon rate with experimental data subtracting an state of the art calculation of the photon yield. The excess photon yield coming from the magnetic field effect shows a good agreement with the above. A more detailed analysis including up to date estimates of the the evolution of the magnetic field [21], with a less stringent set of assumptions and a quantification of the magnetic field induces asymmetry, is being carried out and will be reported elsewhere.

The authors acknowledge useful conversations with M. E. Tejeda-Yeomans. Support for this work has been received in part from PAPIIT-UNAM under grant number IN101515, by CONACyT under grant number 256494 and by the National Research Foundation (South Africa), and the Harry Oppenheimer Memorial Trust OMT Ref. 20242/02.

References

[1] I. Arsene *et al.* (BRAHMS Collaboration), Nucl. Phys. A757, 1 (2005); B. B. Back *et al.*, Nucl. Phys. A **757**, 28 (2005); J. Adams *et al.* (STAR Collaboration), Nucl. Phys. **A757**, 102 (2005); K. Adcox *et al.* (PHENIX Collaboration), Nucl. Phys. **A757**, 184 (2005); P. Jacobs and X. N. Wang, Prog. Part. Nucl. Phys. **54**, 443 (2005).

[2] F. Becattini, J. Phys. Conf. Ser. **527**, 012012 (2014) and references therein.

[3] D. E. Kharzeev, L. D. McLerran and H. J. Warringa, Nucl. Phys. A **803**, 227 (2008).

[4] V. Skokov, A.Y. Illarionov, V. Toneev, Int. J. Mod. Phys. A **24**, 5925 (2009).

[5] V. Voronyuk, V. D. Toneev, W. Cassing, E. L. Bratkovskaya, V. P. Konchakovski, S. A. Voloshin, Phys. Rev. C **83**, 054911 (2011).

[6] A. Bzdak, V. Skokov, Phys. Lett. B **710**, 171-174 (2012).

[7] G. Basar, D. E. Kharzeev, V. Skokov, Phys. Rev. Lett. **109**, 202303 (2012).

[8] G. Basar, D. E. Kharzeev, E. V. Shuryak, Phys. Rev. C **90**, 014905 (2014).

[9] K. Tuchin, Phys. Rev. C **91**, 014902 (2015).

[10] G. Arciniega, P. Ortega, L. Patiño, J. High Energy Phys. **1404**, 192 (2014); S. Y. Wu and D. L. Yang, J. High Energy Phys. **1308**, 032 (2013) K. A. Mamo, J. High Energy Phys. **1308**, 083 (2013).

[11] A. Adare *et al.* [PHENIX Collaboration], Phys. Rev. C **91**, 064904 (2015); J. Adam *et al.* [ALICE Collaboration], Phys. Lett. B **754**, 235 (2016).

[12] A. Adare *et al.* [PHENIX Collaboration], e-Print: arXiv:1509.07758 [nucl-ex]; D. Lohner, *et al.* (ALICE Collaboration), e-Print: arXiv:1212.3995 [nucl-ex].

[13] J.-F. Paquet, C. Shen, G. S. Denicol, M. Luzum, B. Schenke, S. Jeon, C. Gale, e-Print:arXiv:1509.06738 [hep-ph].

[14] H. van Hees, M. He, R. Rapp, Nucl. Phys. A **933**, 256-271 (2015).

[15] O. Linnyk, V. Konchakovski, T. Steinert, W. Cassing, E. L. Bratkovskaya, Phys. Rev. C **92**, 054914 (2015).

[16] For a recent review see C. Shen, e-Print: arXiv:1601.02563 [nucl-th].

[17] A. Ayala. J. D. Castaño-Yepes, C. A. Dominguez, L. A. Hernandez, arXiv:1604.02713 [hep-ph].

[18] J. Schwinger, Phys. Rev. **82**, 664 (1951).

[19] Y. Zhong, C.-B. Yang, X. Cai, S.-Q. Feng, Adv. High Energy Phys. **2014**, 193039 (2014).

[20] W. Broniowski, W. Florkowski, Phys. Rev. **C** 65, 024905 (2002).

[21] L. McLerran and V. Skokov, Nucl. Phys. A **929**, 184-190 (2014); V. Voronyuk, V. D. Toneev, W. Cassing, E. L. Bratkovskaya, V. P. Konchakovski, and S. A. Voloshin, Phys. Rev. C **83**, 054911.

EPJ Web of Conferences **141**, 03001 (2017) DOI: 10.1051/epjconf/201714103001
ISMD 2016

Recent results on forward physics and jets at LHC

Grzegorz Brona[1,a] for the ATLAS and CMS Collaborations

[1] *University of Warsaw, Faculty of Physics, Warsaw, Poland*

Abstract. An overview of the recent results on forward physics from the LHC Run2 is presented. Both CMS and ATLAS results obtained in proton-proton collisions at $\sqrt{s}$ = 13 TeV are shown. Forward energy flow measured at high pseudorapidity, inclusive jets, and jets reconstructed in CMS-CASTOR detector are discussed. In addition, results on the pseudorapidity spectra of charged particles are presented.

1 Forward energy flow

Forward energy flow is an observable that provides an independent measure of the underlying event activity (UE). It can be used to tune the models that simulate this component of an interaction. A good description of the UE is a prerequisity to study hard processes, which lay on top of the UE activity, and to look for new physics phenomena. Moreover it is an important background for precise measurements in QCD and EW domains. The measurement of the UE in new kinematical domains – at high energies and high pseudorapidities, gives also an insight into QCD dynamics at low scales. Finally, results on the forward energy flow can be used as input to the cosmic ray physics modelling.

At CMS [1] two measurements of the forward energy flow at high rapidities at $\sqrt{s}$ =13 TeV are released. The first is performed with the CMS Hadronic Forward (HF) calorimeter within $3.15 < |\eta| < 5.20$, and the CASTOR calorimeter covering $-6.6 < |\eta| - 5.2$ [2], while the second only with CASTOR [3]. Both are done with low luminosity runs.

The analysis with HF and CASTOR is focused on two samples of events: an HF-OR sample defined with a requirement of at least one HF calorimeter tower above 5 GeV at at least one side of CMS, and an HF-AND sample with a requirement of at least one HF tower above 4 GeV at both sides of the CMS detector. The former is an inclusive sample while the latter is a non-single diffractive enhanced sample. The observable in the analysis is a sum over calorimeter towers in a given $|\Delta\eta|$ bin. The results are corrected for pile-up and detector noise. Finally, the energy distributions are corrected back to the particle level with corrections derived from Monte Carlo. The largest uncertainty arrives from the calorimeter global energy scale uncertainty, which in the studied region is in the range 10–17%. In Fig. 1 and 2 energy flow distributions in HF-OR and HF-AND samples are presented, respectively. For the comparison PYTHIA8 with different tunes, and different diffractive component modeling is used. Also predictions of EPOS LHC and QGSJET-II, models for cosmic rays physics,

[a] e-mail: grzegorz.brona@fuw.edu.pl

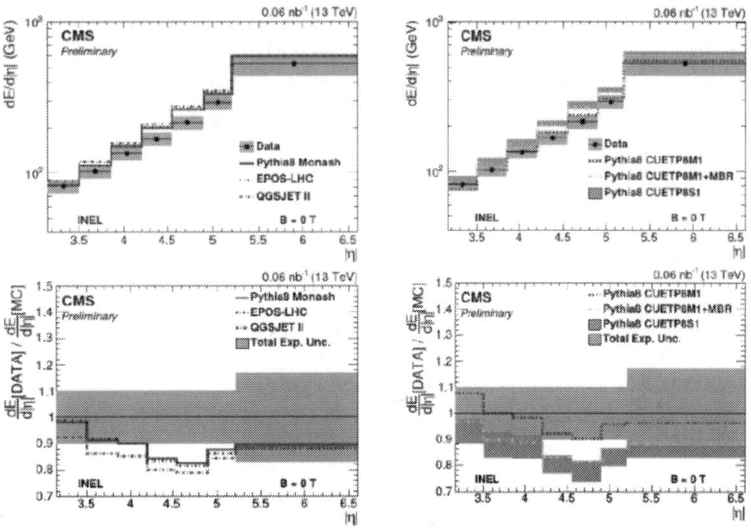

Figure 1. The inelastic forward energy flow distributions. On the left and right the same data points are presented while models differ. On the bottom plots ratios of the results to the predictions are presented [2].

are presented. The best agreement for the HF-OR sample for all the models is obtained in the lowest $|\eta|$ bins. Then for higher HF rapidities models overestimate the data. In the CASTOR bin the agreement is again better. Overall PYTHIA8 CUETP8M1 describes the HF-OR sample best. Predictions with different diffraction models (PYTHIA8 CUETP8M1 with Schuler-Sjöstrand, and with Minimum Bias Rockefeller models) exhibit large variations. For the HF-AND sample the spread between models tends to be smaller. Cosmic ray models are inside the experimental uncertainties, and provide a qualitatively good description. The best agreement among the PYTHIA tunes, is obtained with PYTHIA8 CUETP8M1. The HF-AND and HF-OR distributions do not exhibit significant differences in the spectrum shape.

The second analysis focuses on CASTOR, however the selection of events requires an activity exceeding 5 GeV in at least one HF tower. The CASTOR calorimeter does not have a segmentation in η but it is divided in 14 modules along the z direction and 16 sectors in ϕ. The first two modules are electromagnetic modules, and the rest hadronic. Three observables are defined: total energy recorded in CASTOR per event, total energy deposited in the electromagnetic modules per event, and total energy in the hadronic modules. The dominant uncertainty is again the energy scale uncertainty, 17%. Distributions are corrected to the stable particle level with Monte Carlo models. In Fig. 3 the distributions are plotted. In the first bins diffractive events form a distinctive peak. In the soft part of the spectrum (up to energy deposits of a few hundred GeV) PYTHIA8 and HERWIG tend to overestimate the data. The observables are sensitive to the PYTHIA8 tuning, none of the models and tunes is able to describe all three observables in the whole spectrum of the energy.

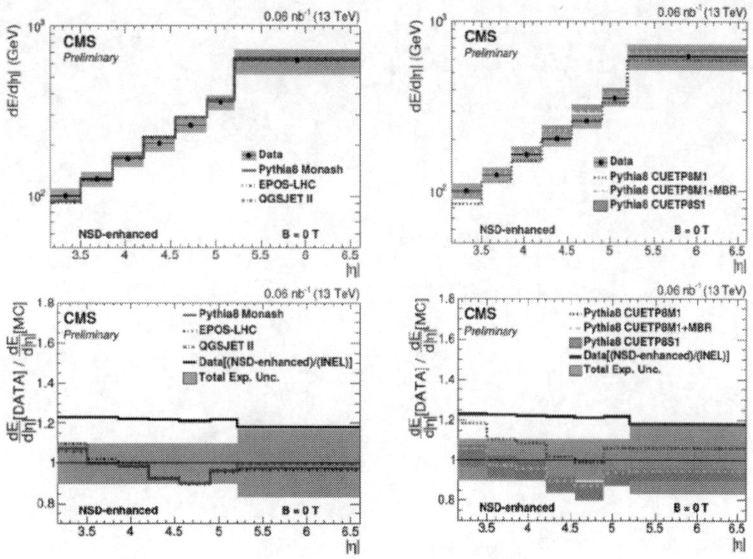

Figure 2. The NSD-enhanced forward energy flow distributions. On the left and right the same data points are presented while models differ. On the bottom plots ratios of the results to the predictions are presented [2].

2 Jets

Both experiments, CMS and ATLAS [4], performed measurements of the inclusive jets cross sections at 13 TeV [5, 6]. The cross sections are obtained in bins of rapidity. In case of CMS seven bins are defined covering a range up to $|y| = 4.7$, and $p_T = 2000$ GeV for most central jets. In ATLAS the range is smaller, up to $y = 3.0$ (six bins) but the coverage in p_T is wider, up to 3200 GeV. The analysis in CMS makes use of the anti-k_T algorithm with two different jet radii R =0.4 and 0.7. The better agreement with the NLO QCD predictions (obtained with NLOJet++) are obtained for a smaller R. In case of ATLAS only $R = 0.4$ is used. The data are compared with NLOJet++ with different PDF sets – CT14, MMHT2014, NNPDF3.0, CT10, HERAPDF2.0 and ABM12. A ratio of theory predictions to the data is presented in Fig. 4. All predictions describe the data overall well. In the forward region at large p_T there is a tendency in the theoretical approach to overestimate the data. Predictions with the ABM12 PDF set have a tendency to underestimate the data in medium and low rapidity bins.

The analysis of jets reconstructed within the CASTOR acceptance is based on 0.212 nb^{-1} of integrated luminosity recorded in a condition of very low pile-up (6% of probability of an interaction per bunch crossing) [7]. Jets are reconstructed with the anti-k_T algorithm with R =0.5 in a range of p_T 3–13 GeV. Very forward jets are sensitive to the low-x gluon PDF, and to non-linear effects. In Fig. 5 the inclusive cross section of jets in CASTOR is plotted versus p_T. Four different sources of systematics are shown. These are: CASTOR energy scale uncertainty (CES), luminosity uncertainty, CASTOR true acceptance (the position of the detector varies during data taking and is known with a precision of 2 mm), and modeling of the unfolding corrections. All tested models are in agreement with the data within the systematic uncertainty band. The spectrum has only a moderate sensitivity to the underlying

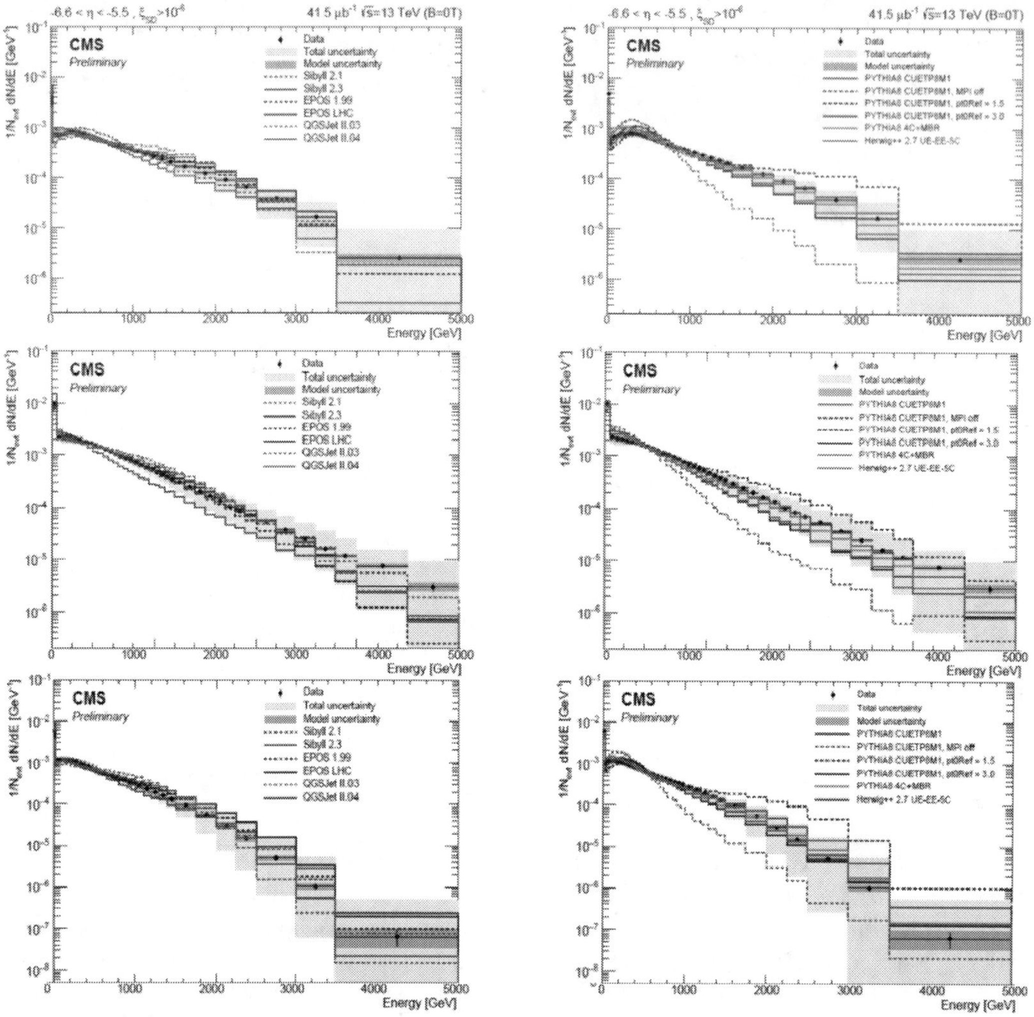

Figure 3. From top to bottom: total energy recorded in CASTOR per event, total energy deposited in the electromagnetic modules per event, and total energy deposited in the hadronic modules. On the left and right plots the same data points are presented while models differ [3].

PDF set of the model (Fig. 5 right). However, switching off the multiple parton interaction mechanism shifts the predicted spectrum significantly, and causes deviation from the data.

3 Charged particles spectra and UE with leading track/jet

Charged particles spectra are measured both by CMS and ATLAS [8–10]. At ATLAS two samples of events with a single primary vertex are selected: with at least 1 (2) tracks within $|\eta| < 2.5$ and with $p_T > 500$ MeV (100 MeV). Then primary charged-particle multiplicities as a function of pseudorapidity are calculated. The distributions are compared to model predictions: the EPOS LHC Monte Carlo fits data best in both samples. To compare with

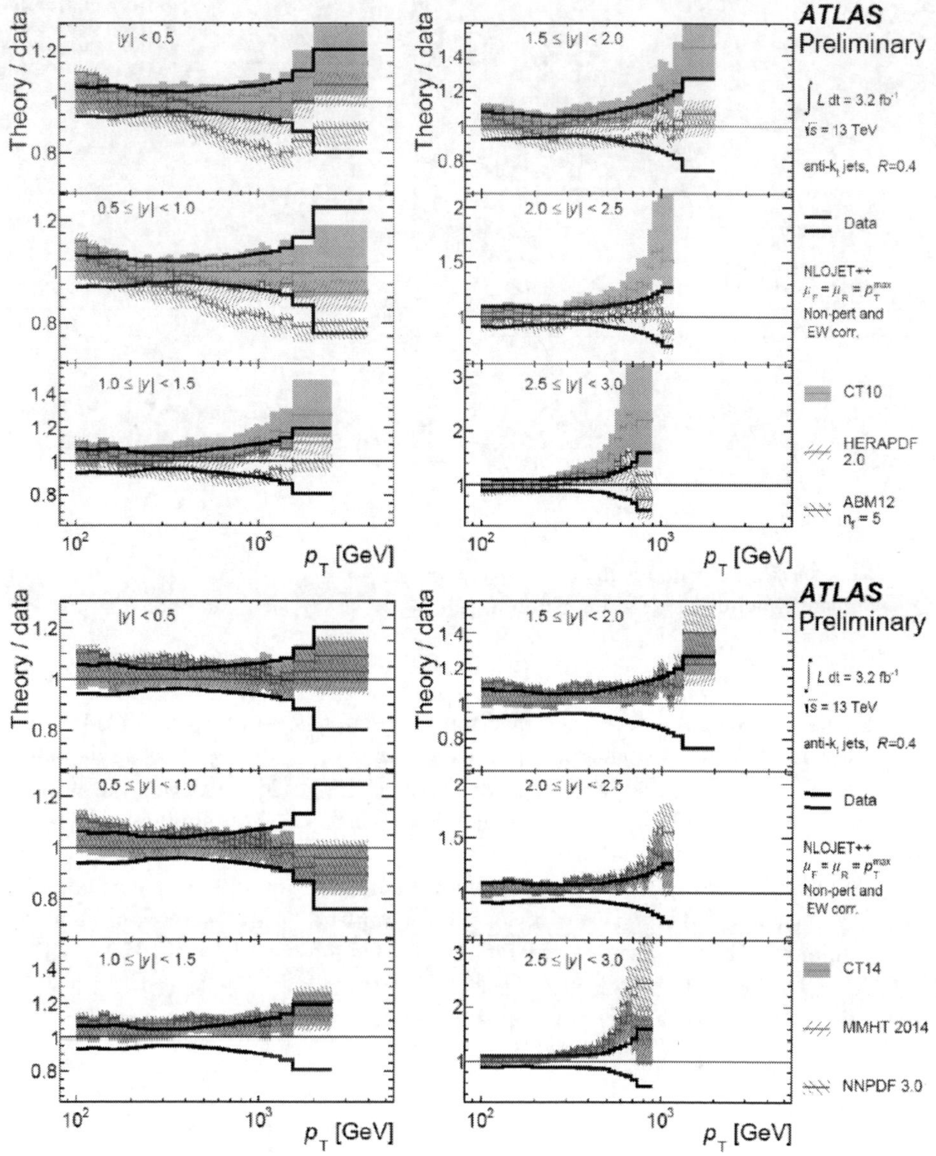

Figure 4. Ratio of the theoretical predictions for the inclusive jets cross section to the data. The plots are obtained for different bins in rapidity. As a model NLOJet++ is used with various PDF sets [6].

previous results the range under study is restricted to $|\eta| < 0.2$ and presented in Fig. 6. A 20% increase of the multiplicity is observed when moving from 7 to 13 TeV center-of-mass energy. All models follow the data while the best description is obtained with EPOS LHC Monte Carlo.

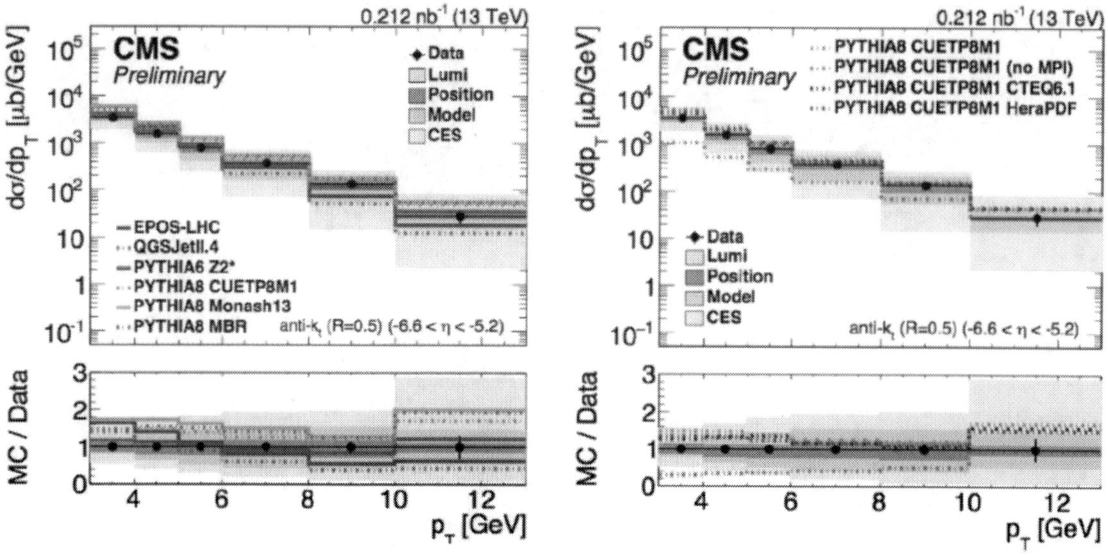

Figure 5. Cross section for the production of jets within $-6.6 < |\eta| < -5.2$. Left and right plots present the same data, while models are different [7].

In case of the CMS analysis, basic selection of events is similar to ATLAS. An event enters the study if at least one charged particle with $p_T > 500$ MeV is registered within $|\eta| < 2.4$ limit. Then three subsamples are defined: 1) inelastic event sample with at least one particle with $E > 5$ GeV in at least one HF detector, 2) non-single-diffractive (NSD) enhanced sample with at least one particle with $E > 5$ GeV in both HF detectors, 3) single diffractive (SD) sample defined in the same way as the inelastic sample but with an additional veto on activity in the second HF. Distributions of the multiplicities are presented in Fig. 7. In the inelastic sample the best agreement with data is obtained with EPOS LHC, HERWIG++ and PYTHIA8 CUETP8M1 models. The last from these models describes also the NSD sample well. The SD distribution is not described by any of the presented models.

Acknowledgements

This work was supported by Polish National Science Center, under contract DEC-2012/07/E/ST2/01406.

References

[1] CMS Collaboration, JINST **3** S08004 (2008)
[2] CMS Collaboration, CMS-FSQ-15-006
[3] CMS Collaboration, CMS-FSQ-16-002
[4] ATLAS Collaboration, JINST 3435 (2008) S08003
[5] CMS Collaboration, Eur. Phys. J. C **76**, 451 (2016)
[6] ATLAS Collaboration, ATLAS-CONF-2016-092

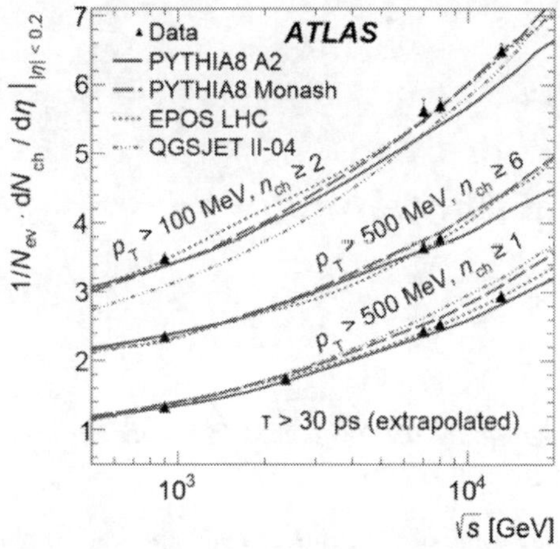

Figure 6. Charged particles multiplicity within $|\eta| < 0.2$ for various energies [10]. The results were extrapolated to include particles with mean lifetime $\tau > 30$ ps.

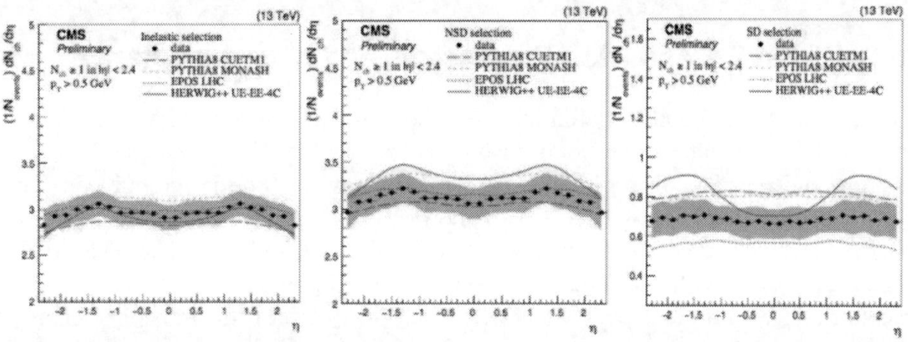

Figure 7. Multiplicity distribution in CMS at $\sqrt{s}$ =13 TeV calculated in three samples: inclusive sample (left), non-single-diffractive enhanced sample (middle), single diffractive sample (right) [8].

[7] CMS Collaboration, CMS-FSQ-16-003
[8] CMS Collaboration, CMS-FSQ-15-008
[9] CMS Collaboration, Phys. Lett. B **758**, 67 (2016)
[10] ATLAS Collaboration, Eur. Phys. J. C **76**, 502 (2016)

Hard diffraction and Photon-induced processes with Proton Tagging at the LHC

Christophe Royon[1,a]

[1] *Department of Physics and Astronomy 1082 Malott,1251 Wescoe Hall Dr. Lawrence, KS 66045-758*

Abstract. We describe some of the measurements that can be performed at the LHC by the TOTEM, CMS and ATLAS collaborations on hard diffraction in order to understand better the structure of the Pomeron. We will also describe the prospects concerning the search for quartic $\gamma\gamma\gamma\gamma$ anomalous couplings.

1 The CMS-TOTEM and ATLAS detectors and their acceptance

The TOTEM and ATLAS collaboration installed some roman pots located at about 200-220 meters from the CMS/ATLAS interaction point on both sides of CMS/ATLAS in order to measure intact protons after collisions [1]. In addition, there were some roman pot detectors located at 147 m but they were moved to 220 m during the 2015 shutdown.

It is possible to collect combined data between CMS and TOTEM [1]. The CMS-TOTEM collaboration aims at combining the data using the vertical roman pots of TOTEM and the CMS detector. The acceptance in (ξ, t) for high $\beta^* \sim 90$m is given in Fig. 1, left (t is the transfered energy squared at the proton vertex and ξ is the proton momentum fraction carried by the Pomeron) and the acceptance for ATLAS is similar. We see that we have a good acceptance in a large domain in ξ and thus in diffractive mass. The luminosity accumulated in special high-β^* is rather low and the high-β^* measurements are restricted to processes with a high enough cross sections. The CMS-TOTEM-Precision Proton Spectrometer (CT-PPS) and the ATLAS Forward Physics (AFP) detectors aim at taking data using the horizontal pots of TOTEM/AFP and the CMS/ATLAS detectors at high luminosity for the standard low-β^* lattice parameters of the LHC. The acceptance is shown in Fig. 1, right. We see that we have a good acceptance for rather high values of ξ up to 0.15, so at high diffractive masses. For these runs, the luminosity is quite high and we are sensitive to processes with low cross section. The situation is thus complementary to the high-β^* running scenario.

2 Measurements at medium luminosities: the Pomeron structure

2.1 Mesaurement at medium luminosities: Hard diffraction

In this section, we will discuss briefly how we can assess the gluon and quark contents in the Pomeron at the LHC, and also the search for BFKL resummation effects.

[a] e-mail: christophe.royon@ku.edu

 DOI: 10.1051/epjconf/201714103002

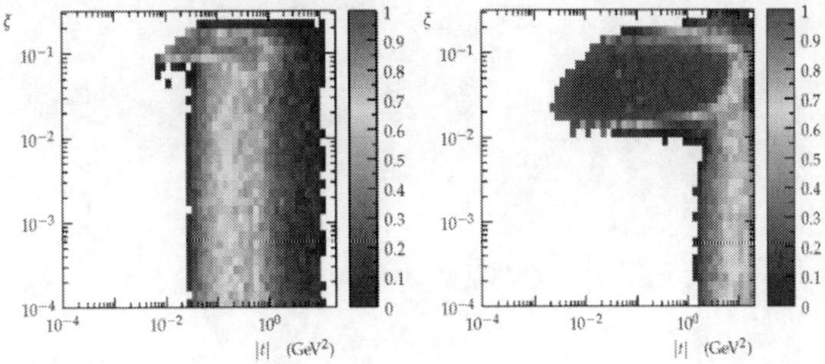

Figure 1. Acceptance in (t,ξ) of the roman pots detectors for high-β^* (left) and low-β^* (right).

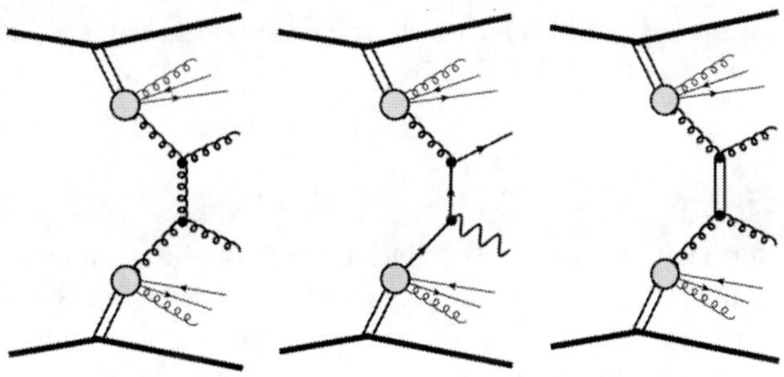

Figure 2. Inclusive diffractive diagrams. From left to right: jet production in inclusive double pomeron exchange, γ+jet production in DPE, jet gap jet events

Inclusive diffraction measurements at the LHC We discuss potential measurements at the LHC that can constrain the Pomeron structure. The Pomeron structure in terms of quarks and gluons has been derived from QCD fits at HERA and it is possible to probe this structure and the QCD evolution at the LHC in a completely new kinematical domain. All the following studies have been performed using the Forward Physics Monte Carlo (FPMC), a generator that has been designed to study forward physics, especially at the LHC [3]. It aims to provide a variety of diffractive processes in one common framework, *i.e.* single diffraction, double pomeron exchange, central exclusive production and two-photon exchange.

Dijet production in double Pomeron exchanges processes and sensitivity to the gluon density in the pomeron

One can first probe if the Pomeron is universal between *ep* and *pp* colliders, or in other words, if we are sensitive to the same object at HERA and the LHC using as an example dijet production in single diffractive and double pomeron exchange at the LHC. It is possible to assess the gluon and quark densities using the dijet and $\gamma + jet$ productions. The different diagrams of the processes that can be studied at the LHC are shown in Fig. 2, namely double pomeron exchange (DPE) production of dijets (left), of γ+jet (middle), sensitive respectively to the gluon and quark contents of the Pomeron, and the jet gap jet events (right).

The dijet production in DPE events at the LHC is sensitive to the gluon density in the Pomeron. In order to quantify how well we are sensitive to the Pomeron structure in terms of gluon density at

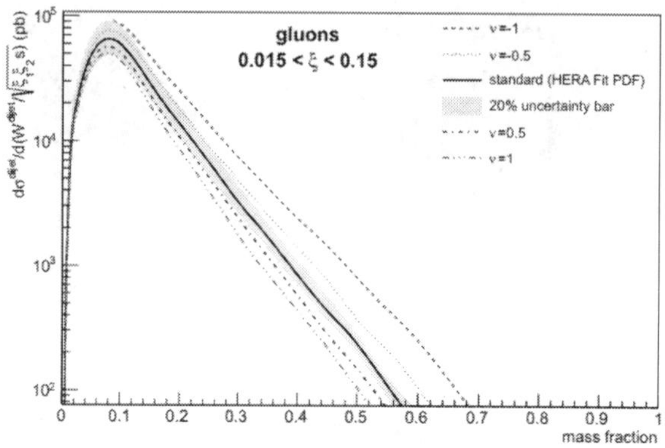

Figure 3. DPE di-jet mass fraction distribution. The different curves correspond to different modifications of the Pomeron gluon density extracted from HERA data (see text).

the LHC, we display in Fig. 3, the dijet mass fraction, the ratio of the dijet mass to the total diffractive mass [4, 5]. The central black line displays the cross section value for the gluon density in the Pomeron measured at HERA including an additional survival probability of 0.03. The yellow band shows the effect of the 20% uncertainty on the gluon density taking into account the normalisation uncertainties. The dashed curves display how the dijet cross section at the LHC is sensitive to the gluon density distribution especially at high β. For this sake, we multiply the gluon density in the Pomeron from HERA by $(1-\beta)^\nu$ where ν varies between -1 and 1. When ν is equal to -1 (resp. 1), the gluon density is enhanced (resp, decreased) at high β. We note that the curves corresponding to the different values of ν are much more separated at high values of the dijet mass fraction, meaning that this observable is indeed sensitive to the gluon density at high β.

Sensitivity to the Pomeron structure in quarks using γ + jet events and W asymmetry

The QCD diffractive fits performed at HERA assumed that $u = d = s = \bar{u} = \bar{d} = \bar{s}$, since data were not sensitive to the difference between the different quark component in the Pomeron. On the contrary, measuring the γ+jet to the dijet cross section ratios as a function of the diffractive mass M allows to distinguish between different assumptions on the quark content of the Pomeron. For instance, varying d/u between 0.25 and 4 leads to a variation of the cross section ratio by a factor 2.5 .

In addition, it is possible to use the W asymmetry in single diffractive W production [6]. Typically, the muon asymmetry is directly sensitive to the quark content of the pomeron and varies by a factor 6 at low ξ between the assumptions $u/d = 2$ or $u/d = 1/2$ for the quark content in the pomeron.

Reggeon contribution at high ξ The dijet diffractive cross section is also sensitive to Reggeon exchanges especially at high ξ. The Reggeon contribution is porrly constrained at HERA and it is important to measure it at the LHC. In Ref. [9], we showed that the Reggeon contribution is of the same order of magnitude as the Pomeron one in single diffractive dijets for $0.1 < \xi < 0.17$.

Jet gap jet production in double Pomeron exchanges processes

The observation of gap between jets in double Pomeron exchange events [8, 10] is ideal to probe the BFKL [7] evolution equation. We predict that the ratio of jet-gap-jet with respect with "standard" dijet production in DPE events will be of the order of 20%, which would represented a good test of the BFKL evolution equation.

2.2 Exclusive processes in diffraction

The advantage of the exclusive diffractive and photon exchange processes is that all particles can be measured in the final state. Both protons can be measured in CMS-TOTEM/CT-PPS and the produced particles (jets, vector mesons, Z boson....) in CMS/ATLAS, and there is no energy losses such as in the pomeron remnants. It is thus possible to reconstruct the properties of the object produced exclusively (via photon and gluon exchanges) from the tagged proton since the system is completely constrained. It is worth mentioning that it is also possible to constrain the background by requesting the matching between the information of the two protons and the produced object, and thus, central exclusive production is a potential channel for beyond standard model physics at high masses.

For instance, the CMS-TOTEM experiment performed extensive studies of possible measurements of exclusive states at high β^*. It is worth mentioning that the search for glueball states and the probe of the low x gluon density down to $x \sim 10^{-4}$ will be possible. It is possible to discover or exclude glueballs at low masses since the 1-10 GeV region in masses can be probed diffractively ($\xi \sim 10^{-4}$ – 10^{-3}), ensuring pure gluonic exchanges. It is then possible to check the $f_0(1500)$ or $f_0(1710)$ glueball candidates (lattice calculations predict a 0 + + glueball at 1.7 GeV with a ~100 MeV uncertainty, favoring the $f_0(1710)$ candidate). The simulation of a possible signal of $f_0(1710) \rightarrow \rho^0\rho^0$ and of the non resonant $\rho^0\rho^0$ background including the CMS tracker performance leading to a 20-30 MeV resolution was performed and a luminosity of $\sim$ 0.06 pb^{-1} is needed for a 7 σ signal. About 0.6 pb^{-1} is needed for decay characterisation In addition, it is possible to perform a spin analysis of $f_0(1710) \rightarrow \rho^0\rho^0 \rightarrow 4\pi$ to determine if $J = 0$ or 2: as an example, a measurement of the polar angle of the $\pi^+\pi^-$ pair for the ρ candidate needs about 5 pb^{-1}

In addition, using the same data set, the measurement of the cross section times branching ratio for the three $\chi_{C,0,1,2}$ states will be performed allowing a comparison with the results of the LHCb measurement.

3 Search for quartic $\gamma\gamma\gamma\gamma$ anomalous coupling events in diffraction

In this section, we will describe briefly the search for $\gamma\gamma\gamma\gamma$ quartic anomalous couplings using proton tagging at the LHC that regained interest after the excess observed in diphoton production by the CMS and ATLAS experiments [11, 12] (the search for $WW\gamma\gamma$ and $ZZ\gamma\gamma$ couplings is given in [13]).

3.1 Theoretical motivations

Four-photon (4γ) interactions through diphoton production via photon fusion with intact outgoing protons are considered. The pure photon dimension-eight operators read

$$\mathcal{L}_{4\gamma} = \zeta_1^\gamma F_{\mu\nu}F^{\mu\nu}F_{\rho\sigma}F^{\rho\sigma} + \zeta_2^\gamma F_{\mu\nu}F^{\nu\rho}F_{\rho\lambda}F^{\lambda\mu} \quad (1)$$

and they can induce the $\gamma\gamma\gamma\gamma$ process, highly suppressed in the SM [14]. We discuss here possible new physics contributions to $\zeta_{1,2}^\gamma$ that can be probed and discovered at the LHC using the forward proton detectors.

Loops of heavy charged particles contribute to the 4γ couplings [14] as $\zeta_i^\gamma = \alpha_{\rm em}^2 Q^4 m^{-4} N c_{i,s}$, where $c_{1,s}$ is related to the spin of the heavy particle of mass m running in the loop and Q its electric charge. These couplings scale as $\sim Q^4$ and are enhanced in presence of particles with large charges. For a 500 GeV vector (fermion) resonance with $Q = 3$ (4), large couplings ζ_i^γ of the order of 10^{-13} – 10^{-14} Gev^{-4} can be reached.

Beyond perturbative contributions to ζ_i^γ from charged particles, non-renor- malizable interactions of neutral particles are also present in common extensions of the SM. Such theories can contain scalar,

Table 1. Number of signal for $Q_{\rm eff} = 4$, $m = 340$ GeV and background events after various selections for an integrated luminosity of 300 fb^{-1} and $\mu = 50$ at $\sqrt{s} = 14$ TeV. Values obtained using the corresponding EFT couplings with and without form factors are also displayed. Excl. stands for exclusive backgrounds and DPE for double pomeron exchange backgrounds.

Cut / Process	Signal (full)	Signal with/without f.f. (EFT)	Excl.	DPE	DY dijet+ pile up	$\gamma\gamma$+ pile up
[$0.015 < \xi_{1,2} < 0.15$, $p_{\rm T1,(2)} > 200$ (100) GeV]	65.	18. (187.)	0.13	0.2	1.6	2968
$m_{\gamma\gamma} > 600$ GeV	64.	17. (186.)	0.10	0	0.2	1023
[$p_{\rm T2}/p_{\rm T1} > 0.95$, $\lvert\Delta\phi\rvert > \pi - 0.01$]	64.	17. (186.)	0.10	0	0	80.2
$\sqrt{\xi_1\xi_2 s} = m_{\gamma\gamma} \pm 3\%$	61.	12. (175.)	0.09	0	0	2.8
$\lvert y_{\gamma\gamma} - y_{pp}\rvert < 0.03$	60.	16. (169.)	0.09	0	0	0

pseudo-scalar and spin-2 resonances that couple to the photon and generate the 4γ couplings by tree-level exchange as $\zeta_i^\gamma = (f_s m)^{-2} d_{i,s}$, where $d_{1,s}$ is related to the spin of the particle.

3.2 Experimental sensitivity to quartic four photon couplings

The $\gamma\gamma\gamma\gamma$ process can be probed via the detection of two intact protons in the forward proton detectors and two energetic photons in the corresponding electromagnetic calorimeters.

The anomalous $\gamma\gamma\gamma\gamma$ process has been implemented in the Forward Physics Monte Carlo (FPMC) generator [3]. The FPMC generator was also used to simulate the background processes giving rise to two intact protons accompanied by two photons, electrons or jets that can mimic the photon signal. Those include exclusive SM production of $\gamma\gamma\gamma\gamma$ via lepton and quark boxes and $\gamma\gamma \to e^+e^-$. The central exclusive production of $\gamma\gamma$ via two-gluon exchange, was simulated using ExHuME. This series of backgrounds is called "Exclusive" in Table 1. FPMC was also used to produce $\gamma\gamma$, Higgs to $\gamma\gamma$ and dijet productions via double pomeron exchange (called DPE background in Table 1 and Fig. 4). Such backgrounds tend to be softer than the signal and can be suppressed with requirements on the transverse momenta of the photons and the diphoton invariant mass. In addition, the final-state photons of the signal are typically back-to-back and have about the same transverse momenta. Requiring a large azimuthal angle $|\Delta\phi| > \pi - 0.01$ between the two photons and a ratio $p_{T,2}/p_{T,1} > 0.95$ greatly reduces the contribution of non-exclusive processes.

Additional background processes include the quark and gluon-initiated production of two photons, two jets and Drell-Yan processes leading to two electrons. The two intact protons arise from pile-up interactions (these backgrounds are called $\gamma\gamma$ + pile-up and DY, dijet + pile-up in Table 1). The pile-up background is further suppressed by requiring the proton missing invariant mass to match the diphoton invariant mass within the expected resolution and the diphoton system rapidity and the rapidity of the two protons to be similar.

The number of expected signal and background events passing respective selections is shown in Table 1 for an integrated luminosity of 300 fb^{-1} for a center-of-mass energy of 14 TeV. Exploiting the full event kinematics with the forward proton detectors allows to completely suppress the background with a signal selection efficiency after the acceptance cuts exceeding 70%. Tagging the protons is absolutely needed to suppress the $\gamma\gamma$ + pile-up events. Further background reduction is even possible

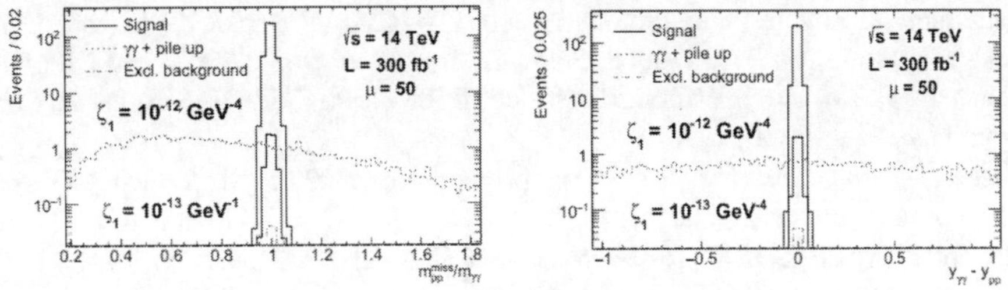

Figure 4. Diphoton to missing proton mass ratio (left) and rapidity difference (right) distributions for signal and backgrounds.

by requiring the photons and the protons to originate from the same vertex that provides an additional rejection factor of 40 for 50 pile-up interactions, showing the large margin on the background suppression. A similar study at a higher pile-up of 200 was performed and led to a very small background. The sensitivity extends up to $7 \cdot 10^{-15}$ GeV^{-4} allowing us to probe further the models of new physics described above.

We also performed a full amplitude calculation in Ref. [11, 14] that avoids the dependence on the choice of form factors needed in order to avoid quadratic divergences of scattering amplitudes. Sensitivities were found to be similar leading to possible discoveries of vector or fermions at high masses and high effective charges.

In this short report, we presented present and potential measurements to be performed by the TOTEM and CMS collaborations related to a better understanding of the pomeron structure in terms of QCD and to exploratory physics where unprecedented sensitivities can be reached on quartic anomalous $\gamma\gamma\gamma\gamma$ couplings (as well as for $\gamma\gamma WW$ and $\gamma\gamma ZZ$ anomalous couplings).

Acknowledgments

The author thanks the organizers for partial support.

References

[1] TOTEM Coll., CERN-LHCC-2014- 020; CMS and TOTEM Coll., CERN-LHCC-2014-021; ATLAS Coll., CERN-LHCC-2011-012.

[2] TOTEM Coll., Europhysics Lett. **101** (2013) 21004; Europhysics Lett. **101** (2013) 21003; Europhysics Lett. **101** (2013) 21002; Europhysics Lett. **96** (2011) 21002; Europhysics Lett. **95** (2011) 41001; Nucl. Phys. B **899** (2015) 527; Phys. Rev. Lett. **111**, 012001 (2013).

[3] M. Boonekamp et al., e-Print: arXiv:1102.2531;

[4] C, Marquet, C. Royon, M. Saimpert, D. Werder, Phys.Rev. D **88** (2013) 7, 074029.

[5] O. Kepka, C. Royon, Phys.Rev. D**76** (2007) 034012; CDF Coll., Phys. Rev. D **77**, (2008) 052004.

[6] A. Chuinard, C. Royon, Rafal Staszewski, preprint arXiv:1510.04218.

[7] V. S. Fadin, E. A. Kuraev, L. N. Lipatov, Phys. Lett. B**60** (1975) 50; I. I. Balitsky, L. N. Lipatov, Sov.J.Nucl.Phys. **28** (1978) 822;

[8] C. Marquet, C. Royon, M. Trzebinski, R. Zlebcik, Phys.Rev. D**87** (2013) 3, 034010; O. Kepka, C. Marquet, C. Royon, Phys. Rev. D**83** (2011) 034036.

[9] C. Marquet, D.E. Martins, A.V. Pereira, M. Rangel, C. Royon, preprint arXiv:1608.05674.
[10] C. Marquet, C. Royon, Phys. Rev. D**79** (2009) 034028; O. Kepka, C. Royon, C. Marquet, R. Peschanski, Eur. Phys .J. C**55** (2008) 259-272; Phys. Lett. B**655** (2007) 236-240; H. Navelet, R. Peschanski, C. Royon, S. Wallon, Phys. Lett. B**385** (1996) 357; H. Navelet, R. Peschanski, C. Royon, Phys. Lett. B366 (1996) 329.
[11] S. Fichet, G. von Gersdorff, C. Royon, preprint : arXiv:1512.05751; S. Fichet, G. von Gersdorff, C. Royon, preprint : arXiv:1601.01712.
[12] CMS Coll., CMS-PAS-EXO-15-004; ATLAS Coll., ATLAS-CONF-2015-081.
[13] E. Chapon, O. Kepka, C. Royon, Phys. Rev. **D81** (2010) 074003; O. Kepka and C. Royon, Phys. Rev. D **78** (2008) 073005.
[14] S. Fichet, G. von Gersdorff, B. Lenzi, C. Royon, M. Saimpert, JHEP 1502 (2015) 165; S. Fichet, G. von Gersdorff, O. Kepka, B. Lenzi, C. Royon, M. Saimpert, Phys.Rev. D**89** (2014) 114004.
[15] S. Fichet, G. von Gersdorff, C. Royon, Phys.Rev.Lett. **116** (2016) no.23, 231801; S. Fichet, G. von Gersdorff, C. Royon, Phys.Rev. D**93** (2016) no.7, 075031.

EPJ Web of Conferences **141**, 03003 (2017) DOI: 10.1051/epjconf/201714103003
ISMD 2016

Hard Diffraction at Colliders

Rafał Staszewski[1,a]

[1] *Institute of Nuclear Physics Polish Academy of Sciences, PL-31342 Kraków, Poland*

Abstract. The phenomenon of hard diffraction is presented. The discussion includes: the present understanding of the process, experimental techniques, past measurements performed at the SPS, HERA and Tevatron colliders, the first LHC data and prospects for new results.

1 Introduction

Diffractive interaction is a process that is governed by strong forces, but in which no quantum numbers are transferred. The most important types of diffractive processes are shown in Fig. 1. The simplest of them is the elastic scattering, where the final state consists of the same particles as the initial state – both particles stay intact. In single diffractive dissociation (single diffraction), one of the particles remain intact, while the other dissociates into a multi-particle state. In double dissociation (double diffraction), both particles dissociate. Finally, in central diffraction both particles stay intact, but the final state consists also of additional particles.

In the Regge theory diffractive interactions are governed by the exchange of the pomeron trajectory, which intercept is close to unity and which is responsible for the growth of the total cross sections with the centre-of-mass energy.

In the quantum chromodynamics the pomeron can be described in the leading order as an exchange of two gluons in a color-signet state. However, at small t values (t is the four-momentum transfer squared and the cross section for diffractive processes is steeply falling with t) the strong coupling constant is not small enough to justify neglecting higher-order diagrams. Therefore, the pomeron should be thought of rather as an exchange of a gluon ladder.

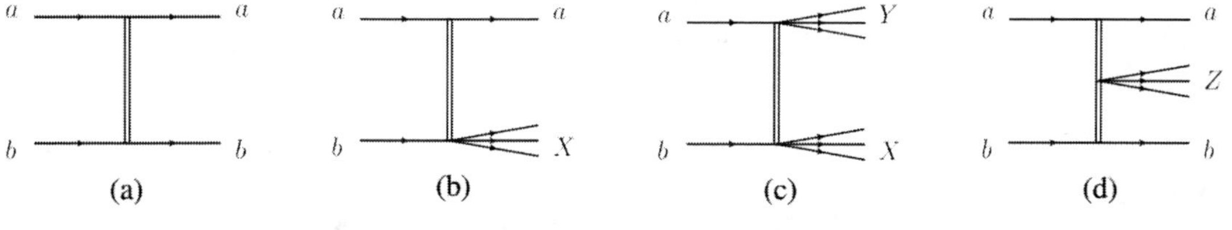

Figure 1. Diagrams of diffractive processes: (a) elastic scattering, (b) single diffractive dissociation (single diffraction), (c) double diffractive dissociation (double diffraction) and (d) central diffraction.

[a]e-mail: rafal.staszewski@ifj.edu.pl

2 Early history

The concept of hard diffraction was born from the assumption that one can consider the single diffractive process as a collision of a hadron with a pomeron emitted by the other hadron [1]. If one assumes that the pomeron consists of partons, then the pomeron–hadron interaction resembles any other hadron–hadron collision. In particular, it is possible that the partons undergo a hard interaction, even though the diffractive exchange (emission of the pomeron) is a soft process. Such a process would lead to a signature with an intact hadron and a hard object (*e.g.* a pair of high-p_T jets) present in the dissociated state.

The above hypothesis was studied first at the CERN SPS collider with $p\bar{p}$ collisions at $\sqrt{s}$ = 630 GeV. The UA8 experiment measured the intact protons and anti-protons, which are scattered at very small angles, with detectors inserted into the accelerator beam pipe with special mechanisms – Roman pots. The jets were detected with the central calorimeter of the UA2 experiment.

The experiment confirmed the existence of hard diffraction [2]. In events with a tagged proton that possessed more than 90% of its initial energy, localised depositions of energy were observed in the calorimeters. These clusters showed typical properties of jets and were similar to the ones observed in non-diffractive events.

3 Mechanism of hard diffraction

Although the hypothesis of existence of hard diffractive events was confirmed, the interpretation in terms of the underlying mechanism was, and still remains, not straightforward. Apart from the picture of the resolved pomeron consisting of partons, other descriptions of hard diffraction are possible.

A hard diffractive interaction can be obtained in a scenario, when the hard object is produced as in any other non-diffractive interaction, but in addition some soft gluon exchanges are presents in the event [3]. These gluons can screen the color-flow present in the hard event. Then, in the full event the overall exchange would be colour-singlet. The process would be considered diffractive, without the need of introducing a pomeron consisting of partons.

4 Diffractive PDFs

Detailed experimental studies of hard diffraction were performed at the HERA *ep* collider. The process of diffractive deeply inelastic scattering (DDIS) is depicted in Fig. 2. It is instructive to discuss the kinematics of this process. The relevant variables are:

- Q^2 – minus the square of the four-momentum exchanged by the electron,
- t – the square of the four-momentum exchanged by the proton (which stays intact),
- W – invariant mass of the photon–proton system,
- M_X – invariant mass of produced particles apart from scattered electron and proton,
- x – fraction of the proton carried by the struck quark,
- $x_{I\!P}$ (sometimes ξ) – fraction of the momentum lost by the proton,
- $\beta = x/x_{I\!P}$ – fraction of the momentum lost by the proton that was carried by the struck quark.

In addition, it is possible to assume a diffractive DIS event in which the proton does not stay intact, but dissociates. Then, additional kinematic variable has to be introduced: M_Y – the mass of the dissociated state.

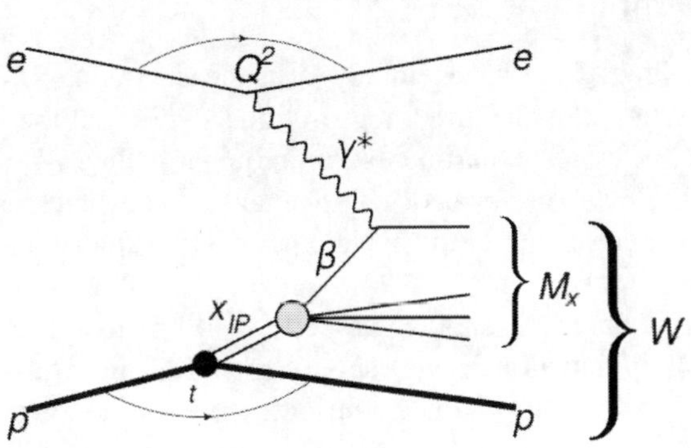

Figure 2. Feynman diagram of diffractive deeply inelastic scattering.

The DDIS process can be thought of as a scattering of the electron off a quark inside a pomeron emitted from the proton. Then, x_{IP} is the fractional momentum carried by the pomeron, β is the fraction of the pomeron momentum carried by the quark and M_X is the photon–pomeron invariant mass.

Similarly to the case of inclusive DIS, it is possible to measure the diffractive structure functions F_2^D. However, it no longer depends only on x and Q^2, but also on the proton kinematics: t and x_{IP}. Since experimentally it is not always possible to detect the intact proton (or the dissociated state), one can also consider the structure functions averaged over proton kinematics. The common naming convention is: $F_2^{D(4)} = F_2^D(x, Q^2, x_{IP}, t)$, $F_2^{D(3)} = \int F_2^{D(4)} \mathrm{d}t$, $F_2^{D(2)} = \int F_2^{D(3)} \mathrm{d}x_{IP}$.

It has been theoretically shown [4] that for DDIS processes the QCD factorisation holds: the cross section for hard processes can be calculated as convolution of the hard matrix element with parton distributions. Therefore, one can introduce the diffractive PDFs, which also depend on x, Q^2, x_{IP} and t. In addition, it is often assumed that these distributions factorise into the pomeron flux and its structure:

$$f_p(x, Q^2, x_{IP}, t) = \Phi(x_{IP}, t) \cdot f_{IP}(\beta, Q^2).$$

Assuming the DPDFs undergo a QCD evolution like the inclusive PDFs, the pomeron structure was extracted from the F_2^D measurements [5].

5 Factorisation breaking

The diffractive PDFs measured at HERA can be used to predict hard diffractive cross sections for other processes. Such predictions have been compared to the measurements performed at the Tevatron accelerator. It turned out that the predictions overshot the measurement by about one order of magnitude. This shows that the factorisation theorem regarding diffractive interactions, which has been proven for DDIS, does not hold for other processes, like $p\bar{p}$ collisions.

The physical mechanism of the factorisation breaking comes from the fact that hadrons have finite size, unlike the highly virtual photon in DIS. Therefore, it is possible that in a hadron–hadron collision several independent (soft) interactions take place. If one of them is diffractive, but others are not, the overall event would not have a diffractive signature.

In order to have a diffractive event, the diffractive interaction must not be accompanied by any other soft interaction. Probability of such a case is often called *rapidity gap survival probability* (the origin of the name will be explained in Section 6). The details of this correction factor, like its energy, kinematics, process dependence, *etc.* are a subject the present research.

6 Experimental techniques

Two experimental techniques have been employed in measurements of diffraction. The first one, already mentioned in Section 2, is the direct registration (*tagging*) of the intact hadron that was scattered diffractively. In accelerator experiments this requires dedicated detectors insertable into the beam chamber. Typically this is performed with Roman–pot mechanisms, but also another solution has been used, where a whole section of the beam pipe was movable and the detectors were fixed to it (Hamburg movable beam pipe).

The other technique is based on another property of diffractive events. When a colour singlet is exchanged, the particle production in the rapidity region of the exchange is strongly suppressed compared the exchange of a coloured object. In diffractive events one may more often find *large rapidity gaps*.

Both above methods have their advantages and disadvantages. The tagging method provides the reconstruction of the scattered hadron momentum, allowing more differential measurements. On the other hand, the kinematic region of the detector acceptance is limited by its position and the magnetic fields of the accelerator. The rapidity-gap method typically uses the already existing calorimetric system of the experiment and does not require dedicated equipment. On the other hand, it does not allow distinguishing between the single-diffractive events and double-diffractive ones, in which one of the dissociation mass is so small that the system escapes the calorimeter acceptance. In addition, this method is limited to large sizes of rapidity gaps, which translates into rather small masses of the dissociated system. High-mass diffraction has gaps of similar size to those present in non-diffractive processes because of statistical fluctuations.

The common experimental problem common for both methods is *pile-up* – a presence of several independent hadron–hadron interactions in a single registered event, which is a result of high luminosity of current hadron colliders. A rapidity gap present in the diffractive event will be filled with particles originating from another, independent interactions. The higher the pile-up, the more difficult it is to observe events with large gaps.

The presence of pile-up affects the tagging method in a different way. A hard non-diffractive event can be accompanied by a soft single-diffractive process that produces a forward intact hadron, which results in a hard-diffractive signature of the whole event. Here, pile-up becomes a source of background.

Pile-up is a big experimental problem, because it limits the instantaneous luminosity in which diffractive events can be studied. This, in turn, translates into limitations on the values of cross sections for processes that can be measured. This is particularly true for the LHC, where the average pile-up multiplicity is of the order of twenty. Apart for special cases of exclusive events, which will be discussed in the next section, diffraction cannot be measured in such conditions and require dedicated runs.

7 Variety of diffractive topologies

As discussed before, diffractive interaction means an exchange of a pomeron. However, this exchange can occur in a number of ways. In the following, different types of diffractive processes will be discussed on the example of the jet production.

If a single pomeron is exchanged leaving one of the hadrons intact, then one deals with the single diffractive jet production. This can be thought of as a collision of the pomeron with the other hadron, which in illustrated in Fig. 3a.

If two pomerons are exchanged and both hadrons stay intact, the process is called central diffraction or a double pomeron exchange and its diagram is presented in Fig. 3b. Fig. 3c presents a similar

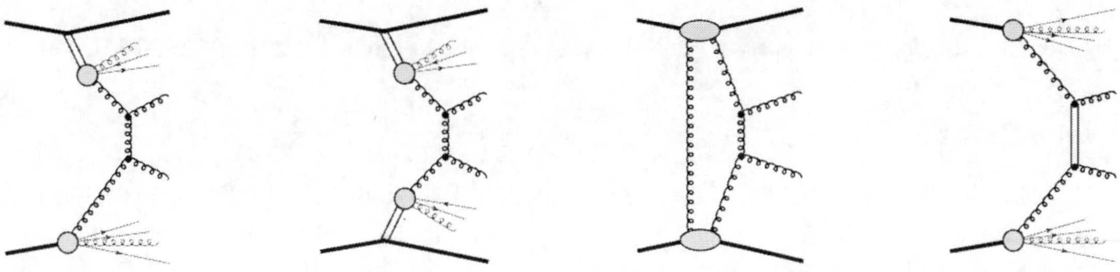

Figure 3. Different types of diffractive jet production processes: (a) single diffraction, (b) central diffraction, (c) central exclusive production, (d) double diffraction (jet–gap–jet).

process, called central exclusive production, which also contains two intact hadrons and a central di-jet system.

The difference between the exclusive and non-exclusive central diffraction is in the presence of the pomeron remnants. Central diffraction can be thought of a collision of two pomerons consisting of partons. One parton from each of the pomerons participates in the hard process producing the jets. However, other pomeron-originating partons may not interact, or interact softly. They are called pomeron remnants, are produced at small angles and usually escape detection. There are no remnants in the exclusive process – the whole energy taken from the initial-state hadrons goes into the production of the central state.

The exclusive processes are a very special types of hadron–hadron interaction. For them, using the detectors that can measure diffractively scattered hadrons, it is possible to measure all particles of the final state. They are very interesting also from the phenomenological point of view, since they can be calculated directly in QCD, without the need of introducing the diffractive PDFs. The experimental detection of these processes is a challenge because of their very small cross sections. However, the energy–momentum conservation introduces very strong correlations between the centrally produced state and the forward hadrons. This allows their measurement even in pile-up-contaminated environment.

Fig. 3a presents a diagram for the jet–gap–jet process. Here, unlike in all above cases, the pomeron exchange is a part of the hard process. The colour singlet is exchanged between the jets. In this process both hadrons dissociate, so the tagging method cannot be used. The large rapidity gap will be present between the jets.

8 The Large Hadron Collider

At the time of writing this article LHC has already provided first measurements on hard diffraction. The diffractive contributions have been observed for the jet and electroweak bosons [6–8]. All these measurements are based on the rapidity gap method and their precision is rather limited.

Two LHC experiments, ATLAS and CMS, are now equipped with systems of detectors that are used for tagging diffractively scattered protons. It is expected that they will provide much more detailed and precise measurements of hard diffraction in various topologies. The forward physics programme for the LHC including feasibility studies for many processes can be found in the report [9] prepared by a dedicated Forward Physics and Diffraction Working Group.

Acknowledgements

This work was supported in part by Polish National Science Centre grant 2012/05/B/ST2/02480.

References

[1] G. Ingelman and P. E. Schlein, Phys. Lett. **152B**, 256 (1985). doi:10.1016/0370-2693(85)91181-5
[2] R. Bonino *et al.* [UA8 Collaboration], Phys. Lett. B **211**, 239 (1988). doi:10.1016/0370-2693(88)90840-4
[3] A. Edin, G. Ingelman and J. Rathsman, Phys. Lett. B **366**, 371 (1996) doi:10.1016/0370-2693(95)01391-1 [hep-ph/9508386].
[4] J. C. Collins, Phys. Rev. D **57**, 3051 (1998) Erratum: [Phys. Rev. D **61**, 019902 (2000)] doi:10.1103/PhysRevD.61.019902, 10.1103/PhysRevD.57.3051 [hep-ph/9709499].
[5] A. Aktas *et al.* [H1 Collaboration], Eur. Phys. J. C **48**, 715 (2006) doi:10.1140/epjc/s10052-006-0035-3 [hep-ex/0606004].
[6] CMS Collaboration, Eur. Phys. J. C **72**, 1839 (2012).
[7] CMS Collaboration, Phys. Rev. D **87**, no. 1, 012006 (2013).
[8] ATLAS Collaboration, Phys. Lett. B **754**, 214 (2016).
[9] K. Akiba *et al.* 2016 J. Phys. G: Nucl. Part. Phys. 43 110201.

DOI: 10.1051/epjconf/201714103004

Effects of diffraction in pp and pA collisions

Gösta Gustafson[1,a]

[1] *Dept. of Astronomy and Theoretical Physics, Lund University*
Solvegatan 14A, 22362 Lund, Sweden

Abstract. Diffractive excitation is a large fraction of the pp cross section, also at high energies. Diffraction has been described by multi-Regge diagrams, or in the Good-Walker formalism as a result of fluctuations. The two formalisms are, however, just different sides of the same phenomenon. The dipole cascade formalism in impact parameter space is well suited to describe diffractive excitation including effects of saturation. Diffractive excitation is also an important effect in pA scattering, where the Glauber formalism has been used to estimate the number of NN subcollisions and of "wounded" nucleons. Diffractive excitation has here been either neglected or included in a simplified way, not including excitation of target nucleons. In this talk we discuss how these effects can be included with the help of the dipole cascade model DIPSY.

1 Introduction. The role of pertubation theory and unitarity constraints

Hera has shown that the parton density at small x grows rapidly $\sim 1/x^{1.3}$, as predicted by the perturbative BFKL pomeron. For pp collisions this implies a very large probability for gluon-gluon subcollisions, which implies that unitarity constraints are very important. These constraints lead to saturation of the gluon density, and suppression of partons with $k_\perp < Q_s^2$. This may explain why models based on multiple perturbative partonic subcollisions (like Pythia [1, 2]) are very successful at high energies. One can then ask: if perturbative physics dominates, is it then possible to calculate the result from basic principles, without input pdf's?

Unitarity constraints and saturation are most easily described in impact parameter space. Rescattering is represented by a convolution in $\mathbf{k}_\perp$-space, which simplifies to a product in $\mathbf{b}$-space. The small size of $\mathrm{Re}\, A_{el}^{pp}$ indicates that the pp interaction is driven by absorption. If the absorption probability in Born approximation equals $2F(b)$, the optical theorem and the eikonal approximation gives the result: $d\sigma_{el}/d^2b = T^2 = (1 - e^{-F})^2$, $d\sigma_{tot}/d^2b = 2T = 2(1 - e^{-F})$.

2 Dipole cascade evolution

Mueller's dipole cascade model [3, 4] is a formulation of LL BFKL evolution in impact parameter space. A colour charge is always screened by an accompanying anticharge, forming a colour dipole. The dipole emits bremsstrahlung gluons coherently, which splits the dipole in two dipoles. The new

[a]e-mail: gosta.gustafson@thep.lu.se

dipoles split repeatedly, developing into a cascade. When two cascades collide, s-channel unitarity is restored in the eikonal approximation.

The *Lund dipole cascade model, DIPSY* [5–7], is a generalization of Mueller's model. Thus it is based on BFKL evolution, but includes a set of important corrections:

1) Important non-leading effects in BFKL evolution. (The most essential are related to energy conservation and running α_s.)

2) Saturation from pomeron loops in the evolution. (Not included by Mueller or in the BK equation.)

3) Confinement, which also implies t-channel unitarity.

4) It is implemented in the DIPSY MC it gives also fluctuations and correlations.

5) It can be applied to collisions between electrons, protons, and nuclei.

3 Fluctuations and diffractive excitation

3.1 Good–Walker formalism

A projectile with a substructure may be diffractively excited to a different mass eigenstate. This can be described in the Good–Walker formalism as the result of fluctuations [8]. Assume that the projectile is a linear combination of diffractive eigenstates, Φ_n, with definite eigenvalues T_n. The elastic amplitude is then given by $\langle\Psi_{in}|T(b)|\Psi_{in}\rangle$, where the average is taken over both projectile and target states. The differential total and elastic cross sections are then given by

$$d\sigma_{tot}/d^2b = 2\langle T\rangle, \quad d\sigma_{el}/d^2b = \langle T\rangle^2. \tag{1}$$

The total diffractive cross section, including elastic scattering, is given by $\langle T^2\rangle$. Diffractive excitation is obtained subtracting the elastic, and thus given by the fluctuations $\langle T^2\rangle - \langle T\rangle^2$. For a fluctuating projectile scattering against a fluctuating target, the single and dubble excitations are obtained by separate averages over projectile and target states, below denoted by subscripts p and t respectively. Thus the cross sections for single excitation of the projectile and of the target, and for double excitation are given by:

$$d\sigma_{SD,p}/d^2b = \langle\langle T\rangle_t^2\rangle_p - \langle T\rangle_{p,t}^2; \tag{2}$$
$$d\sigma_{SD,t}/d^2b = \langle\langle T\rangle_p^2\rangle_t - \langle T\rangle_{p,t}^2; \tag{3}$$
$$d\sigma_{DD}/d^2b = \langle T^2\rangle_{p,t} - \langle\langle T\rangle_t^2\rangle_p - \langle\langle T\rangle_p^2\rangle_t + \langle T\rangle_{p,t}^2. \tag{4}$$

3.2 Diffractive eigenstates

The parton cascades discussed in sec. 2 can come on shell through interaction with a target. The BFKL evolution has a stochastic nature with large fluctuations, and following Hatta *et.al* [9] we assume that these cascades represent the diffractive eigenstates in high energy collisions. (A similar approach was presented by Miettinen and Pumplin [10].) In ref. [11] it is also shown that the fluctuations in the BFKL dynamics imply that this Good–Walker formalism actually agrees with the more commonly used triple-pomeron analysis. The Good–Walker formalism has, however, the advantage that it is more easy to account for saturation effects.

Fig. 1 shows a comparison between DIPSY and CDF results for single diffraction *vs* max M_X^2 [12]. It is also possible to calculate diffractive final states, and fig. 1 shows the result for $dn_{ch}/d\eta$ in $p\bar{p}$ collisions [13]. Note in particular that the MC is here only tuned to $\sigma_{\rm tot}$ and $\sigma_{\rm el}$, with no new parameter. (The MC result is too large for high η due to the lack of valence quarks in the model.)

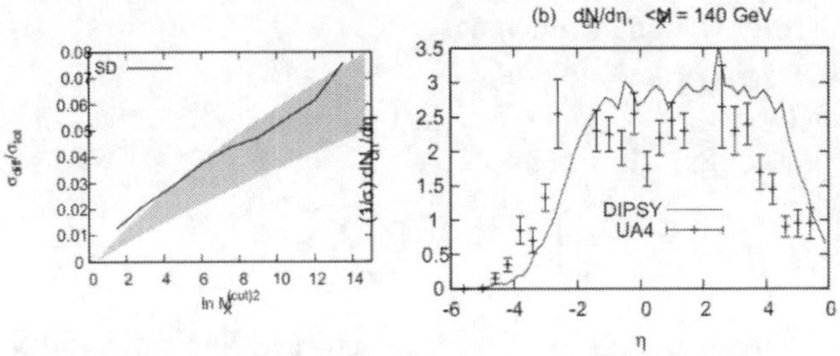

Figure 1. *Left*: $\int dM_X^2\, d\sigma_{SD}/dM_X^2$ for $M_X < M_X^{(cut)}$. The shaded area is an estimate of CDF results. *Right*: $dn_{ch}/d\eta$ in $p\bar{p}$ collisions at 546 GeV and $\langle M_X \rangle$= 140 GeV. Data from UA4 [14].

4 Collisions with nuclei

In collisions with nuclei, understanding the initial state is essential for the interpretation of collective final state effects. Studies of pA collisions can then be valuable to test models for the initial state in AA. DIPSY here gives a full partonic picture for the initial state in collisions with nuclei, in form of a dense gluon soup, and the model accounts for saturation within the cascades, finite size effects, and for correlations and fluctuations in the partonic state.

4.1 The Glauber model and Gribov corrections

The Glauber model is frequently used in analyses of experimental data, *e.g.* for estimating centrality or the number of "wounded" nucleons and binary NN collisions. In **b**-space rescattering is given by a product of amplitudes. For a projectile proton at impact parameter **b**, hitting a nucleus with nucleon positions $\mathbf{b}_\nu$ ($\nu = 1, \ldots, A$), the amplitude $T = 1 - S$ will then be determined by the relation

$$S^{(pA)}(\mathbf{b}) = \prod_{\nu=1}^{A} S^{(pp,\nu)}(\mathbf{b} - \mathbf{b}_\nu). \tag{5}$$

Gribov pointed out that diffractive excitation gives significant corretions to Glauber's formulae [15]. In pA collisions the projectile is frozen in the same state, k, during the passage through the nucleus, while the target nucleons are in different, uncorrelated states l_ν. In the Good–Walker formalism the elastic pA scattering amplitude is then given by the average

$$\langle T^{(pA)}(\mathbf{b}) \rangle = 1 - \langle \langle \prod_{\nu} \langle \{1 - T_{k,l_\nu}^{(pp,\nu)}(\mathbf{b} - \mathbf{b}_\nu)\} \rangle_{l_\nu} \rangle_{\mathbf{b}_\nu} \rangle_k, \tag{6}$$

with $d\sigma_{tot}^{pA}/d^2b = 2\langle T^{(pA)}(\mathbf{b})\rangle$ and $d\sigma_{el}^{pA}/d^2b = \langle T^{(pA)}(\mathbf{b})\rangle^2$. Note that the target average is over the pp amplitude to first power, but the average over projectile states contains also higher powers of $T^{(pp)}$.

4.2 Specification of wounded nucleons

Białas and coworkers noted that the central particle density in pA collisions is approximately determined by the number of participating, or wounded, nucleons: $\frac{dN^{pA}}{d\eta} \approx \frac{1+N_w^t}{2}\frac{dN^{pp}}{d\eta}$ [16, 17].

In most applications all inelastically interacting nucleons are treated as wounded nucleons, also those who are diffractively excited. This may be OK for forward observables, where these nucleons contribute to particle production. If the excited mass has a distribution $\sigma_{SD}/dM_X^2 \sim dM_X^2/(M_X^2)^{1+\epsilon}$ excited nucleons contribute significantly to central particle production at LHC only if ϵ is very small. For the (weakly) favoured value $\epsilon \approx 0.1$ [18] their contribution is suppressed by a factor $\sim 1/2$, and for larger ϵ-values even more. We thus conclude that depending on the observable under consideration, wounded nucleons ought to correspond either to non-diffractive, "absorbed", nucleons or to inelastic nucleons including diffraction ("inclusively wounded" nucleons).

i) Wounded nucleons = absorbed nucleons (inel. non diffr.)

The absorption probability is given by $d\sigma_{abs}/d^2b = 1 - S^2$. As the S-matrix factorizes, this is also the case for S^2, and the absorptive cross section is given by

$$d\sigma^{pA}_{abs}/d^2b = \langle 1 - \prod_\nu (S^{(pp,\nu)})^2 \rangle = 1 - \langle \prod_\nu \langle \{1 - T^{(pp,\nu)}(\mathbf{b} - \mathbf{b}_\nu)\}^2 \rangle_{l_\nu} \rangle_k. \tag{7}$$

Note that this expression includes the target average of $T^{(pp)}$ squared, taken to different powers n before averaging over projectile states.

ii) Inclusively wounded nucleons

Here we get

$$d\sigma^{pA}_{winc}/d^2b = 1 - \langle \prod_\nu \{1 - \langle T^{(pp,\nu)}(\mathbf{b} - \mathbf{b}_\nu)\rangle^2_{l_\nu}\} \rangle_k. \tag{8}$$

Just like the total and elastic cross sections, this expression includes only the target average of $T(\mathbf{b})$ to first power.

4.3 Simple approximations

Many analyses of experimental data use simple approximations for the pp amplitude:

i) Black disc model: $T^{pp}(b) = \theta(R - b)$

Here the amplitude depends on a single parameter R, and it gives the result $\sigma^{pp}_{abs} = \sigma^{pp}_{el} = \sigma^{pp}_{tot}/2$. Thus diffraction is totally neglected. If R is chosen to reproduce σ^{pp}_{inel}, then σ^{pA}_{tot} will be overestimated, while if it is choosed to reproduce σ^{pp}_{tot}, σ^{pA}_{inel} will be correspondingly underestimated [19].

ii) Gray disc model: Projectile is absorbed with probability a, for $b < R$

This model has two parameters, which makes it possible to reproduce *e.g.* σ^{pp}_{tot} and σ^{pp}_{el}. It is, however, not possible to distinguish SD_{target}, SD_{proj}, and DD, and therefore it is not possible to take separate averages over projectile and target states, as in eqs. (6-8) (see also ref. [19]).

4.4 The model by Strikman and coworkers

This model, sometime called the Glauber–Gribov (GG) model [20, 21], accounts for a fluctuating projectile, but not for fluctuations in the target. If the fluctuating pp total cross section, averaged over target states, is denoted $\hat{\sigma}_{tot}$, *i.e.* $\hat{\sigma}_{tot} = 2\int d^2b \langle T^{(pp)}(b) \rangle_{targ}$, then the pp total cross section is given by the average over projectile states: $\sigma^{(pp)}_{tot} = \langle \hat{\sigma}_{tot} \rangle_{proj}$. The distribution in values for $\hat{\sigma}_{tot}$ is assumed to have the following form:

$$\frac{dP}{d\hat{\sigma}_{tot}} = \rho \frac{\hat{\sigma}_{tot}}{\hat{\sigma}_{tot} + \sigma^{tot}_0} \exp\left\{-\frac{(\hat{\sigma}_{tot}/\sigma^{tot}_0 - 1)^2}{\Omega^2}\right\}. \tag{9}$$

Here Ω is a parameter determining the fluctuations, related to $\sigma^{(pp)}_{SD}$, σ^{tot}_0 is fixed from $\sigma^{(pp)}_{tot}$, and ρ is a normalization constant.

i) Wounded nucleons = absorbed nucleons

Define the fluctuating absorptive pp cross section:

$$\hat{\sigma}_{abs} = \int d^2b \langle \{2T^{(pp)}(b) - T^{(pp)2}(b)\} \rangle_{targ}; \tag{10}$$

$$\sigma^{(pp)}_{abs} = \langle \hat{\sigma}_{abs} \rangle_{proj}. \tag{11}$$

DOI: 10.1051/epjconf/201714103004

The same functional form is used as for σ_{tot} in eq. (9), but the width parameter Ω need not be the same. Note that σ_0^{abs} ought to be adjusted to reproduce $\sigma^{pp}_{inel\,ND}$, but is often tuned to $\sigma^{pp}_{inel\,tot}$.

ii) Wounded nucleons include target excitations

Here $\hat{\sigma}_{abs}$ should be exchanged for $\hat{\sigma}_{wincl}$ obtained by replacing $\langle T^{(pp)2}(b)\rangle_{targ}$ by $\langle T^{(pp)}(b)\rangle^2_{targ}$ in eq. (10).

In fig. 2 we compare the result from DIPSY with the analytic form in eq. (9), with averages and widths tuned to the DIPSY cross sections. We note that the DIPSY results have a more asymmetric form with a longer tail to large cross sections, which can be well fitted by a log-normal distribution.

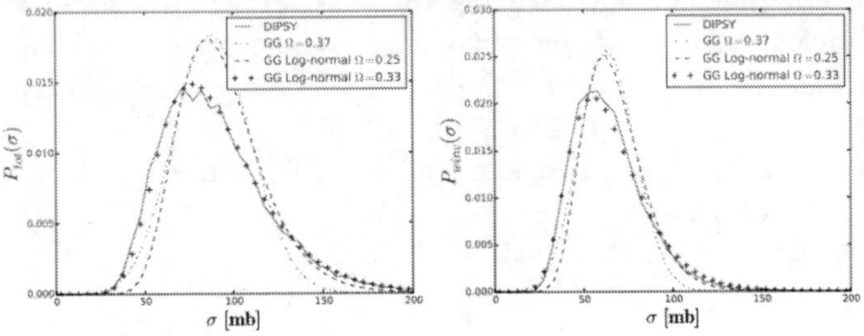

Figure 2. Prob. distrib. for the total (left) and the wounded incl. (right) cross sections. The solid line shows the DIPSY result, and the dotted line the GG form with the same average and width. Plus signs and dashed line are log-normal fits.

4.5 Final states

The old Fritiof model [22] was inspired by the wounded nucleon idea, and it worked quite well in the fixed target energy range. We have now constructed an updated version, called FritiofP8, in form of a toy model incorporating diffractive excitation using Pythia8. If different branches of a proton cascade interact via gluon exchange with two target nucleons, the result will be one dipole chain stretching in rapidity from the proton up to a point where it branches off to two chains, continuing to the two target nucleons. The result will be the same as one chain stretched over the full rapidity range, plus one chain stretched between the second target nucleon and the branching point, which can be placed randomly in rapidity. The result of the secondary chain is then similar to the contribution from a diffractively excited nucleon, as generated by the Pythia8 MC. This picture can be directly generalized to a case, where the proton interacts absorptively with several target nucleons. If diffractive target nucleons are produced with a mass distribution $d\sigma/dM_X^2 \sim 1/M_X^2$, which is the default in Pythia8, then the contribution from secondary absorbed and diffractively excited target nucleons will be very similar.

The FritiofP8 model is implemented in a MC [23], simulating final states obtained from the initial states given by DIPSY. Results for $\sum E_\perp$ in the forward (nucleus) direction is shown in fig. 3 *left*, which do agree very well with results from Atlas [24]. For illustration we also show result from a model called "Absorptive", which assumes that all wounded target nucleons contribute like a pp collision at full energy. This model is similar to the "G-Pythia" model used by Alice. We note that for large $\sum E_\perp$ this model is almost a factor 10 above FritiofP8 and the data. Fig. 3 *right* shows FritiofP8 results for the η-distribution in central pPb collisions, compared with the absorptive model and Atlas data [24]. We see that also here the FritiofP8 model agrees quite well with the data.

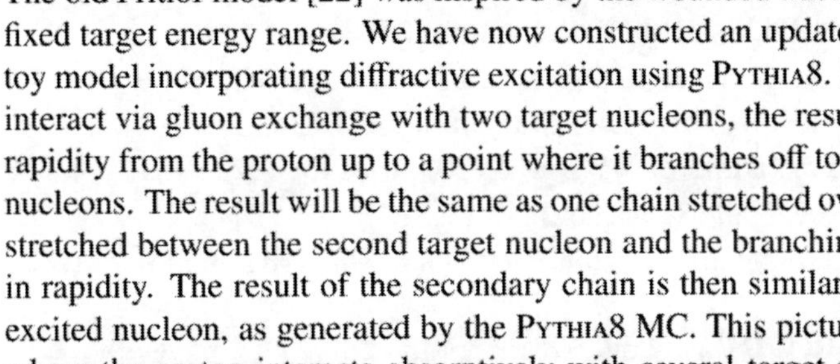

Figure 3. *Left*: $\sum E_\perp$ in forward direction from FritiofP8 (dashed) compared with the "Absorptive" model (solid). *Right*: η-distribution in central collisions from FritiofP8 (dashed), compared with the "Absorptive" model (solid) and with data from Atlas [24].

5 Conclusions

Saturation suppresses low-$p_\perp$ gluons, which implies that high energy hadronic collisions are dominantly perturbative.

The DIPSY dipole cascade model is based on BFKL dynamics with non-leading corrections and saturation. Implemented in a MC it includes correlations and fluctuations. It gives a fair description of DIS and pp data, with no input pdf's, and can give the initial condition in pA and AA collisions.

The Glauber model is frequently used in experimental analyses. Gribov pointed out the importance of diffractive scattering, but this is frequently not taken into account, or treated in an improper way.

A significant fraction of the interacting target nucleons are diffractively excited, and their contribution depends on the observable under study. For observables which obtain small contributions from diffractively excited target nucleons, the distribution in the GG model should be normalized to $\sigma_{inel.ND} \approx 2/3\,\sigma_{inel}$.

A simple model based on Fritiof, called FritiofP8, which combines DIPSY and Pythia8, works rather well for min bias final states in pA collisions.

References

[1] T. Sjostrand, M. van Zijl, Phys. Lett. **B188**, 149 (1987)
[2] T. Sjöstrand, S. Ask, J.R. Christiansen, R. Corke, N. Desai et al. (2014), 1410.3012
[3] A.H. Mueller, Nucl. Phys. **B415**, 373 (1994)
[4] A.H. Mueller, B. Patel, Nucl. Phys. **B425**, 471 (1994), hep-ph/9403256
[5] E. Avsar, G. Gustafson, L. Lönnblad, JHEP **07**, 062 (2005), hep-ph/0503181
[6] E. Avsar, G. Gustafson, L. Lönnblad, JHEP **01**, 012 (2007), hep-ph/0610157
[7] C. Flensburg, G. Gustafson, L. Lönnblad, JHEP **1108**, 103 (2011), 1103.4321
[8] M.L. Good, W.D. Walker, Phys. Rev. **120**, 1857 (1960)
[9] Y. Hatta, E. Iancu, C. Marquet, G. Soyez, D. Triantafyllopoulos, Nucl.Phys. **A773**, 95 (2006), hep-ph/0601150
[10] H.I. Miettinen, J. Pumplin, Phys. Rev. **D18**, 1696 (1978)
[11] G. Gustafson, Phys. Lett. **B718**, 1054 (2013), 1206.1733
[12] C. Flensburg, G. Gustafson, JHEP **1010**, 014 (2010), 1004.5502
[13] C. Flensburg, G. Gustafson, L. Lönnblad, JHEP **1212**, 115 (2012), 1210.2407
[14] D. Bernard et al. (UA4), Phys.Lett. **B166**, 459 (1986)
[15] V.N. Gribov, Sov. Phys. JETP **29**, 483 (1969), [Zh. Eksp. Teor. Fiz.56,892(1969)]
[16] A. Białas, M. Bleszynski, W. Czyz, Nucl. Phys. **B111**, 461 (1976)
[17] A. Bialas, J. Phys. **G35**, 044053 (2008)
[18] G. Aad et al. (ATLAS), Eur. Phys. J. **C72**, 1926 (2012), 1201.2808
[19] G. Gustafson, L. Lönnblad, A. Ster, T. Csörgő, JHEP **10**, 022 (2015), 1506.09095
[20] B. Blaettel, G. Baym, L.L. Frankfurt, H. Heiselberg, M. Strikman, Phys. Rev. **D47**, 2761 (1993)
[21] M. Alvioli, M. Strikman, Phys. Lett. **B722**, 347 (2013), 1301.0728
[22] B. Andersson, G. Gustafson, B. Nilsson-Almqvist, Nucl. Phys. **B281**, 289 (1987)
[23] C. Bierlich, G. Gustafson, L. Lönnblad (2016), 1607.04434
[24] G. Aad et al. (ATLAS), Eur. Phys. J. **C76**, 199 (2016), 1508.00848

Parton Distribution Functions and the role of forward region data

Voica Radescu[1,a]

[1] *University of Oxford, Denys Wilkinson Building Keble Road, Oxford OX1 3RH, U.K.*

Abstract. In the era of searching for new discoveries at the LHC, it is crucial to achieve a higher level of precision in understanding the proton structure which will allow for unambiguous interpretations of the high energy, luminous data ahead. The knowledge of proton's constituents comes mainly from the deep inelastic scattering data at HERA, complemented by the measurements from the fixed target, Tevatron, and now increasingly precise more data from LHC. A road-map that marks the most recent developments from the past and present experiments sensitive to the proton constituents will be presented here with an emphasis on the recent results using LHC data from the forward region.

1 Motivation

It is a most exciting phase of particle physics where there are many crucial questions which point to the deficiencies of the Standard Model (SM) that need to be answered. These questions can be addressed by the long term plan of the LHC running in different stages for achieving ever higher energies and luminosity. This poses new challenges for the experimental set-up and for the understanding of quark-gluon dynamics of the incoming proton beam particles at the LHC in the search for new particles or new interaction effects. The goal is first to facilitate the discovery of new physics by comparing measurements with the expectations driven either by the current SM hypothesis or by testing various beyond-SM scenarios. Precision is crucial to control the dominant uncertainties in order to observe and later analyse possible deviations from expectations. On the experimental side, the current measurements are now available at sub percent precision (e.g. W, Z at 7 TeV, Z p_T at 8 TeV), while the last decade has seen an enormous progress in achieving percent level precision on the theory calculations. Many of these are now available up to the next-to-next-to leading order (NNLO) in the truncated perturbative series of the strong coupling. In the following, an overview of the progress on improving the current limiting factor on precision, which arises from the limited knowledge of the proton's constituents, is presented.

2 Parton Distribution Functions

Proton's Parton Distribution Functions (PDFs) are understood, at the leading order approximation in QCD, as the probability of finding a parton of a given flavour that carries a fraction x of the total proton's momentum. Once QCD corrections are included, PDFs become scheme dependent. Their shape and normalisation are very different for each flavour, reflecting the different underlying dynamics that

[a] e-mail: voica.ana.maria.radescu@cern.ch

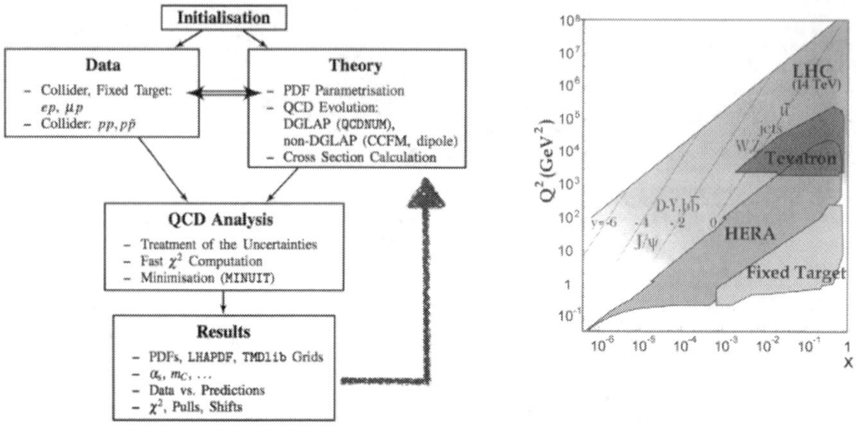

Figure 1. A schematic illustration of a PDF extraction mechanism as used by the xFitter [4] platform is shown on the left hand side. A kinematic coverage in x versus scale of experimental data [10] sensitive to PDFs is shown on the right hand side.

determine them, e.g. the valence quarks peak at the high x, while the gluon and sea quarks dominate at low x.

The PDFs cannot be directly calculated by QCD, however their evolution with the scale is predicted by perturbative QCD via evolution equations [1, 2]. The extraction of PDFs relies on the interplay between the precision of data versus theory. It invokes the factorisation concept where the hadronic cross section can be viewed as a convolution between calculable partonic cross sections (short distance process) and non-calculable part (long distance process).

2.1 Methodology in Extracting the PDFs

The following general steps are involved in the extraction of PDFs:

1. Parametrise PDFs at a starting scale;
2. Evolve to the scale corresponding to a data point;
3. Calculate the cross section;
4. Compare the calculated cross section with the data via χ^2 function;
5. Minimize χ^2 with respect to the PDF parameters.

There are many available choices for each of the above steps. A more detailed schematics of PDF extraction is also shown in figure 1. For step (1) there are multiple options for functional forms to be used to parametrise PDFs. For step (2), different methodologies can be used to evolve the PDFs at the starting scale, using collinear or k_T ordered evolutions. For step (3), there are various theoretical schemes to account for the heavy quark masses. Since the computational time is lengthy, there are also various fast grid techniques to speed up the calculations. For step (4), accounting for different sources of experimental uncertainties can be dealt by using either the covariance matrix or nuisance parameter approaches. And finally for step (4) there are different ways to minimize and extract the fit parameters, such as using data driven regularisation methods or using the MINUIT package [3].

Currently, there are several open-source software codes available, starting from the xFitter [4], which opened the ways of QCD-fit open-source code sharing, followed by the codes from various

	CT14	MMHT14	NNPDF3.0	HERAPDF2.0	ABM12(ABMP)	CJ15	JR14
HQ scheme	VFNS (ACOT-χ)	VFNS (TR opt)	VFNS (FONLL)	VFNS (TR opt)	FFNS Run mc (ABM)	VFNS (ACOT)	FFNS (JR)
order in pQCD	LO, NLO, NNLO	LO, NLO, NNLO	LO, NLO, NNLO	LO, NLO, NNLO	NNLO	NLO	NLO, NNLO
α(Mz)	fixed(fitted)	fixed (fitted)	fixed	fixed	fitted	fixed	fitted
α(Mz) LO α(Mz) NLO α(Mz) NNLO	0.1300 0.1180 (0.117) 0.1180 (0.115)	0.1350 0.1180 (0.1201) 0.1180 (0.1172)	0.1180 0.1180 0.1180	0.1300 0.1180 0.1180	- - 0.1132	- 0.118 -	- 0.1158 0.1136
Nr param.	Pol. Bernst. 28	Pol. Cheb. 25	NN (259)	Pol. 14	Pol. 24	Pol. 22	Pol.25
PDF assumptions	ubar/dbar=1(x->0) u/d=1 (x->0)	s-sbar=fit. dbar-ubar=fit.	dbar-ubar=fit	ubar=dbar (x->0) sbar=0.67*dbar	s=sbar dbar-ubar=fit	dv/uv=const s+sbar=k(ubar+dbar)	dbar-ubar=fit
Stat. treatm.	Hessian 2 stages: Δχ2=100 90% CL region	Hessian Δχ2 Dynamical (68% CL)	Monte Carlo (68% CL)	Hessian Δχ2=1 (68% CL)	Hessian Δχ2=1 (68% CL)	Hessian Δχ2=1 (68% CL)	Hessian Δχ2=1 (68% CL)
Q2min	2	2	3.5	3.5	2.5	1.69	2
References	arXiv:1506.07443	arXiv:1412.3989	arXiv:1410.8849	arXiv:1506.06042	arXiv:1310.3059	arXiv:1212.1702	arXiv:1403.1852

Figure 2. The list of known PDF differences. The analyses differ in many areas: different treatment of theories with quarks masses, inclusion of various data sets and account for possible tensions, different assumption on values of strong couplings, different assumptions in procedure (parametrisation, corrections).

groups: APFEL [5] , ALPOS [6], OPENQCDRAD [7, 8], QCDNUM [9], just to mention the main ones in use.

On the data side, there has been a persistent experimental effort over the last 40 years by both fixed-target and collider experiments around the world to provide constraints and accurate data. The kinematic coverage is illustrated in figure 1. The cleanest probe of the proton structure is the deep inelastic scattering experiments, where HERA has provided its final word on extracting PDFs in shape of HERAPDF2.0 [11]. However, complementary information is provided by the Drell-Yan processes at the collider experiments from Tevatron and the abundant data from the LHC.

2.2 Current PDF Groups

The extraction of PDFs is subject to many choices which lead to the formation of different analysis groups to extract global PDFs. The main list of differences comprises from: choice of data selection, choice of data treatment (corrections, uncertainties), various theory calculations for each process, e.g. formalism, automation, assumptions, parametrisation of PDFs and fit methodology, treatment of uncertainties (from data to theory). The most current and active PDF groups to date are: CT14 [12], MMHT14 [13], NNPDF3.0 [14], HERAPDF2.0 [11], CJ15 [15], ABM12 [16], JR14 [17] as well as dedicated PDFs produced by studies carried out by the LHC experiments, xFitter and PROSA [19]. Figure 2 summarises the known differences among these PDF groups.

Most of these differences have been addressed by benchmark exercises to assess the true differences in the methodologies of the groups [20, 21]. The level of precision for these benchmarked PDFs (based on General-Mass-Variable-Flavour-Number-Schemes) reaches below 10% in the bulk x region of $10^{-3} - 10^{-1}$, however, outside this region, the level of uncertainties escalate considerably, as shown in figure 3 for the gluon distribution. With increased mass ranges accessible by the LHC for discovery search, the PDF uncertainties increase considerably (especially for the gluon initiated processes), as illustrated in figure 3. This motivates strongly the need to improve the uncertainties for the high mass reaches.

There are also nuclear PDF groups, such as nCTEQ [22], HKN [23], EPS [24], DSSZ[25]. However, the analysis of nuclear data to extract nuclear PDFs relies on isospin symmetry and the assumption that bound proton PDFs obey the same evolution equations and sum rules as the free proton PDFs. There have been recent results from the nCTEQ group which analysed the W and Z production from LHC on proton-lead interaction with the power to provide extra constraint on light u,d quarks [18].

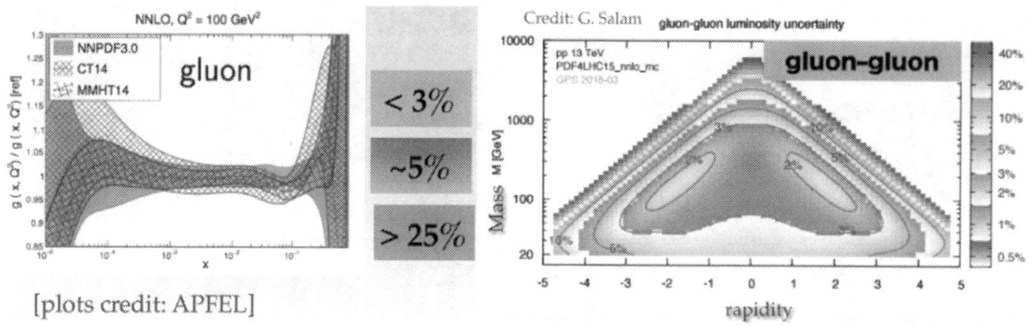

Figure 3. On the left hand side, the gluon relative uncertainties as function of x is shown for NNPDF3.0, CT14 and MMHT14 groups (plot credit APFEL). The panel on the right hand side demonstrates that with increased mass range (vertical axis) as function of the rapidity (horisontal axis), the PDF uncertainties increase considerably for the gluon-gluon parton luminosity distributions (plot credit G. Salam).

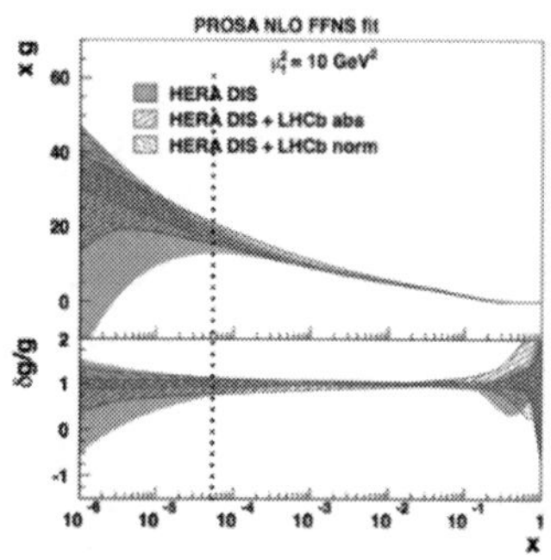

Figure 4. Figure shows the impact of the LHCb heavy quark production data on gluon PDF as performed by the PROSA collaboration.

3 Impact of the LHC measurements

The LHC provides an extended kinematic range in x by its three experiments ATLAS, CMS and LHCb. There is an interesting complementarity offered by the LHCb measurements, which can be exploited with its forward detector design accessing the low x region. It has been shown [19] that the access to the low x region reduces the uncertainties on gluon and sea quarks, as shown illustratively in figure 4. Moreover, the low x kinematics of the LHCb can be linked to the neutrino physics coverage, for which the main background for astrophysical neutrinos is generally the flux of neutrinos from the decays of charm mesons in cosmic ray collisions in the atmosphere. Therefore, the heavy quark production data from LHC could validate calculations of the prompt neutrino flux.

The LHC data can provide not only PDF discrimination by confronting theory with data, but also PDF improvement by using specific processes aimed at constraining PDFs. For gluons the inclusive jets, dijets and trijets measurements target the medium to large x sensitivity to PDFs, as well as the top pair production is an interesting observable for large x region. The transverse momentum of the Z distribution is particularly interesting as it can be very precisely measured experimentally (below percent level). For the quarks, the W and Z rapidity spectra are the main channels to constrain them at medium x. Interesting measurements are also the low and high mass Drell-Yan distributions. With

DOI: 10.1051/epjconf/201714103005

the boost of availability of the NNLO calculations, a re-analysis of many measurements are worth to be pursued to further constrain the PDFs.

4 Latest developments for PDFs

The fixed order calculations do not always work in describing data. A classical example is the Z transverse momentum distribution [26]. The predictions suffer from this lack of a consistent formalism applicable from small to large scales and there are efforts to merge the fixed order perturbative calculations with the parton showering or soft gluon resummation approaches.

The transverse momentum distributions, which can be viewed as adding extra dimensions to the PDFs, k_T dependent PDFs, are introduced for a proper simulation of parton showers [27]. This requires generalisation of QCD factorisation with an explicit dependence on transverse momentum and polarisations. They also obey the evolution equations which generalise the ordinary Renormalisation Group Equations of collinear PDFs. Another approach is to extract PDFs using rather resumed calculations than the pure fixed order calculations which is under development [28].

At the LHC the QED effects start to bring an important contributions due to access to high scales processes, therefore it starts to become relevant implementation of the combined QCD and QED evolution to extract a complete set of PDFs including photon PDFs. An active development is ongoing in all of the PDF groups.

5 Summary

The PDFs are very important as they still limit our knowledge of cross sections whether SM or beyond SM. Enormous progress achieved in pushing towards percent level precision on theory and experimental measurements. On the theory side there has been a boost due to the availability of the NNLO state-of-the-art calculations, the inclusion of the electro-weak effects, the release of special-case PDFs based on resumed calculations, parton showering, intrinsic charm. This has been complemented by the development of the advanced statistical methods (Monte Carlo replica, reweighting, profiling). On the data side, there is a pressing demand in pushing for precision measurements for constraining PDFs and measurement from clean processes are expected to bring a decisive impact on PDFs and flavour separation. The precise measurements should be presented with de-convoluted correlation information to help cross-calibrate systematic uncertainties. The next big goal of the LHC is to search for hints of new physics beyond the Standard Model and this can only be achieved if we can control better all our free parameters.

References

[1] V.N. Gribov and L.N. Lipatov, Sov. J. Nucl. Phys. 15, 438 (1972); L.N. Lipatov, ibid 20, 95 (1975); G. Altarelli and G. Parisi, Nucl. Phys. B126, 298 (1977); Yu.L. Dokshitzer, Sov. Phys. JETP 46, 641 (1977).

[2] E. A. Kuraev, L. N. Lipatov and V. S. Fadin, Sov. Phys. JETP 44, 443 (1976) [Zh. Eksp. Teor. Fiz. 71, 840 (1976)]; E. A. Kuraev, L. N. Lipatov and V. S. Fadin, Sov. Phys. JETP 45, 199 (1977) [Zh. Eksp. Teor. Fiz. 72, 377 (1977)]; I. I. Balitsky and L. N. Lipatov, Sov. J. Nucl. Phys. 28, 822 (1978) [Yad. Fiz. 28, 1597 (1978)].

[3] F. James and M. Roos, Comput. Phys. Commun. **10**, 343 (1975). doi:10.1016/0010-4655(75)90039-9

[4] S. Alekhin *et al.*, Eur. Phys. J. C **75**, no. 7, 304 (2015) doi:10.1140/epjc/s10052-015-3480-z [arXiv:1410.4412 [hep-ph]].

[5] V. Bertone, S. Carrazza and J. Rojo, Comput. Phys. Commun. **185**, 1647 (2014) doi:10.1016/j.cpc.2014.03.007 [arXiv:1310.1394 [hep-ph]].

[6] D. Britzger *et al.*, [http://desy.de/~britzger/alpos/].

[7] S. Alekhin *et al.*, [https://www-zeuthen.desy.de/~alekhin/OPENQCDRAD/].

[8] S. Alekhin, J. Blumlein and S. Moch, Phys. Rev. D **86**, 054009 (2012) doi:10.1103/PhysRevD.86.054009 [arXiv:1202.2281 [hep-ph]].

[9] M. Botje, Comput. Phys. Commun. **182**, 490 (2011) doi:10.1016/j.cpc.2010.10.020 [arXiv:1005.1481 [hep-ph]].

[10] C. Patrignani et al. (Particle Data Group), Chin. Phys. C, 40, 100001 (2016).

[11] H. Abramowicz *et al.* [H1 and ZEUS Collaborations], Eur. Phys. J. C **75**, no. 12, 580 (2015) doi:10.1140/epjc/s10052-015-3710-4 [arXiv:1506.06042 [hep-ex]].

[12] S. Dulat *et al.*, Phys. Rev. D **93**, no. 3, 033006 (2016) doi:10.1103/PhysRevD.93.033006 [arXiv:1506.07443 [hep-ph]].

[13] L. A. Harland-Lang, A. D. Martin, P. Motylinski and R. S. Thorne, Eur. Phys. J. C **75**, no. 5, 204 (2015) doi:10.1140/epjc/s10052-015-3397-6 [arXiv:1412.3989 [hep-ph]].

[14] R. D. Ball *et al.* [NNPDF Collaboration], JHEP **1504**, 040 (2015) doi:10.1007/JHEP04(2015)040 [arXiv:1410.8849 [hep-ph]].

[15] J. F. Owens, A. Accardi and W. Melnitchouk, Phys. Rev. D **87**, no. 9, 094012 (2013) doi:10.1103/PhysRevD.87.094012 [arXiv:1212.1702 [hep-ph]].

[16] S. Alekhin, J. Blumlein and S. Moch, Phys. Rev. D **89**, no. 5, 054028 (2014) doi:10.1103/PhysRevD.89.054028 [arXiv:1310.3059 [hep-ph]].

[17] P. Jimenez-Delgado and E. Reya, Phys. Rev. D **89**, no. 7, 074049 (2014) doi:10.1103/PhysRevD.89.074049 [arXiv:1403.1852 [hep-ph]].

[18] A. Kusina, Proceedings DIS2016, [http://pos.sissa.it/archive/conferences/265/026/DIS2016_026.pdf].

[19] PROSA collaboration; [https://prosa.desy.de/].

[20] J. Butterworth *et al.*, J. Phys. G **43**, 023001 (2016) doi:10.1088/0954-3899/43/2/023001 [arXiv:1510.03865 [hep-ph]].

[21] A. Accardi *et al.*, Eur. Phys. J. C **76**, no. 8, 471 (2016) doi:10.1140/epjc/s10052-016-4285-4 [arXiv:1603.08906 [hep-ph]].

[22] K. Kovarik *et al.*, Phys. Rev. D **93**, no. 8, 085037 (2016) doi:10.1103/PhysRevD.93.085037 [arXiv:1509.00792 [hep-ph]].

[23] M. Hirai, S. Kumano and T.-H. Nagai, Phys. Rev. C **76**, 065207 (2007) doi:10.1103/PhysRevC.76.065207 [arXiv:0709.3038 [hep-ph]].

[24] K. J. Eskola, H. Paukkunen and C. A. Salgado, JHEP **0904**, 065 (2009) doi:10.1088/1126-6708/2009/04/065 [arXiv:0902.4154 [hep-ph]].

[25] D. de Florian, R. Sassot, P. Zurita and M. Stratmann, Phys. Rev. D **85**, 074028 (2012) doi:10.1103/PhysRevD.85.074028 [arXiv:1112.6324 [hep-ph]].

[26] G. Aad *et al.* [ATLAS Collaboration], Eur. Phys. J. C **76**, no. 5, 291 (2016) doi:10.1140/epjc/s10052-016-4070-4 [arXiv:1512.02192 [hep-ex]].

[27] R. Angeles-Martinez *et al.*, Acta Phys. Polon. B **46**, no. 12, 2501 (2015) doi:10.5506/APhysPolB.46.2501 [arXiv:1507.05267 [hep-ph]].

[28] M. Bonvini, QCD LHC 2016 contribution.

Gluon polarization measurements and the possible role of diffractive process in the transverse single spin asymmetry measurements in RHIC-PHENIX

Itaru Nakagawa[1,2,a] for the RHIC Spin Collaboration

[1] *RIKEN, 2-1 Hirosawa, Wako, Saitama, 351-0198, Japan*
[2] *Riken-BNL Research Center, 20 Pennsylvania Avenue Bldg.510A, Upton, NY 11973, U.S.A.*

Abstract. Two selected topics from the latest RHIC spin results are discussed here. For the transversely polarized spin program, an unexpectedly large single spin asymmetry in the very forward neutron production observed in polarized proton + nucleus collisions at $\sqrt{s}$ = 200 GeV is discussed in this document. For the longitudinal program, the latest highlights from the measurements on the gluon spin components of the proton spin is discussed. After a decade of continuous efforts to hunt for the gluon polarization, the RHIC collaboration is about to catch the tail of the experimental evidence that gluon carries substantially large portion of the proton spin.

1 Introduction

Proton is known as a strongly interacting, relativistic bound state of quarks and gluons (referred to as partons) in quantum chromodynamics (QCD). The goal of the RHIC spin collaboration is to address the open questions on the proton's spin structure which are necessary to understand the fundaments of QCD. See reference [1] for extensive discussion what have achieved and what will be explored in the future as the spin program at RHIC. The polarized proton-proton collider at RHIC provides an unique opportunity to study the spin structure of proton. The latest highlights from the RHIC spin program are selectively introduced in this document. The spin program is classified into two main programs based on the spin orientation of incident proton spin orientation with respect to its momentum direction either transverse or longitudinal. The main focus of this document discussed in section 2 is about the quite striking atomic mass dependence in the single spin asymmetry of the very forward (almost zero degree) neutron production recently observed by the PHENIX experiment. The asymmetry observed in p + p was well reproduced by the interference between one pion exchange (OPE) and a_1-Reggeon. However this hadronic interference predicts only moderate dependence on the atomic mass number. I will introduce the possible mechanism which has been overlooked such as the ultra peripheral collision (UPC). For longitudinal program, the latest highlights from the measurements on the gluon spin component of the proton spin is discussed in sections 3, respectively.

[a] e-mail: itaru@riken.jp

2 The Very Forward Neutron Asymmetry

The transverse single spin asymmetry (TSSA) of very forward (almost zero degree) was first discovered[2] at RHIC in transversely polarized proton+proton collision at $\sqrt{s}$ = 200 GeV. Neutrons were measured by a zero-degree calorimeter (ZDC) [3] with a position-sensitive shower-maximum detector (SMD). One ZDC module is composed of Cu-W alloy absorbers with PMMA-based communication grade optical fibers and corresponds to 1.7 nuclear interaction lengths. A single photomultiplier collects Cherenkov light via optical fibers. Three ZDC modules are located in series (5.1 nuclear interaction lengths) at 18 m away from the collision point, covering 10 cm in the transverse plane. The detector covers the psuedorapidity range of $6.8 \leq \eta \leq 8.8$ where the neutron emission angle with respect to the incident proton direction is limited to less than 2.2 mrad. Relatively large energy neutrons are selected in the analysis, i.e. the neutron energy cut was applied at the energy fraction of incident proton $x_{\rm F} = E_{\rm n}/E_{\rm p} \geq 0.4$. The production mechanism of the such a neutron is driven by soft process and of which cross section is well described by one pion exchange (OPE) model. The corresponding four momentum transfer can be defined $0.02 \leq -t \leq 0.5$ (GeV/c)^2 at under the assumption of OPE.

The single spin asymmetries for such a forward neutron turned out to be large[2]. Although the OPE model was quite successful to reproduce the cross section data, the predicted asymmetry within the OPE framework appeared to be very tiny and far underestimates the data[4]. The forward neutron asymmetry is formulated as

$$A_{\rm N} = \phi_{\rm flip}\phi_{\rm non-flip} \sin\delta \quad (1)$$

where $\phi_{\rm flip}$ ($\phi_{\rm non-flip}$) is spin flip (spin non-flip) amplitude between incident proton and out-going neutron, and δ is the relative phase between these two amplitudes. Although the OPE can contribute to both spin flip and non-flip amplitudes, resulting $A_{\rm N}$ is small due to the small relative phase. The decent amplitude can be generated only by introducing the interference between spin flip π exchange and spin non-flip a_1-Reggeon exchange which has large phase shift in between[4].

In year 2015, the first attempt was made to collide the transversely polarized proton and nucleus in RHIC at $\sqrt{s}$ = 200 GeV. In order to explore the evolving origin of the asymmetry as a function of atomic mass number (A), the forward (p-going side) neutron asymmetries were measured for p+Au and p+Al collisions. Since the proton beam is unpolarized at LHC, the measurement is the high energy frontier of the polarized proton and nucleus collision. The existing polarized proton+nucleus experiments are carried out in one order of magnitude lower $\sqrt{s}$, for instance, at RHIC[6] and Fermi Lab.[5] using polarized proton beam and fixed nuclear targets.

The asymmetry analysis basically followed the same procedures and conditions as previous p+p measurements. The measurement in p+p collision system was repeated in year 2015 and resulting asymmetry was consistent with previous measurements. The ZDC acceptance was segmented into eight azimuthal bins and evaluated left-right raw asymmetries ϵ_N^{exp} defined by the Eqn. (2) in the each bin independently.

$$\epsilon_N^{raw}(\phi) = \frac{N^+(\phi) - N^-(\phi)}{N^+(\phi) + N^-(\phi)} \quad (2)$$

where $N^+(\phi)$ and $N^-(\phi)$ are number of neutrons observed in the given azimuthal ϕ bin when protons are polarized up and down, respectively.

Fig.1 shows preliminary results of forward neutron $A_{\rm N}$ measurements in 2015. The red points are the results of forward neutron $A_{\rm N}$ inclusive measurements. Surprisingly, they show unexpectedly strong mass number (A) dependence. The asymmetry even flips the sign from p+p, p+Al to p+Au.

The p+Au data point shows magnificently large A_N of about 0.18 which is three times larger than that of p+p in absolute amplitude. The existing π and a_1-Reggeon interference framework predicts only moderate evolution as growing A and does not have any mechanism to flip the sign of A_N in any p+A collision systems.

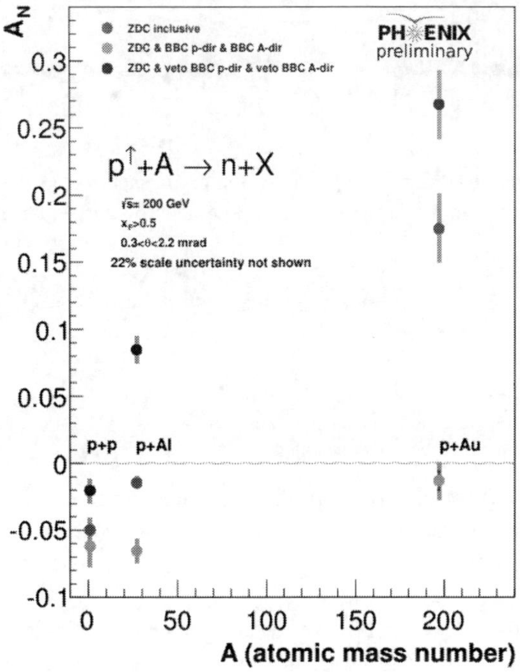

Figure 1. (Observed forward neutron A_N in transversely polarized proton-nucleus collisions. Data points are A=1, A=27, and A=197 are results of p+p, p+Al, and p+Au, respectively. Red, Blue and Green data points are neutron inclusive, neutron + BBC veto, and BBC tagged events, respectively

More interestingly, another drastic dependence of A_N was observed in correlation measurements in addition to the inclusive neutron. In these measurements, another out-going charged particle was either tagged or vetoed within the acceptance of the beam-beam counter (BBC) in both North and South arms which covers $3.1 \leq |\eta| \leq 3.9$. The BBCs cover such a limited acceptance, but the resulting asymmetries behaved remarkably contradicts. Once BBC hits (BBC tagging) are required in both arms (green data points), the drastic behavior of inclusive A_N is vanished and no flipping sign was observed between p+p and p+Au. On the contrary, the asymmetries are pushed even more positive for p+Al and p+Au data points once no hits in BBC are required (BBC vetoed) as represented by blue data points.

It was pointed out the importance of the electro-magnetic (EM) interaction effect in the existing fixed target experiments [5] and [6], which were carried out in small t, one may need to consider the effect here, although it was ignored in p+p data. Due to the smallness of the four momentum transfers of the present kinematics, i.e. $-t \leq 0.5$ (GeV/c)^2, the EM interaction may play a role which becomes increasingly important in large atomic number nucleus. The Coulomb field of the nucleus becomes rich source of exchanging photons between the polarized proton. This is known as the ultra-peripheral collision (UPC) in heavy ion collider experiments. In the UPC process, there is no charge exchange at the collision vertex unlike π or a_1 meson exchange. The forward neutron in the final state then can be produced through the diagrams shown in Fig. 2.

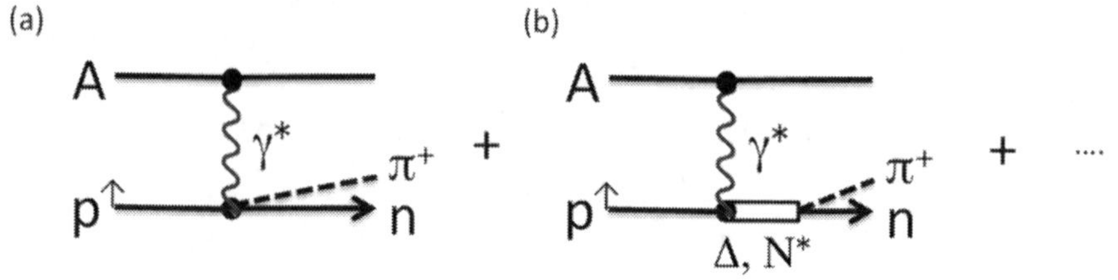

Figure 2. Feynman diagrams of UPC process to leave the high-z neutron in the final state to the proton going directions. (a) threshold photo pion production (b) Δ or N^* excitation and its decay into n+π^+ channel.

The description of A_N is thus extended from Eqn. (1) to Eqn. 4, which includes not only hadronic but also EM amplitudes:

$$\begin{aligned} A_N &= \phi^{\rm had}_{\rm flip}\phi^{\rm had}_{\rm non-flip}\sin\delta_1 + \phi^{\rm EM}_{\rm flip}\phi^{\rm had}_{\rm non-flip}\sin\delta_2 \\ &+ \phi^{\rm had}_{\rm flip}\phi^{\rm EM}_{\rm non-flip}\sin\delta_3 + \phi^{\rm EM}_{\rm flip}\phi^{\rm EM}_{\rm non-flip}\sin\delta_4 \end{aligned} \quad (3)$$

where 'EM' and 'had' stand for electromagnetic and hadronic interactions, and $\delta_1 \sim \delta_4$ are relative phases, respectively. The second and the third terms are known as Coulomb nuclear interference (CNI), which is observed to cause $< 5\%$ asymmetry of elastic scattering in p + p, and p + C processes [6].

According to MC study [7], the neutron and its counter part π^+ via UPC process is substantially boosted towards the proton beam direction and therefore the fragmenting π^+ are mostly emitted in even higher rapidity region than BBC. Only small fraction of π^+ are detected by BBC. Thus EM processes are suppressed in the BBC tagging events while enhanced in the BBC vetoed events. As a consequence, one may draw a hypothesis that the moderate A_N evolution of the BBC tagging data are consistent with the prediction based on π and a_1-Reggeon interference scenario because the data are dominated by the hadronic amplitude. Although it is not trivial to predicts the sign of the A_N caused by EM amplitude, the observed growing feature of A_N for BBC vetoed events makes sense if EM amplitude guides to opposite sign of hadronic interactions at least qualitatively. Nevertheless it is not known at all at this moment why EM should have opposite sign nor should have such a large amplitude or large relative phases between hadronic/EM spin-flip and spin-nonflip amplitudes to produce such the large asymmetries. It remains mystery.

3 Gluon Polarization

Gluon polarization within the proton is probed through the measurements of double helicity asymmetry, A_{LL} in various production channels involving gluon dominated hard scattering processes. A_{LL} is defined as the difference between cross sections for colliding bunches with the same helicity and opposite helicity, divided by the sum of them. The most statistically sufficient channels are inclusive π^0 and jet production. Shown in Fig. 3 are inclusive π^0 and jet measurements by PHENIX and STAR experiments, respectively. The polarized gluon distribution function $\Delta g(x, Q^2)$ is extracted by fitting data based on a next-to-leading order (NLO) pQCD framework[12][13]. These global fitting models are confirmed to reproduce inclusive cross sections. As a consequence of the integration of so determined $\Delta g(x, Q^2 = 10\text{GeV})$ within the range of $x > 0.05$ is calculated to be $0.20^{+0.05}_{-0.07}$ at 90 % C.L. It implies the gluon carries significant fraction of proton spin 1/2 (40% in the central value). The experimental efforts to extend the measured region to lower x region where the majority of gluon exists is now underway.

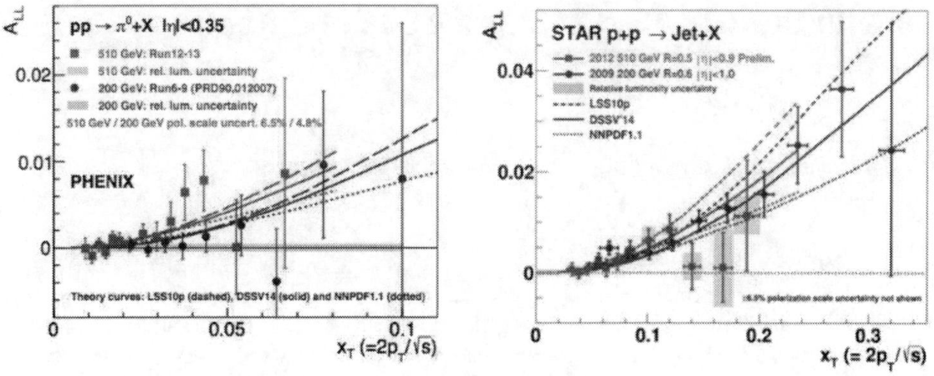

Figure 3. (left) $A_{\rm LL}$ vs. x_T for π^0 production at mid-rapidity with the point-to-point uncertainties in 200 GeV (blue circles)[8] and 510 GeV (red squares)[9] p+p collisions, compared to NLO predictions for three recent NLO global analyses [12–14] (blue curves for 200 GeV and red curves for 510 GeV). The gray/gold bands give the correlated systematic uncertainties. for inclusive π^0 measured in PHENIX at $\sqrt{s} = 510$ GeV. Theory curve is DSSV. (right) $A_{\rm LL}$ vs. xT for inclusive jet production at mid-rapidity in 200 GeV (blue circles) [6] and 510 GeV (red squares)[11] p+p collisions, compared to NLO predictions for three recent NLO global analyses [12–14] (blue curves for 200 GeV and red curves for 510 GeV).

References

[1] RHIC The Cold Nuclear Plan for 2017 to 2023, nucl-ex, arXiv:1602.03922 and the references therein.

[2] Y. Fukao et al., Phys. Lett. B 650, 325 (2007), A. Adare et al.: Phys. Rev. D 88, 032006 (2013).

[3] C. Adler, A. Denisov, E. Garcia, M. J. Murray, H. Strobele, and S. N. White, Nucl. Instrum. Methods Phys. Res., A 470, 488 (2001).

[4] B. Z. Kopeliovich, I. K. Potashnikova, and Ivan Schmidt: Phys. Rev. Lett. 64, 357 (1990).

[5] D. C. Carey et al.: Phys. Rev. D 79, 094014 (2009).

[6] I. G. Alekseev et al.: Phys. Rev. D 79, 094014 (2009).

[7] G. Mitsuka: Eur. Phys. J. C 75:614 (2015).

[8] PHENIX Collaboration, Phys.Rev. D90 (2014) 072008.

[9] A. Adare et al. (PHENIX Collaboration), Phys. Rev. D 93, 011501R (2016) .

[10] STAR Collaboration, Phys. Rev. L115, 092002 (2015).

[11] Z. Chang, Proceeding Spin-2014, arXiv:1512.05400.

[12] D. de Florian, R. Sassot, M. Stratmann, and W. Vogelsang, Phys. Rev. Lett. 113, 012001 (2014).

[13] The NNPDF Collaboration: E. R. Nocera, R. D. Ball, St. Forte, G. Ridolfi, and J. Rojo, Nucl. Phys. B 887 (2014) 276.

[14] E. Leader, A.V. Sidorov, and D.B. Stamenov, Phys. Rev. D 82 (2010) 114018.

Measurements of the elastic, inelastic and total pp cross sections with the ATLAS, CMS and TOTEM detectors

Simon Holm Stark[1,a]

[1] *Niels Bohr Institute, Blegdamsvej 17, 2100 Copenhagen OE, Denmark*

Abstract.
The elastic ($\sigma_{\rm el}$), inelastic ($\sigma_{\rm inel}$), and total ($\sigma_{\rm tot}$) cross sections are related through the differential elastic cross section using the optical theorem. ATLAS and TOTEM have used forward detectors to measure the elastic scattering at $\sqrt{s}$ = 8 TeV. The results for the total cross section are $\sigma^{\rm TOTEM}_{\rm tot,\, 8\, TeV} = 101.7 \pm 2.9$ mb and $\sigma^{\rm ATLAS}_{\rm tot,\, 8\, TeV} = 96.07 \pm 0.92$ mb and for the inelastic cross section $\sigma^{\rm TOTEM}_{\rm inel,\, 8\, TeV} = 74.7 \pm 1.7$ mb and $\sigma^{\rm ATLAS}_{\rm inel,\, 8\, TeV} = 71.73 \pm 0.71$ mb. Alternatively, the inelastic cross section can be measured in a fiducial region and extrapolated to full phase-space using theoretical models. The results from ATLAS and CMS at $\sqrt{s}$ = 13 TeV are $\sigma^{\rm CMS}_{\rm inel,\, 13\, TeV} = 71.3 \pm 3.5$ mb and $\sigma^{\rm ATLAS}_{\rm inel,\, 13\, TeV} = 79.3 \pm 2.9$ mb.

1 Introduction

The elastic ($\sigma_{\rm el}$), inelastic ($\sigma_{\rm inel}$) and total ($\sigma_{\rm tot}$) pp cross sections are fundamental quantities which cannot be calculated with perturbative QCD since processes with low momentum transfer are involved. Therefore, measurements of the cross sections provide a probe of the models used to describe the non-perturbative regime of QCD.

The optical theorem gives a relation between $\sigma_{\rm tot}$ and $\sigma_{\rm el}$:

$$\sigma^2_{\rm tot} = \frac{16\pi(\hbar c)^2}{1+\rho^2} \left.\frac{{\rm d}\sigma_{\rm el}}{{\rm d}t}\right|_{t=0}, \qquad \rho = \frac{{\rm Re}\,[F_{\rm el}(t)]}{{\rm Im}\,[F_{\rm el}(t)]},$$

where t is the four-momentum transfer. Therefore, a measurement of the differential cross section down to small angles gives not only the elastic cross section by integration, but also the total cross section from which the inelastic cross section can be deduced.

This paper is divided in two parts. The first part describes the measurement of $\sigma_{\rm el}$, $\sigma_{\rm inel}$ and $\sigma_{\rm tot}$ at $\sqrt{s}$ = 8 TeV from the differential elastic cross section with the forward detectors TOTEM and the ATLAS-ALFA subdetector. The second part describes the measurement of $\sigma_{\rm inel}$ at $\sqrt{s}$ = 13 TeV using the central detectors ATLAS and CMS in a fiducial region followed by extrapolation to full phase-space.

[a] e-mail: smortens@nbi.dk

2 Elastic, inelastic and total cross section at $\sqrt{s} = 8$ TeV

The elastic, inelastic and total cross sections are determined by ATLAS [1] and TOTEM [2–4] by measuring the differential elastic cross section as a function of the four-momentum transfer t. The four-momentum transfer can be calculated from the beam momentum of the LHC and the scattering angle of the protons. The smaller the values of t that can be measured, the smaller the extrapolation uncertainty on σ_{tot}. Therefore, both TOTEM and the ALFA subdetector of ATLAS use tracking detectors about 230 meters from the interaction point (IP) and inside Roman Pots in order to go as close as 1 mm from the beam center. There are two stations on each side of the IP allowing both a position and a local angle measurement. From the knowledge of the settings for the LHC magnet between the IP and the detectors, the scattering angle can be inferred. In addition, TOTEM has two tracking telescopes close to the IP used to measure the inelastic rate, thereby allowing a luminosity independent use of the optical theorem. The inelastic rate measurement is similar to the ATLAS and CMS approach in section 3.

The data are taken in special LHC runs with low pile-up below 0.1 and high β^* collision optics since, the minimum accessible t-value is inversely proportional to β^*. Both ATLAS and TOTEM have published results with $\beta^* = 90$ m and TOTEM also with $\beta^* = 1$ km.

The elastic events are selected when a track is reconstructed in all four stations. Furthermore, due to the back-to-back signature of an elastic event, the tracks are required to fulfill certain correlation criteria between inner-outer station and left-right side of the IP. The correlations are smeared by detector resolution and beam divergence and the boundaries are determined by simulation. The background contribution comes from beam halo protons and protons from diffractive processes that occur by happenstance. The background is determined from data by events where the two protons have an identical reconstructed vertical scattering angle. It is assumed that this will happen just as often as the real background with opposite signs. The background level is for both ATLAS and TOTEM below 0.12%. Also the track reconstruction inefficiency is data driven by determining the number of elastic events with a track in less than four stations. The t-resolution is influenced by detector resolution and beam divergence. The relative t-resolution is determined by MC to be better than 10 % and data is corrected by unfolding. The acceptance correction is also determined by simulation.

The differential elastic cross section is a superposition of the strong interaction amplitude and the Coulomb amplitude. Adding in quadrature gives

$$\frac{d\sigma_{\mathrm{el}}}{dt} \propto \frac{G^4(t)}{|t|^2} + \sigma_{\mathrm{tot}}^2(1+\rho^2)\cdot\exp(-B|t|) - \frac{\sigma_{\mathrm{tot}}G^2(t)}{|t|}\left[\sin(\phi(t)) + \rho\cos(\phi(t))\right]\cdot\exp\left(\frac{-B|t|}{2}\right),$$

where G is the proton form factor, B is the nuclear slope assuming a purely exponential fall off and ϕ is the phase between between the Coulomb and nuclear interaction. The first term in the equation is the Coulomb part, the second is the nuclear part, and the third is the interference term. Figure 1 shows a simulation of the contributions from the three terms in the differential elastic cross section.

The unfolded differential cross section is fitted with σ_{tot} and B as free parameters. A lower bound on the fit range is determined by requiring a certain acceptance level in the detectors. An upper bound is set to ensure that the exponential decay is a valid approximation. At $\beta^* = 90$ m, the lower bound is about $|t| \sim 0.01$ GeV^{-2}, hence the sensitivity to the interference and Coulomb terms is absent and ρ must be taken from global model extrapolations. At $\beta^* = 1$ km (only TOTEM), the lower bound is $|t| = 6 \cdot 10^{-4}$ GeV^{-2} which also allows a measurement of ρ.

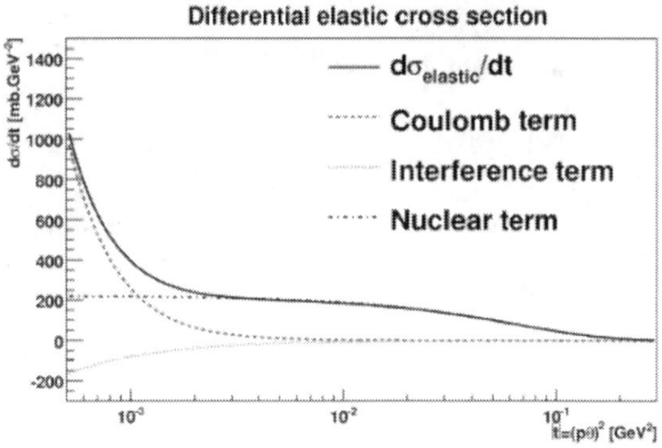

Figure 1. Simulation of the differential elastic cross section with σ_{tot} = 100 mb, B = 18 GeV^{-2} and ρ = 0.13.

2.1 Results

The results are shown in Table 1. The difference in σ_{tot} between ATLAS and TOTEM at β^* = 90 m corresponds to 1.9σ. Figure 2 shows the results along with measurements at other center-of-mass energies and a fit to data. The increase of the cross sections with energy has already been observed at lower $\sqrt{s}$ and is still evident at $\sqrt{s}$ = 8 TeV.

Table 1. Results for the elastic, inelastic and total pp cross sections at $\sqrt{s}$ = 8 TeV using elastic scattering and the optical theorem.

	ATLAS (β^* = 90 m)	TOTEM (β^* = 90 m)	TOTEM (β^* = 1 km)
σ_{tot}	96.07 ± 0.92	101.7 ± 2.9	102.9 ± 2.3
σ_{el}	24.33 ± 0.39	27.1 ± 1.4	-
σ_{inel}	71.73 ± 0.71	74.7 ± 1.7	-

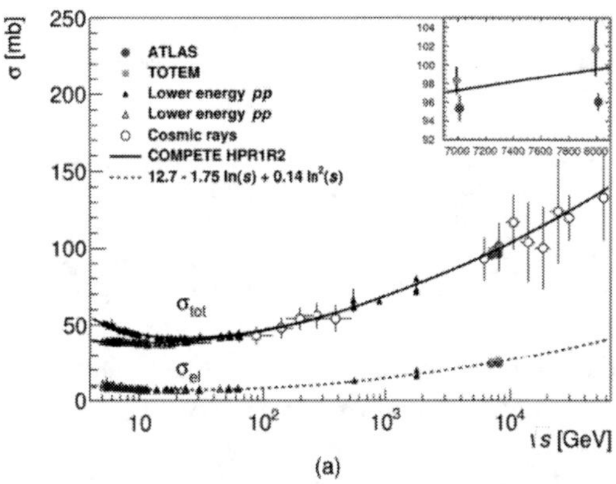

Figure 2. The total and elastic cross sections as a function of center-of-mass energy. The fit has not been updated with the ATLAS $\sqrt{s}$ = 8 TeV result.

The assumption of a purely exponential decay of the elastic nuclear amplitude was investigated by TOTEM. Figure 3 shows the differential elastic cross section with respect to a reference function using a pure exponential. Only the t-dependent systematic uncertainties, i.e. without normalization, are important for this measurement. The purely exponential form is excluded by TOTEM by more than 7σ. ATLAS had t-dependent uncertainties that were too great to perform a similar test. The beam energy is the dominating contribution, and whereas TOTEM used a 0.1% uncertainty, ATLAS used a

0.65% uncertainty. Instead, ATLAS observed the effect of different *B*-parametrizations and measured the RMS value of σ_{tot} to be 0.28 mb.

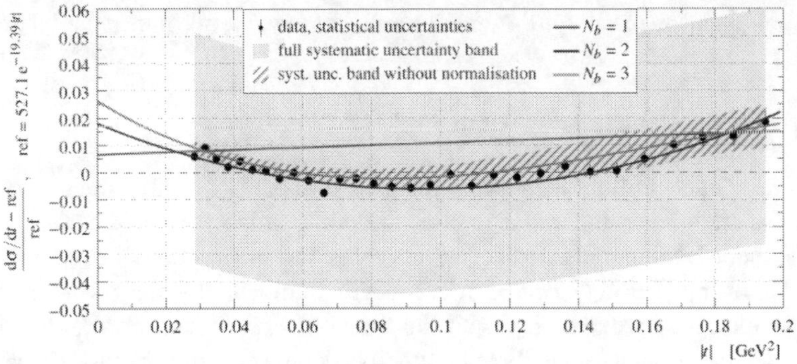

Figure 3. Differential elastic cross section relative to a reference exponential. The colored lines are fits to data where a polynomial of order N_b is used for the nuclear slope.

From the $\beta^* = 1$ km data, TOTEM was able to measure the ρ-parameter to be:

$$\rho = 0.12 \pm 0.03\,,$$

which is compatible with a model extrapolation from lower energies giving $\rho = 0.140 \pm 0.007$.

3 Inelastic cross section at $\sqrt{s} = 13$ TeV

The strategy for determining the inelastic cross section at $\sqrt{s} = 13$ TeV used by ATLAS [5] and CMS [6] is to measure the inelastic cross section in a fiducial region and then use theoretical models to extrapolate to the full phase-space. The fiducial inelastic cross section is determined from the number of observed collisions corrected for background, pile-up, detector efficiencies and luminosity.

3.1 Experimental setup

ATLAS uses thin plastic scintillation counters called the minimum-bias trigger scintillators (MBTS). They are placed at $z = \pm 3.6$ m on both sides of the IP and have a pseudo-rapidity coverage of $2.07 < |\eta| < 3.86$. CMS uses the hadron forward calorimeters (HF) and the CASTOR calorimeter, which cover $3.0 < |\eta| < 5.2$ and $-6.6 < \eta < -5.2$, respectively. The HFs are built from iron absorbers and quartz fiber for read out. CASTOR has tungsten for absorption and quartz plates for read out. Data was taken in special LHC runs with low pile-up (< 0.5) in order to minimize the effect of multiple collisions.

3.2 Results

The results for the fiducial inelastic cross section for CMS are:

$$\sigma_{\text{inel}}^{\text{fid,HF}} = 65.8 \pm 0.8(\text{exp.}) \pm 1.8(\text{lum.})\ \text{mb}\,,$$
$$\sigma_{\text{inel}}^{\text{fid,HF+CASTOR}} = 66.9 \pm 0.4(\text{exp.}) \pm 2.0(\text{lum.})\ \text{mb}\,.$$

For ATLAS, the result is:

$$\sigma_{\text{inel}}^{\text{fid}} = 68.1 \pm 0.6(\text{exp.}) \pm 1.3(\text{lum.})\ \text{mb}\,.$$

Since the fiducial region in ATLAS and in CMS using only HF is the same, the ATLAS $\sigma^{\rm fid}_{\rm inel}$ and the CMS $\sigma^{\rm fid,HF}_{\rm inel}$ are directly comparable and they agree to within less than one standard deviation.

For each model, an extrapolation factor from fiducial to full phase-space can be found. CMS extrapolates the HF+CASTOR fiducial result using the average of the extrapolation factor for all their models as the nominal value and then taking the maximum difference between the models for systematics. The result is:

$$\sigma^{\rm CMS}_{\rm inel} = 71.3 \pm 0.5({\rm exp.}) \pm 2.1({\rm lum.}) \pm 2.7({\rm ext.})\ {\rm mb}\ .$$

ATLAS tunes the theoretical models before using them for the extrapolation of the fiducial cross section to full phase-space. The fraction of the inelastic cross section coming from diffractive processes, f_D, differs between the theoretical models. However, it is related to the fraction of events with hits in ATLAS on only one side of the IP, R_{SS}, since such events are dominated by single diffractive processes. Figure 4 shows the measured value of R_{SS} and the functional dependence on f_D for the different models from which the amount that f_D needs to be tuned is deduced. After the models have been tuned using f_D, the predicted hit multiplicity in the MBTS for the different models are compared to data. Only the PYTHIA8 DL and PYTHIA8 MBR models give a reasonable description of the data, hence only these are used by ATLAS in the extrapolation. ATLAS constrains the model dependence

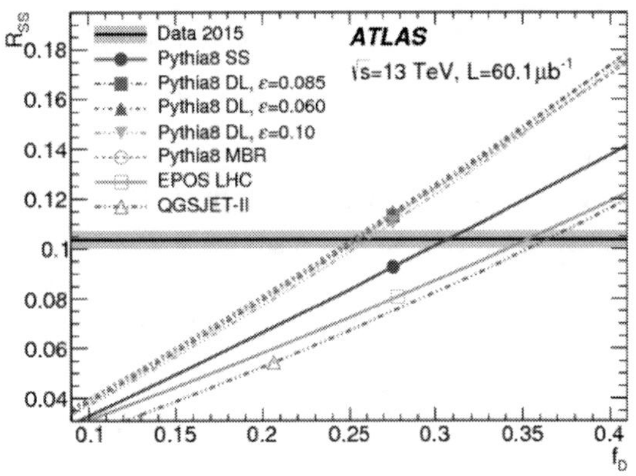

Figure 4. The relation between the ratio of single-sided events (R_{SS}) and the fraction of diffractive processes in inelastic scattering (f_D) for different models. The default value of f_D in each model is shown with a marker.

in the extrapolation using a previous and precise measurement of $\sigma_{\rm inel}$ at $\sqrt{s} = 7$ TeV done with the ALFA subdetector [7] using the same strategy as described in Section 2. The full phase-space inelastic cross section is calculated as

$$\sigma_{\rm inel,\,13\,TeV} = \sigma^{\rm fid}_{\rm inel,\,13\,TeV} + (\sigma^{\rm ALFA}_{\rm inel,\,7\,TeV} - \sigma^{\rm fid}_{\rm inel,\,7\,TeV}) \times \frac{\sigma^{\rm MC,\,fid}_{\rm inel,\,13\,TeV}}{\sigma^{\rm MC,\,fid}_{\rm inel,\,7\,TeV}}$$

with PYTHIA8 DL as the nominal model and PYTHIA8 MBR used as a systematic uncertainty. The result is:

$$\sigma^{\rm ATLAS}_{\rm inel} = 79.3 \pm 0.6({\rm exp.}) \pm 1.3({\rm lum.}) \pm 2.5({\rm ext.})\ {\rm mb}\ .$$

The results are summarized in Figure 5 and compared to the models used by CMS. Both CMS and the first ATLAS result tend to be lower than expected from the model predictions for both the fiducial and the full phase-space inelastic cross sections. An update of the ATLAS result shows better agreement with the model expectations.

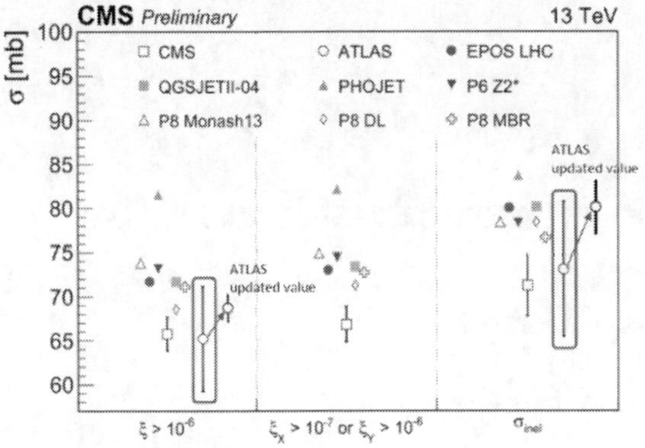

Figure 5. The inelastic cross section in different phase space regions compared to the models used by CMS.

4 Conclusion

The elastic, inelastic and total cross sections at $\sqrt{s}$ = 8 TeV have been measured by ATLAS and TOTEM using the differential elastic cross section. The total cross section is observed to increase with center-of-mass energy. TOTEM has excluded a purely exponential decay of the nuclear elastic amplitude by more than 7σ. Furthermore, they have measured $\rho = 0.12 \pm 0.03$, which is in good agreement with a model extrapolation from lower energies.

The inelastic cross section at $\sqrt{s}$ = 13 TeV has been measured by ATLAS and CMS. Theoretical models were used to extrapolate the fiducial cross section to full phase-space. For several models, ATLAS has observed deviations from data in the fraction of diffractive processes and in hit multiplicities in the detector. The predicted cross section agrees well with the ATLAS measurement, whereas the CMS value is lower than anticipated. By comparison of the ATLAS and CMS $\sqrt{s}$ = 13 TeV results wrt. the ATLAS and TOTEM $\sqrt{s}$ = 8 TeV results, one can observe that there has been an increase in the inelastic cross section wrt. $\sqrt{s}$ in the case of ATLAS, but not in the case of CMS.

References

[1] ATLAS Collaboration, Phys. Lett. B **761**, 158-178 (2016)
[2] TOTEM Collaboration, Phys. Rev. Lett. **111**, 012001 (2013)
[3] TOTEM Collaboration, Nucl. Phys. B **899**, 527-546 (2015)
[4] TOTEM Collaboration, CERN-PH-EP-2015-325
[5] ATLAS Collaboration, arXiv:1606.02625
[6] CMS Collaboration, CMS-PAS-FSQ-15-005
[7] ATLAS Collaboration, Nucl. Phys. B **889**, 486-548 (2014)

EPJ Web of Conferences **141**, 03008 (2017) DOI: 10.1051/epjconf/201714103008
ISMD 2016

Recent results on diffractive and forward physics at HERA

Yuji Yamazaki[1,a] on behalf of the H1 and ZEUS Collaborations

[1] *Rokkodai-cho 1-1, Nada, Kobe, 657-8501 Japan*

Abstract.
Recent studies on diffractive processes in ep scattering at HERA are reviewed. Heavy quark production in diffractive processes was studied by tagging a D^* meson in diffractive DIS events. The cross section for events with a prompt photon in the final state was also measured. The data are well reproduced by models using diffractive parton densities determined from inclusive diffractive DIS cross sections. Two studies were also made on exclusive vector meson production. The cross-section ratio $\sigma(\psi(2S))/\sigma(\psi(1S))$, which reflects the difference in transverse shape of the charmonium wave functions, are compared to various models. Moreover, the diffractive ρ^0 photoproduction with a forward neutron, which has two vertices with small-p_T recoil, were measured to study factorisation properties. The data show some indication of re-scattering.

1 Introduction

HERA, the only $e^{\pm}p$ collider, has provided opportunities to study high-energy electron-proton collisions beyond the electroweak scale. The main process, the Deep-Inelastic Scattering (DIS), probes the internal quark structure of the proton via exchanged virtual photons (γ^*). The point-like nature of the virtual photon with $Q^2 \gg \Lambda_{\rm QCD}$, where Q^2 is the negative mass of the virtual photon, ensures that the photon probes the quarkonic structure of the proton.

The DIS process can also be interpreted as collisions of $\gamma^* p$, which allows discussion of the DIS and photoproduction processes in the same framework; the DIS process can be considered as a special case of $\gamma^* p$ collisions where the Q^2 of the γ^* is large ($\gg \Lambda_{\rm QCD}$). In the photoproduction process, $Q^2 \approx 0$ and no hard scale is present if there is no high-p_T particle in the final state and the process is described by the exchange of particles. In processes where a hard scale is present, individual partons are visible and the cross sections can be calculated in perturbative QCD.

Diffractive processes at HERA are characterised as a colourless exchange between the virtual photon and the proton, where the photon system dissociates into a hadronic system X while the proton may remain intact or may dissociate to a system Y. The the momentum exchanged at the $p \to Y$ vertex, t, is typically small ($-t \le \Lambda_{\rm QCD}$) unless the mass of the system Y is large. There it is expected that the exchanged state at the proton vertex is particle-like e.g. pions, Reggeons or "Pomerons" ($\mathbb{P}$). It may be, however, that the exchange can be explained by partonic colourless states e.g. (dressed) two gluons and can be interpreted by pQCD calculations.

[a]e-mail: yamazaki@phys.sci.kobe-u.ac.jp

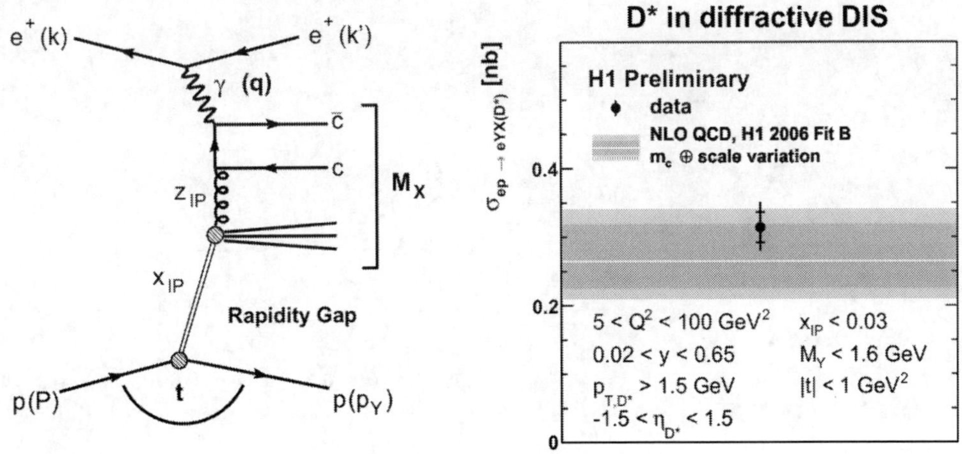

Figure 1. (left) A typical diagram of charm quark pair production in photon-dissociative diffractive processes in *ep* scattering. (right) The charm quark production cross section in comparison to the NLO QCD calculation using H1 2006 Fit B diffractive PDF [1].

HERA allows quantitative investigations to be made of the partonic content of such exchanges by measuring the cross section behaviour of Diffractive DIS (DDIS), which is defined as DIS processes with diffractive final states (experimentally defined as a presence of large rapidity gap, observation of the fast-going proton or neutron, etc.). The gluon density in the exchange can also be extracted through the scaling violation behaviour of the DDIS cross sections in Q^2 (for example in Ref. [1]), provided that the factorisation holds between the γ^* vertex and other part of the DDIS diagram. The factorisation was theoretically proven for DDIS but not for other processes, such as diffraction in photoproduction or hadron-hadron collisions, even with a presence of a hard scale. The factorisation breaking may lead to "absorption" or re-scattering between two diffractive systems. The cross sections are then suppressed from the theoretical predictions based on the diffractive parton densities extracted from DDIS. Such behaviour was first largely advertised by the Tevatron measurements on dijet production in diffraction [2]. There are several observations also at HERA, e.g. photoproduced dijets [3] as well as the t−distribution in forward neutron production [4].

In the following sections, four recent results are presented. D^* production in diffractive DIS [5] and prompt photon production in diffractive photoproduction [6] are compared to the models based on the diffractive PDFs. The production ratio $\psi(2S)/\psi(1S)$ in diffractive DIS [7] is compared to various models with approaches based on perturbative QCD in Section 4. Exclusive production of ρ^0 in photoproduction with a fast neutron [8] is reviewed in Section 5 in the context of the absorptive phenomena.

 DOI: 10.1051/epjconf/201714103008

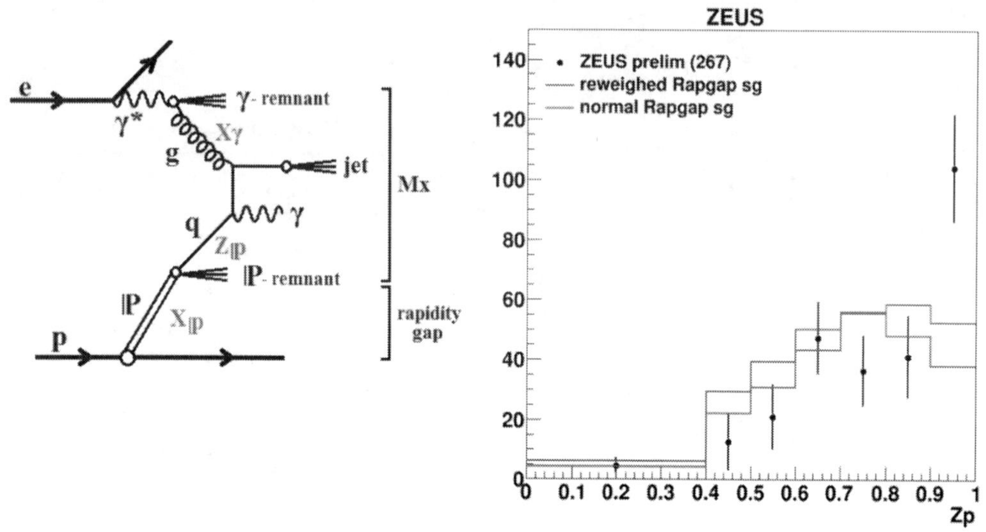

Figure 2. (left) An example of diagrams where a prompt photon is produced with a quark jet in the resolved diffractive photoproduction process in *ep* scattering. (right) The $z_\mathbb{P}$ dependence of the cross sections with a prompt photon, a jet and a large rapidity gap in photoproduction.

2 D^* production in diffractive DIS

Charm quarks in diffractive DIS are predominantly produced through the so-called boson-gluon fusion diagram (Fig. 1(left)). The cross sections with a D^* in diffractive DIS are therefore sensitive to the gluonic content of the diffractive exchange.

The cross section was measured with 281pb^{-1} of *ep* collisions by the H1 Collaboration in the deep-inelastic regime ($5 < Q^2 < 100\text{GeV}^2$) with a presence of a large rapidity gap from the edge of the detector in the proton direction, hence selecting either events with proton staying intact or dissociated to a small mass state Y.

The cross section (Fig. 1(right)) shows good agreement with the NLO prediction within the uncertainty if using the diffractive PDF H1 2006 Fit B [1] extracted from the scaling violation of the inclusive diffractive DIS data. This indicates that the factorisation holds in diffractive DIS. Differential cross sections are also measured as a function of other variables described in Fig. 1(right). All the distributions are well reproduced by the NLO calculation, providing a cross-check on the diffractive parton densities, in particular the gluon density.

3 Prompt photon production in diffractive photoproduction

This measurement requires an isolated photon and a jet in the photon dissociation system in addition to the presence of a large rapidity gap in photoproduction events. The advantage of a photon + a jet over dijet events is (a) less sensitive to the QCD radiation and (b) more sensitivity on quark content of the diffractive exchange as the prompt photon couples only to quarks (Fig. 2(left)).

The fraction of parton momentum involved in the hard scattering with respect to the two incoming "particles", for the photon (x_γ) and for the diffractive exchange ($z_\mathbb{P}$), can be reconstructed using the momentum of the photon and jet. The events are further classified as "direct" and "resolved" events using the reconstructed x_γ. In this analysis, $x_\gamma > 0.8$ is defined as direct-enriched region and the rest

 DOI: 10.1051/epjconf/201714103008

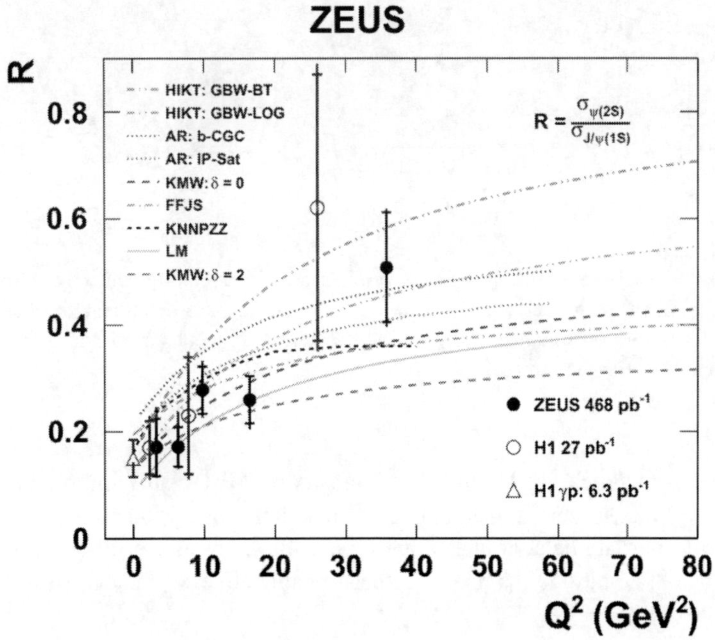

Figure 3. The cross section ratio $\sigma(\psi(2S))/\sigma(\psi(1S))$ as a function of Q^2 in comparison to various models.

is resolved-enriched. The fraction of the resolved and direct processes in the RAPGAP model was reweighted so as to describe the data.

Theoretical prediction by RAPGAP reproduces well various distributions, such as E_T and η of the photon and jets. The only exception is $z_{\mathbb{P}}$ distribution (Fig. 2(right)). It fails to describe the peak near $z_{\mathbb{P}} = 1$ observed in the data. This may indicate that the model lacks some diagrams like two partons in the diffractive exchange with very asymmetric configuration i.e. one of the two partons carries almost the entire momentum of the diffractive exchange. Such diagrams are higher-twist contribution and not included in the framework of the QCD fit to diffractive DIS cross sections.

4 The production ratio $\psi(2S)/\psi(1S)$ in diffractive DIS

This process measures the cross section ratio of the quasi-elastic charmonium production $ep \to e\psi Y$ where Y is either a proton or a state with small mass. The ratio $\sigma(\psi(2S))/\sigma(\psi(1S))$ is expected to have a strong dependence in Q^2, which is originated from different wave function shape of $1S$ and $2S$ onium states. The wave function of the $1S$ state is always positive, while the function for the $2S$ state has a node at around 0.35 fm.

The vector boson is generated from a virtual photon dissociated into a $q\bar{q}$ state. The transverse profile of the $q\bar{q}$ state wave function is similar to that of the charmonium for low Q^2 ($Q^2 \simeq m_\psi^2$). In this case, the convolution of the two wave functions, $q\bar{q}$ state and the $\psi(2S)$ state, may become small since the wave function changes the sign along the transverse position. The $q\bar{q}$ state, however, becomes much smaller than the ψ wave function for $Q^2 \gg m_\psi$ and it becomes likely that the two quarks probe similar values of wave functions. No such cancellation is expected in the convolution in this case. This predicts that the cross section ratio is to increase with Q^2.

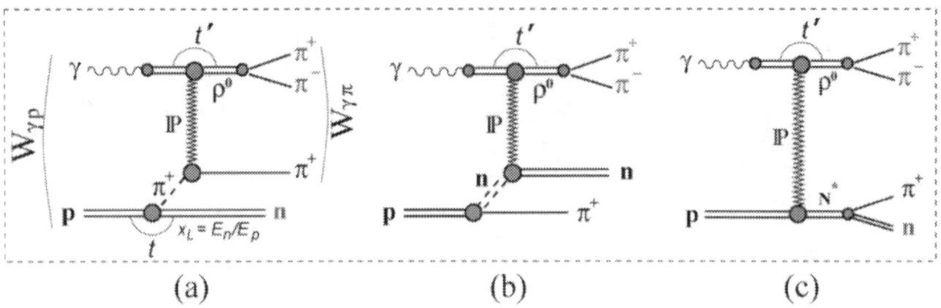

Figure 4. Diagrams for processes contributing to ρ^0 production with a leading neutron at HERA. Three processes are considered as signal: (a) pion exchange (b) neutron exchange and (c) a neutron produced from the decay an excited nucleon or from a non-resonant $\pi + n$ state. The event with a neutron produced from dissociation of the system Y is considered as background (d).

The data in Fig. 3 was measured with full luminosity of HERA runs and greatly improves the precision of the previous measurement. The data confirms qualitatively the above-explained expectation. Looking into detail, the data behaviour is consistent with many of the models although some of the models are excluded. The ratio is flat against other variables like W, the centre-of-mass energy of the $\gamma^* p$ system, and $|t|$.

5 Exclusive ρ^0 photoproduction with a fast neutron

Exclusive ρ^0 production with a fast neutron is a process without the explicit presence of a hard scale since both the outgoing neutron and ρ^0 do not carry high transverse momentum. It is therefore appropriate to describe the process in terms of particle exchange. For events with large $x_L = E_n/E_p$, the dominant process is one-pion exchange (OPE) (Fig. 4 (a)). This allows the diffractive production of $\gamma\pi$ scattering to be studied, especially by comparing to the results from exclusive production of ρ^0 without a tagged neutron.

There are other two other diagrams (Fig. 4 (b) and (c)), which do not leave any significant activity in the forward part of the detector and are considered as signal in this analysis. The neutrons through the proton dissociation system Y is considered as background and was subtracted.

The shape of the measured neutron spectrum is well reproduced by choosing appropriate "pion flux" models among those proposed (see the Ref.[8] for more detail). The production rate is, however, suppressed by about a factor two. The cross section ratio $\sigma_{\gamma\pi\to\rho^0\pi}/\sigma_{\gamma p\to\rho^0 p}$ was measured 0.25 ± 0.06, which is far less than that expected from the additive quark model (~ 0.6). Similar suppression was observed in inclusive neutron production[10]. This may be an indication of absorptive correction.

The exponential slope parameters of the leading neutron b_n determined from the p_T dependence $\sigma(x_L, p_T) = A\exp(b_n p_T^2)$ was measured at HERA for inclusive neutron production [9] and compared to models with various pion flux, giving constraint to models (Fig. 5(right)). A similar measurement was made for the exclusive ρ^0 production and is shown in Fig. 5. The behaviour is quite different from the inclusive measurement at high x_L. This may be a further indication of the presence of re-scattering, which predicts to suppress the events with a neutron with large p_T.

References

[1] H1 Collaboration, A. Aktas et al., Eur. Phys. J. **C48**, 715 (2006).

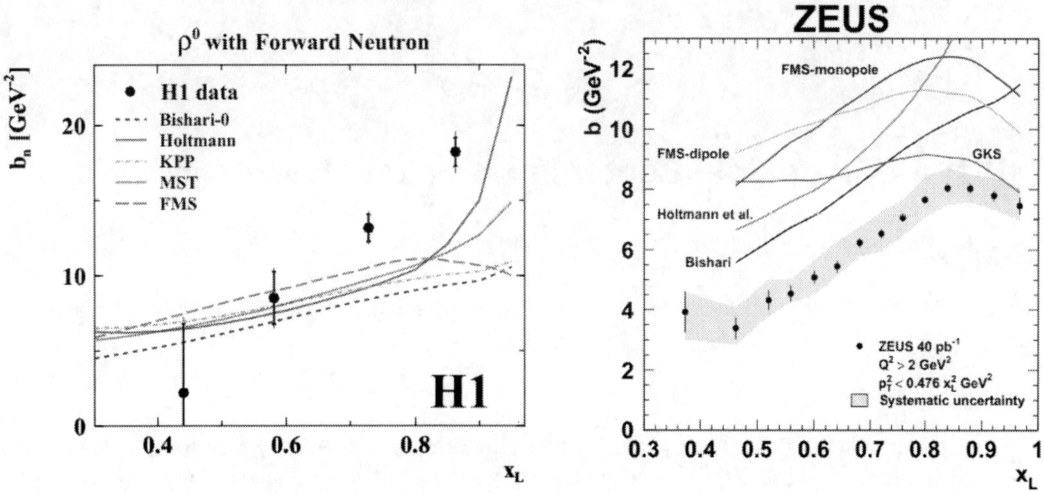

Figure 5. (left) The exponential slope parameter b_n for the exclusive ρ^0 production. (right) The corresponding parameters for inclusive DIS with a leading neutron[9].

[2] CDF Collaboration, T. Affolder et al., Phys. Rev. Lett. **84**, 5043 (2000).

[3] H1 Collaboration, A. Aktas et al., Eur. Phys. J. **C51**, 549 (2007).

[4] ZEUS Collaboration, S. Chekanov et al., Nucl. Phys. **B776**, 1 (2007).

[5] H1 Collaboration, H1prelim-16-011, available under: `http://www-h1.desy.de/h1/www/publications/htmlsplit/H1prelim-16-011.long.html`.

[6] ZEUS Collaboration, ZEUS-prel-15-001, available under: `https://inspirehep.net/record/1368789?ln=en`

[7] ZEUS Collaboration, H. Abramowicz et al., Nucl. Phys. **B909**, 934 (2016).

[8] H1 Collaboration, V. Andreev et al., Eur. Phys. J. **C76**, No. 1, 41 (2016).

[9] ZEUS Collaboration, S. Chekanov et al, Nucl. Phys. **B776**, 1 (2007).

[10] ZEUS Collaboration, S. Chekanov et al, Nucl. Phys. **B637**, 3 (2002); H1 Collaboration, F.D. Aaron et Al, Eur. Phys. J **C68**, 381 (2010).

GSDC: A Unique Data Center in Korea for HEP research

Sang-Un Ahn[1,a]

[1] *Korea Institute of Science and Technology Information, 245 Daehak-ro Yuseong-gu 34141 Daejeon, South Korea*

Abstract. Global Science experimental Data hub Center (GSDC) at Korea Institute of Science and Technology Information (KISTI) is a unique data center in South Korea established for promoting the fundamental research fields by supporting them with the expertise on Information and Communication Technology (ICT) and the infrastructure for High Performance Computing (HPC), High Throughput Computing (HTC) and Networking. GSDC has supported various research fields in South Korea dealing with the large scale of data, e.g. RENO experiment for neutrino research, LIGO experiment for gravitational wave detection, Genome sequencing project for bio-medical, and HEP experiments such as CDF at FNAL, Belle at KEK, and STAR at BNL. In particular, GSDC has run a Tier-1 center for ALICE experiment using the LHC at CERN since 2013. In this talk, we present the overview on computing infrastructure that GSDC runs for the research fields and we discuss on the data center infrastructure management system deployed at GSDC.

1 Introduction

Nowadays, we can access to e-mails, files and documents from anywhere at anytime via web[1] and apps not only on laptop or desktop but also on mobile devices. This is true when we use commercial services such as Google Docs[2] or Apple iCloud[3] in which our e-mails, files and documents are always accessible when we are connected to the Internet. Also it is true in research fields where IT infrastructure is crucial to store and process the large scale of data produced by the instruments and to make the data accessible to stakeholders. These can be done thanks to so called Data Centers and we have benefited from them for many years without knowing their existence. This paper discusses about Data Center based on practical experience of operations of Global Science experimental Data hub Center (GSDC) at Korea Institute of Science and Technology Information (KISTI) in terms of which a Data Center should satisfy for highly available infrastructure and introduces GSDC as an example of the highly available infrastructure for IT operations.

2 Data Center for Big Data

Recently Data Centers have been built for facilitating the process of Big Data in business. The trends in business have shifted to provide services and products tailored for individual customers based

[a] e-mail: sahn@kisti.re.kr

upon their personal interests. The information related to each individual's interest remains to his or her Internet activities via mobile devices and personal computers, for example, keywords entered in search engines, web pages or images clicked, and articles to read. The amount of these information is too big to be dealt with a mainframe or a conventional supercomputer so that the Data Center's role became crucial. Also, Data Mining[4] on these scattered information in order to extract a few specific keywords or items, and even more complicated pattern recognition techniques such as machine-learning[5] or deep-learning[6] are important in the same regards.

In research fields, development of high resolution instrumentation for better results has lead to the huge production of experimental data. Usually the amount of produced data is far beyond of the capability of a researcher's workstation or a moderate cluster in his or her laboratory. For example, four giant experiments built at European Organization for Nuclear Research (CERN)[7] to exploit the Large Hadron Collider (LHC)[8] – ALICE[9], ATLAS[10], CMS[11], and LHCb[12] – have produced a hundred petabytes of data since the start of operations in 2009[13]. A hierarchical model[14] for distribution of the data produced from the experiments was introduced at the design step of the LHC and it has evolved to the formation of the intercontinental collaboration of Data Centers around the world, which is called the Worldwide LHC Computing Grid[15].

A Data Center can be defined as a facility used to house computer systems and associated components, such as telecommunications and storage systems. It generally includes redundant or backup power supplies, redundant data communications connections, environmental controls (e.g. air conditioning, fire suppression) and various security devices. Large Data Centers are industrial scale which consumes electricity as much as a small town. The keywords for Data Center are housing computer systems, redundancy or backup, power efficiency and security.

3 Highly Available IT Infrastructure

The principle goal of Data Center in both business and research fields is to preserve data – user's e-mails, files and documents or invaluable experimental data produced from the giant instrumentation – safely. In order for Data Center to achieve it, a fault-tolerant system architecture – in other words, redundancy – is mandatory among the required design features[16]. Nowadays, IT operations are a crucial aspect of most organizational operations around the world. If a IT infrastructure for a service has a single server connected with a single switch or router, and one of them goes failure, the entire IT service will be stopped. Having backup to keep IT service alive helps avoid the worst scenario, i.e. business continuity relies on the reliability of IT infrastructure which is the key for the business success.

There are two folds to have IT infrastructure highly available: not allowing a single point of failure in the system architecture and building a fast recovery mechanism. No single point of failure can be achieved by having all critical points in pairs. For example, each server can have dual power supplies in which it prevents from accidental power cut from one source. For a cluster, servers can be connected to two or more physical switches in order to have a backup uplink for networking. In a scale of Data Center, data stored in a Data Center can be replicated to the other sites so that the access to the data should be available all the time unless all sites are closed.

Recovery cycle for a IT service is carried out as following: 1) malfunctioning or glitches are occurred 2) they are detected by monitoring system and alarms are raised 3) for a critical case, a specific service has to be rolled back to its original state in which it can be achieved by automated provisioning and configuration management 4) the IT service comes back online. Figure 1 describes the recovery cycle for a IT service. In this cycle, automated provisioning and configuration management coupled with alarm system are crucial to have a fast recovery system for IT services. To have a backup for

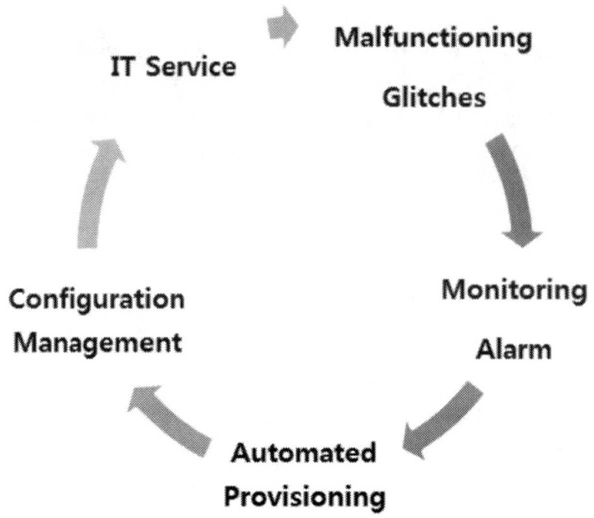

Figure 1. Recovery cycle for a IT service in case that malfunctioning or glitches are occurred. Automated provisioning and configuration management are crucial in this cycle.

Table 1. Summary of specification of computing resources provided by GSDC for experiments

Experiment	Type	CPU	Disk	Tape
ALICE	Grid	3,000	1.5 PB	1.5 PB
CMS	Non-Grid	500	500 TB	-
Belle II	Grid	300	80 TB	-
LIGO	Non-Grid	576	150 TB	-
RENO	Non-Grid	192	250 TB	-
Genome	Non-Grid	192	250 TB	-

the data or meta-data in the service is mandatory. Virtualization[18] and High Availability add-on[19] provides more options for reliable and agile IT infrastructure.

4 Global Science experimental Data hub Center

GSDC is a government funding project to promote the fundamental research in South Korea by providing IT infrastructure. In the first phase of its project, GSDC focused on collider physics conducted at CERN, FNAL[20] and KEK[21]. Now in the second phase, GSDC expand its services to domestic region where requiring large amount of storage and computing power. The final goal is to become the National Data Center for fundamental research. The roadmap of GSDC is described in figure 2.

GSDC has supported officially six experiments: ALICE, CMS, Belle II[22], LIGO[23], RENO[24] and Genome project. The specification of computing resources provided for each experiment is summarized in table 1. To support each experiment, a certain level of complexity is required to build essential IT services. For example, User Interface (UI) to give an access point to users, Computing Element (CE) to process user's tasks, Storage Element (SE) to store and manage data.

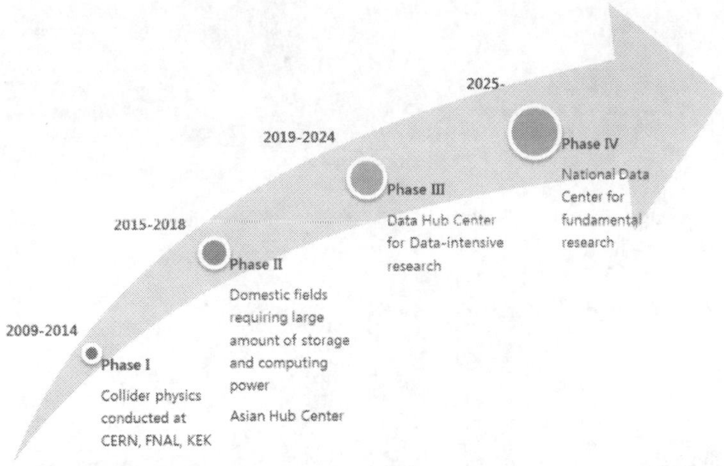

Figure 2. A roadmap of GSDC project. Now it is in the second phase and the final goal is to become the National Data Center for fundamental research in 10 years.

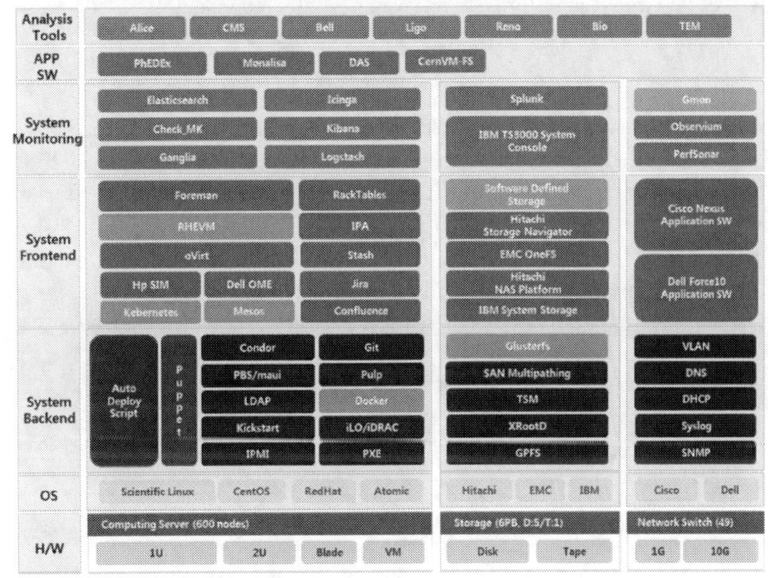

Figure 3. GSDC System Administration Software Stack

However, these are like the tip of iceberg which means that the services provided to users, i.e. UI or CE, are only the seen part of the whole IT infrastructure. A complicated system administration software stack shown in figure 3 is in behind the scene and the computing and storage resources are provided upon the stack. For example, remote control on hardware can be done via IPMI tools, and installation of Operating System and required packages for a specific service can be done via provisioning tool. For provisioning and configuration management, GSDC introduced Foreman[25] and Puppet[26]. Most of all services are profiled in Puppet code written in Ruby. Puppet is coupled

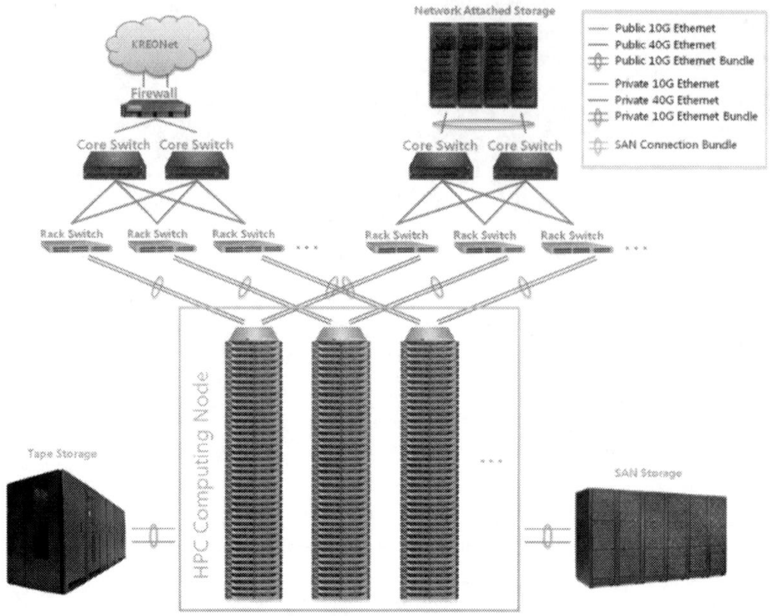

Figure 4. GSDC System Architecture for Network which is fully redundant.

with Foreman so that a server can be defined to run a specific service at the provisioning step. The entire infrastructure of GSDC is monitored by various kind of tools, such as Elasticsearch coupled with Logstash to query service status and it is visualized by Kibana[27]. Storage system and network are monitored by other tools, for example, Splunk[28] and Observium[29]. For Grid systems, perfSONAR[30] services are enabled to monitor network status among sites periodically. The system architecture for GSDC network is fully redundant as shown in figure 4.

It is worth to point out that the entire software stack for system administration is based upon open-sourced projects. It requires a high skillset to deploy and configure in order to satisfy production level of service.

5 Conclusion

Recently Data Centers are built for processing and storing the large scale of data. Highly available IT infrastructure is the key of the success of the business and fundamental research. GSDC is the sole Data Center in South Korea which government funds for fundamental research, and it is targeting the most reliable Data Center in the world.

Acknowledgement

This work was supported by the National Research Foundation of Korea (NRF) through contract N-16-NM-CR01 and the Program of Construction and Operation for Large-scale Science Data Center (K-16-L01-C06).

References

[1] T. Berners-Lee, M. Fischetti and M. L. Foreword By-Dertouzos, *Weaving the Web: The original design and ultimate destiny of the World Wide Web by its inventor* (HarperInformation, 2000)
[2] Google Docs, http://docs.google.com/
[3] Apple iCloud, http://www.icloud.com/
[4] M. J. Berry and G. Linoff, *Data mining techniques: for marketing, sales, and customer support* (John Wiley & Sons, 1997)
[5] C. Andrieu, N. De Freitas, A. Doucet and M. I. Jordan, Machine learning **50**, 5-43 (2003)
[6] Y. LeCun, Y. Bengio and G. Hinton, Nature **521**, 436-444 (2015)
[7] European Organization for Nuclear Research, http://www.cern.ch/
[8] L. Evans and P. Bryant, Journal of Instrumentation **3**, S08001 (2008)
[9] The ALICE Collaboration, Journal of Instrumentation **3**, S08002 (2008)
[10] The ATLAS Collaboration, Journal of Instrumentation **3**, S08003 (2008)
[11] The CMS Collaboration, Journal of Instrumentation **3**, S08004 (2008)
[12] The LHCb Collaboration, Journal of Instrumentation **3**, S08005 (2008)
[13] C. Sutton, CERN Courier **53**, 6 (2013)
[14] Models of Networked Analysis at Regional Centres for LHC Experiments (MONARC), http://monarc.web.cern.ch/MONARC/
[15] The Worldwide LHC Computing Grid (WLCG), http://wlcg.web.cern.ch/
[16] L. A. Barroso, J. Clidaras, and U. Hölzle, Synthesis lectures on computer architecture **8**, 1-154 (2013)
[17] M. J. Earl, *Information Management* (Oxford University Press, Oxford, UK, 1996) 456-484.
[18] P. Barham et al, ACM SIGOPS Operating Systems Review **37**, 164-177 (2003)
[19] Red Hat High Availability add-on Overview, https://access.redhat.com/documentation/en-US/Red_Hat_Enterprise_Linux/7/html/High_Availability_Add-On_Overview/index.html
[20] Fermi National Accelerator Laboratory, http://www.fnal.gov/
[21] High Energy Accelerator Research Organization, http://www.kek.jp/
[22] Belle II Experiment, https://www.belle2.org/
[23] LIGO Scientific Collaboration, http://ligo.org/
[24] Reactor Experiment for Neutrino Oscillation, http://reno01.snu.ac.kr/~reno/wiki/index.php/Main_Page
[25] Foreman, https://theforeman.org/
[26] Puppet, https://puppet.com/
[27] Elastic Stack, https://www.elastic.co/products/
[28] Splunk, https://www.splunk.com/
[29] Observium, http://www.observium.org/
[30] perfSONAR, http://www.perfsonar.net/

Search for the QCD Critical Point with Fluctuations of Conserved Quantities in Relativistic Heavy-Ion Collisions at RHIC

Xiaofeng Luo[1,a]

[1] *Institute of Particle Physics and Key Laboratory of Quarks & Lepton Physics (MoE), Central China Normal University, Wuhan, China, 430079*

Abstract. Fluctuations of conserved quantities are predicted to be sensitive observables to probe the QCD phase transition and critical point in heavy-ion collisions. Experimentally, various cumulants (up to fourth order) of net-proton (proxy for net-baryon (B)), net-kaon (proxy for net-strangeness (S)), and net-charge (Q) multiplicity distributions of Au+Au collisions at $\sqrt{s_{NN}}$ = 7.7 ~ 200 GeV has been measured by the STAR and PHENIX experiments from the first phase of beam energy scan program at the Relativistic Heavy Ion Collider. In this paper, i will discuss those experimental results and the corresponding physics implications.

1 Introduction

The Quantum Chromodynamics (QCD) phase structure can be displayed in the two dimensional phase diagram (temperature, T vs. baryon chemical potential, μ_B). Finite temperature Lattice QCD calculations has shown that at zero μ_B ($\mu_B = 0$) region, it is a crossover transition between hadronic phase and quark-gluon phase. At large μ_B region, the QCD based models predicted that the QCD phase transition is of the first order and there should exist a so called QCD Critical Point (CP) as the endpoint of the first order phase boundary [1]. Due to sign problem at finite μ_B region, it is difficult to precisely determine the location of the CP or even its existence [2, 3]. One of the main goals of the Beam Energy Scan (BES) program at the Relativistic Heavy Ion Collider (RHIC) is to search for the CP in heavy-ion collisions experiment. Experimental confirmation of the existence of the CP will be an excellent verification of QCD theory in the non-perturbative region and a milestone of exploring the QCD phase structure. To vary the temperature (T) and baryon chemical potential (μ_B) of the nuclear matter created in heavy-ion collisions [4], physicist has tuned the Au+Au nuclear colliding energies from high to low values, this allows us to access a broad region of the QCD phase diagram.

Theretically, fluctuations of conserved quantities, such as net-baryon (B), net-charge (Q) and net-strangeness (S), have been long time predicted to be sensitive to the correlation length [5–7] and directly connected to the susceptibilities computed in the first principle Lattice QCD calculations [8, 9]. Thus, it can serve as a powerful observables to study the phase transition and search for the CP in heavy-ion collisions. During the first phase of the RHIC BES (2010 to 2014), the STAR experiment has measured the cumulants (up to the fourth order) of net-proton (proton minus anti-proton number, proxy of net-baryon [10]) , net-charge and net-kaon multiplicity distributions in Au+Au collisions at

[a] e-mail: xfluo@mail.ccnu.edu.cn

$\sqrt{s_{NN}}$= 7.7, 11.5, 14.5, 19.6, 27, 39, 62.4 and 200 GeV. In those energies, the data of 14.5 GeV is taken in the year 2014, 19.6, 27 GeV are taken in the year 2011, and the other energies are collected in the years 2010. In this paper, we will discuss the experimental results from the STAR experiment for the fluctuations of conserved quantities. The corresponding physics implications will be discussed.

The paper is organized as follows: In the second section, we will give a brief discussion of the techniques used in the moment analysis. The experimental results and discussion will be presented in the third section. Finally, we will give a summary and outlook.

2 Experimental Observables

2.1 Definition

Particle multiplicity fluctuations can be characterized by the cumulants of the multiplicity distributions. It can be calculated as

$$C_1 = \langle N \rangle, C_2 = \left\langle (\delta N)^2 \right\rangle, C_3 = \left\langle (\delta N)^3 \right\rangle, C_4 = \left\langle (\delta N)^4 \right\rangle - 3\left\langle (\delta N)^2 \right\rangle^2 \tag{1}$$

Where N is particle or net-particle number and the $\langle N \rangle$ is average over events, $\delta N = N - \langle N \rangle$. With the definition of cumulants, we can also define mean (M), variance (σ^2), skewness (S) and kurtosis (κ) as:

$$M = C_1, \sigma^2 = C_2, S = \frac{C_3}{(C_2)^{\frac{3}{2}}}, \kappa = \frac{C_4}{(C_2)^2} \tag{2}$$

In addition, the moments product $\kappa\sigma^2$ and $S\sigma$ can be expressed in terms of the ratios of cumulants:

$$\kappa\sigma^2 = \frac{C_4}{C_2}, S\sigma = \frac{C_3}{C_2}, \sigma^2/M = \frac{C_2}{C_1} \tag{3}$$

The ratios of cumulants are independent on system volume. The statistical errors of those cumulants and cumulants ratios are estimated by the Delta theorem [11, 12]. In general, the statistical errors strongly depend on the shape of the distributions, especially the width. For gaussian distributions, the statistical errors of cumulants (C_n) can be approximated as $error(C_n) \propto \sigma^n/(\sqrt{N}\epsilon^n)$, where σ is the measured width of the distribution, N represents the number of events and ϵ is the particle detection efficiency.

2.2 Data Analysis Techniques

Experimentally, one cannot directly measure the collision centrality and/or initial collision geometry of the system of two nuclei. The collision centrality is usually determined by comparing the measured particle multiplicity with the Glauber Monte Carlo simulations. It is denoted as a percentage value (for e.g. 0-5%, 30-40%,...70-80%) for a collection of events to represent the fraction of the total cross section. This in general can cause two undesirable effects in the multiplicity fluctuations within finite centrality bin. One is the so called centrality bin width effect, which is caused by volume variation within a finite centrality bin size and the other one is centrality resolution effect depending on the number of particles used in the centrality definitions. Detector effects such as finite detection efficiency will dilute the origin signals and proper methods are needed to be applied to recover the signals. Finally, statistical errors estimation for the high order fluctuations is not straightforward and very complicated. Although it is difficult measurement, a set of analysis techniques were built up to measure the various order fluctuations of conserved quantities in heavy-ion collisions. Those include: (1) Novel centrality determination to account for the effects of centrality resolution and auto-correlation [13]. (2) Centrality bin width correction [13, 14]. (3) Efficiency correction and error estimation [11, 12, 15, 16]. Those techniques are very important to address the background effects and extract the real physics fluctuation signals from the observables. The centrality bin width correction is to address the volume fluctuations effects on the higher order moments of conserved quantities distributions within a finite width centrality bin. It is also called centrality bin width effect. This effect needs to be eliminated, as an artificial centrality dependence could be introduced by it. To do this, the centrality bin width correction is applied to calculate the various moments of particle multiplicities distributions in one wide centrality bin. Experimentally, the smallest centrality bin is determined by a single value of particle multiplicity. Experimental measurements are usually analysed for a wider centrality bin (a range of particle multiplicity), such as $0-5\%$,$30-40\%$,...70-80%,etc., to reduce statistical errors. To eliminate the centrality bin width effects, we calculate the various order cumulants (C_n) for each single particle multiplicity within one wider centrality bin and weighted averaged by the corresponding number of events in that multiplicity.

$$C_n = \sum_{r=N_1}^{N_2} \omega_r C_n^r \tag{4}$$

where the n_r is the number of events for multiplicity value r and the corresponding weight for the multiplicity r, $\omega_r = n_r / \sum_{r=N_1}^{N_2} n_r$. N_1 and N_2 are the lowest and highest multiplicity values for one centrality bin. For more details, one can see [13].

3 Experimental Results

The STAR experiment published the energy dependence of cumulants (up to fourth order) of net-proton [17] and net-charge [18] multiplicity distributions in Au+Au collisions at $\sqrt{s_{NN}}$= 7.7, 11.5, 19.6, 27, 39, 62.4 and 200 GeV. In the CPOD2014 and QM2015 conference [19–21], the STAR experiment reported the new net-proton fluctuation results of wider transverse momentum coverage ($0.4 < p_T < 2$ GeV/c) by identifying proton and anti-proton with time of flight detector. Meanwhile, the preliminary results of cumulants of net-kaon distributions in Au+Au collision at $\sqrt{s_{NN}}$= 7.7, 11.5, 14.5, 19.6, 27, 39, 62.4 and 200 GeV are shown in the QM 2015 conference [21].

Figure 1 panel (a), (c), (d) show the energy dependence of $\kappa\sigma^2$ of net-charge, net-kaon and net-proton multiplicity distributions in Au+Au collisions measured by the STAR experiment for two centralities (0-5% and 70%-80%) at $\sqrt{s_{NN}}$= 7.7, 11.5, 14.5 19.6, 27, 39, 62.4 and 200 GeV. Fluctuations of net-charge in Au+Au collisions measured by the PHENIX experiment are shown in the panel (b) of Fig. 1. It shows that the net-charge and net-kaon fluctuations measured by the STAR experiment are with larger statistical errors than the errors of net-proton fluctuations. This is because the statistical errors of various order cumulants depend on the width (σ) of the multiplicity distributions (($error(C_n) \propto \sigma^n/(\sqrt{N}\epsilon^n)$)) and the net-charge distributions are much wider than net-proton and net-kaon. On the other hand, due to decay of kaons, the efficiency of kaon (~ 40%) is much lower than proton (~ 80%), this also leads to larger statistical errors for net-kaon fluctuations. The Skellam (Poisson) expectations shown in the figure as dashed lines with unity value reflect a system of totally uncorrelated, statistically random particle production. It predicts the $\kappa\sigma^2$ to be unity for Skellam expectations as well as in the hadron resonance gas model. For the net-proton results, a clear non-monotonic energy dependence is observed for $\kappa\sigma^2$ in the most central (0-5%) Au+Au collisions with a minimum around 19.6 and 27 GeV. This non-monotonic behavior cannot be described by various

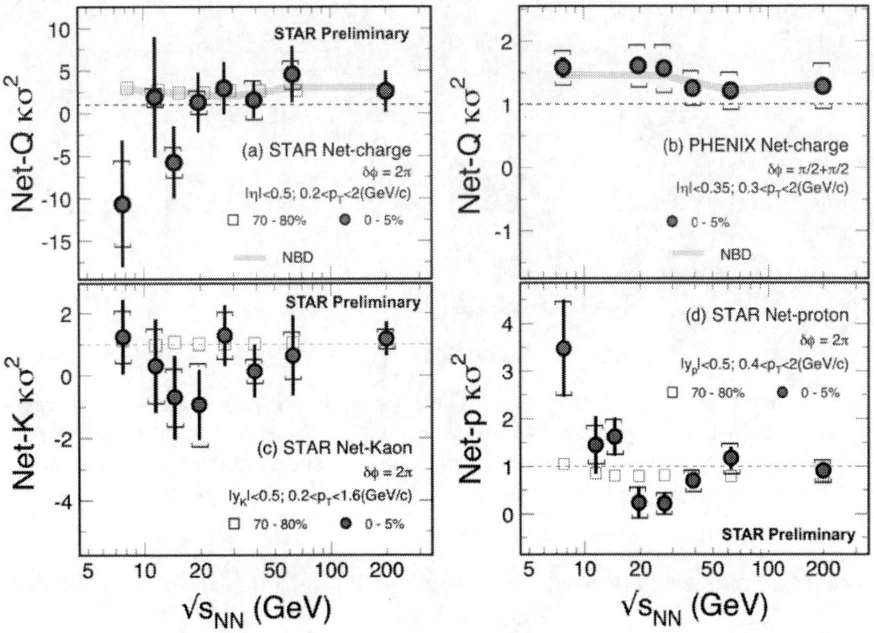

Figure 1. (Color online) The STAR measured energy dependence of $\kappa\sigma^2$ of net-proton, net-charge (top left) and net-kaon distributions in Au+Au collisions at $\sqrt{s_{\mathrm{NN}}}$= 7.7, 11.5, 14.5, 19.6, 27, 39, 62.4 and 200 GeV. The net-charge fluctuations measured by the PHENIX experiment in Au+Au collisions at $\sqrt{s_{\mathrm{NN}}}$= 7.7, 19.6, 27, 39, 62.4 and 200 GeV are shown in top right panel.The statistical and systematical error are shown in bars and brackets, respectively.

model calculations [22, 23]. At energies above 39 GeV, the results are close to Skellam expectation. For the STAR net-charge and net-kaon results, we did not observe non-monotonic behavior for $\kappa\sigma^2$ within current statistics. The expectations from negative binomial distribution can better describe the net-charge and net-kaon data than the Poisson (Skellam) expectations. The PHENIX net-charge results show smaller errors than the one measured by the STAR experiment. This is because the PHENIX detector has much smaller acceptance than the STAR detector and thus the width of the net-charge distributions measured by the PHENIX experiment is much narrower. On the other hand, the resonance decay effects are not excluded in the current experimental measurements of fluctuations of net-proton, net-kaon and net-charge multiplicity. However, based on the hadron resonance gas model calculation [24], the decay effects for net-proton $\kappa\sigma^2$ is small and at 2% level. But for the net-charge, the decay effects are large. The statistical error of cumulants ($\Delta(C_n)$) are related to the width of the distribution as $\Delta(C_n) \sim O(\sigma^n)$ [11, 12]. Thus, the wider is the distribution, the larger are statistical errors for the same number of events. It is predicted that the coupling strength of pions and kaons to the critical fluctuation is much smaller than protons. Thus, the fluctuation of net-protons will be more sensitive to the critical fluctuations than the net-charge and net-kaon [6].

4 Detector upgrades and beam energy scan phase II at RHIC

A second phase of the beam energy scan (BES-II) at RHIC has been planed in the years 2019-2020 and focusing on energies below $\sqrt{s_{\mathrm{NN}}}$=20 GeV [25]. This is to collect more number of events to confirm the trends observed in the BES-I. It will allow us to draw a solid conclusion and have more complete physics picture from various experimental measurements. The long beam bunches and stochastic

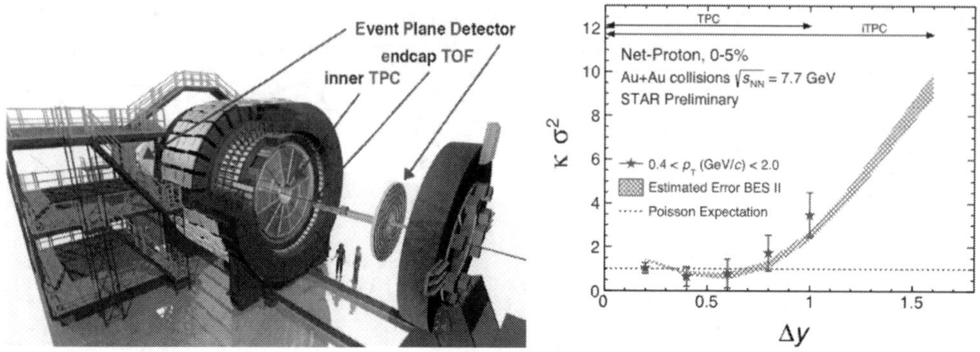

Figure 2. (Color online) Left: Upgrades of the STAR detector for the second phase of beam energy scan at RHIC. Right: Rapidity coverage dependence of the $\kappa\sigma^2$ of net-proton distribution in 0-5% central Au+Au collisions at $\sqrt{s_{NN}}$=7.7 GeV. The blue band shows the expecting trend and statistical error for net-proton $\kappa\sigma^2$ at BES-II. For this analysis, the rapidity coverage can be extend to $|y| < 0.8$ with iTPC upgrades.

electron cooling technique will be used to accelerate gold beams. It will increase the luminosity by a factor of 5-15 compared to the BES-I. Since the luminosity will decrease as decreasing the colliding energy, the increasing of the luminosity is much more necessary and important at low energies, such as 7.7 GeV. To study the QCD phase structure at high baryon density, operating the STAR detector at fixed target mode has been also proposed. In the BES-II, fixed target mode Au+Au collisions allow us to have energy coverage from $\sqrt{s_{NN}}$=3 GeV (μ_B=720 MeV) up to 7.7 GeV. In Fig.2 left, the inner TPC (iTPC) of STAR is to be upgraded to improve the energy loss resolution and can extend the pseudo-rapidity coverage from $|\eta| < 1$ to $|\eta| < 1.5$ [26]. It is also planed to install a end cap Time-of-Flight (eTOF) detector at the west end of the STAR TPC to extend the PID capability in the forward region [27]. The iTPC upgrade is very important to search for the criticality and study the dynamical evolution of the fluctuations by looking at the rapidity acceptance dependence for the fluctuations of the conserved quantities [28]. In the forward and backward region of STAR detector, a new Event Plane Detector (EPD) will be also built and used to replace the Beam-Beam Counters (BBC) detector for centrality and event plane determination, which can suppress the auto-correlations. In Fig.2 right, the blue band is the extrapolating from current measurements by assuming a power law behavior induced by critical fluctuations ($\kappa\sigma^2 \propto N^3$ [29]).

5 Summary

The STAR and PHENIX experiments at RHIC have measured the fluctuation of net-proton, net-kaon, and net-charge multiplicity in Au+Au collisions at $\sqrt{s_{NN}}$= 7.7, 11.5, 14.5, 19.6, 27, 39, 62.4 and 200 GeV. Different measurements/experiments are affected by acceptance, kinematic cuts, resonance decays, and other dynamical effects differently. To search for the signature of the QCD critical point, one needs to do careful studies. Non-monotonic energy dependence of $\kappa\sigma^2$ of the net-proton distributions is observed in the 0 ~ 5% most central Au+Au collisions measured by the STAR experiment. This non-monotonic behavior cannot be described by various transport model calculations. The net-charge and net-kaon results are consistent with the binomial and Poisson expectations with current statistics. Experimentally, the second phase of the beam energy scan at RHIC has been planed in 2019-2020 with upgraded detectors and increased luminosity to explore the phase structure focusing on energies below 20 GeV with high precision.

Acknowledgement

The work was supported in part by the MoST of China 973-Project No.2015CB856901, NSFC under grant No. 11575069, 11221504.

References

[1] K. Rajagopal, F. Wilczek, *"At the Frontier of Particle Physics / Handbook of QCD"*, Vol. 3 (World Scientific, 2001)
[2] Z. Fodor and S. Katz, JHEP 0203, **014** (2002); 0404, **050** (2004); R. V. Gavai, S. Gupta, Phys. Rev. D **71**, 114014 (2005); **78**, 114503 (2008).
[3] R. V. Gavai, arXiv:1404.6615.
[4] M. M. Aggarwal *et al.* (STAR Collaboration), arXiv: 1007.2613
[5] M. A. Stephanov, Phys. Rev. Lett. **102**, 032301 (2009).
[6] C. Athanasiou *et al.*, Phys. Rev. D **82**, 074008 (2010).
[7] M. A. Stephanov, Phys. Rev. Lett. **107**, 052301 (2011); B. Schaefer, M. Wagner, Phys. Rev. D **85**, 034027 (2012).
[8] S. Gupta, X. Luo, B. Mohanty, H. G. Ritter, N. Xu, Science **332**, 1525 (2011).
[9] R. V. Gavai and S. Gupta, Phys. Lett. B **696**, 459 (2011).
[10] Y. Hatta and M. A. Stephanov, Phys. Rev. Lett. **91**, 102003 (2003).
[11] X. Luo, J. Phys. G: Nucl. Part. Phys. **39**, 025008 (2012).
[12] X. Luo, Phys. Rev. C **91**, 034907 (2015).
[13] X. Luo, J. Xu, B. Mohanty, N. Xu, J. Phys. G **40**, 105104 (2013).
[14] X. Luo (for the STAR Collaboration), J. Phys.: Conf. Ser. **316**, 012003 (2011).
[15] Adam Bzdak and Volker Koch, Phys. Rev. C **86**, 044904 (2012).
[16] Adam Bzdak and Volker Koch, Phys. Rev. C **91**, 027901 (2015).
[17] L. Adamczyk *et al.*, (STAR Collaboration), Phys. Rev. Lett. **112**, 032302 (2014); Phys. Rev. Lett. **113**, 092301 (2014);
[18] L. Adamczyk *et al.*, (STAR Collaboration), Phys. Rev. Lett. **113**, 092301 (2014).
[19] X. Luo (for the STAR Collaboration), PoS(CPOD2014)019, 2015. [arXiv: 1503.02558]
[20] X. Luo, Nucl. Phys. A **956**, 75 (2016). [arXiv: 1512.09215].
[21] J. Thaeder, (for the STAR Collaboration), Nucl. Phys. A **956**, 320 (2016). [arXiv:1601.00951]
[22] J. Xu, S. L. Yu, F. Lui, X. Luo, Phys. Rev. C **94**, 024901(2016).
[23] S. He, X. Luo, Y. Nara, S. Esumi, N. Xu, Phys. Lett. B **762**, 296 (2016).
[24] P. Garg *et al.*, Phys. Lett. B **726**, 691 (2013).
[25] STAR Note 0598: BES-II whitepaper: http://drupal.star.bnl.gov/STAR/starnotes/public/sn0598
[26] STAR Note 0619: iTPC proposal: http://drupal.star.bnl.gov/STAR/starnotes/public/sn0619
[27] L. Adamczyk *et al.*, STAR Collaboration, arXiv:1609.05102
[28] Masakiyo Kitazawa, Nucl. Phys. A **942**, 65 (2015).
[29] Bo Ling and Mikhail A. Stephanov, Phys. Rev. C **93**, 034915 (2016). [arXiv: 1512.09125].

Recent hard probe measurements from STAR

Fuqiang Wang[1,a] (for the STAR Collaboration)

[1] *Department of Physics and Astronomy, Purdue University, West Lafayette, IN 47907, USA*

Abstract. High momentum transfer processes can be reliably calculated by perturbative quantum chromodynamics (pQCD). Products of those hard processes, high transverse momentum particles, jets, and heavy quarks (hard probes), interact with the quark-gluon plasma (QGP) created in relativistic heavy-ion collisions. Properties of the QGP can be studied by comparing hard probe measurements in heavy-ion collisions to those in p+p collisions or from pQCD calculations. I will present recent results from STAR on hard probe measurements, including a wide range of observables: heavy flavor, dijet transverse momentum imbalance, hadron-jet correlations, γ-jet correlations, and dihadron correlations. I will discuss what we have learned from these measurements.

1 Introduction

The primordial universe was immersed by the most elementary particles: quarks, gluons, leptons, and photons. This extremely high energy density state–the quark-gluon plasma (QGP)–is created in today's laboratory by relativistic heavy-ion collisions [1–3]. The question scientists want to answer, among others, is how the fundamental theory of quantum chromodynamics (QCD) behaves in the high density medium, such as the QGP, and what novel phenomena may emerge. Two main ways to study the QGP are to measure how it dissembles, much like how the universe expanded and cooled down, and how energetic quarks and gluons (hard probes) interact with it and fragment into showers of collimated hadrons (jets). Hard probes are a valuable tool to study the QGP in nuclear (AA) collisions because their initial production can be reliably calculated by perturbative QCD (pQCD) or measured in proton+proton (pp) collisions. The effect of their interactions with the QGP can, therefore, be studied by comparing AA collisions to pp collisions and/or pQCD calculations.

This proceedings article reviews hard probe results from the STAR experiment at RHIC. In particular, the following hard probe results are reviewed: heavy quark production and collective flow, dijet transverse momentum (p_T) imbalance, hadron-jet correlations, γ-hadron correlations, and dihadron correlations.

2 Heavy flavor probe

One central question is how the QGP quickly thermalizes from the initial non-equilibrium state of heavy-ion collisions. Heavy quarks, referred to as heavy flavor (HF), are a sensitive probe to this question because they, being heavy, do not thermalize easily.

[a] e-mail: fqwang@purdue.edu

Figure 1, left panel, shows the nuclear modification factor, R_{AA}, the ratio of production rate of D^0 mesons (containing a heavy flavor charm quark) in Au+Au collisions to that in pp collisions properly normalized by the number of binary nucleon-nucleon collisions [4]. D^0 production is strongly suppressed at large p_T, a result of charm quark energy loss due to interactions with the QGP.

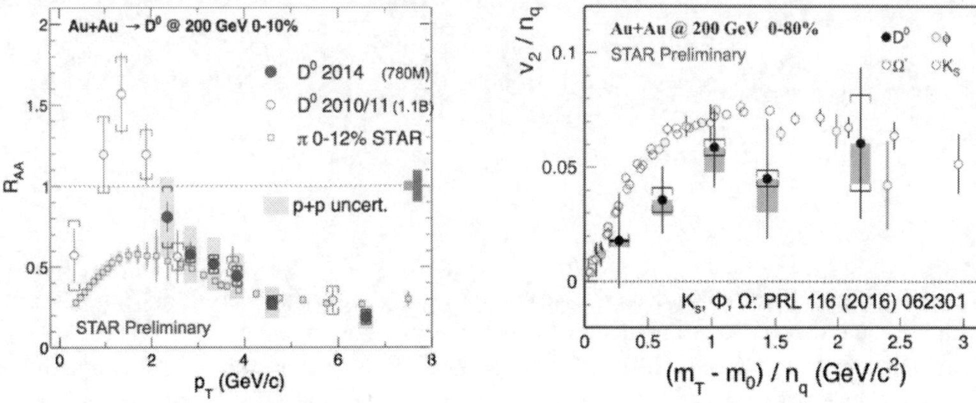

Figure 1. (Color online) Nuclear modification factor R_{AA} [4] (left) and elliptic anisotropy (right) [5] of D^0 mesons in Au+Au collisions, compared to those of other particles.

Figure 1, right panel, shows the elliptic azimuthal anisotropy of D^0 mesons [5], compared to that of other hadrons. This azimuthal anisotropy is generated by interactions of charm quarks with the QGP. Large anisotropy values are generally taken as evidence of strong interactions and low viscosity hydrodynamics. The comparable magnitude of D^0 anisotropy to that of light hadrons' would suggest that heavy quarks are as thermalized as light quarks, which implies that the QGP is highly thermalized. Recent transport studies [6] and other approaches suggest that hydrodynamics may not be a unique explanation for the anisotropy.

3 Dijet transverse momentum imbalance

Jets are produced back-to-back in azimuth by hard scatterings of partons, initially moving, in leading order QCD, along the longitudinal directions. Higher order effects of QCD, e.g. nucleon Fermi motion in transverse direction inside the nucleus and initial-state gluon radiation of partons in the nucleons, render a non-strict back-to-back balance of dijet p_T. Interactions with the QGP produce additional p_T imbalance between leading and subleading jets of the dijet system. Comparison of the p_T imbalance, quantified by the $A_j \equiv \frac{p_T^{\rm lead}-p_T^{\rm sublead}}{p_T^{\rm lead}+p_T^{\rm sublead}}$ variable, between AA and pp collisions will reveal the strength of jet-medium interactions.

Figure 2 shows the A_j distributions in AA and pp collisions [7] by the red solid and open symbols, respectively. The A_j distribution in AA collisions is shifted to larger values than that in pp collisions. This shift is the effect of additional p_T imbalance due to jet interactions with the QGP.

Some words are in order here on the details of jets measured in experiments. Jets are reconstructed by clustering final-state particles in a cone of given size $R = \sqrt{\Delta\eta^2 + \Delta\phi^2}$ where $\Delta\eta$ is spread in pseudorapidity (η) and $\Delta\phi$ is that in azimuth. There are several jet reconstruction algorithms; the jet results in this article are obtained by the anti-k_T algorithm which emphasizes high-p_T particles [8]. In Fig. 2, $R = 0.4$ is used. Not all energy of the initial parton is reconstructed within the finite size

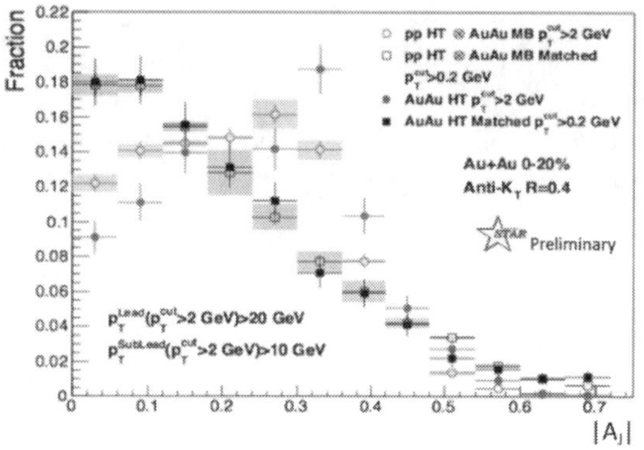

Figure 2. (Color online) Dijet p_T imbalance in central AA and pp collisions [7]. Jets are reconstructed by the anti-k_T algorithm with $R = 0.4$, and particle $p_T > 2\text{GeV}/c$ (red points). Results with the same dijets but including $p_T > 0.2\text{GeV}/c$ particles in jet energy are shown in black.

because of out-of-cone radiation. In addition, a lower $p_T > 2$ GeV/c cutoff is imposed in order to reduce background. The broadened A_j distribution in AA indicates energy loss (energy is shifted from high p_T to low p_T) and/or broadening (energy is shifted out of cone).

Once jets are reconstructed with the fragment $p_T > 2$ GeV/c cut, this p_T cut can be relaxed or removed to include jet energies residing in low-p_T particles. The black filled and open points in Fig. 2 show the A_j results with a relaxed $p_T > 0.2$ GeV/c (dictated by the detector aperture) for AA and pp collisions, respectively. The two distributions are consistent with each other, suggesting that jet-medium interactions result in jet softening, but not broadening beyond $R = 0.4$. For smaller R cone size used in jet reconstruction, however, dijet A_j distributions are found not to match between AA and pp (broader in *AA* than pp) even with $p_T > 0.2$ GeV/c. This indicates that jet-medium interactions do result in jet broadening, besides softening, but only within $R = 0.4$. Note that the dijets shown in Fig. 2 are required to satisfy $p_T^{\text{lead}} > 20$ GeV/c and $p_T^{\text{sublead}} > 10$ GeV/c, relatively large for RHIC energy of 200 GeV. At the LHC it is found that jets are broadened to significantly larger R at low p_T [9].

4 Hadron-jet correlations

Jet-medium interactions can be studied in more detail by hadron-jet correlations where the hadron, called the trigger, is required to be at high p_T. Because of energy loss and the expeditiously dropping jet production with increasing p_T, those high-p_T trigger hadrons are predominantly produced by hard scattering near the "cylindrical" surface of the collision zone and directed outward. The back-to-back jet partners are mostly directed inward, having to traverse the entire size of the collision zone.

Figure 3 shows that the charged hadron jet p_T spectra for $R = 0.3$ and $R = 0.5$ [10]. The jet area A_{jet} is required to be larger than 0.20 and 0.65, respectively, to reduce fake jets. All charged particles of $p_T > 0.2$ GeV/c are include in jet reconstruction (anti-k_T). In heavy-ion collisions, with essentially no p_T cut imposed on particles, fake jet rate is large. In this analysis fake jets are reconstructed by a mixed-event technique, where one and only one track is sampled from each event of similar characteristics (centrality, event plane, etc.). The mixed-event technique yields a good description of fake jets in kinematic ranges where all reconstructed jets are known to be fake. The mixed-event fake jets have been subtracted for the spectra in Fig. 3. In addition, the jet energy has been corrected for underlying event contribution by subtracting ρA_{jet} where ρ is the medium area density of the background particle energy, and unfolded for underlying event background fluctuations and particle momentum resolution.

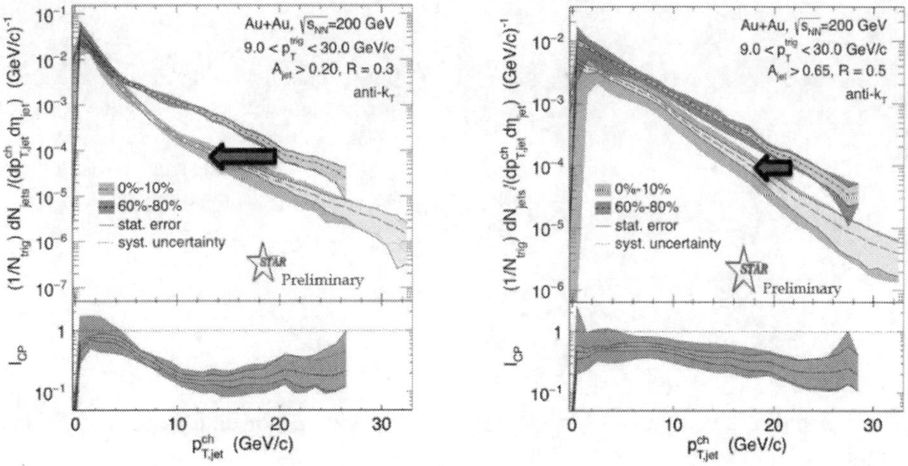

Figure 3. (Color online) Hadron-jet correlations [10]: recoil jet p_T spectra in peripheral and central Au+Au collisions for $R = 0.3$ and $R = 0.5$ (upper) and I_{AA}, ratio of jet spectra in central to peripheral collisions (lower).

In Fig. 3, the trigger hadron p_T range is 9-30 GeV/c. Both the peripheral and central jet spectra are shown. The spectra ratio of central to peripheral collisions is shown in the lower panels of Fig. 3. The central spectrum is suppressed at large p_T indicating jet energy loss in medium. The suppression can be viewed as the result of a horizontal shift to the left of the spectra from peripheral to central collisions. The shift indicates the amount of energy loss. The shift is smaller for larger R at RHIC, and for $R = 0.5$, is smaller at RHIC than LHC [11].

5 γ-hadron correlations

Reconstructed jet energy does not precisely reflect the parent parton energy or the hard scattering momentum transfer. Jet studies always involve data-theory comparisons where jets are reconstructed in the exactly same way in real data and theory calculations. Gamma-jet correlations, on the other hand, offer a direct measure of the hadron scattering momentum transfer and a gauge of the recoil jet energy. Gamma-hadron correlations, a prestep toward the golden γ-jet measurements, have been widely studied.

Figure 4 shows I_{AA}, the ratio of jet fragmentation functions in AA collisions to pp collisions, where $z_T \equiv p_T^{\rm hadron}/p_T^{\gamma}$ [12]. The γ-hadron data are compared to the π^0-hadron correlation data where the $z_T \equiv p_T^{\rm hadron}/p_T^{\pi^0}$ for the latter. The trigger p_T^{γ} and $p_T^{\pi^0}$ are restricted within 12-20 GeV/c and the hadron p_T is required to be larger than 1.2 GeV/c. The recoil jets from π^0 or γ are both strongly suppressed. The suppression is less significant at low z_T (low associated hadron p_T). With the current statistical precision, the γ-hadron and π^0-hadron correlation data are compatible. Note, however, that (1) the energies of the recoil jets in the two cases are different because the π^0 takes only portion of the jet energy while the γ closely reflects the recoil jet energy, thus the z_T definitions are different for these two measurements; (2) the π^0-hadron correlations are biased to surface hard scatterings and the γ-hadron correlations are unbiased (except for possible bias introduced by the $p_T^{\rm hadron} > 1.2$ GeV/c cut); and (3) the recoil jet from the γ is mostly quark jet and that recoil the π^0 is mostly gluon jet.

 DOI: 10.1051/epjconf/201714104002

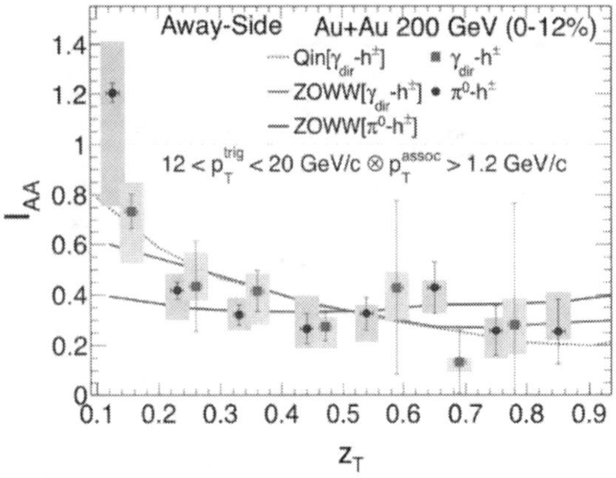

Figure 4. (Color online) γ-hadron correlations [12]: ratio of the recoil jet "fragmentation" function in Au+Au collisions to that in pp collisions. The data are compared to those from π^0-hadron correlations and theoretical calculations.

Figure 4 also shows comparisons to theoretical calculations. The calculations and data are consistent within uncertainties.

6 Dihadron correlations

Jet-medium interactions shift energy from high-p_T particles to low-p_T particles. The bulk of jet-medium interaction effects is contained in low-p_T particles. Jet-correlated low p_T have been most difficult to measure because of the overwhelming background from the underlying event. The azimuthal distribution shape of this background is "arbitrary" because of the possible azimuthal harmonics of all orders due to initial state geometry fluctuations. Subtraction of this background suffers from large systematic uncertainties.

A novel method has been recently devised by requiring large p_T recoil from a relatively high-p_T trigger particle and subtracting anisotropic underlying event background using data itself [13]. Figure 5 shows the Gaussian width of the away-side correlation peak from this method as a function of centrality [13]. The width increases with centrality; the increase can be due to the increasing nuclear k_T broadening with centrality, broadening due to jet-medium interactions, or both. To shed more light, Fig. 5 right panel shows the widths in peripheral and central collisions and the quadratic difference of the widths as a function of p_T. Because the change in the nuclear k_T effect from peripheral to central collisions does not depend on hadron p_T, the increase of the width difference with p_T indicates the effect of jet-medium interactions; the effect seems larger for high p_T correlated particles.

7 Summary

STAR has made a variety of hard probe measurements. All hard probes are suppressed in heavy-ion collisions compared to pp collisions and pQCD calculations. Lost energy by hard probes are shifted to lower p_T and larger angles. Heavy quarks flow, almost as strongly as light quarks. These measurements show strong medium effects; more quantitative knowledge about the medium requires rigorous theory and model comparisons to the data.

References

[1] K. Adcox et al. (PHENIX Collaboration), Nucl.Phys. **A757**, 184 (2005), `nucl-ex/0410003`
[2] J. Adams et al. (STAR Collaboration), Nucl.Phys. **A757**, 102 (2005), `nucl-ex/0501009`

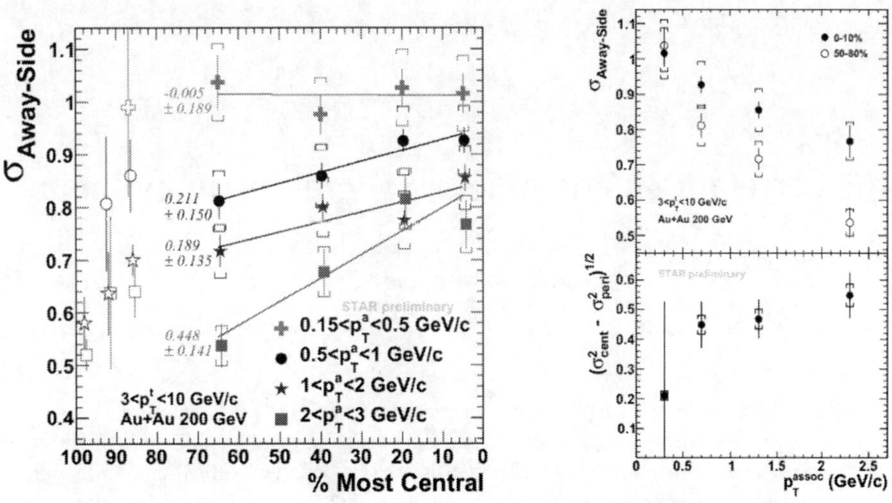

Figure 5. (left) Away-side dihadron correlation peak width as a function of centrality for various associated particle p_T ranges. (right) Away-side peak width in peripheral and central collisions (upper) and the quadratic difference of the widths (lower) as a function of associated particle p_T.

[3] B. Muller, J. Schukraft, B. Wyslouch, Ann.Rev.Nucl.Part.Sci. **62**, 361 (2012), 1202.3233
[4] G. Xie (STAR), Nucl. Phys. **A956**, 473 (2016), 1601.00695
[5] M.R. Lomnitz (STAR), Nucl. Phys. **A956**, 256 (2016), 1601.00743
[6] L. He, T. Edmonds, Z.W. Lin, F. Liu, D. Molnar, F. Wang, Phys. Lett. **B753**, 506 (2016), 1502.05572
[7] L. Adamczyk et al. (STAR) (2016), 1609.03878
[8] M. Cacciari, G.P. Salam, G. Soyez, JHEP **0804**, 063 (2008), 0802.1189
[9] S. Chatrchyan et al. (CMS), Phys. Rev. **C84**, 024906 (2011), 1102.1957
[10] P.M. Jacobs, A. Schmah (STAR), Nucl. Phys. **A956**, 641 (2016), 1512.08784
[11] J. Adam et al. (ALICE), JHEP **09**, 170 (2015), 1506.03984
[12] N.R. Sahoo (STAR), Nucl. Phys. **A956**, 621 (2016), 1512.08782
[13] K. Jiang (STAR Collaboration), J. Phys. Conf. Ser. **636**, 012011 (2015)

EPJ Web of Conferences **141**, 04003 (2017)
ISMD 2016 DOI: 10.1051/epjconf/201714104003

Multi-jet correlations and colour coherence phenomena

Jason S.H. Lee[1,a] on behalf of the CMS Collaboration

[1] *University of Seoul, South Korea*

Abstract. In multi-jet events, colour coherence was shown to play a important role in the topology of these events. In modern Monte Carlo generators, the colour coherence effect has become an inherent feature, such that the colour coherence effect can no longer be disentangled in a simple manner. This study looks at the multi-jet correlations by comparing Monte Carlo generators in Parton Shower dominant and Matrix Element dominant regions.

1 Introduction

When a quark pair is produced, one of the jets can radiate a gluon. Due to quantum chromodynamics (QCD), in the fragmentation phase, the outgoing partons interact with each other. This phenomena was first described by the LUND string model [1], predicting a suppression of particle production in the region between the two quark jets due to destructive interference and an enhancement of particle production in the region between the quark and gluon jets. This colour coherence effect was initially observed in e^+e^- collisions by several experiments at e^+e^- colliders (PETRA, PEP and LEP), leading to the discovery of the gluon.

In hadron collisions, constituents are also coloured making the three jets colour connected to the proton constituents. The Tevatron experiments showed that the variable 'β', the ratio of the difference in pseudorapidity over the difference in azimuthal angle between the second and third jet, is sensitive to colour coherence [2]. The second jet is the jet with the second highest transverse momentum (p_T) and the third jet is most likely to have been the radiated gluon. They showed good agreement with data for parton shower Monte Carlo (MC) generators with the colour coherence effects implemented via angular ordering. They tuned PYTHIA [3] with the angular ordering on and off to highlight this effect.

Section 2 summaries the some of the QCD jets measurement, from the CMS collaboration [4] at the CERN LHC, which are directly and indirectly affected by the colour coherence effect.

This has led to the colour coherence effect becoming an inherent feature of modern MC generators, such that the colour coherence effect can no longer be disentangled in a simple manner. To study the multi-jet correlations, it was suggested by the PYTHIA authors that a more practical test on MC generators would be to test the Parton Shower (PS) dominant and Matrix Element (ME) dominant regions. Section 3 compares four MC generators, one with only PS calculations, one with only ME calculations, and two with both PS and ME calculations.

[a] e-mail: jason.lee@cern.ch

2 Jet Measurement by CMS

CMS uses particle flow objects as inputs for the anti-kt jet clustering algorithm [5], with the radius parameter set to R = 0.5 in Run1, R = 0.4 for Run2.

The di-jet azimuthal decorrelation is sensitive to the radiation of additional jets as measured in Run1 in [6]. In the region where the di-jet azimuthal separation is large, $\Delta\phi_{di-jet} > 2\pi/3$, it was shown to be well modelled by the MC generators. However, for the region where the di-jet azimuthal separation is small, $\Delta\phi_{di-jet} \approx \pi/2$, differences started to show up between data and MC as the azimuthal separation decreased. Here, smaller azimuthal separation is due to the production of additional jets and this region is indirectly sensitive to colour coherence.

In the Run1 measurement of the jet charge [7], the jet charge is measured in three directions, in the direction of the transverse momentum, longitudinal to jet axis and transverse to jet axis. A 'κ' factor was introduced to give more weight to low momentum tracks within the jet for small 'κ'. The jet charge defined in the direction transverse to the jet axis together with a small 'κ' would be most sensitive to angular correlation effects like colour coherence. And, unsurprisingly, this region showed the biggest discrepancies between data and MC.

Similar to the Tevatron measurement, CMS produced a colour coherence result from Run1 data [8] using the same 'β' variable. The leading order MC generators were shown to have a poor description of the data in forward region. And similarly to the Tevatron measurement, a comparison to data with colour coherence effects turned on and off was shown. Turning on the colour coherence effects with PYTHIA shows better agreement with data compared to the case when colour coherence is turned off. Currently, only a very old version of PYTHIA can turn off colour coherence effects.

3 Parton Shower vs Matrix Element

As the colour coherence effect became an inherent feature of MC generators, in the newer MC generators, the colour coherence effect can no longer be disentangled in a simple manner. It was suggested that a more practical test on MC generators would be to test the performance in a Parton Shower (PS) dominant and a Matrix Element (ME) dominant regions. This study used three MC generators, PYTHIA8[3], MADGRAPH[9] and POWHEG BOX[10–12], to make three jet events with only PS, only ME and both ME + PS calculations. PYTHIA8 (version 8.212) is the standard tool for the generation of events in high-energy collisions which includes soft and hard QCD processes and hard QCD processes using standard 2 $\rightarrow$ 2 processes. Initial state and final state radiations are implemented by parton shower (PS) models which is based on dipole style p_T-ordered evolution. PYTHIA8 was generated using the latest CUETP8M1 tune, which is based on Monash with NNPDF2.34LO parton distribution function. Both MADGRAPH (version 5.2.4) and POWHEG (version 2) are ME based generator and both using the PYTHIA8 for the PS For MADGRAPH, the QCD processes 2 $\rightarrow$ 3 was done at leading order (LO) calculation, while POWHEG is next leading order (NLO). Three jet events where also generated using POWHEG only, with PS turned off in PYTHIA8, so that the events would only be generated with ME calculations. Four samples were made to study the PS dominant and ME dominant regions:

- PYTHIA8 (PS only)
- POWHEG + PYTHIA8 without PS (ME only)
- POWHEG + PYTHIA8 with PS (ME + PS)
- MADGRAPH + PYTHIA8 (ME + PS)

 DOI: 10.1051/epjconf/201714104003

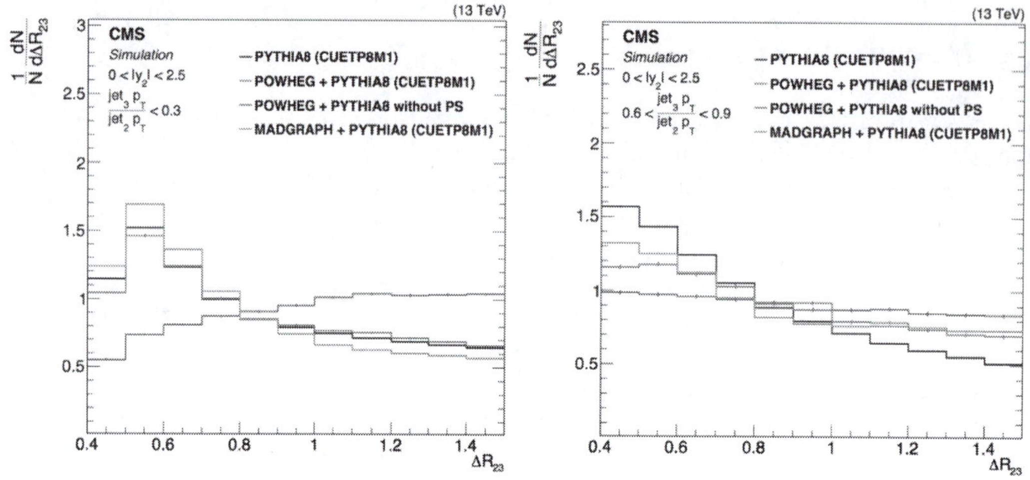

Figure 1. ΔR_{23} distributions for $p_{T3}/p_{T2} < 0.3$ PS dominant (left) and high $0.6 < p_{T3}/p_{T2} < 0.9$ ME dominant (right) regions. PYTHIA8 is PS only, POWHEG + PYTHIA8 without PS is ME only and both POWHEG + PYTHIA8 and MADGRAPH + PYTHIA8 have ME + PS.

This MC study looked at events with only three jets, all with $p_T > 30$ GeV and $\eta < 2.5$. A much higher p_T requirement on the leading jet of 510 GeV was used so that it can be triggered in real CMS data. Two variables were used to separate and study the PS dominant and ME dominant regions. The angular separation between the second and third p_T ordered jets, $\Delta R_{23} = \sqrt{(\Delta\eta)^2 + (\Delta\phi)^2}$, where $\Delta\eta = \eta_3 - \eta_2$ and $\Delta\phi = \phi_3 - \phi_2$. The second variable is the p_T ratio of the third jet over the second jet, p_{T3}/p_{T2}. A soft third jet should be PS dominated, while a hard third jet should be ME dominated. Soft third jet will have a low p_{T3}/p_{T2} and small angular separation ΔR_{23}, while a hard third jet will have p_{T3}/p_{T2} near 1 and larger angular separation ΔR_{23}.

Figure 1 shows the ΔR_{23} distributions for low $p_{T3}/p_{T2} < 0.3$ and high $0.6 < p_{T3}/p_{T2} < 0.9$ events. In the low p_{T3}/p_{T2} region, the third jet is soft and PS is expected to be dominant. POWHEG + PYTHIA8 without PS is ME only and the distribution is very different to all the other samples which have PS. PYTHIA8 agrees well with POWHEG + PYTHIA8 and MADGRAPH + PYTHIA8, even though the latter two also have additional ME contributions. In the high p_{T3}/p_{T2} region, the third jet is hard and ME is expected to be dominant. Although it isn't as striking, it is clearly visible that PYTHIA8 isn't as flat as the other samples with ME. This effect is greater at high ΔR_{23}, where POWHEG + PYTHIA8 and MADGRAPH + PYTHIA8 are close to POWHEG + PYTHIA8 without PS.

Figure 2 shows the p_{T3}/p_{T2} distributions for small $0.4 < \Delta R_{23} < 1.0$ and large $1.0 < \Delta R_{23} < 1.5$ events. In the small ΔR_{23} region, the third jet has a small angular separation from the second jet and PS is expected to be dominant. PYTHIA8 agrees very well with MADGRAPH + PYTHIA8, while both POWHEG samples seem to be different. In the large ΔR_{23} region, the third jet has a large angular separation from the second jet and ME is expected to be dominant. PYTHIA8 seems to disagree with all the others, with POWHEG + PYTHIA8 and MADGRAPH + PYTHIA8 closer to POWHEG + PYTHIA8 without PS.

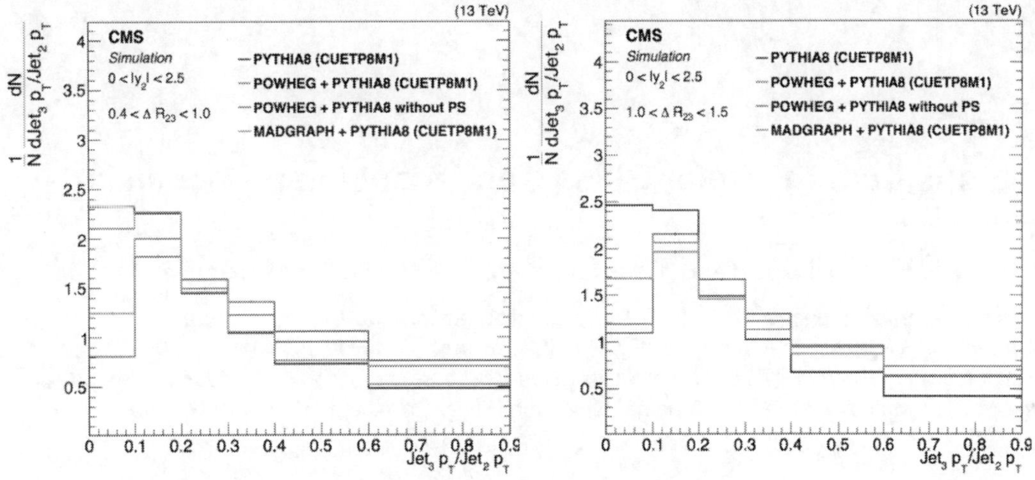

Figure 2. p_{T3}/p_{T2} distributions for small $0.4 < \Delta R_{23} < 1.0$ PS dominant (left) and large $1.0 < \Delta R_{23} < 1.5$ ME dominant (right) regions. PYTHIA8 is PS only, POWHEG + PYTHIA8 without PS is ME only and both POWHEG + PYTHIA8 and MADGRAPH + PYTHIA8 have ME + PS.

4 Summary

As shown by the CMS measurements on jets, multi-jet correlations like colour coherence still needs to be understood better. These results show that the three jet events can be separated into ME and PS dominated regions. Further optimisations could be done to separate the ME and PS even further. Comparisons to data by looking at these regions will make it much easier to understand where the generators are poorly performing.

References

[1] B. Andersson, G. Gustafson, B. Soderberg, Z. Phys. **C20**, 317 (1983)
[2] B. Abbott et al. (D0), Phys. Lett. **B414**, 419 (1997), hep-ex/9706012
[3] T. Sjöstrand, S. Ask, J.R. Christiansen, R. Corke, N. Desai, P. Ilten, S. Mrenna, S. Prestel, C.O. Rasmussen, P.Z. Skands, Comput. Phys. Commun. **191**, 159 (2015), 1410.3012
[4] S. Chatrchyan et al. (CMS), JINST **3**, S08004 (2008)
[5] M. Cacciari, G.P. Salam, G. Soyez, JHEP **04**, 063 (2008), 0802.1189
[6] CMS Collaboration, CMS-PAS-SMP-14-015 (2015)
[7] CMS Collaboration, CMS-PAS-SMP-15-003 (2016)
[8] S. Chatrchyan et al. (CMS), Eur. Phys. J. **C74**, 2901 (2014), 1311.5815
[9] J. Alwall, M. Herquet, F. Maltoni, O. Mattelaer, T. Stelzer, JHEP **06**, 128 (2011), 1106.0522
[10] P. Nason, JHEP **11**, 040 (2004), hep-ph/0409146
[11] S. Frixione, P. Nason, C. Oleari, JHEP **11**, 070 (2007), 0709.2092
[12] S. Alioli, P. Nason, C. Oleari, E. Re, JHEP **06**, 043 (2010), 1002.2581

EPJ Web of Conferences **141**, 04004 (2017) DOI: 10.1051/epjconf/201714104004
ISMD 2016

Photons from the Color Glass Condensate in p+A collisions

Sanjin Benić[1,2], Kenji Fukushima[2], Oscar Garcia-Montero[3], and Raju Venugopalan[4]

[1] *Physics Department, Faculty of Science, University of Zagreb, Zagreb 10000, Croatia*
[2] *Department of Physics, The University of Tokyo, 7-3-1 Hongo, Bunkyo-ku, Tokyo 113-0033, Japan*
[3] *Institut für Theoretische Physik, Universität Heidelberg, Philosophenweg 16, 69120 Heidelberg, Germany*
[4] *Physics Department, Brookhaven National Laboratory, Bldg. 510A, Upton, NY 11973, USA*

Abstract. We report on a first NLO computation of photon production in p+A collisions at collider energies within the Color Glass Condensate framework, significantly extending previous LO results. At central rapidites, our result is the dominant contribution and probes multi-gluon correlators in nuclei. At high photon momenta, the result is directly sensitive to the nuclear gluon distribution. The NLO result contains two classes of processes, the annihilation process and the process with $q\bar{q}$ pair together with a photon in the final state. Using the McLerran-Venugopalan model, we show full numerical results for the photon spectrum coming from the annihilation process.

1 Introduction

Nuclei at high energies are governed by the dynamics of saturated gluons described within the Color Glass Condensate (CGC) effective theory [1–3]. The CGC introduces a new dynamical scale, called the saturation scale Q_s, characterizing gluon correlations in nuclei with $k_\perp \lesssim Q_s$. The CGC description should well describe the physics already at the energy reach of the RHIC and the LHC, where typically $x \sim 10^{-2}$ and $x \sim 10^{-3}$ at central rapidity is expected, respectively. Anticipating new experimental results on photon spectrum in p+A collisions from RHIC and LHC we aim at a detailed theoretical description that would help constrain fundamental multi-gluon correlators as predicted by the CGC.

The LO contribution for photon production is valence quark bremsstrahlung (see leftmost diagram in Fig. 1). The CGC result for the $O(\alpha_e)$ cross section was calculated by Gelis and Jalilian-Marian in [4] through which the photon spectrum becomes sensitive to the two-point gluon correlator of the nucleus. This is a good description for central rapidities at low energies or, alternatively, in forward kinematics, where the valence quarks dominate the proton wave function. When the center of mass energy is increased, the gluon component of the proton contributes significantly to photon production in the central rapidity region.

Working within the dilute-dense approximation, appropriate for the p+A kinematics, we investigate NLO processes with a gluon in place of the valence quark being emitted from the proton. In Fig. 1 (middle and right) the gluon emits a quark-antiquark pair. The pair either annihilates back to a photon (class II process) or goes to the final state (class III process). The class II process was calculated in [5], while the class III process was calculated in [6]. The inclusive rate in both cases is of the order $O(\alpha\alpha_s)$. To compare with valence quark bremsstrahlung we must multiply with the respective

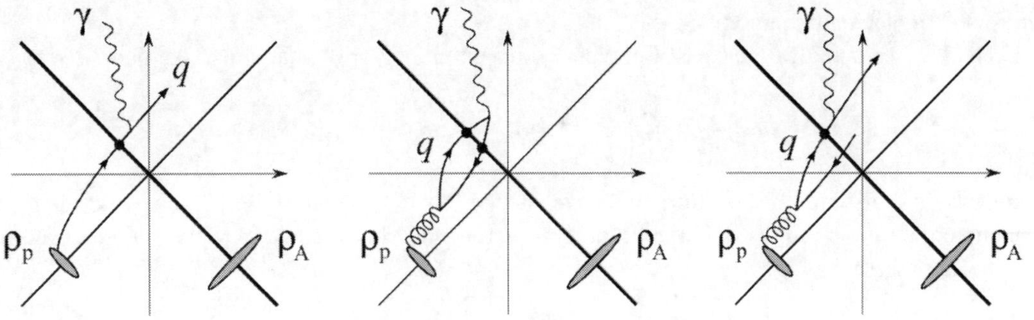

Figure 1. Leftmost diagram belongs to the leading order contribution giving the rate of $O(\alpha)$. The middle (class II) and the rightmost (class III) processes are typical examples of next-to-leading order contributions resulting in rates of $O(\alpha\alpha_s)$.

distribution functions for quarks f_q and for gluons f_g inside the proton. We consider the high energy regime where the new processes become dominant as $f_q \ll \alpha_s f_g$.

2 Calculation of the diagrams

We solve the classical Yang-Mills equations with the dilute-dense approximation for the color sources of the proton ρ_p and of the nucleus ρ_A: $\rho_p \ll \rho_A$. This allows for a systematic computation of the gluon field in powers of ρ_p [7, 8]. We denote the resulting $O(\rho_p^{0,1})$ gluon field as $\mathcal{A}^\mu_{(0,1)}(x)$. In the $\mathcal{A}^+ = 0$ gauge the field of the nucleus alone, $\mathcal{A}^\mu_{(0)}$, is

$$\mathcal{A}^\mu_{(0)}(x) = -g\delta^{\mu-}\delta(x^+)\frac{1}{\partial_\perp^2}\rho_A(\boldsymbol{x}_\perp). \tag{1}$$

The field $\mathcal{A}^\mu_{(1)}$ is given in Fourier space as [8, 9]

$$\mathcal{A}^\mu_{(1)}(q) = \frac{ig\,\delta^{\mu i}q^i_\perp}{(q^+ + i\epsilon)(q^- - i\epsilon)}\frac{\rho_p(\boldsymbol{q}_\perp)}{\boldsymbol{q}_\perp^2} + \frac{ig}{q^2 + iq^+\epsilon}\int_{\boldsymbol{k}_{1\perp}}\int_{\boldsymbol{x}_\perp} e^{i(\boldsymbol{q}_\perp - \boldsymbol{k}_{1\perp})\cdot\boldsymbol{x}_\perp}\, C^\mu(q,\boldsymbol{k}_{1\perp})\, U(\boldsymbol{x}_\perp)\,\frac{\rho_p(\boldsymbol{k}_{1\perp})}{\boldsymbol{k}_{1\perp}^2}, \tag{2}$$

where

$$C^+(q,\boldsymbol{k}_{1\perp}) = 0, \quad C^-(q,\boldsymbol{k}_{1\perp}) = \frac{-2\boldsymbol{k}_{1\perp}\cdot(\boldsymbol{q}_\perp - \boldsymbol{k}_{1\perp})}{q^+ + i\epsilon}, \quad \boldsymbol{C}_\perp(q;\boldsymbol{k}_{1\perp}) = \frac{\boldsymbol{q}_\perp\,\boldsymbol{k}_{1\perp}^2}{(q^+ + i\epsilon)(q^- + i\epsilon)} - 2\boldsymbol{k}_{1\perp}, \tag{3}$$

and $U(\boldsymbol{x}_\perp)$ is the Wilson line with $\mathcal{A}^\mu_{(0)}$ in the adjoint representation

$$U(\boldsymbol{x}_\perp) = \mathcal{P}_+ \exp\left[ig\int_{-\infty}^{\infty} dz^+ \mathcal{A}^-_{(0)}(z^+,\boldsymbol{x}_\perp)\cdot T\right]. \tag{4}$$

The quark propagator $S_{(0)}(x,y)$ is given by the Dirac equation in the $\mathcal{A}^\mu_{(0)}$ background [10] as

$$\begin{aligned} S_{(0)}(x,y) \equiv S_F(x-y) &+ i\theta(x^+)\theta(-y^+)\int_z \delta(z^+)[\tilde{U}(\boldsymbol{z}_\perp) - 1]S_F(x-z)\gamma^+S_F(z-y) \\ &- i\theta(-x^+)\theta(y^+)\int_z \delta(z^+)[\tilde{U}^\dagger(\boldsymbol{z}_\perp) - 1]S_F(x-z)\gamma^+S_F(z-y), \end{aligned} \tag{5}$$

where $\tilde{U}(\boldsymbol{x}_\perp)$ is the Wilson line in the fundamental representation, that is, with the fundamental generators t^a, replacing the adjoint ones T^a in Eq. (4).

The LSZ formula for the class II process with a single flavor of charge e in the quark loop is

$$\mathcal{M}_\lambda(\boldsymbol{k}_\gamma) = eg \int_{xy} e^{-ik_\gamma \cdot x} \mathrm{Tr}[\not{\epsilon}_\lambda(\boldsymbol{k}_\gamma) S_{(0)}(x,y) \mathcal{A}_{(1)}(y) S_{(0)}(y,x)] \,, \tag{6}$$

where $\boldsymbol{k}_\gamma$ is the photon momentum. We have checked that (6) satisfies the photon Ward identity. To calculate the rate we square the amplitude and perform a color average over the sources. The result can be summarized as [5]

$$\frac{dN}{d^2\boldsymbol{k}_{\gamma\perp} d\eta_{k_\gamma}} = \frac{\alpha_e\,\alpha_s}{16\pi^8} \frac{N_c}{N_c^2-1} \int_0^1 dx\,dx' \int_{\boldsymbol{y}_\perp \boldsymbol{y}'_\perp \boldsymbol{z}_\perp \boldsymbol{z}'_\perp} e^{i\boldsymbol{k}_{\gamma\perp}\cdot \boldsymbol{r}_\perp} S(\boldsymbol{y}_\perp, \boldsymbol{z}_\perp, \boldsymbol{y}'_\perp, \boldsymbol{z}'_\perp)$$
$$\times \int_{\boldsymbol{k}_{1\perp}} e^{i\boldsymbol{k}_{1\perp}\cdot \boldsymbol{r}_\perp} \varphi_p(\boldsymbol{k}_{1\perp}) \left[(\hat{z}_\perp - \hat{\boldsymbol{y}}_\perp)\cdot(\hat{z}'_\perp - \hat{\boldsymbol{y}}'_\perp)\Psi_1\Psi_1'^* + \Psi_2\Psi_2'^* + 2(\hat{z}_\perp - \hat{\boldsymbol{y}}_\perp)\cdot \hat{\boldsymbol{k}}_{1\perp}\Psi_1\Psi_2'^*\right]. \tag{7}$$

where

$$S(\boldsymbol{y}_\perp, \boldsymbol{z}_\perp, \boldsymbol{y}'_\perp, \boldsymbol{z}'_\perp) \equiv \frac{1}{N_c}\left\langle \mathrm{Tr}_c\left[\tilde{U}(\boldsymbol{y}_\perp) T_F^a \tilde{U}^\dagger(\boldsymbol{z}_\perp)\right] \mathrm{Tr}_c\left[\tilde{U}(\boldsymbol{z}'_\perp) T_F^a \tilde{U}^\dagger(\boldsymbol{y}'_\perp)\right]\right\rangle \,, \tag{8}$$

$\varphi_p(\boldsymbol{k}_{1\perp})$ is the unintegrated gluon distribution in the proton, $\boldsymbol{r}_\perp = (1-x)\boldsymbol{y}_\perp + x\boldsymbol{z}_\perp - (1-x')\boldsymbol{y}'_\perp - x'\boldsymbol{z}'_\perp$. The quantities $\Psi_{1,2}$ and $\Psi'_{1,2}$ are functions of all the integration variables, and detailed expression are to be found in [5].

We have expanded the Wilson lines to recover the perturbative calculation. We have found that the $O(\rho_p\rho_A)$ contribution vanishes as it should by charge conjugation. The lowest order non-vanishing contribution is $O(\rho_p\rho_A^2)$. We have explicitly checked that this is UV finite as it should by gauge invariance. Taking the zero quark mass limit is completely safe in this amplitude as there are no collinear singularities. Finally, we have checked that the term in the last line of Eq. (7) has a well defined $\boldsymbol{k}_{1\perp} \to 0$ limit so that our result can be brought in a collinear factorized form on the proton side.

Turning to the class III process, in addition to the photon, the final state contains a quark with momentum $\boldsymbol{q}$ and a antiquark with momentum $\boldsymbol{p}$. The LSZ formula for the amplitude is

$$\mathcal{M}_\lambda(\boldsymbol{k}_\gamma, \boldsymbol{q}, \boldsymbol{p}) = ieg \int_{xyz} e^{-ik_\gamma\cdot x - iq\cdot y - ip\cdot z} \bar{u}(\boldsymbol{q})(i\overrightarrow{\not{\partial}}_y - m)[S_{(0)}(y,w)\mathcal{A}_{(1)}(w)S_{(0)}(w,x)\not{\epsilon}_\lambda(\boldsymbol{k}_\gamma)S_{(0)}(x,z)$$
$$+ S_{(0)}(y,x)\not{\epsilon}_\lambda(\boldsymbol{k}_\gamma)S_{(0)}(x,w)\mathcal{A}_{(1)}(w)S_{(0)}(w,z)](i\overleftarrow{\not{\partial}}_z + m)v(\boldsymbol{p}) \,. \tag{9}$$

Inserting the quark propagator $S_{(0)}(x,y)$ and the gluon field $\mathcal{A}^\mu_{(1)}(x)$ we find [5]

$$\mathcal{M}_\lambda(\boldsymbol{k}_\gamma, \boldsymbol{q}, \boldsymbol{p}) = -eg^2 \int_{\boldsymbol{k}_\perp \boldsymbol{k}_{1\perp}} \int_{\boldsymbol{x}_\perp \boldsymbol{y}_\perp} \frac{\rho_p^a(\boldsymbol{k}_{1\perp})}{\boldsymbol{k}_{1\perp}^2} e^{i\boldsymbol{k}_\perp\cdot \boldsymbol{x}_\perp + i(\boldsymbol{k}_{\gamma\perp}+\boldsymbol{q}_\perp+\boldsymbol{p}_\perp-\boldsymbol{k}_\perp-\boldsymbol{k}_{1\perp})\cdot \boldsymbol{y}_\perp}$$
$$\times \bar{u}(\boldsymbol{q})\epsilon_{\lambda\mu}(\boldsymbol{k}_\gamma)\left\{T_g^\mu(\boldsymbol{k}_{1\perp})U(\boldsymbol{x}_\perp)^{ba}t^b + T_{q\bar{q}}^\mu(\boldsymbol{k}_\perp, \boldsymbol{k}_{1\perp})\tilde{U}(\boldsymbol{x}_\perp)t^a\tilde{U}^\dagger(\boldsymbol{y}_\perp)\right\}v(\boldsymbol{p}) \,. \tag{10}$$

where the explicit expressions for the functions $T_g^\mu(\boldsymbol{k}_{1\perp})$ and $T_{q\bar{q}}^\mu(\boldsymbol{k}_\perp, \boldsymbol{k}_{1\perp})$ can be found in [5]. We have checked that (10) satisfies the photon Ward identity and also that the leading twist result satisfies the gluon Ward identities. In general, the result for the amplitude is gauge dependent. We have made the calculation in the $\mathcal{A}^+ = 0$ and also in the $\partial_\mu\mathcal{A}^\mu = 0$ gauge [6, 11]. Interestingly, after some algebraic manipulation, the two results for the amplitude, within this particular gauge choices, become identical.

The total cross section for $\bar{q}q\gamma$ production is found as [6]

$$\frac{d\sigma}{d^2\boldsymbol{k}_{\gamma\perp}d\eta_{k_\gamma}d^2\boldsymbol{q}_\perp d\eta_q d^2\boldsymbol{p}_\perp d\eta_p} = \frac{\alpha_e\alpha_s^2}{256\pi^8 C_F}\int_{\boldsymbol{k}_{1\perp}\boldsymbol{k}_{2\perp}} (2\pi)^2\delta^{(2)}(\boldsymbol{k}_{\gamma\perp}+\boldsymbol{q}_\perp+\boldsymbol{p}_\perp-\boldsymbol{k}_{1\perp}-\boldsymbol{k}_{2\perp})\frac{\varphi_p(\boldsymbol{k}_{1\perp})}{\boldsymbol{k}_{1\perp}^2\boldsymbol{k}_{2\perp}^2}$$
$$\times\Bigg\{\int_{\boldsymbol{k}_\perp\boldsymbol{k}'_\perp}\mathrm{Tr}[(\not{q}+m)T^\mu_{q\bar{q}}(\boldsymbol{k}_{1\perp},\boldsymbol{k}_\perp)(m-\not{p})\gamma^0 T^\dagger_{q\bar{q}\mu}(\boldsymbol{k}_{1\perp},\boldsymbol{k}'_\perp)\gamma^0]\phi_A^{q\bar{q},q\bar{q}}(\boldsymbol{k}_\perp,\boldsymbol{k}_{2\perp}-\boldsymbol{k}_\perp;\boldsymbol{k}'_\perp,\boldsymbol{k}_{2\perp}-\boldsymbol{k}'_\perp)$$
$$+\int_{\boldsymbol{k}_\perp}\mathrm{Tr}[(\not{q}+m)T^\mu_{q\bar{q}}(\boldsymbol{k}_{1\perp},\boldsymbol{k}_\perp)(m-\not{p})\gamma^0 T^\dagger_{g\mu}(\boldsymbol{k}_{1\perp})\gamma^0]\phi_A^{q\bar{q},g}(\boldsymbol{k}_\perp,\boldsymbol{k}_{2\perp}-\boldsymbol{k}_\perp;\boldsymbol{k}_{2\perp}) + \mathrm{h.c.}$$
$$+\,\mathrm{Tr}[(\not{q}+m)T^\mu_g(\boldsymbol{k}_{1\perp})(m-\not{p})\gamma^0 T^\dagger_{g\mu}(\boldsymbol{k}_{1\perp})\gamma^0]\phi_A^{g,g}(\boldsymbol{k}_{1\perp})\Bigg\}. \tag{11}$$

The cross section depends on three different generalized unintegrated gluon distribution functions of the nucleus $\phi_A^{q\bar{q},q\bar{q}}$, $\phi_A^{q\bar{q},g}$ and $\phi_A^{g,g}$. They correspond to different Wilson line correlators in the fundamental and in the adjoint representation identical to the ones found for $q\bar{q}$ production [12]. The inclusive photon production is found by integrating over the quark and antiquark phase space.

Expanding the Wilson lines we find a contribution at $O(\rho_p\rho_A)$ (leading twist). This is the $k_\perp$-factorized photon production as it would be obtained by a pQCD computation. Taking the collinear limit on the proton ($\boldsymbol{k}_{1\perp}\to 0$) and on the nucleus ($\boldsymbol{k}_{2\perp}\to 0$) side, the result can be expressed in terms of gluon distribution functions of the proton and the nucleus and the inclusive cross section is given as

$$\frac{d\sigma}{d^2\boldsymbol{k}_{\gamma\perp}d\eta_{k_\gamma}} = \frac{1}{16}\int_0^\infty\frac{dq^+}{q^+}\frac{dp^+}{p^+}\int_{\boldsymbol{q}_\perp\boldsymbol{p}_\perp}(2\pi)^2\delta^{(2)}(\boldsymbol{k}_{\gamma\perp}+\boldsymbol{q}_\perp+\boldsymbol{p}_\perp)x_p f_{g,p}(x_p,Q^2)x_A f_{g,A}(x_A,Q^2)|\mathcal{M}_{gg\to q\bar{q}\gamma}|^2 . \tag{12}$$

From this result we get a clear relation between the photon spectrum and the gluon distribution in the nuclei.

In the limit of $\boldsymbol{k}_{\gamma\perp}\to 0$ the amplitude $\mathcal{M}_{gg\to q\bar{q}\gamma}$ factorizes in accordance to the soft photon theorem to a radiative part and the remaining, non-radiative part, responsible for $q\bar{q}$ production. The radiative part gives an infrared enhancement to the inclusive cross section as $1/\boldsymbol{k}_{\gamma\perp}^2$.

3 Numerical results

Here we discuss the numerical computation of the class II process. The general analytical expression (7) is valid for a single massive quark flavor. For simplicity we will take the limit of zero quark mass.

The respective Wilson line correlator occurring in the class II process is referred to in the literature as the inelastic quadrupole [13]. Interestingly, within the large N_c approximation, the inelastic quadrupole does not factorize. This means that in principle one must handle a simultaneous integration over all the coordinates of the quadrupole.

For the numerical calculations we assume the case of a large nucleus, that is $A \gg 1$, in which case approximate translational invariance in the transverse plane holds. This turns one transverse plane integration into a volume factor πR_A^2, where R_A is the nuclear radii. For the calculation of the inelastic quadrupole we use the McLerran Venugopalan model $\langle\rho_A^a(\boldsymbol{x}_\perp)\rho_A^b(\boldsymbol{y}_\perp)\rangle \equiv g^2\delta^{ab}\mu_A^2\delta^{(2)}(\boldsymbol{x}_\perp-\boldsymbol{y}_\perp)$ [14–16]. The saturation scale in the MV model is defined as

$$Q_s^2 \equiv \frac{N_c^2-1}{4N_c}g^4\mu_A^2 . \tag{13}$$

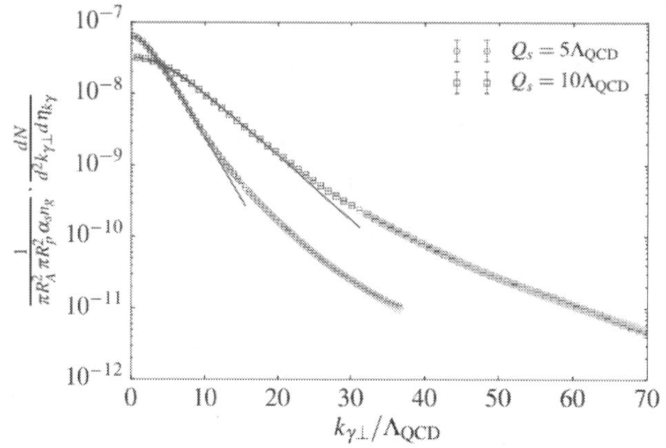

Figure 2. Photon spectrum from the class II diagram. The results are shown for a single quark flavor and in the chiral limit.

Finally, we determine $\varphi_p(\boldsymbol{k}_{1\perp})$ from the MV model for the proton as $k_{1\perp}^2 \varphi_p(\boldsymbol{k}_{1\perp}) = \pi(N_c^2 - 1) g^4 \mu_p^2 \pi R_p^2$.

The numerical integration is optimized through a combination of the MISER Monte Carlo algorithm and a Quasi Discrete Hankel Transform algorithm, for details see [6]. The computation is performed for two different values of the saturation scale $Q_s = 5\Lambda_{\rm QCD}$ and $Q_s = 10\Lambda_{\rm QCD}$. The numerical results for the transverse momentum spectrum of the photon is shown on Fig. 2, where we factorize the transverse proton density fluctuation parameter as $n_g \equiv (N_c^2 - 1) g^4 \mu_p^2/(4N_c)$.

The parametrization of the soft part of the spectrum (up to $k_{\gamma\perp} \sim 2Q_s$) is possible by an exponential as $\exp\left(-\sqrt{k_{\gamma\perp}^2 + (0.5Q_s)^2}/0.5Q_s\right)$, while semi-hard part with $k_{\gamma\perp} \gtrsim 2Q_s$ is well fitted by a power law tail $\left(\log(k_{\gamma\perp}/Q_s)\right)^{1.5}/k_{\gamma\perp}^{5.6}$. The thin lines on Fig. 2 are the exponential fit, while the thick lines are the power law fit.

4 Conclusions

We have performed a detailed analytical calculation of NLO photon rates in p+A collisions. The NLO photon rates probe multi-gluon correlations in the nuclei and as such substantiate and complement related investigations using hadron production [12, 17–21]. Our results will allow a direct study of the nuclear gluon distribution function, as well as the higher-twist corrections thereoff, in a systematic way. Furthermore, they may help resolve an experimental puzzle [22] concerning the excess of soft photons in comparison to the predictions by the Low-Burnett-Kroll soft photon theorem. Relative to hadrons, photons are free from hadronization uncertainties and dedicated photon measurements would go a long way towards exploring the physics of gluon saturation.

Acknowledgments

S. B. was supported by the European Union Seventh Framework Programme (FP7 2007-2013) under grant agreement No. 291823, Marie Curie FP7-PEOPLE-2011-COFUND NEWFELPRO Grant No. 48. K. F. was supported by MEXT-KAKENHI Grant No. 15H03652 and 15K13479. R. V. is supported under DOE Contract No. DE-SC0012704. This material is also based upon work supported by the

U.S. Department of Energy, Office of Science, Office of Nuclear Physics, within the framework of the TMD Topical Collaboration. S. B. acknowledges HZZO Grant No. 8799 at Zagreb University for computational resources.

References

[1] E. Iancu, R. Venugopalan, *The Color glass condensate and high-energy scattering in QCD*, in *In *Hwa, R.C. (ed.) et al.: Quark gluon plasma* 249-3363* (2003), hep-ph/0303204
[2] J. Jalilian-Marian, Y.V. Kovchegov, Prog. Part. Nucl. Phys. **56**, 104 (2006), hep-ph/0505052
[3] F. Gelis, E. Iancu, J. Jalilian-Marian, R. Venugopalan, Ann. Rev. Nucl. Part. Sci. **60**, 463 (2010), 1002.0333
[4] F. Gelis, J. Jalilian-Marian, Phys. Rev. **D66**, 014021 (2002), hep-ph/0205037
[5] S. Benic, K. Fukushima (2016), 1602.01989
[6] S. Benic, K. Fukushima, O. Garcia-Montero, R. Venugopalan (2016), 1609.09424
[7] J.P. Blaizot, F. Gelis, R. Venugopalan, Nucl. Phys. **A743**, 13 (2004), hep-ph/0402256
[8] K. Fukushima, Y. Hidaka, Nucl. Phys. **A813**, 171 (2008), 0806.2143
[9] F. Gelis, Y. Mehtar-Tani, Phys. Rev. **D73**, 034019 (2006), hep-ph/0512079
[10] L.D. McLerran, R. Venugopalan, Phys. Rev. **D59**, 094002 (1999), hep-ph/9809427
[11] O. Garcia-Montero, *Prompt photon production in pA collisions as a probe of many-body gluodynamics at high energies*, Master's Thesis, Ruprecht-Karls-Universität Heidelberg (2016)
[12] J.P. Blaizot, F. Gelis, R. Venugopalan, Nucl. Phys. **A743**, 57 (2004), hep-ph/0402257
[13] F. Dominguez, C. Marquet, B.W. Xiao, F. Yuan, Phys. Rev. **D83**, 105005 (2011), 1101.0715
[14] L.D. McLerran, R. Venugopalan, Phys. Rev. **D49**, 2233 (1994), hep-ph/9309289
[15] L.D. McLerran, R. Venugopalan, Phys. Rev. **D49**, 3352 (1994), hep-ph/9311205
[16] L.D. McLerran, R. Venugopalan, Phys. Rev. **D50**, 2225 (1994), hep-ph/9402335
[17] H. Fujii, F. Gelis, R. Venugopalan, Nucl. Phys. **A780**, 146 (2006), hep-ph/0603099
[18] H. Fujii, K. Watanabe, Nucl. Phys. **A915**, 1 (2013), 1304.2221
[19] Z.B. Kang, Y.Q. Ma, R. Venugopalan, JHEP **01**, 056 (2014), 1309.7337
[20] Y.Q. Ma, R. Venugopalan, H.F. Zhang, Phys. Rev. **D92**, 071901 (2015), 1503.07772
[21] Y.Q. Ma, R. Venugopalan, Phys. Rev. Lett. **113**, 192301 (2014), 1408.4075
[22] A. Belogianni et al., Phys. Lett. **B548**, 129 (2002)

EPJ Web of Conferences **141**, 05001 (2017) DOI: 10.1051/epjconf/201714105001
ISMD 2016

Collective flow measurements at RHIC energies

ShinIchi Esumi[1,a]

[1] *Division of Physics, Faculty of Pure and Applied Sciences, University of Tsukuba, Center for Integrated Research in Fundamental Science and Engineering (CiRfSE), Tenno-dai 1-1-1, Tsukuba, Ibaraki 305-8571, Japan*

Abstract. Recent experimental results on collective flow measurements from relativistic heavy-ion collider (RHIC) are presented and discussed to study high-temperature and high-density quark-nuclear matter, Quark Gluon Plasma (QGP) especially focusing on bulk properties, such as freeze-out parameters, temperature, chemical potential, collective expansion, azimuthal event anisotropy measurements. Their relations to the various correlation and fluctuation studies are also discussed, including initial geometrical and E- and B-field conditions as well as possible collective flow evolution that could even be developed in small systems. Current results and understandings from the beam energy scan program (BES) and future plans are discussed and reviewed.

1 Baryon density, directed flow and initial fields

Identified particle yields and ratios are used to extract chemical freeze-out temperature T_{ch} and Baryon chemical potential μ_B based on thermal models with chemical freeze-out assumptions, as shown in Fig.1 (left), which confirms that the high temperature system indeed moves towards the high Baryon density region by reducing the colliding beam energy [1]. The thermal kinetic freeze-out parameters T_{kin} and transverse radial expansion velocity β_T has been extracted by Blast wave model based on hydrodynamical expansion with a common kinetic freeze-out for all particle species. Fig.1 (right) shows that the system cools down further more by larger radial expansion with increasing the centrality and beam energy [2].

Directed flow gradient vs pseudo-rapidity η of anti-proton (top), proton (middle) and net-proton (bottom) are shown in Fig.2 (left) as a function of beam energy, the negative slope of the net-proton v_1 as a function of η might indicate the softening of the equation of state, which could be related to a sudden change of the system pressure close to the critical point [3]. Asymmetric colliding system Cu+Au has provided a finite directed flow at mid-rapidity as shown in Fig.2 (right), where the small difference of v_1 between positive and negative charged particles could be attributed by the initial electric field parallel to the reaction plane inside the participant region expected in the asymmetric collision system [4].

Very strong magnetic fields are expected to be formed in early stage of non-central collisions because of fast moving positive beam ions passing close by. Possible charge asymmetry effects from the strong B-field via 3-particle correlation (like- and un-like signed pair correlation with respect to the

[a] e-mail: esumi.shinichi.gn@u.tsukuba.ac.jp

DOI: 10.1051/epjconf/201714105001

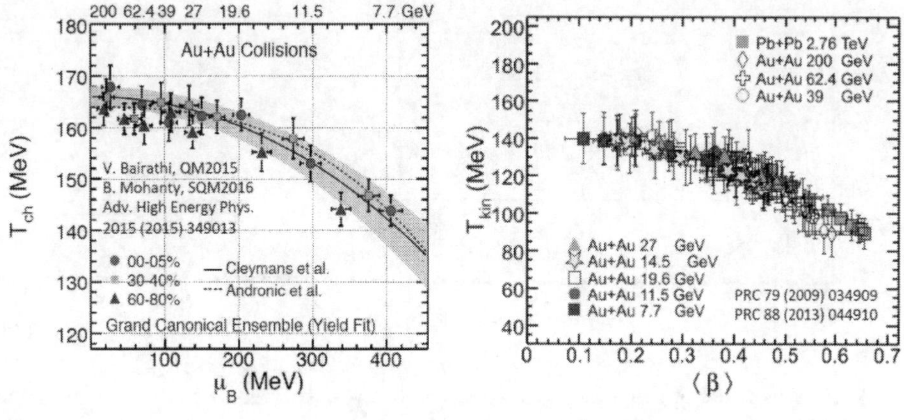

Figure 1. Chemical (left) and kinetic (right) thermal freeze-out parameters as functions of beam energy and centrality [1, 2].

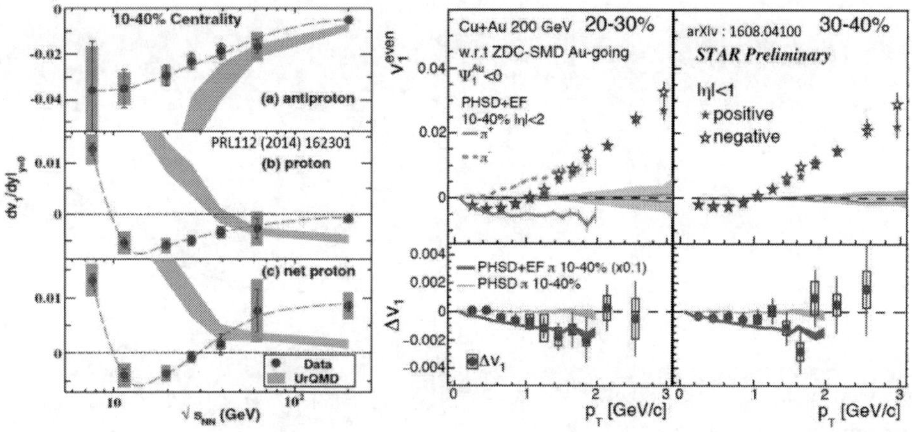

Figure 2. Directed flow v_1 slope vs η for anti proton, proton and net-proton as a function of colliding beam energy in Au+Au collisions (left) [3]. Charge dependence of directed flow v_1 vs p_T in asymmetric Cu+Au collisions (right) [4].

reaction plane) are shown in Fig.3 (left) [5] and v_2 differences between negative and positive charged pion are shown in Fig.3 (right) as a function of charge asymmetry A_{ch} of the event [6], where these observations are consistent with the expectations from the chiral magnetic effects and waves [5, 6].

2 Elliptic, higher order harmonic flow and correlations

Triggered by the ridge observation in high-multiplicity p+p collisions at LHC energies, flow and correlation studies in small systems; p+A, d+A and ^{3}He+A collisions are carried out to investigate the collective phenomena in small systems at RHIC energies. Fig.4 shows the elliptic and triangular event anisotropy measurements in high multiplicity (central) events in these small collision systems,

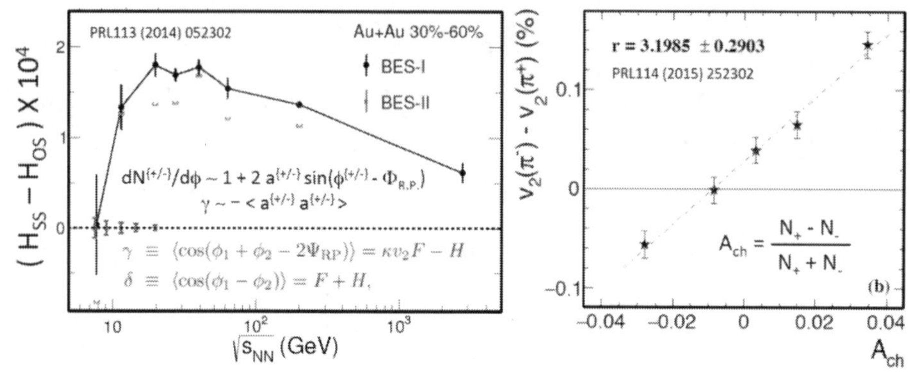

Figure 3. Possible signature of chiral magnetic effects and waves observed as charge asymmetry in 3-particle correlation (left) [5] and as charge dependent v_2 (right) [6].

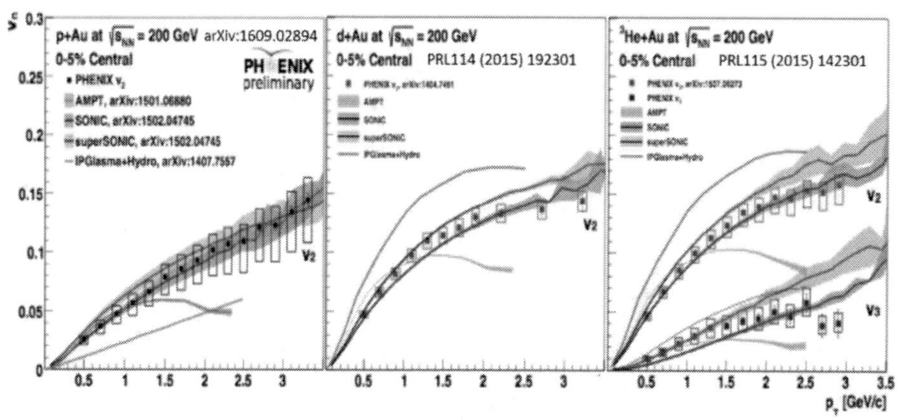

Figure 4. Event anisotropy v_2 and v_3 measurements in small collision systems, p+Au (left), d+Au (middle) and ^{3}He+Au (right) collisions at 200 GeV [7–9].

where results are remarkably well reproduced by hydrodynamical models and transport simulations [7–9].

The higher order event anisotropy measurements for identified particles are shown in Fig.5 (left), where the hydrodynamics-like mass dependence at low P_T region and quark coalescence-like constituent quark number dependence at intermediate P_T region are confirmed also for the higher event anisotropy [10]. Triangular event anisotropy v_3 normalized by the system energy density; $(dN/d\eta)/(N_{part}/2)$ are shown as a function of colliding beam energy in Fig.5 (right), where non-monotonic variation is observed around $\sqrt{s_{NN}}$ = 10-30 GeV [11], where this non-monotonic behavior seems to be occurred at similar beam energy region as v_1 slope in Fig.2 and as charge asymmetry from B-field in Fig.3.

Higher order event plane angles are utilized to study HBT quantum interferometry correlation in order to measure the geometrical size, duration time as well as the shape of the emission source of the colliding zone. The extracted sideward R_s and outward R_o radius parameters are shown in Fig.6

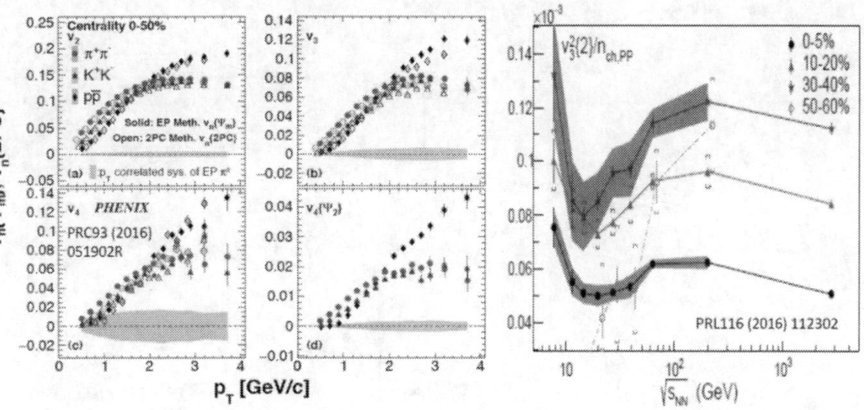

Figure 5. Higher order event anisotropy measurements $v_2(\Psi_2)$, $v_3(\Psi_3)$, $v_4(\Psi_4)$ and $v_4(\Psi_2)$ for identified particles $\pi^{+,-}$, $K^{+,-}$, p and $\overline{p}$ (left) [10]. Beam energy dependence of v_3 normalized by the system energy density (right) [11].

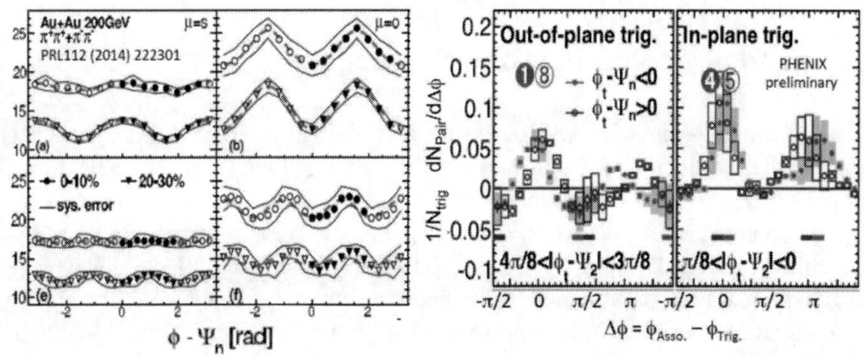

Figure 6. Azimuthal angle dependence of extracted HBT radius parameters R_s^2 and R_o^2 with respect to the 2nd and 3rd order event planes (left) [12]. Trigger angle dependence of di-hadron azimuthal correlation with respect to the 2nd order event plane (right) [13].

(left) as a function of relative pair angle from the 2nd and the 3rd order event planes, the measured eccentricity $\epsilon_{2,\text{final}}$ at the end of freeze-out is found to be smaller (closer to the isotropic distribution) than the initial eccentricity $\epsilon_{2,\text{initial}}$, while the triangularity $\epsilon_{3,\text{final}}$ seems to be almost zero or slightly reversed at the end of freeze-out, because of the very strong radial expansion [12]. Di-hadron correlation yields per trigger are shown in Fig.6 (right) by selecting trigger particle at in-plane or out-of-plane directions including left/right dependent trigger selections with respect to the 2nd order event plane, where the correlation yields and shapes are clearly different for various trigger angle cases, which could be caused by path-length dependence of jet modification as well as flow-jet interplay coming from jet-medium interaction inside the expanding medium [13].

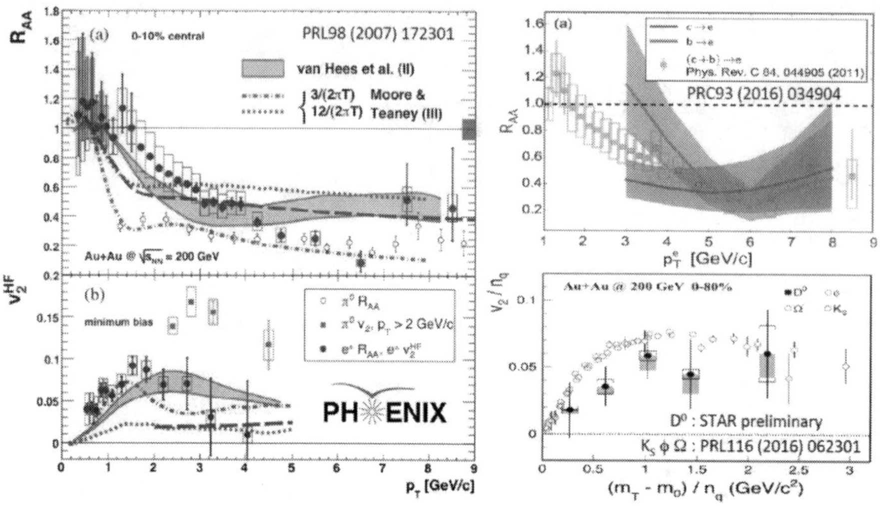

Figure 7. Heavy flavor electron suppression R_{AA} and elliptic event anisotropy v_2 at RHIC (left) [14]. Charm and bottom separated electron R_{AA} (top-right) [15] and reconstructed D^0 meson v_2 compared with strangeness hadrons (bottom-right) [16].

3 Heavy quarks, photon flow and fluctuations

Heavy quark flow and suppression have been observed to be rather significant at RHIC energies, which leads to the conclusion of strongly-interacting Quark Gluon Plasma (sQGP) formation at RHIC as shown in Fig.7 (left) [14]. Recent improved measurements especially in terms of charm and bottom quark separation are seen in Fig.7 (right panels) using new vertex detectors in both PHENIX experiment with Silicon Vertex Detector (VTX) and STAR experiment with Heavy Flavor Tracker (HFT), where the charm and bottom separated heavy flavor electron R_{AA} is shown in top-right panel of Fig.7 and the identified D^0 meson v_2 is shown in bottom-right panel of Fig.7. Possible hits of heavy quark mass dependence of flow and suppression could be seen [15, 16].

Direct photon production at RHIC energy have been confirmed to be as a penetrating probe (R_{AA}~1 and v_2~0) at high p_T region above 5-6 GeV/c [17], while the enhanced photon production at lower p_T region are considered as thermal photon radiation from early stage of hot matter [18] as shown in Fig.8 (left). Therefore, it has been very much surprising to observe quite large magnitude of elliptic and triangular event anisotropies v_2 and v_3 at low p_T region below 4 GeV/c as shown in Fig.8 (middle- top and bottom panels), which is somewhat contradicting to have such large $v_{2,3}$ for early photons emitted before the strong expansion, where this needs further investigations [19]. Fluctuation of conserved quantity is expected to be sensitive to the critical point as an end point of 1st order phase transition, because a sudden change of correlation length is expected close to the critical point. Higher order cumulant ratio of net-proton distribution (ratio of 4th-order over 2nd-order cumulant) is shown as a function of beam energy [20], where there might be some possible critical fluctuations at similar beam energy region again around 10-30 GeV, although the statistical and systematic errors of the measurements are comparable with the magnitude of the observed possible critical fluctuation. This is the main reason why the beam energy scan phase 2 program will be pursued in the next 3-5 years at RHIC.

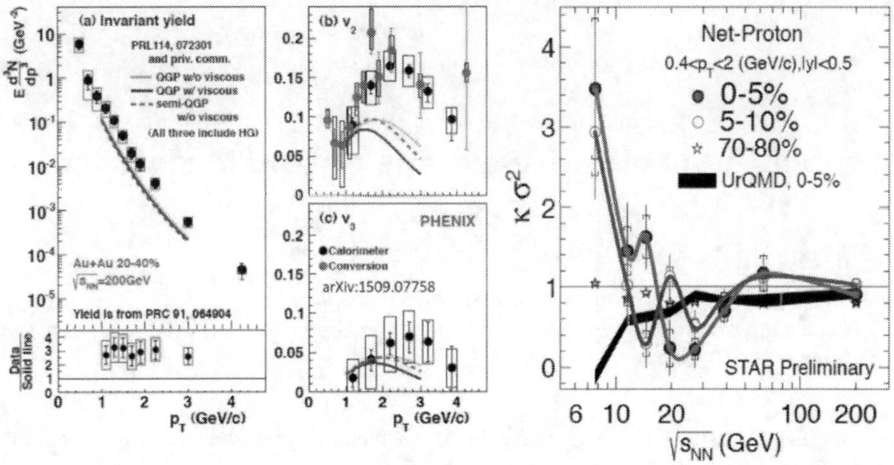

Figure 8. Low p_T direct photon yield and azimuthal anisotropy $v_{2,3}$ (left) [19]. Cumulant ratio of 4th-order over 2nd-order ($\kappa\sigma^2$) of net-proton distribution as a function of beam energy (right) [20].

References

[1] S. Chatterjee et al., Adv. High Energy Phys. **2015**, 349013 (2015); V. Bairathi, QM2015; B. Mohanty, SQM2016

[2] B. I. Abelev et al. (STAR Collaboration), Phys. Rev. **C79**, 034909 (2009); B. Abelev et al. (ALICE Collaboration), Phys. Rev. **C88**, 044910 (2013)

[3] L. Adamczyk et al. (STAR Collaboration), Phys. Rev. Lett. **112**, 162301 (2014)

[4] L. Adamczyk et al. (STAR Collaboration), arXiv:1608.04100

[5] L. Adamczyk et al. (STAR Collaboration), Phys. Rev. Lett. **113**, 052302 (2014)

[6] L. Adamczyk et al. (STAR Collaboration), Phys. Rev. Lett. **114**, 252302 (2015)

[7] C. Aidala et al. (PHENIX Collaboration), arXiv:1609.02894

[8] A. Adare et al. (PHENIX Collaboration), Phys. Rev. Lett. **114**, 192301 (2015)

[9] A. Adare et al. (PHENIX Collaboration), Phys. Rev. Lett. **115**, 142301 (2015)

[10] A. Adare et al. (PHENIX Collaboration), Phys. Rev. **C93**, 051902R (2016)

[11] L. Adamczyk et al. (STAR Collaboration), Phys. Rev. Lett. **116**, 112302 (2016)

[12] A. Adare et al. (PHENIX Collaboration), Phys. Rev. Lett. **112**, 222301 (2014)

[13] T. Todoroki for the PHENIX Collaboration, Nucl. Phys. **A904-905**, 495c (2013)

[14] A. Adare et al. (PHENIX Collaboration), Phys. Rev. Lett. **98**, 172301 (2007)

[15] A. Adare et al. (PHENIX Collaboration), Phys. Rev. **C93**, 034904 (2016)

[16] L. Adamczyk et al. (STAR Collaboration), Phys. Rev. Lett. **116**, 062301 (2016)

[17] S. Afanasiev et al. (PHENIX Collaboration), Phys. Rev. Lett. **109**, 152302 (2012); A. Adare et al. (PHENIX Collaboration), Phys. Rev. Lett. **109**, 122302 (2012)

[18] A. Adare et al. (PHENIX Collaboration), Phys. Rev. **C91**, 064904 (2015)

[19] A. Adare et al. (PHENIX Collaboration), arXiv:1509.07758

[20] X. Luo for the STAR Collaboration), QM2015

Interplay between collective expansion and Mach cone

Tetsufumi Hirano[1,a], Yasuki Tachibana[2,1,b], and Michito Okai[1,c]

[1] *Department of Physics, Sophia University, Tokyo 102-8554, Japan*
[2] *Institute of Particle Physics and Key Laboratory of Quark and Lepton Physics (MOE), Central China Normal University, Wuhan, 430079, China*

Abstract. By using a hybrid dynamical model which describes space-time evolution of the bulk medium, (mini-)jet propagation and interactions between medium and (mini-)jets, we study hydrodynamic responses to (mini-)jet propagation in high energy nuclear collisions. When an energetic jet traverses the bulk matter, it loses its energy into the matter and forms a Mach-cone like structure. On the other hand, the bulk matter expands radially due to pressure gradient. As a result, there happens an interplay between radial expansion and the Mach cone. We discuss possible phenomena and observables related with this in asymmetric gamma-jet events. We also discuss phenomena in which many mini-jets propagate the bulk matter at once in an event and calculate higher harmonics of azimuthal angle distribution.

1 Introduction

One of the important findings in ultra-relativistic heavy ion collisions at the LHC energy is a clear signal of jet quenching, namely, a phenomenon in which hard partons lose their energies during traversing a dense medium. In an asymmetric di-jet event, a few hundred GeV jet is dragged by a dense medium with temperature of the order of a few hundred MeV [1]. This kind of striking phenomenon attracts a lot of theoretical and experimental interests [2].

Dynamics of how the jet loses its energy has been focused in the conventional theoretical analyses. On the other hand, we focus on where the lost energy and momentum go in the medium and, in particular, how the medium responses to the energy and momentum deposited from the jet. To investigate them, we developed a fluid + jet model [3] in which we solve relativistic hydrodynamic equations for space-time evolution of the quark gluon plasma (QGP) fluid with energy and momentum source terms from jet energy loss.

In our previous study, we analysed asymmetric di-jet events at the LHC energy by using this model and found energy/momentum transport away from the jet axis results from the hydrodynamic response to the jet propagation [3]. In the present analysis, we study asymmetric γ-jet events to obtain more information about the hydrodynamic responses to jet propagation. Further details on this topic can be also found in Ref. [4]. Furthermore, we also study an event in which many mini-jets traverse the QGP medium at once. These mini-jets are expected to disturb bulk evolution of the QGP fluid [5].

[a] e-mail: hirano@sophia.ac.jp
[b] e-mail: yasuki.tachibana@mail.ccnu.edu.cn
[c] e-mail: michito0605@eagle.sophia.ac.jp

To see this, we analyse higher order harmonics of the azimuthal angle distribution of final hadrons in low and intermediate transverse momentum regions.

Throughout this paper, we use a natural unit, $\hbar = c = k_{\rm B} = 1$.

2 Model

Relativistic hydrodynamics works quite well in description of space-time evolution of the bulk QCD matter in relativistic heavy ion collisions. We employ a fluid + jet model to study hydrodynamic responses to propagation of jets.

We solve relativistic hydrodynamic equations with source terms

$$\partial_\mu T^{\mu\nu} = \sum_i J_i^\nu, \tag{1}$$

where $T^{\mu\nu}$ is energy momentum tensor of the fluid and J^ν is four vector of the source term. We neglect dissipative effects of the fluid for simplicity, then the energy momentum tensor can be written as

$$T^{\mu\nu} = eu^\mu u^\nu - P(g^{\mu\nu} - u^\mu u^\nu). \tag{2}$$

Here e is energy density, P is pressure, u^μ is four flow velocity and $g^{\mu\nu} = \mathrm{diag}(1,-1,-1,-1)$ is Minkowski metric. As for the equation of state $P = P(e)$, we employ a recent result from lattice QCD simulations [6].

We assume jets lose their energy and momentum, p_i^μ, and that the deposited energy and momentum are quickly equilibrated at the position of jet partons, $\vec{x}_i(t)$, in the QGP fluid. Thus we parametrise the source term as

$$J_i^\mu = -\frac{dp_i^\mu}{dt}\delta^{(3)}\left(\vec{x} - \vec{x}_i(t)\right). \tag{3}$$

Since the purpose of the present study is to investigate hydrodynamic responses to jet propagation, we employ a rather simple parametrised formula for energy loss of jet partons. The amount of energy loss per unit length is assumed to be proportional to parton density of the QGP fluid and, in turn, the cube of temperature in an ideal gas limit. Thus the energy loss formula can be

$$\frac{dp_i^0}{dt} = -\left(\frac{T(t,\vec{x}_i(t))}{T_0}\right)^3 \left.\frac{dE}{dl}\right|_0, \tag{4}$$

where a reference temperature is $T_0 = 0.5$ GeV and the amount of energy loss per unit length at that temperature is $dE/dl|_0 = 15$ GeV/fm in the case of γ-jet events and $dE/dl|_0 = 5$ GeV/fm in many mini-jets cases. The former value leads to reproduce the nuclear modification factor for jets at $p_{\rm T}^{\rm jet} \sim 100$ GeV at the LHC energy [7].

Once we set initial conditions, we are able to describe space-time evolution of the QGP fluids by solving Eq. (1) numerically. In the actual simulations, we employ the Milne coordinates ($\tau = \sqrt{t^2 - z^2}$, $\eta_{\rm s} = (1/2)\ln[(t+z)/(t-z)]$) in the longitudinal direction and the Cartesean coordinates in the transverse plane. In the following, we suppose very central ($b = 0$ fm) Pb+Pb collisions at $\sqrt{s_{\rm NN}} = 2.76$ TeV. Initial entropy density profile can be factrised into the transverse profile $\rho_\perp$ and the longitudinal profile H as

$$s(\tau_0, \vec{x}_\perp, \eta_{\rm s}) = \rho_\perp(\vec{x}_\perp)H(\eta_{\rm s}), \tag{5}$$

where the initial time is taken to be $\tau_0 = 0.6$ fm. By using the optical version of the Glauber model, we estimate the transverse profile as

$$\rho_\perp(\vec{x}_\perp) = \frac{C}{\tau_0}\left[(1-\alpha_{\rm hard})\frac{n_{\rm part}(\vec{x}_\perp)}{2} + \alpha_{\rm hard}n_{\rm coll}(\vec{x}_\perp)\right]. \tag{6}$$

The longitidinal profile can be parametrised as

$$H(\eta_s) = \exp\left[-\frac{(|\eta_s| - \eta_{\rm flat}/2)^2}{2\sigma_\eta^2}\theta\left(|\eta_s| - \eta_{\rm flat}/2\right)\right]\theta(Y_{\rm beam} - |\eta_s|). \tag{7}$$

Here $n_{\rm part}$ and $n_{\rm coll}$ are the number density of participants and that of binary collisions, respectively. $C = 41.4$ is a parameter to control the overall normalisation of the multiplicity at midrapidity. A fraction of hard components is $\alpha_{\rm hard} = 0.08$. The rapidity region of Bjorken-like longitudinally boost-invariant flat profile [8] is $\eta_{\rm flat} = 3.8$ and the width of half Gaussians to smoothly connect the profile to vacuum near the beam rapidity, $Y_{\rm beam}$, is $\sigma_\eta = 3.2$. Initial transverse flow is set to vanish and longitudinal flow velocity is taken from Bjorken's solution $u^{\eta_s}(\tau_0) = 0$ [8]. Initial profile and its parameters are taken from Ref. [9].

We solve Eq. (1) all the way down to $T = 0$ MeV. However, we assume hadrons are kinetically and thermally frozen in a hypersurface Σ defined by $T(\tau, \vec{x}_\perp, \eta_s) = T_{\rm f} = 145$ MeV. We calculate momentum distributions of hadrons by using the Cooper-Frye formula [10]. Here we neglect contribution from resonance decays throughout this paper.

In the analysis of γ-jet events, the initial jet $p_{\rm T}$ spectrum can be parametrised as

$$\frac{dN}{dp_{\rm T}^{\rm jet}d\eta} \propto \frac{1}{p_0}\left(\frac{p_0}{p_{\rm T}^{\rm jet}}\right)^\alpha. \tag{8}$$

Here $p_0 = 205$ GeV and $\alpha = 6.43$ are chosen to fit the data in p-p collisions at $\sqrt{s_{\rm NN}} = 2.76$ TeV [7]. Neglecting higher order corrections, we suppose a photon and a jet are emitted in the opposite directions ($\phi_p = 0$ for a photon and $\phi_p = \pi$ for a jet) with the same transverse momentum. When we analyse many mini-jet propagation on top of bulk QGP fluid evolution, we utilise an event generator PYTHIA [11] to generate mini-jet spectra. The number of mini-jet at midrapidity with $p_{\rm T}^{\rm minijet} > 2$ GeV is estimated to be $\sim$ 600. Here each binary collision in a Pb+Pb collision is simulated with PYTHIA and the number of binary collisions at some transverse point is estimated by using the Glauber model. We put all these mini-jets into the source term (3) in this analysis. Spatial distribution of the jet production points in the transverse plane is assume to be proportional to the number density of binary collision. These mini-jets move along their eikonal paths at the speed of light until either their energy decreases to 2 GeV or temperature of the fluid at the position of them goes down to $T = 160$ MeV. In the analysis of mini-jet propagation, freezeout temperature is taken to be $T_{\rm f} = 160$ MeV.

3 Results

3.1 γ-jet events

We first show results in γ-jet events at the LHC energy in this subsection. We calculate azimuthal angle distribution of charged pions at midrapity $\eta = 0$ in γ-jet events in Pb+Pb collisions at $b = 0$ fm. Trigger conditions are $100 < p_{\rm T}^{\rm jet} < 110$ GeV for jets and $110 < p_{\rm T}^{\gamma} < 120$ GeV for photons. To see the effect of jet propagation, we subtract an azimuthal angle distribution without any jet propagation as a background:

$$\left\langle \Delta\frac{dN_{\pi^\pm}}{d\phi_p d\eta}\bigg|_{\eta=0}\right\rangle = \left\langle \frac{dN_{\pi^\pm}}{d\phi_p d\eta}\bigg|_{\eta=0}\right\rangle - \frac{dN_{\pi^\pm}}{d\phi_p d\eta}\bigg|_{\eta=0,\text{ no jet}}. \tag{9}$$

Here $\langle\cdots\rangle$ is event average and there is no contribution from jet fragmentation. Figure 1 (left) shows difference of azimuthal angle distribution of charged pions between with and without jet propagation

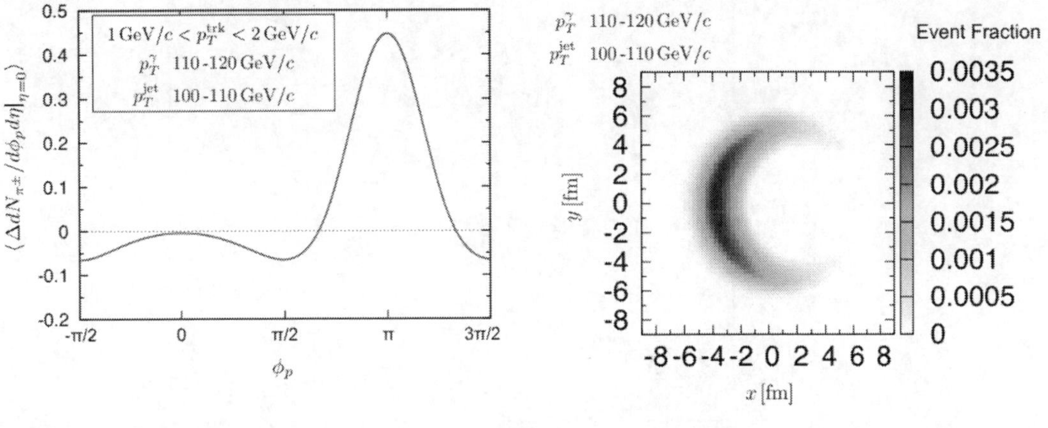

Figure 1. (Left) Difference of azimuthal angle distribution of charged pions between with and without jet propagation. Triggered jet (γ) transverse momentum is $100 < p_T^{\rm jet} < 110$ GeV ($110 < p_T^{\gamma} < 120$ GeV). Transverse momentum of charged pions are in $1 < p_T < 2$ GeV. (Right) Distribution of production point of a photon and a jet under this trigger condition.

as defined in Eq. (9). A clear peak located at $\phi_p = \pi$ results from deposition of energy and momentum from a jet to a fluid. Note that a jet and a photon are emitted in the opposite directions: $\phi_p = 0$ for a photon and $\phi_p = \pi$ for a jet. On the other hand, two dips appear at $\phi_p \sim \pm\frac{\pi}{2}$. This is a consequence of interplay between Mach cone propagation and radial expansion. A dip around $\phi_p \sim \frac{\pi}{2}$ $(-\frac{\pi}{2})$ is originated from jets produced in the positive (negative) y regions. When jets propagate in the negative x direction, the Mach cone is supposed to be composed of two lateral wakes propagating in the direction at $\pm(\frac{\pi}{2} + \theta)$, where θ is Mach angle. However, Mach cone structure is shoved by radial expansion and consequently is tilted slightly. As the action-reaction law, the lateral wake of the Mach cone propagating in the direction at $\frac{\pi}{2} + \theta$ prevents the matter from expanding in the direction at $\phi_p \sim \frac{\pi}{2}$ when jets start their propagation from negative y regions. A similar phenomenon happens when jets starts from positive y regions. This is the reason why particle emission is suppressed in these two directions. These results indicate one has a chance to obtain information about the production points of jets.

3.2 many mini-jets traversing QGP fluid

We utilise the QGP fluid + jet model when multiple mini-jets traverse the QGP fluid and deposit their energies and momenta. The bulk QGP itself expands isotropically in the transverse plane if one neglects the source terms in Eq. (1) since it starts from a perfect circular shape in this study. Nevertheless, many mini-jets are expected to disturb transverse expansion of the QGP fluids due to random momentum direction of each mini-jet in the transverse plane. To see this effect, we calculate flow coefficients of azimuthal angle distributions of charged pions in very central Pb+Pb collisions at $\sqrt{s_{\rm NN}} = 2.76$ TeV. Here we employ the event plane method to analyse Fourier coefficients of azimuthal angle distributions. For the i-th event, we first calculate the event plane angle

$$\tan n\Psi_n^i = \frac{\langle \sin n\phi_p \rangle_i}{\langle \cos n\phi_p \rangle_i}. \tag{10}$$

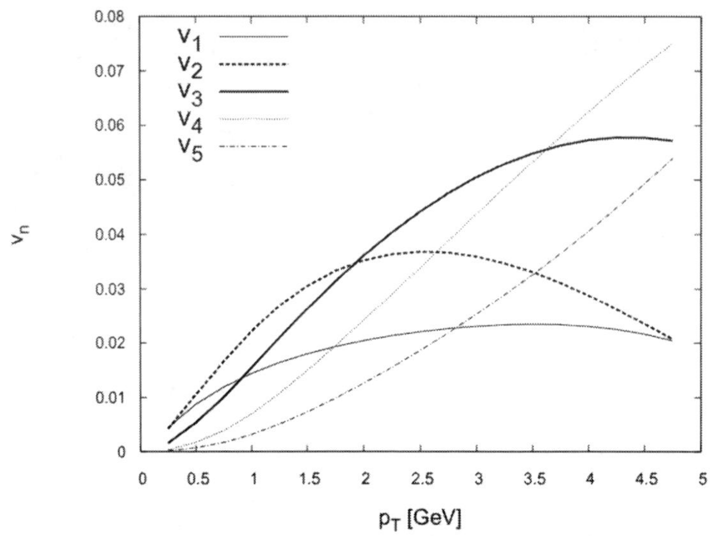

Figure 2. Fourier coefficients (n = 1–5) of the azimuthal angle distributions of charged pions in central Pb+Pb collisions when ~ 600 minijets traverse the smooth QGP matter at midrapidity.

In a single event, an angle bracket in Eq. (10) means

$$\langle O \rangle_i = \frac{\int O \frac{dN^i}{dp_\mathrm{T} d\phi_p} dp_\mathrm{T} d\phi_p}{\int \frac{dN^i}{dp_\mathrm{T} d\phi_p} dp_\mathrm{T} d\phi_p}. \tag{11}$$

When we calculate the event plane angle above, integration with respect to p_T is performed in the whole p_T region at midrapidity. Averaging over all events, we obtain

$$v_n = \frac{1}{N_\mathrm{ev}} \sum_{i=1}^{N_\mathrm{ev}} \langle \cos n(\phi_p - \Psi_n^i) \rangle_i. \tag{12}$$

Our preliminary results of p_T dependences of flow coefficients with the limited number of events, $N_\mathrm{ev} = 30$, are shown in Fig. 2. Since the bulk QGP does not generate any anisotropic flow, finite v_n are attributed solely to minijet propagation. Although statistics is limited, we see considerable amount of flow coefficients. In the low p_T region, v_2 is the largest among them. On the other hand, in the intermediate p_T region ($2 < p_\mathrm{T} < 3$ GeV), v_3 is larger than the others. It would be interesting to see how large the disturbing effect due to minijet propagation is when the initial profile of the bulk QGP fluctuates, which will be discussed in the future publication.

4 Summary

In this study, we utilised the QGP fluid+jet model to investigate hydrodynamic responses to (mini-)jet propagation in central Pb+Pb collisions at the LHC energy. We found one clear peak and two dips in azimuthal distributions of charged pions with respect to jet axis in asymmetric γ-jet events under some trigger condition. The peak results from energy and momentum deposition by a jet, while two dips appear as interplay between Mach cone propagation and radial flow. We also found multiple minijet propagation changes bulk QGP expansion from radially to anisotropically due to random directions of minijet momenta. This can be quantified by flow coefficients of azimuthal angle distributions. In addition to conventional initial fluctuations of bulk matter profile, it can be regarded as a new source of anisotropic flow.

References

[1] S. Chatrchyan *et al.* (CMS Collaboration), Phys. Rev. C **84**, 024906 (2011)
[2] G. Y. Qin and X. N. Wang, *Quark-Gluon Plasma 5* (World Scientific, Singapore, 2016) 309
[3] Y. Tachibana and T. Hirano, Phys. Rev. C **90**, 021902 (2014)
[4] Y. Tachibana and T. Hirano, Phys. Rev. C **93**, 054907 (2016)
[5] M. Schulc and B. Tomášik, Phys. Rev. C **90**, 064910 (2014)
[6] S. Borsányi, Z. Fodor, C. Hoelbling, S. D. Katz, S. Krieg and K. K. Szabó, Phys. Lett. B **730**, 99 (2014)
[7] G. Aad *et al.* (ATLAS Collaboration), Phys. Rev. Lett. **114**, 072302 (2015)
[8] J. D. Bjorken, Phys. Rev. D **27**, 140 (1983)
[9] T. Hirano, P. Huovinen and Y. Nara, Phys. Rev. C **84**, 011901 (2011)
[10] F. Cooper and G. Frye, Phys. Rev. D **10**, 186 (1974)
[11] T. Sjöstrand *et al.*, Comput. Phys. Commun. **191**, 159 (2015)

EPJ Web of Conferences **141**, 05003 (2017) DOI: 10.1051/epjconf/201714105003
ISMD 2016

Longitudinal correlations in the initial stages of ultra-relativistic nuclear collisions

Wojciech Broniowski[2,3,a] and Piotr Bożek[1,b]

[1] *AGH University of Science and Technology, Faculty of Physics and Applied Computer Science, 30-059 Cracow, Poland*
[2] *The H. Niewodniczański Institute of Nuclear Physics, Polish Academy of Sciences, 31-342 Cracow, Poland*
[3] *Institute of Physics, Jan Kochanowski University, 25-406 Kielce, Poland*

Abstract. In this talk we review some of our results for the longitudinal correlations in connection with recent experimental data, in particular for the *torque* effect and for the two-particle correlations in pseudorapidity, $C(\eta_1, \eta_2)$. The model framework involves event-by-event fluctuations of the initial conditions, followed with 3D viscous hydrodynamic evolution and statistical hadronization.

The phenomenon of collectivity allows for simple understanding of very basic features found in ultra-relativistic nuclear collisions (for a review and literature see, e.g., [1]), such as the harmonic flow, the emergence of the near-side ridge, or even the transverse-momentum fluctuations [2, 3].

In the long-range rapidity correlations, the phenomenon is manifest through the shape-flow transmutation depicted in Fig. 1: If the transverse sections in the initial state are similar in the forward (F) and backward (B) directions, then the corresponding harmonic flow (directions of the principal axes, magnitude) are also similar, up to decorrelation effects from fluctuations. Of course, the similarity of the F and B shapes must come out from the early production mechanism, for instance flux-tubes spanned along a large rapidity range. Thus, actually, the starting point of the approach relies on an exact similarity of the transverse fireball shapes in distant F and B directions, which would lead to perfect collimation of the flow axes.

The torque effect, cartooned in Fig. 2, quantifies departure from this perfect collimation. As originally proposed in Ref. [4] in the framework of the wounded-nucleon model [5], the decollimation happens as a combination of two features:

- First, there are event-by-event fluctuations in the number of the F and B going participants (left part of Fig. 2).
- Second, the participants must have asymmetric emission profiles (relating to the deposition of the initial entropy) in the rapidity, with a preference that a participant shines preferably in its forward direction [6].

These lead to the event-by-event torque, as shown in the right part of Fig. 2, appearing independently for subsequent Fourier components.

[a]e-mail: Wojciech.Broniowski@ifj.edu.pl
[b]e-mail: Piotr.Bozek@fis.agh.edu.pl

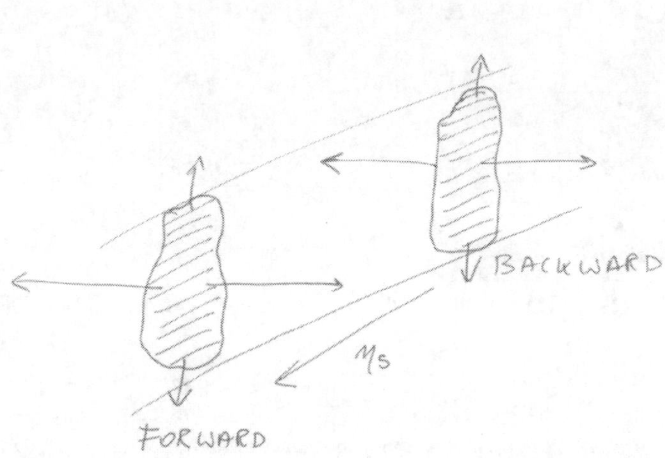

Figure 1. Collimation of flow at very distant longitudinal separations, following from the shape-flow transmutation.

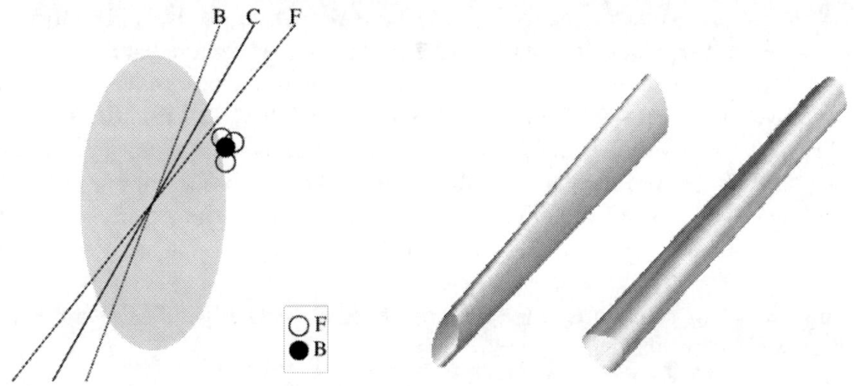

Figure 2. The torque effect. The fluctuations (left) in the number of forward- (F) and backward-going (B) participants, together with asymmetric emission profiles in rapidity, lead to event-by-event fluctuations of the event-plane angle (right). [4]

Apart for the decollimation due to the initial fluctuations, which are the object of our interest, there are also "unwanted" final-stage fluctuation resulting from the statistical nature of hadronization. These may be eliminated through the use of appropriate measures, for instance the 3-bin correlator proposed by the CMS Collaboration [7]:

$$r_n(\eta^a,\eta^b) = \frac{V_{n\Delta}(-\eta^a,\eta^b)}{V_{n\Delta}(\eta^a,\eta^b)}. \tag{1}$$

$$V_{n\Delta}(\eta^a,\eta^b) = \langle\langle e^{in(\phi_1-\phi_2)}\rangle\rangle = \langle\langle e^{in(\psi_n(\eta_a)+\phi_1'-\psi_n(\eta_b)-\phi_2')}\rangle\rangle \simeq \langle e^{in[\psi_n(\eta_a)-\psi_n(\eta_b)]}\rangle\langle\langle e^{in\phi_1'-in\phi_2'}\rangle\rangle.$$

The azimuthal angles of the measured particles, ϕ_1 and ϕ_2, are obtained in a certain reference frame, whereas $\psi_n(\eta_a)$ and $\psi_n(\eta_b)$ are the event-plane angles of the fireball in bins centered around η_a and η_b.

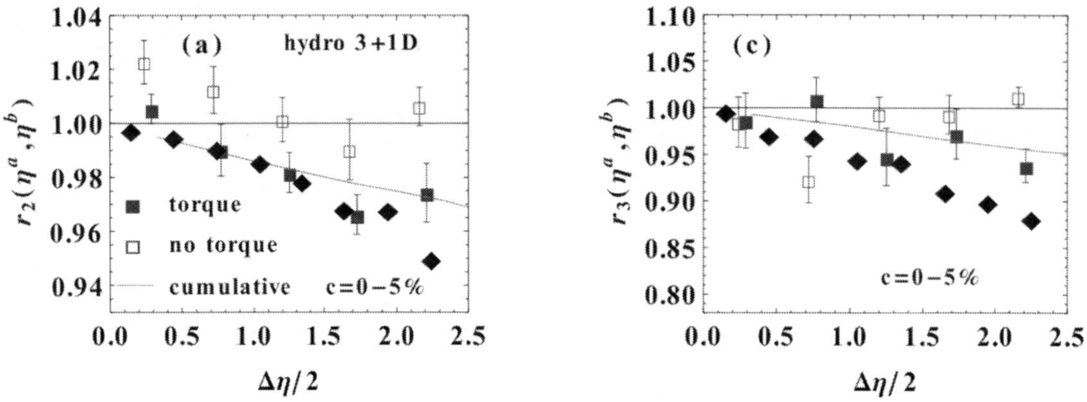

Figure 3. The torque effect for Pb+Pb collisions at $\sqrt{s_{NN}} = 2.76$ TeV, expressed through the 3-bin CMS measure $r_n(\eta^a, \eta^b)$, plotted as a function of $\Delta\eta/2 \equiv \eta_a$ with the reference bin $4 < \eta_b < 5$. Our event-by-event model simulations [8] qualitatively reproduce the CMS data [7] when the ingredients leading to torque are incorporated (in particular, the asymmetric particle emission profile from a given wounded nucleon). The effect is significantly larger for the triangular flow (right) than for the elliptic flow (left).

The angles ϕ'_1 and ϕ'_2 are evaluated relative to $\psi_n(\eta_a)$ and $\psi_n(\eta_b)$, respectively. The forward reference bin $4, 4 < \eta_b < 5$ is well separated from the two measurement bins centered at η_a and $-\eta_a$.

The departure of r_n from unity, seen in Fig. 3, is a measure of the event-plane angle decorrelation, i.e., the torque. We note that our simulations, based on GLISSANDO [9, 10], 3D viscous hydrodynamics [11], and THERMINATOR [12, 13], lead to fair agreement with the data when the two conditions itemized above are incorporated (filled points labeled "torque" in Fig. 3). Without asymmetric production profiles there is no effect (empty squares labeled "no torque"). The solid lines in Fig. 3 show the calculation where many THERMINATOR events for the same initial condition are averaged (cumulative events) to improve statistics.

Interestingly, the CMS measurements for the p-Pb collisions [7] indicates the necessity of an additional decorrelation mechanism:

- Fluctuating strings.

This is because in p-A collisions the transverse shape of the fireball is determined by the positions of the participants in A. If the corresponding strings span all the relevant range in pseudorapidity, the transverse shape of the fireball is the same at all pseudorapidities (see Ref. [14] for a detailed discussion). Adding fluctuating end-points of the strings [15, 16] (see the left part of Fig. 4) leads to proper description of the data.

In Fig. 5 we summarize the results for the slope parameter introduced by the CMS collaboration,

$$r_n(\eta_a, \eta_b) = e^{-2F_n^{\eta}\eta_a}. \tag{2}$$

We note that from general considerations [14] the slope for the $n = 4$ harmonic is very simply related to the slope for $n = 2$:

$$F_4^{\eta}/4 \simeq F_2^{\eta}. \tag{3}$$

Next, we discuss the interpretation of the recent ATLAS measurements [17, 18] of the two-particle correlation function in pseudorapidity, $C_N(\eta_1, \eta_2)$ (for details see Ref. [19]). A typical model result

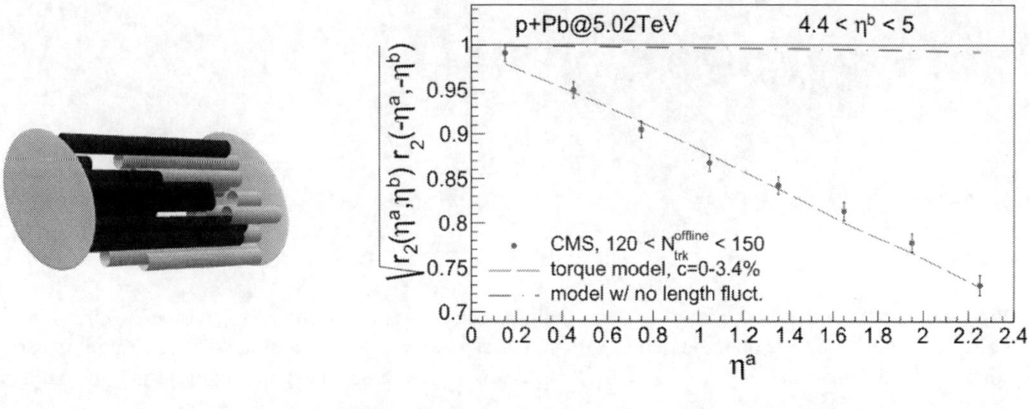

Figure 4. The torque effect for p-Pb. The decorrelation mechanism from string breaking (left) is crucial for the explanation of the CMS data [7]. For strings spanning all the pseudorapidity range there is no decorrelation (dash-dotted line in the right part). The model calculation is carried out with GLISSANDO [10].

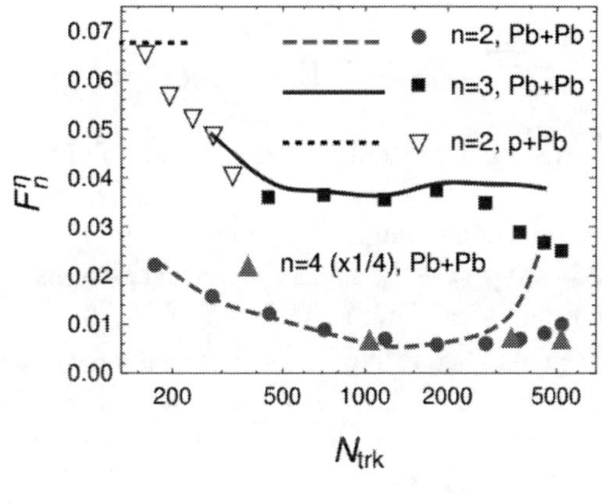

Figure 5. Slope in η_a of the r_n decorrelation measure for Pb-Pb and p-Pb collisions. The data are from Ref. [7], and the model results are from GLISSANDO [10]. We note fair agreement, except for the highest multiplicity events. Note that relation (2) is well satisfied by the data.

is shown in the left part of Fig. 6, where we see that the inclusion of the string fluctuations enhances the correlations and builds-up a characteristic ridge structure along $\eta_1 = \eta_2$. Quantitative measures are conveniently provided with the a_{nm} coefficients defined in Refs, [20, 21]. In our simple model the corresponding formulas are analytic [19]. We introduce $N_+ = N_A + N_B$, $N_- = N_A - N_B$, where $N_{A,B}$ denotes the number of sources associated with the given nucleus. We obtain the non-vanishing terms

DOI: 10.1051/epjconf/201714105003

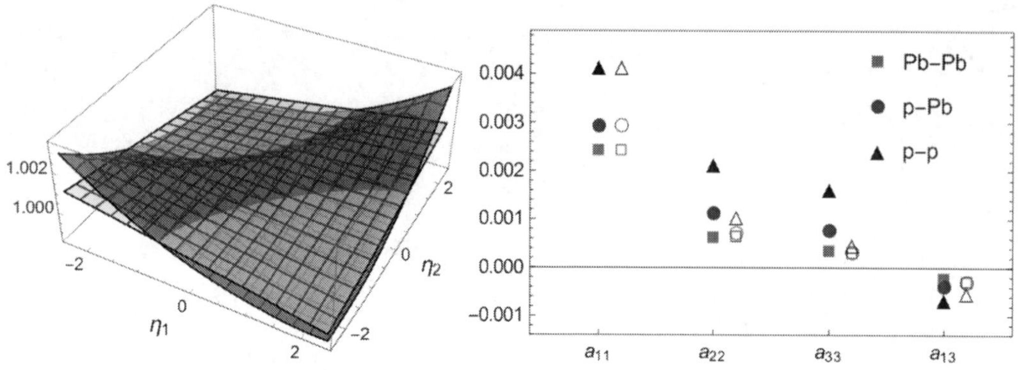

Figure 6. Left: Model prediction for the normalized correlation function $C_N(\eta_1, \eta_2)$ for Pb-Pb collisions at the LHC for $c = 30-40\%$. The lighter color surface corresponds to the calculation without string fluctuations. The surface with an elongated ridge (darker color) corresponds to the case incorporating the string fluctuations. Right: the a_{nm} coefficients

(for $n, m > 0$)

$$a_{nn} = \frac{\mathrm{var}(N_-)/\langle N_+\rangle + (1-r)s(\omega) - r}{6\langle N_+\rangle}\frac{Y^2}{y_b^2}\delta_{n1} + r\frac{s(\omega)+1}{(2n-1)(2n+3)\langle N_+\rangle}\frac{Y}{y_b}, \quad (4)$$

$$a_{n,n+2} = a_{n+2,n} = -r\frac{s(\omega)+1}{2(2n+3)\sqrt{(2n+1)(2n+5)}\langle N_+\rangle}\frac{Y}{y_b},$$

where $r = 0$ or $r = 1$ corresponds to absent or present string fluctuations, respectively, y_b denotes the rapidity of the beam, $[-Y, Y]$ is the range of the acceptance bin, and $s(\omega)$ is the square of the scaled standard deviation of the fluctuating distribution of strength of the sources (such additional fluctuations are indeed needed to obtain the multiplicity distributions in the Glauber approach). The results for the a_{nm} coefficients are shown in the right part of Fig. 6.

With the help of experimental data one may carry out an interesting analysis. The observed number of charged hadrons is proportional to the number of sources, $N_{\mathrm{ch}} = cN_+$. We may then obtain c from matching $a_{11}^{\mathrm{exp}} = a_{11}^{\mathrm{mod}}$. This yields $N_{\mathrm{ch}} = 4.7N_+$. In the CMS acceptance range $\Delta\eta = 4.8$, hence we have $dN_{\mathrm{ch}}/d\eta \simeq 1 \times N_+$. On the other hand, from the multiplicity data $dN_{\mathrm{ch}}/d\eta \simeq (3-4) \times N_W$ and $dN_{\mathrm{ch}}/d\eta \simeq 1.3 \times Q_W$ [22], where N_W indicates the number of wounded nucleons and Q_W the number of wounded quarks. We note that the number inferred from the matching of a_{11} suggests the wounded quark picture, or more generally, a picture with ~ 3 constituents (hot-spots) in the nucleon.

We conclude this talk by showing two photographs alluding to the collectivity paradigm (left part of Fig. 7), occasionally questioned during the meeting. The surfers motion is strongly collimated even if they are 2 miles away, due to the underlying collective wave. On the other hand, the non-flow events, while present, are quite rare (right part of Fig. 7).

Research supported by the Polish National Science Centre grants 2015/17/B/ST2/00101, 2012/06/A/ST2/00390, and 2015/19/B/ST2/00937.

Figure 7. Collectivity of flow (left) and rare non-flow events (right).

References

[1] U. Heinz, R. Snellings, Ann.Rev.Nucl.Part.Sci. **63**, 123 (2013), 1301.2826
[2] W. Broniowski, M. Chojnacki, L. Obara, Phys. Rev. **C80**, 051902 (2009), 0907.3216
[3] P. Bożek, W. Broniowski, Phys. Rev. **C85**, 044910 (2012), 1203.1810
[4] P. Bożek, W. Broniowski, J. Moreira, Phys. Rev. **C83**, 034911 (2011), 1011.3354
[5] A. Białas, M. Błeszyński, W. Czyż, Nucl. Phys. **B111**, 461 (1976)
[6] A. Białas, W. Czyż, Acta Phys. Polon. **B36**, 905 (2005), hep-ph/0410265
[7] V. Khachatryan et al. (CMS), Phys. Rev. **C92**, 034911 (2015), 1503.01692
[8] P. Bożek, W. Broniowski, A. Olszewski, Phys.Rev. **C91**, 054912 (2015), 1503.07425
[9] W. Broniowski, M. Rybczyński, P. Bożek, Comput. Phys. Commun. **180**, 69 (2009), 0710.5731
[10] M. Rybczyński, G. Stefanek, W. Broniowski, P. Bożek, Comput. Phys. Commun. **185**, 1759 (2014), 1310.5475
[11] P. Bożek, Phys. Rev. **C85**, 034901 (2012), 1110.6742
[12] A. Kisiel, T. Tałuć, W. Broniowski, W. Florkowski, Comput. Phys. Commun. **174**, 669 (2006), nucl-th/0504047
[13] M. Chojnacki, A. Kisiel, W. Florkowski, W. Broniowski, Comput. Phys. Commun. **183**, 746 (2012), 1102.0273
[14] P. Bożek, W. Broniowski, Phys. Lett. **B752**, 206 (2016), 1506.02817
[15] S.J. Brodsky, J.F. Gunion, J.H. Kuhn, Phys. Rev. Lett. **39**, 1120 (1977)
[16] A. Białas, M. Jeżabek, Phys. Lett. **B590**, 233 (2004), hep-ph/0403254
[17] G. Aad et al. (ATLAS Collaboration) (2015), ATLAS-CONF-2015-020
[18] G. Aad et al. (ATLAS Collaboration) (2015), ATLAS-CONF-2015-051
[19] W. Broniowski, P. Bożek (2015), 1512.01945
[20] A. Bzdak, D. Teaney, Phys.Rev. **C87**, 024906 (2013), 1210.1965
[21] J. Jia, S. Radhakrishnan, M. Zhou (2015), 1506.03496
[22] P. Bożek, W. Broniowski, M. Rybczyński, Phys. Rev. **C94**, 014902 (2016), 1604.07697

Rescuing the nonjet (NJ) azimuth quadrupole from the flow narrative

Thomas A. Trainor[1]

[1]*CENPA 354290 University of Washington, Seattle, Washington, USA*

Abstract. According to the flow narrative commonly applied to high-energy nuclear collisions a cylindrical-quadrupole component of 1D azimuth angular correlations is conventionally denoted by quantity v_2 and interpreted to represent elliptic flow. Jet angular correlations may also contribute to v_2 data as "nonflow" depending on the method used to calculate v_2, but 2D graphical methods are available to insure accurate separation. The nonjet (NJ) quadrupole has various properties inconsistent with a flow interpretation, including the observation that NJ quadrupole centrality variation in A-A collisions has no relation to strongly-varying jet modification ("jet quenching") in those collisions commonly attributed to jet interaction with a flowing dense medium. In this presentation I describe isolation of quadrupole spectra from pt-differential $v_2(p_t)$ data from the RHIC and LHC. I demonstrate that quadrupole spectra have characteristics very different from the single-particle spectra for most hadrons, that quadrupole spectra indicate a common boosted hadron source for a small minority of hadrons that "carry" the NJ quadrupole structure, that the narrow source-boost distribution is characteristic of an expanding thin cylindrical shell (strongly contradicting hydro descriptions), and that in the boost frame a single universal quadrupole spectrum (Lévy distribution) on transverse mass m_t accurately describes data for several hadron species scaled according to their statistical-model abundances. The quadrupole spectrum shape changes very little from RHIC to LHC energies. Taken in combination those characteristics strongly suggest a unique *nonflow* (and nonjet) QCD mechanism for the NJ quadrupole conventionally represented by v_2.

1 Introduction

The flow narrative, believed by some to be of central importance to high-energy nuclear collisions [1, 2], is based primarily on v_2 data interpreted to represent elliptic flow – azimuthal modulation of radial flow of a locally-thermalized bulk medium in non-central A-A collisions. Earlier versions referred to more-central A-A collisions at higher collision energies. More recently, "collectivity" in small systems (p-p, p-A) has been claimed as well, based on certain LHC data [3, 4]. However, evidence strongly contradicting the flow narrative has accumulated over the past ten years: Differential analysis of p_t spectra from 200 GeV Au-Au collisions reveals no evidence for radial flow – the blast-wave model said to measure radial flow responds instead to a predicted and observed strong jet contribution to spectra [5]. Measurements of p_t-integral nonjet (NJ) v_2, based on model fits to 2D angular correlations that exclude a jet contribution ("nonflow") [6], reveal v_2 systematics uncorrelated with "jet quenching" [7] contradicting claims for a dense bulk medium [8, 9]. Equivalent systematic v_2 trends are observed for high-multiplicity A-A collisions and for p-p collisions down to negligible particle densities [10]. There is no evidence for a QCD phase transition or changing equation of state.

In this talk I present *quadrupole spectra* inferred from $v_2(p_t, b)$ data that further contradict the flow narrative. Quadrupole spectra from recent LHC Pb-Pb data are compared to those from previous analysis of 200 GeV Au-Au data [11]. Quadrupole spectra reveal a hadron source boost incompatible with Hubble expansion of a flowing bulk medium, and the spectrum shape is very different from the single-particle (SP) spectrum for most final-state hadrons.

2 Quadrupole Spectrum Definition

p_t-differential v_2 is defined as the ratio of the quadrupole ($m = 2$) Fourier amplitude of the event-wise azimuth-dependent SP spectrum to the azimuth-averaged SP spectrum.

$$v_2(p_t, b) = \frac{V_2(p_t, b)}{\bar{\rho}_0(p_t, b)} = \frac{V_2\{2D\}(p_t) + \text{jet contribution}}{(N_{part}/2)S_{NN}(p_t) + N_{bin} r_{AA}(p_t, b) H_{NN}(p_t)}. \tag{1}$$

As defined that ratio may include two jet contributions: (a) jet-related angular correlations in the numerator and (b) SP spectrum hard component $H_{AA}(p_t, b) \equiv r_{AA}(p_t, b) H_{NN}(p_t)$ in the denominator. The v_2 definition assumes that almost all hadrons "carry" the quadrupole correlation component and therefore are described by the same SP spectrum that should then cancel in the v_2 ratio. The NJ quadrupole Fourier amplitude, assuming source-boost azimuth distribution $\Delta y_t(\phi_r, b) = \Delta y_{t0}(b) + \Delta y_{t2}(b)\cos(2\phi_r)$, can be expressed as [11]

$$V_2(y_t, b) = \frac{1}{2\pi}\int_{-\pi}^{\pi} d\phi \, \bar{\rho}_0(y_t, b, \phi_r)\cos(2\phi_r) \approx p_t' \frac{\Delta y_{t2}(b)}{2T_2} \bar{\rho}_2[y_t, b; \Delta y_{t0}(b)], \tag{2}$$

where ϕ_r is ϕ relative to a reference angle, p_t' is p_t in the boost frame, transverse rapidity is $y_t \equiv \ln[(p_t + m_t)/m_h]$ and $\bar{\rho}_2[y_t, b; \Delta y_{t0}(b)]$ is the spectrum for those hadrons carrying the quadrupole correlation component, which may or may not be equivalent to $\bar{\rho}_0(y_t, b)$.

3 Ideal Hydro

The following simple system illustrates some implications of Eq. (2). The source-boost distribution, which should be broad for Hubble expansion of a bulk medium, is assumed to be a single fixed value Δy_{t0} for each collision system. The quadrupole spectrum $\bar{\rho}_2[y_t, b; \Delta y_{t0}(b)]$ is assumed to coincide with SP spectrum $\bar{\rho}_0[p_t, b; \Delta y_{t0}(b)]$ including centrality-dependent radial flow [12] measured by $\Delta y_{t0}(b)$. In that case $v_2(p_t) \approx p_t'(\Delta y_{t2}/2T_2)$ and $v_2(p_t)/p_t \propto p_t'/p_t$. Figure 1 (a) shows p_t' in the boost frame vs lab p_t for three hadron species. Figure 1 (b) shows ratio p_t'/p_t vs y_t (with proper hadron mass) as a *universal curve* common to all hadron species, with shape determined solely by fixed source boost $\Delta y_{t0} = 0.6$. Figure 1 (c) shows panel (a) adjusted to anticipate 200 GeV $v_2(p_t)$ data, and panel (d) shows the equivalent for ratio $v_2(p_t)/p_t$. In each case viscous-hydro predictions for 200 GeV Au-Au collisions (dotted curves) [13] shown for comparison exhibit striking differences from the "ideal-hydro" example. This exercise illustrates "mass ordering" at lower p_t in the conventional plotting format of panel (a). The quantitative source-boost distribution that hydro theory actually predicts (broad or narrow?) is more easily tested in the format of panel (d) than in (a) or (c).

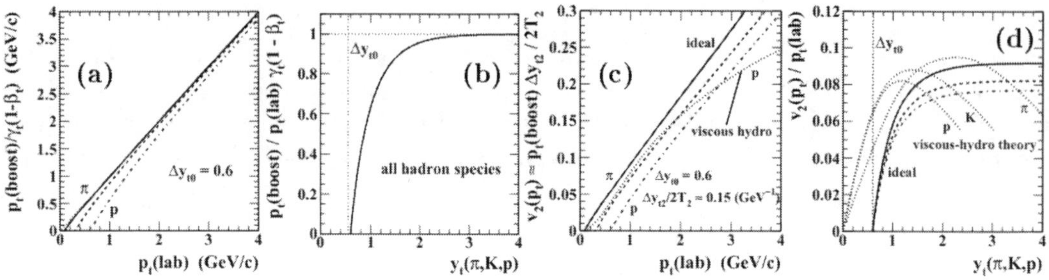

Figure 1. "Ideal-hydro" kinematic trends assuming single value of source boost Δy_{t0} and quadrupole spectrum equal to SP spectrum [11]. Viscous-hydro theory trends [13] are included for comparison.

4 200 GeV Au-Au Quadrupole Spectra

Quadrupole spectra can be inferred from $v_2(p_t, b)$ data in a few steps. Figure 2 (a) shows 0-80% central $v_2(p_t)$ data for three hadron species in the conventional plotting format [14, 15]. "Mass ordering" below 1.5 GeV/c is said to indicate a hydro mechanism. Curves representing "ideal hydro" cross the top edge. A viscous-hydro theory curve for protons is indicated by R [13]. Curves passing through data at higher p_t are explained below. Figure 2 (b) shows the same data in the form $v_2(p_t)/p_t(\text{lab})$ vs y_t with proper mass for each hadron species. The ideal-hydro curves go to a constant value for higher y_t, and the data trends for three hadron species share a common zero intercept that can be identified with fixed source boost $\Delta y_{t0} \approx 0.6$. Figure 2 (c) confirms that the (dotted) viscous-hydro theory curve is strongly falsified by Lambda (or proton) data. Figure 2 (d) shows the quadrupole Fourier coefficient as the product $V_2(y_t) = \bar{\rho}_0(y_t)v_2(y_t)$ obtained via SP spectra for three hadron species [11].

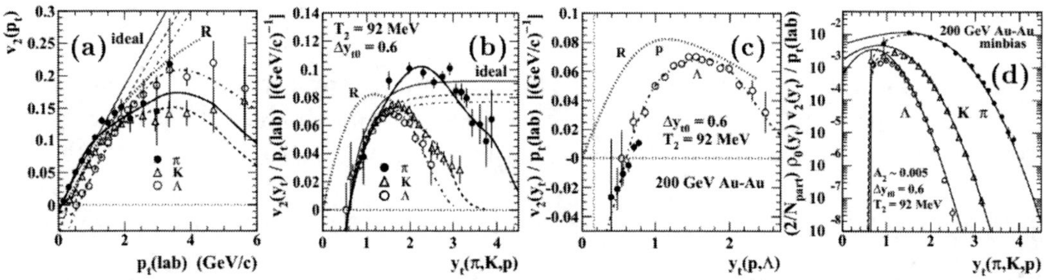

Figure 2. $v_2(p_t)$ data for three hadron species from 0-80% 200 GeV Au-Au collisions [14, 15] processed to obtain $V_2(y_t) = \bar{\rho}_0(y_t)v_2(y_t) \propto p_t'(y_t)\,\bar{\rho}_2(y_t; \Delta y_{t0})$ quadrupole spectra as in Ref. [11].

To obtain final quadrupole spectra from the data in Fig. 2 (d) three more steps are required. The spectra are multiplied by $p_t(\text{lab})/p_t(\text{boost})$ determined exactly by inferred $\Delta y_{t0} = 0.6$, shifted left by Δy_{t0} from lab frame y_t to boost frame y_t' and finally transformed to boost-frame m_t' with the appropriate Jacobian. The result is shown in Figure 3. The spectra, rescaled by their statistical-model abundances relative to pions [16], lie on a common locus (solid curve) with slope parameter $T_2 = 92$ MeV and Lévy exponent $n_2 = 14$ dramatically different from the hadron SP spectrum with $T_0 = 145$ MeV and $n_0 = 12$ [5]. The solid curve, back-transformed by reversing the sequence, gives the curves passing through data in Fig. 2.

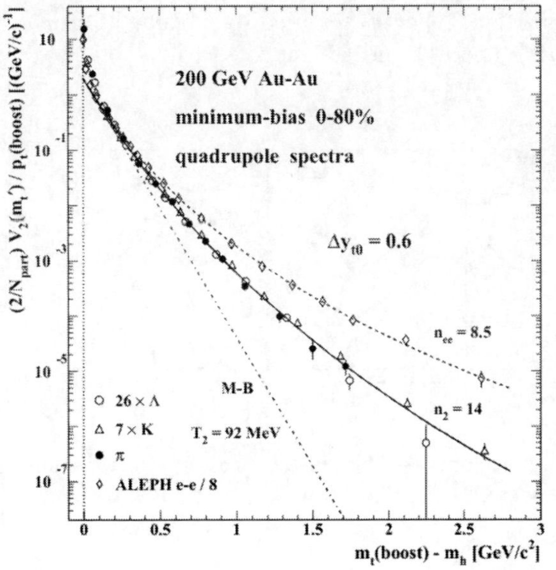

Figure 3. Quadrupole spectra from 200 GeV Au-Au collisions for three hadron species (points in legend) rescaled by statistical-model abundances relative to pions (factors noted in the legend) plotted on m_t' in the boost frame. The solid curve is a Lévy distribution with $T_2 = 92$ MeV (dash-dotted curve) and exponent $n_2 = 14$ very different from Au-Au SP spectrum values $T_0 = 145$ MeV and $n_0 = 12$ [5]. The dashed curve and data are derived from the fragment p_t spectrum (perpendicular to the dijet axis) for 91 GeV LEP dijets [17] also exhibiting $T \approx 90$ MeV.

5 2.76 TeV Pb-Pb Quadrupole Spectra

Figure 4 (a,b) shows $v_2(p_t, b)$ data vs p_t for pions and protons from seven centralities of 2.76 TeV Pb-Pb collisions [18]. 200 GeV $v_2(p_t, b)$ data are presented in Ref. [19] for comparison. Figure 4 (c) shows proton data plotted as in Fig. 2 (b) but rescaled by p_t-integral values $v_2(b)$ [20]. The dashed curve is the 200 GeV equivalent. The 2.76 TeV data reveal significant variation of source boost $\Delta y_{t0}(b)$ with centrality. Figure 4 (d) shows the same data with all centralities boosted (shifted on y_t) to coincide with the 30-40% data. Within uncertainties all centralities follow the same locus that coincides also with the 200 GeV dashed curve.

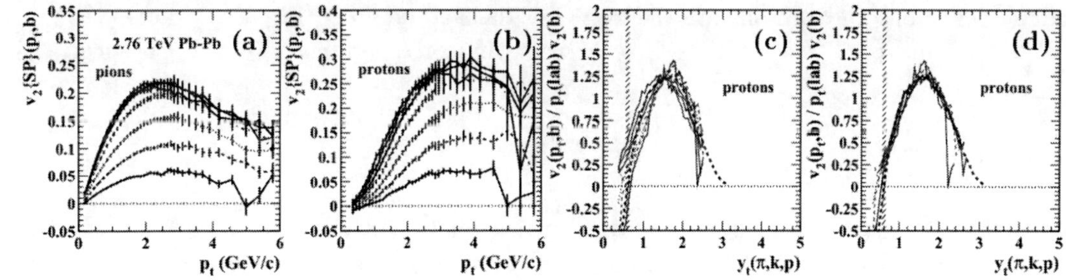

Figure 4. $v_2(p_t, b)$ data vs p_t for pions and protons from 2.76 TeV Pb-Pb collisions [18].

Figure 5 (a,b) shows SP spectra for pions and protons from p-p collisions (open points) and from 30-40% central 2.76 TeV Pb-Pb (solid dots) as the average of 0-5% and 60-80% central collisions with two issues [21]: (a) The Pb-Pb pion spectrum normalized by $N_{part}/2$ must be divided by factor 1.65 to coincide with the *p-p* spectrum at lower p_t in accord with 200 GeV data [5]. (b) The Pb-Pb proton spectrum normalized by $N_{part}/2$ (solid dots) is divided by the same factor but falls substantially below (factor 2) the *p-p* spectrum at lower p_t suggesting significant uncorrected inefficiency in that interval. Figure 5 (c) shows the equivalent of Fig. 2

(d) for four hadron species. The dashed curves through those data are the dashed curves for 200 GeV $v_2(p_t, b)/p_t$(lab) multiplied by the Pb-Pb SP spectra rescaled by $1/1.65$. Figure 5 (d) shows those data multiplied by factor $p_t(\text{lab})/p_t(\text{boost})$ and transformed to the boost frame (shifted left by $\Delta y_{t0} = 0.6$).

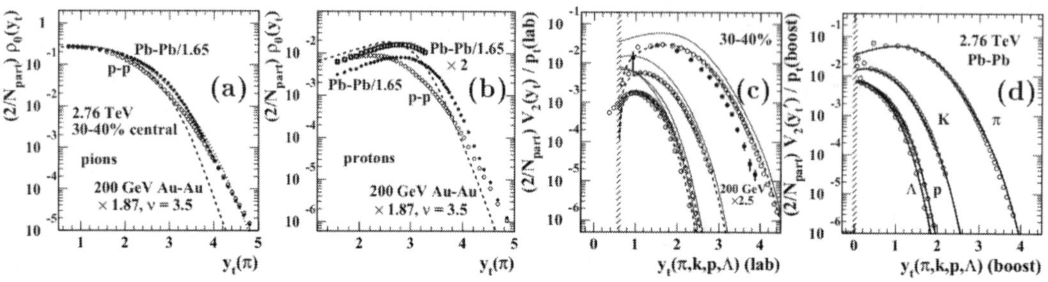

Figure 5. SP spectra for pions (a) and protons (b) from 30-40% central 2.76 TeV Pb-Pb collisions (solid points). Quadrupole spectra for the same system in the lab frame (c) and boost frame (d).

Figure 6 shows data and curves in Fig. 5 (d) transformed to m_t' in the boost frame via the proper Jacobian and rescaled by statistical-model abundances relative to pions (points in legend). The pion data fall on the bold solid curve with $T_2 = 94$ MeV and $n_2 = 12$. The 200 GeV quadrupole spectrum data (inverted solid triangles and thin solid curve) are shown multiplied by factor 2.5 expected from observed energy scaling of v_2 and $\bar{\rho}_0$. Slope parameters T_2 are not significantly different, but the Lévy exponent decreases significantly at the higher energy consistent with the trend for SP spectrum soft component $\hat{S}_0(m_t)$ [21, 22]. Just as at RHIC energies there is a great difference between quadrupole spectra and SP spectra. SP spectrum soft component $\hat{S}_0(m_t')$ for 2.76 TeV *p-p* collisions plotted on m_t' is shown as the dashed curve. The spectrum for protons (and Lambdas) falls substantially below the bold solid curve for $m_t' < 0.7$ GeV/c^2, consistent with the spectrum result in Fig. 5 (b). The dotted curves for kaons and protons are the 200 GeV dashed curves in Fig. 5 (c) derived from the solid curve in Fig. 3 processed in this case with the corresponding 2.76 TeV SP spectra.

6 Summary

Quadrupole spectra for identified hadrons, which may involve only a small fraction of final-state particles that actually "carry" the quadrupole component, can be extracted from p_t-differential $v_2(p_t, b)$ data by a simple sequence of transformations given the availability of matching single-particle hadron spectra. The sequence leads to determination of a hadron source-boost distribution consistent with a single value Δy_{t0} (for a given collision system) that is inconsistent with the broad distribution expected for Hubble expansion of a flowing bulk medium. Quadrupole spectra at 2.76 TeV are remarkably similar to those for 200 GeV and very different from single-particle hadron spectra, contradicting a basic assumption of the flow narrative that almost all hadrons must participate in such flows. These new results, combined with related trends observed over the past ten years, strongly suggest that the nonjet azimuth quadrupole does not represent a hydrodynamic flow. The NJ quadrupole may instead be the manifestation of a QCD mechanism similar to QED antenna radiation.

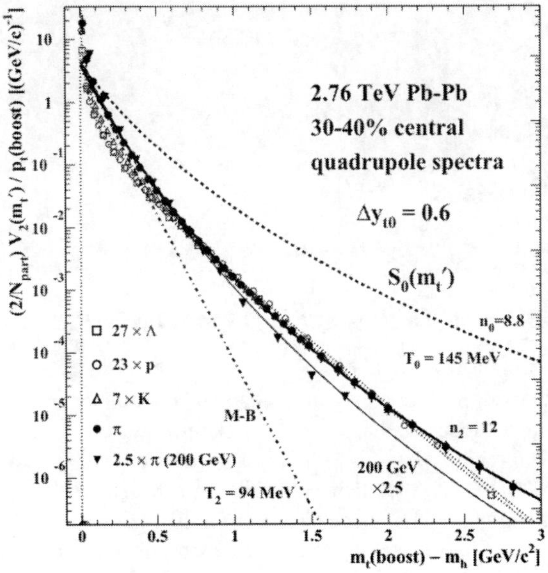

Figure 6. Quadrupole spectra from 2.76 TeV Pb-Pb collisions for four hadron species (points in legend) rescaled by statistical-model abundances relative to pions (factors noted in the legend) plotted on m_t' in the boost frame. The solid curve is a Lévy distribution with $T_2 = 94$ MeV (dash-dotted curve) and exponent $n_2 = 12$, very different from p-p SP spectrum soft component $\hat{S}_0(m_t')$ (dashed curve) values $T_0 = 145$ MeV and $n_0 = 8.8$ [5]. The 200 GeV quadrupole spectrum multiplied by anticipated energy-scaling factor 2.5 is plotted as the thin solid curve and inverted solid triangles

References

[1] D. Teaney, J. Lauret and E. V. Shuryak, Phys. Rev. Lett. **86**, 4783 (2001).
[2] M. Gyulassy and L. McLerran, Nucl. Phys. A **750**, 30 (2005).
[3] CMS Collaboration, JHEP **1009**, 091 (2010) , arXiv:1009.4122.
[4] T. A. Trainor and D. T. Kettler, Phys. Rev. C **84**, 024910 (2011).
[5] T. A. Trainor, Int. J. Mod. Phys. E **17**, 1499 (2008).
[6] D. T. Kettler (STAR collaboration), Eur. Phys. J. C **62**, 175 (2009).
[7] J. Adams *et al.* (STAR Collaboration), Phys. Rev. Lett. **91**, 172302 (2003).
[8] T. A. Trainor, J. Phys. G **37**, 085004 (2010).
[9] T. A. Trainor, D. T. Kettler, D. J. Prindle and R. L. Ray, J. Phys. G **42**, 025102 (2015).
[10] T. A. Trainor and D. J. Prindle, Phys. Rev. D **93**, no. 1, 014031 (2016).
[11] T. A. Trainor, Phys. Rev. C **78**, 064908 (2008).
[12] B. I. Abelev *et al.* (STAR Collaboration), Phys. Rev. C **79**, 034909 (2009).
[13] P. Romatschke and U. Romatschke, Phys. Rev. Lett. **99**, 172301 (2007).
[14] C. Adler *et al.* (STAR Collaboration), Phys. Rev. C **66**, 034904 (2002).
[15] J. Adams *et al.* (STAR Collaboration), Phys. Rev. Lett. **92**, 052302 (2004).
[16] F. Becattini, M. Gazdzicki and J. Sollfrank, Eur. Phys. J. C **5**, 143 (1998).
[17] D. Buskulic *et al.* [ALEPH Collaboration], Z. Phys. C **55**, 209 (1992).
[18] B. B. Abelev *et al.* (ALICE Collaboration), JHEP **1506**, 190 (2015).
[19] D. T. Kettler, D. J. Prindle and T. A. Trainor, Phys. Rev. C **91**, 064910 (2015).
[20] K. Aamodt *et al.* (ALICE Collaboration), Phys. Rev. Lett. **105**, 252302 (2010)
[21] T. A. Trainor, arXiv:1603.01337 [hep-ph].
[22] B. B. Abelev *et al.* (ALICE Collaboration), Phys. Lett. B **736**, 196 (2014).

Saturation and geometrical scaling

Michal Praszalowicz[1,a]

[1] *M. Smoluchowski Institute of Physics, Jagiellonian University, Lojasiewicza 11, 30-348 Krakow, Poland*

Abstract. We discuss emergence of geometrical scaling as a consequence of the non-linear evolution equations of QCD, which generate a new dynamical scale, known as the saturation momentum: Q_s. In the kinematical region where no other energy scales exist, particle spectra exhibit geometrical scaling (GS), *i.e.* they depend on the ratio p_T/Q_s, and the energy dependence enters solely through the energy dependence of the saturation momentum. We confront the hypothesis of GS in different systems with experimental data.

1 Introduction

In this report we present a concise analysis of GS, slightly extended with respect to the presentation given at the XLVI International Symposium on Multiparticle Dynamics. One can find more details in the original publications [1]– [4] and in the recent conference proceedings [5], that cover the same topics.

In QCD we have basically two sets of evolution equations that describe the change of parton densities with decreasing resolution scale $1/Q^2$ – DGLAP equations, or with growing energy (or equivalently with decreasing Bjorken x) – BFKL equation. In both cases the number of partons, or more precisely the number of gluons, is growing rapidly with the evolution variable. In the BFKL case however (since the average transverse size of gluons is fixed), we enter a regime where the partonic system is not dilute and the linear evolution breaks down. A modified BFKL equation that includes the non-linear terms is known as the Balitsky-Kovchegov (BK) equation [6]. One of the consequences of the nonlinearities is the emergence of the so called saturation scale of the form [7, 8]:

$$Q_s^2(x) = Q_0^2(x/x_0)^{-\lambda}. \tag{1}$$

Munier and Peschanski [8] draw an analogy between the BK equation and the time evolution of the wave front $u(t,z)$ in one dimensional space variable z:

$$\frac{\partial}{\partial t}u(t,z) = O(\partial/\partial z) \tag{2}$$

where $O(\partial/\partial z)$ is a non-linear differential operator corresponding in QCD to the BK kernel. For a wide class of operators O and initial conditions for u, wave front u converges asymptotically to the *traveling wave*:

$$u(x,z) \to u(z - v_c t), \tag{3}$$

[a] e-mail: michal@if.uj.edu.pl

i.e. to the fixed front that moves rigidly with the critical velocity v_c. In the QCD context time corresponds to the logarithm of Bjorken x: $t = \ln(x_0/x)$ where x_0 is a constant corresponding to $t = 0$, and $z = \ln(p_T^2/Q_0^2)$. Translating the argument of the travelling wave to the QCD variables gives:

$$z - v_c t = \ln\left(\frac{p_T^2}{Q_0^2}\right) - v_c \ln\left(\frac{x_0}{x}\right) = \ln\left(\frac{p_T^2}{Q_0^2(x/x_0)^{-v_c}}\right). \tag{4}$$

Therefore the travelling wave corresponds to the scaling solution with the saturation momentum given by Eq. (1). In the following we shall check whether GS is present in different pieces of high energy data.

2 Deep inelastic scattering (DIS)

Geometrical scaling was first introduced in the context of DIS for $F_2(x)/Q^2$ [9]. In Fig. 1 we plot $F_2(x)/Q^2$ as a function of Q^2 (left panel) and in terms of the scaling variable $\tau = Q^2/Q_s^2(x)$ for $\lambda = 0.329$ (right panel) for the combined HERA data [10]. Points of different colors correspond to different Bjorken x's. We see from Fig. 1 that DIS data scale very well with some exception in the right part of Fig. 1.b. These points, however, correspond to large Bjorken x's where GS is supposed to break.

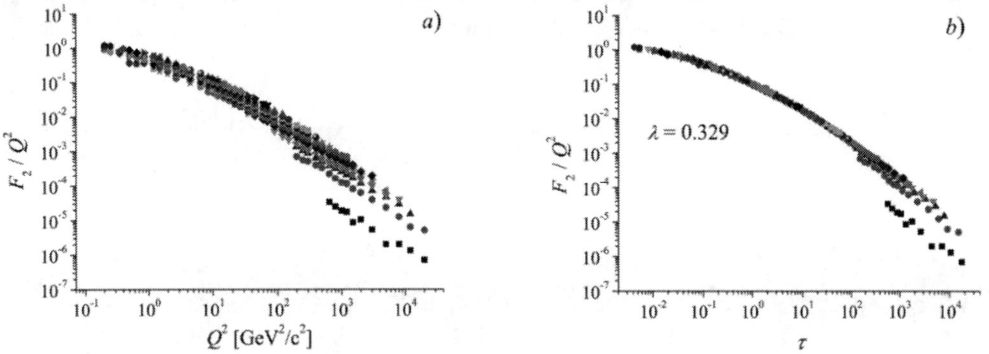

Figure 1. Combined DIS data [10] for F_2/Q^2. Different points forming a wide band as a function of Q^2 in the left panel correspond to different Bjorken x's. They fall on a universal curve when plotted in terms of τ (right panel). (Figure from the first paper of Ref. [1]).

Since the reduced cross-section in ep scattering (which is essentially proportional to F_2/Q^2) is given as a convolution of the virtual photon wave function and an unintegrated gluon distribution of the proton $\varphi_p\left(\vec{k}_T^2, x\right)$, the fact the DIS data scale implies that $\varphi_p\left(k_T^2, x\right) = \varphi_p\left(k_T^2/Q_s^2(x)\right)$.

3 Inelastic p_T spectra at the LHC

The cross-section for not too hard gluon production in pp collisions can be described in the k_T–factorization approach by the formula [11]:

$$\frac{d\sigma}{dyd^2p_T} = \frac{3\pi\alpha_s}{2p_T^2} \int d^2\vec{k}_T\, \varphi_p\left(\vec{k}_T^2, x_1\right) \varphi_p\left((\vec{k} - \vec{p})_T^2, x_2\right) \tag{5}$$

where φ_p denotes the unintegrated gluon distribution. Here $x_{1,2} = e^{\pm y} p_{\mathrm{T}}/\sqrt{s}$. In the following we shall assume that produced gluons are in the mid rapidity region ($y \simeq 0$), hence both Bjorken x's of colliding gluons are equal $x_1 \simeq x_2$ (denoted in the following as x).

Unintegrated gluon distributions have dimension of transverse area. We shall assume that – following our discussion of DIS – that they depend only on the ratio of the transverse momentum over the saturation scale

$$\varphi_p\left(k_{\mathrm{T}}^2, x\right) = S_\perp \phi\left(k_{\mathrm{T}}^2/Q_{\mathrm{s}}^2(x)\right) \tag{6}$$

where ϕ is a *dimensionless* function. In the case of DIS $S_\perp = \sigma_0$ is the dipole-proton cross-section for large dipoles [12]. In the case of heavy ion collisions $S_\perp$ is the transverse size of an overlap of two large nuclei for a given centrality class. In both cases one can assume that $S_\perp$ is energy independent (or weakly dependent). In that case $d^2\vec{k}_{\mathrm{T}}$ integration in (5) leads to

$$\frac{d\sigma}{dyd^2p_{\mathrm{T}}} = S_\perp^2 \mathcal{F}(\tau) \tag{7}$$

where $\tau = p_{\mathrm{T}}^2/Q_{\mathrm{s}}^2(x)$ is a scaling variable and $\mathcal{F}(\tau)$ is a function related to the integral of ϕ_p's. We follow here the parton-hadron duality hypothesis [13], assuming that the charged particle spectra are on the average identical to the gluon spectra.

In Fig. 2 we plot ALICE pp data [14] in terms of p_{T} (left panel) and in terms of the scaling variable τ (right panel) for $\lambda = 0.32$. We see that three different curves from the left panel in Fig. 2 overlap over some region if plotted in terms of $\sqrt{\tau}$. The exponent for which this happens over the largest interval of τ is $\lambda = 0.32$ [4], which is the value compatible with our model independent analysis of the DIS data [1].

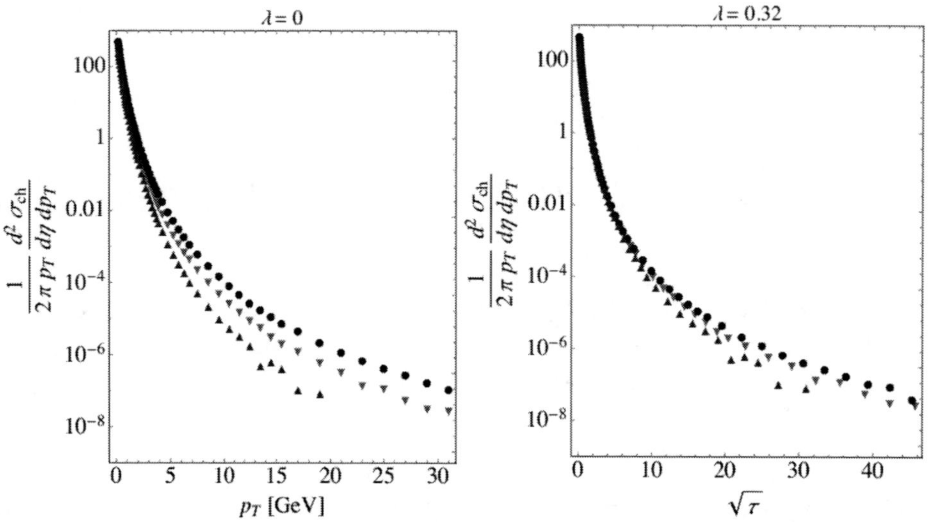

Figure 2. Data for pp scattering from ALICE [14] plotted in terms of p_{T} and $\sqrt{\tau}$. Full (black) circles correspond to $W = 7$ TeV, down (red) triangles to 2.76 TeV and up (blue) triangles to 0.9 TeV. (Figure from Ref. [4].)

In order to illustrate the method of adjusting λ, we plot in Fig. 3 ratios of the cross-sections at 7 TeV to 2.76 and 0.9 TeV. Proximity of both ratios to unity for $\lambda = 0.32$ is the sign of GS for p_{T} up to 4.25 GeV/c [4].

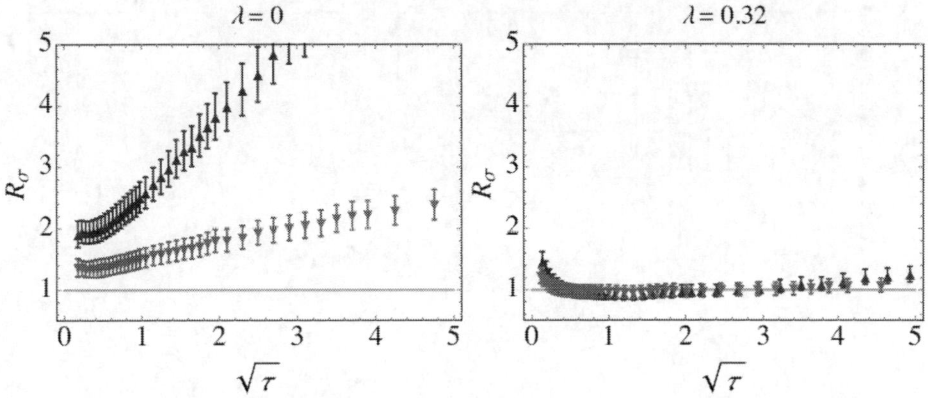

Figure 3. Ratios of the cross-sections at 7/2.76 TeV – down (red) triangles and 7/0.9 TeV – up (blue) triangles, for $\lambda = 0$ (left) and 0.32 (right). (Figure from Ref. [4].)

It has been argued previously [2] that GS should hold for multiplicities, rather than for the cross-sections. This would be true if the relation between the two was energy independent. This may be the case in HI or p*A* collisions where we trigger on some $S_\perp$ by selecting the centrality classes with given number of participants, but it is not true in the case of the inelastic pp scattering:

$$\frac{dN}{dyd^2p_\mathrm{T}} = \frac{1}{\sigma^{\mathrm{MB}}(W)}\frac{d\sigma}{dyd^2p_\mathrm{T}} = \frac{S_\perp^2}{\sigma^{\mathrm{MB}}(W)}\mathcal{F}(\tau) \tag{8}$$

where the minimum bias cross-section $\sigma^{\mathrm{MB}}(W) \neq S_\perp$ is energy-dependent. Nevertheless the multiplicity does scale, however for different value of the exponent ($\lambda = 0.22 - 0.24$), and over a smaller range of p_T. This is discussed in more detail in Ref. [4].

4 Electron-positron scattering

One of the obvious questions concerning pp scattering is whether the observed scaling is indeed a consequence of gluon saturation or whether it appears due to the properties of final state radiation that takes place before hadronization (note that by referring to the parton-hadron duality we have ignored possible effects of the QCD fragmentation functions). The best way to address this issue is to perform analogical analysis of the hadronic p_T spectra in e^+e^- collisions where no GS is expected. To this end we shall use spectra published by TASSO collaboration [15] at $W = \sqrt{s} = 14$, 22, 35 and 44 GeV.[1] One might argue that these energies are too small to reach a firm conclusion on GS in e^+e^-. However, in Ref. [16] we have analyzed pp scattering at even lower energies (6 – 17 GeV) finding positive evidence of GS in the mid rapidity region.

In Fig. 4 ratios of multiplicity spectra

$$R(W) = \left.\frac{1}{\sigma_{tot}}\frac{d\sigma}{2\pi p_\mathrm{T} dp_\mathrm{T}}\right|_{44\ \mathrm{GeV}} \Big/ \left.\frac{1}{\sigma_{tot}}\frac{d\sigma}{2\pi p_\mathrm{T} dp_\mathrm{T}}\right|_{W} \tag{9}$$

for the remaining three values of W are plotted in terms of the scaling variable $\sqrt{\tau}$ for $\lambda = 0$ (note that in this case $\sqrt{\tau} = p_\mathrm{T}$) and for $\lambda = 0.32$. In the left panel w see similar behaviour of R as in the pp

[1] The author is grateful to J. Chwastowski for pointing this reference.

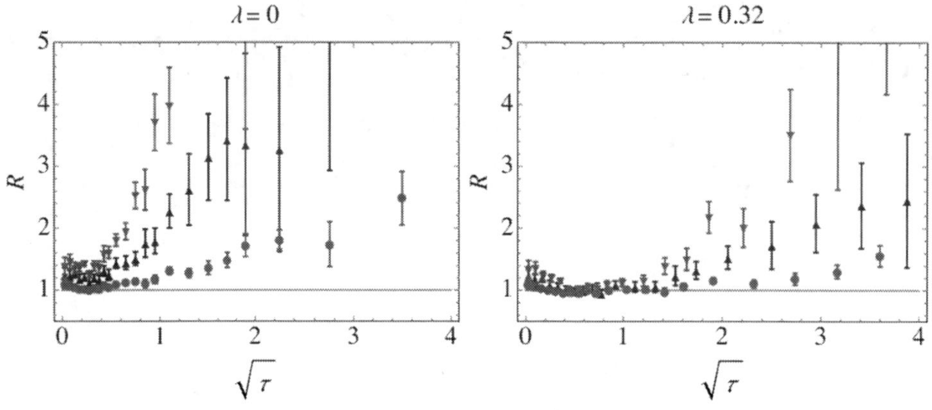

Figure 4. Ratios $R(W)$ of multiplicity spectra for $W = 14$ (down-green triangles), 22 (up-blue triangles) and 35 GeV (red-full circles) as functions of p_T (left panel) and $\sqrt{\tau}$ for $\lambda = 0.32$ (right panel).

case (see Fig. 3.a), however in the right panel where we plot R for $\lambda = 0.32$ no GS is seen (compare with Fig. 3.b).

5 Summary

In this short note we have argued that geometrical scaling is clearly seen in the DIS and pp data. In the latter case an observable that scales is the differential cross-section rather than the multiplicity distribution. Phenomenological analysis analogical to the one described here has been used to look for the effects of GS in pA and heavy ion collisions and in $\langle p_T \rangle$ correlation with multiplicity [2]–[5]. Most recently effects of the fluctuations of the saturation momentum in the pA collisions have been studied in Ref. [17].

Scaling violations that emerge in the kinematical limit when one of the Bjorken x's in (5) is close to unity, have been observed in Ref. [16]. Here, in Sect. 4, we have presented a new evidence that GS is not present in the spectra of particles produced in e^+e^- collisions. Although this has been theoretically expected, conclusions drawn from Fig. 4 support our main hypothesis that geometrical scaling seen in the experimental data is due to the saturation effects in the initial state of the colliding hadrons.

Ackowledgements

The author wants to thank the organizers for an invitation and for putting up together this successful meeting. This work was supported by the Polish NCN grant 2014/13/B/ST2/02486.

References

[1] M. Praszalowicz, T. Stebel, JHEP **1303**, 090 (2013) and JHEP **1304**, 169 (2013).
[2] L. McLerran, M. Praszalowicz, Acta Phys. Pol. B **41**, 1917 (2010) and Acta Phys. Pol. B **42**, 99 (2011).
[3] M. Praszalowicz, Acta Phys. Pol. B **42**, 1557 (2011) and arXiv:1205.4538 [hep-ph].

[4] M. Praszalowicz, A. Francuz, Phys. Rev. D **92**, 074036 (2015) .
[5] M. Praszalowicz, Acta Phys. Polon. Supp. **9**, 395 (2016).
[6] I. Balitsky, Nucl. Phys. B **463**, 99 (1996);
Y. V. Kovchegov, Phys. Rev. D **60**, 034008 (1999) and Phys. Rev. D **61**, 074018 (2000).
[7] A. H. Mueller and D. N. Triantafyllopoulos, Nucl. Phys. B **640**, 331 (2002).
[8] S. Munier, R. B. Peschanski, Phys. Rev. Lett. **91**, 232001 (2003) and Phys. Rev. D **69**, 034008 (2004).
[9] A. Stasto, K. Golec-Biernat, J. Kwiecinski, Phys. Rev. Lett. **86**, 596 (2001).
[10] C. Adloff *et al.* [H1], Eur. Phys. J. C **21** 33 (2001);
S. Chekanov *et al.* [ZEUS], Eur. Phys. J. C **21**, 443 (2001) .
[11] L. V. Gribov, E. M. Levin, M. G. Ryskin, Phys. Lett. B **100**, 173 (1981).
[12] K. J. Golec-Biernat, M. Wusthoff, Phys. Rev. D **59**, 014017 (1998) and Phys. Rev. D **60**, 114023 (1999).
[13] Ya.I. Azimov, Yu.L. Dokshitzer, V.A. Khoze and S.I. Troian, Z. Phys. C **27**, 65 (1985);
Yu. L. Dokshitzer, V. A. Khoze and S. I. Troyan, J. Phys. G **17**, 1585 (1991) ;
V.A. Khoze and W. Ochs, Int. J. Mod. Phys. A **12**, 2949 (1997) .
[14] B. B. Abelev *et al.* [ALICE Collaboration], Eur. Phys. J. C **73**, 2662 (2013).
[15] W. Braunschweig *et al.* [TASSO Collaboration], Z. Phys. C **47**, 187 (1990). doi:10.1007/BF01552339
[16] M. Praszalowicz, Phys. Rev. D **87**, 071502 (2013).
[17] L. McLerran and M. Praszalowicz, Annals Phys. **372**, 215 (2016).

Recent heavy flavor measurements from PHENIX at RHIC

Takashi Hachiya[1,a] for the PHENIX collaboration
[1] *RIKEN BNL Research Center*

Abstract. Heavy flavor is a useful probe to study the property of the strongly coupled quark gluon plasma created in high-energy heavy-ion collisions. They are produced in the early stage of the collisions, and then propagate though QGP. Therefore, they carry information of QGP. PHENIX measures the separated bottom and charm production using the precise tracking by the silicon vertex detector installed at mid-rapidity and forward rapidity. In this article, we present the recent results of separated charm and bottom modification in Au+Au and Cu+Au collisions, and charmonia production in p+Al and p+Au collisions.

1 Introduction

Hadrons carrying heavy quarks, bottoms and charms are crucial probes to study the nature of quark gluon plasma (QGP) created in high energy heavy ion collisions. Because of their large masses ($M_b \approx 4.2$ GeV/c^2 and $M_c \approx 1.3$ GeV/c^2), heavy quarks are mainly produced via parton-parton collisions at the initial stage of the reaction. Once produced, heavy quarks then interact with QGP when they traverse it. Then, they suffer the QGP effects such as energy loss, flow effect and dissociation of their pairs. Therefore, the modification of the heavy quark production is sensitive to the property of QGP. Heavy quarks are also modified by cold nuclear matter (CNM) effects. To disentangle QGP effects from CNM effects, it is necessary to study the heavy flavor production in $p+p$, p+A and A+A collisions.

PHENIX studied a variety of open and closed heavy favor productions in $p + p$ and Au+Au collisions. We previously measured a strong suppression and a substantial flow of heavy flavors in the measurement of single electrons from semi-leptonic decays of bottom and charm hadrons[1, 2]. The results are not well described by the expectation of energy loss via gluon radiation such as the dead cone effect[3]. PHENIX $\psi(2S)$ measurement in central d+Au collisions at mid-rapidity showed stronger suppression than the J/ψ but similar suppression in peripheral collisions[4]. The sequential suppression of $\psi(2S)$ and J/ψ can not be explained by the nuclear absorption.

A Barrel and Forward Silicon Vertex Tracker (VTX and FVTX) were installed in PHENIX in 2011 and 2012. VTX (FVTX) covers $|y| < 1.2$ ($1.2 < |y| < 2.2$) in rapidity and almost 2π (full) in azimuth. These silicon detectors provide precise displaced tracking and better momentum resolution. The precise tracking enables to measure the bottom and charm components separately in the single electron measurement by using the difference of their decay lengths. Better mass resolution allows to separate J/ψ and $\psi(2S)$ in the measurement of lepton pairs from their decays. In this article, we

[a] e-mail: hachiya@rcf.rhic.bnl.gov

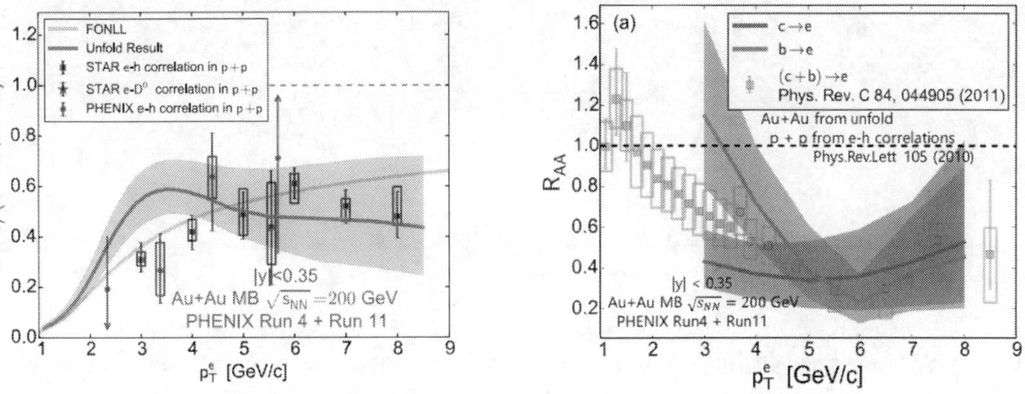

Figure 1. (Left) The bottom electron fraction in minimum bias Au + Au collisions at $\sqrt{s_{NN}}$ = 200 GeV[5], which are compared with the FONLL calculation[8]. Also the $p + p$ result at same energy from PHENIX[6] and STAR[7] are shown. (Right) R_{AA} for electrons from separated bottom and charm hadron decays as well as inclusive heavy flavor electrons.

report on separated bottom and charm yields in mid and forward rapidity in heavy ion collisions and J/ψ and $\psi(2S)$ production in p+A collisions.

2 Separated bottom and charm yields in Au + Au collisions

We utilize the difference of the decay lengths of B and D mesons ($c\tau_{B^0} \approx 455\ \mu m$ and $c\tau_{D^0} \approx 123\ \mu m$) in order to separate bottom and charm components in the single electron measurement. Electrons are identified using a ring-imaging Čerenkov detector and the energy-momentum matching. VTX measures the distance of closest approach (DCA) of charged tracks from the collision vertex. The DCA distribution of inclusive electrons contains a large amount of backgrounds that originate from non-desired physics processes such as Dalitz decays of π^0 and η, photon conversions as well as detector effects such as mis-identified hadrons and high multiplicity background. Those backgrounds are estimated and subtracted using the event driven method and the PHENIX detector simulations. The heavy flavor signals in the DCA distribution are extracted, and the bottom and charm components are separated by the unfolding method based on Bayesian inference technique. The analysis procedure is described in detail[5].

Figure. 1 (left) shows a fraction of electrons from bottom decays to inclusive heavy flavors. The red curve and pink band represent the central value and systematic uncertainties. The bottom electron fraction is compared with the $p + p$ results[6, 7] and FONLL calculation[8]. It is found that the Au + Au result shows a apparent peak at $p_T \sim 3$ GeV/c but $p + p$ results increase smoothly with p_T and is consistent with the FONLL within uncertainties.

The nuclear modification factor (R_{AA}) for electrons from bottom and charm hadron decays are calculated using the bottom electron fraction in Au + Au and $p + p$, and R_{AA} of inclusive heavy flavor electrons[2], as shown in the right panel of Fig. 1. We find that electrons from bottom decays are less suppressed than that from charm decays for $p_T < 4$ GeV/c and similarly suppressed for higher p_T. This result is qualitatively consistent with the expectations from the mass dependence of energy loss[3, 9].

DOI: 10.1051/epjconf/201714105006

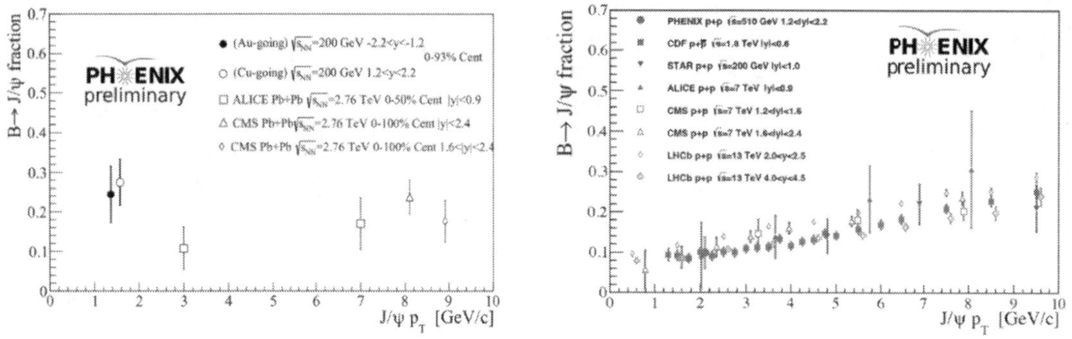

Figure 2. The fraction of $B \rightarrow J/\psi$ relative to inclusive J/ψ in minimum bias Cu + Au collisions at $\sqrt{s_{NN}}$ = 200 GeV (left) and $p + p$ collisions at $\sqrt{s_{NN}}$ = 510 GeV (right) The Cu+Au result are compared with the LHC results.

3 Open bottom measurement from $B \rightarrow J/\psi$ decays in Cu + Au collisions

J/ψ from B meson decays is a direct measurement of open bottom production. PHENIX measure the signal via $B \rightarrow J/\psi \rightarrow \mu\mu$ at forward and backward rapidity in 200 GeV Cu+Au collisions. J/ψ is reconstructed from unlike sign di-muon pairs. DCA of single muons from J/ψ decays is measured by FVTX. The DCA distribution is asymmetric for the B meson decays because the negative side of the DCA has better detector acceptance, while prompt J/ψ from the collision vertex is symmetric. The difference of their shapes in the DCA distribution is used to identify muons from the $B \rightarrow J/\psi$ decays. The DCA distribution is fit with the template shapes of $B \rightarrow J/\psi$ and prompt J/ψ components to extract the B decay signal. Their template shapes are determined by PYTHIA + PHENIX detector simulation.

Figure 2 shows the preliminary result of the fraction of $B \rightarrow J/\psi$ decays relative to inclusive J/ψ in 200 GeV Cu+Au (left) and 510 GeV $p + p$ (right) collisions. In the left panel, the fraction is consistent for Cu going and Au going direction but larger than the CMS and ALICE results[10, 11]. The difference is attributed to a smaller suppression of prompt J/ψ at the LHC energy. On the other hand, the $p + p$ result in the right panel follows the same trend with the world data. The nuclear modification factor R_{CuAu} of $B \rightarrow J/\psi$ is calculated using these fractions and the R_{CuAu} of inclusive J/ψ. Here, we assume that the fraction in 200 GeV $p + p$ collisions is 0.1 based on the world average data. Figure 3 shows R_{CuAu} of $B \rightarrow J/\psi$. The R_{CuAu} of $B \rightarrow J/\psi$ are consistent with unity within uncertainties for both Cu and Au going direction. The result indicates that the bottom suffers smaller suppression at low p_T at RHIC energy. In the comparison, it is found that the R_{CuAu} follows the same trend with the LHC results, where the suppression is less at low p_T and more at high p_T.

4 Closed heavy flavor in p + A collisions : J/ψ and ψ(2S)

The J/ψ and ψ(2S) productions in p + A collisions are a good tool to study the CNM effect. In the CNM effect, the initial state effects such as gluon shadowing and Cronin effect should be similar for both J/ψ and ψ(2S), while the final state effects such as nuclear breakup effect are different because of their binding energies. PHENIX studies the J/ψ and ψ(2S) production in 200 GeV $p + p$, p+Al and p+Au collisions at forward and backward rapidity.

The double ratio of ψ(2S) to J/ψ yield in p+A collisions relative to the same ratio in p+p collisions is used to quantify the modification of the production. Figure 4 shows the preliminary result of the

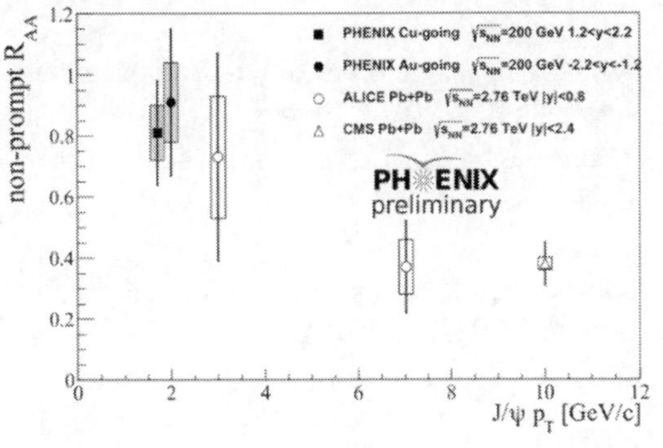

Figure 3. R_{CuAu} for $B \to J/\psi$ is compared with the LHC results

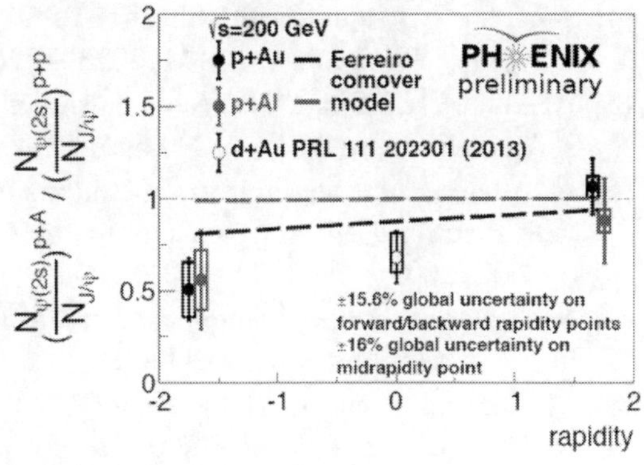

Figure 4. The double ratio of ψ(2S) to J/ψ in 200GeV p+Al and p+Au collisions. The model calculation qualitatively describe the data.

double ratio in p+Al and p+Au collisions. The previous result at mid-rapidity in d+Au collisions is also plotted for the comparison. We find that the double ratio at forward rapidity, positive side, is consistent with unity within uncertainty. This indicates that the nuclear modifications at forward rapidity are comparable for J/ψ and ψ(2S). However, the double ratio at backward rapidity, negative side, shows ψ(2S) is more suppressed than J/ψ with a factor of 2. Including the result at mid rapidity in d+Au collisions, the rapidity dependence of the double ratios is clearly seen. We compared with the model including the interaction with the co-moving hadrons[12], and find that the model qualitatively describes the result.

5 Summary

PHENIX measured open and closed heavy flavor productions in Au+Au, Cu+Au, p+A collisions. VTX and FVTX enables us to separate the bottom and charm contributions in the single electron measurement in Au+Au collisions and to identify J/ψ from B meson decays in Cu+Au collisions. The better mass resolution by FVTX provides the separation of J/ψ and ψ(2S) in p+A collisions. For the first time, we found the less suppression of electrons from bottom decays than that from charm decays for $3 < p_T < 4$ GeV/c and similar suppression for higher p_T. The first result of $B \to J/\psi$ decays also shows the small suppression at low p_T in Cu+Au collisions. The sequential suppression of ψ(2S) and J/ψ is found at backward rapidity, which is described by the comoving hadron model. These results from the heavy flavor measurement shed a light on the QGP property and the CNM effect from different dimensions. We collected large amount of Au + Au and $p + p$ data in 2014 to 2016. These new dataset should provide definitive measurements of open and closed heavy flavor productions.

References

[1] A. Adare *et al.* Energy Loss and Flow of Heavy Quarks in Au + Au Collisions at $\sqrt{s_{NN}} = 200$ GeV *Phys. Rev. Lett.* **98** 172301 (2007)

[2] A. Adare *et al.* Heavy-quark production in $p + p$ and energy loss and flow of heavy quarks in Au + Au collisions at $\sqrt{s_{NN}} = 200$ GeV *Phys. Rev.* C **84**, 044905 (2011)

[3] Y. L. Dokshitzer and D. E. Kharzeev *Phys. Lett.* B 519, 199 (2001)

[4] A. Adare *et al.* Nuclear Modification of Ψ', χ_c, and J/Ψ Production in d + Au Collisions at $\sqrt{s_{NN}}$ = 200GeV *Phys. Rev. Lett.* **111**, 202301 (2013)

[5] A. Adare *et al.* Single electron yields from semileptonic charm and bottom hadron decays in Au + Au collisions at $\sqrt{s_{NN}}$ = 200 GeV *Phys. Rev.* C **93** 034904 (2016)

[6] A. Adare *et al.* Measurement of Bottom versus Charm as a Function of Transverse Momentum with Electron-Hadron Correlations in $p + p$ Collisions at $\sqrt{s}$ = 200 GeV *Phys. Rev. Lett.* 103 082002 (2009)

[7] M. Aggarwal *et al.* Measurement of the Bottom contribution to non-photonic electron production in $p + p$ collisions at $\sqrt{s}$ = 200 GeV *Phys. Rev. Lett.* 105 202301 (2010)

[8] M. Cacciari, P. Nason, and R.Vogt, QCD Predictions for Charm and Bottom Quark Production at RHIC *Phys. Rev. Lett.* **95** 122001 (2005)

[9] A. Adil and I. Vitev Collisional dissociation of heavy mesons in dense QCD matter *Phys. Lett.* B 649 139–146 (2007)

[10] S. Chatrchyan *et al.* Suppression of non-prompt J/ψ, prompt J/ψ, and Υ(1S) in PbPb collisions at $\sqrt{s_{NN}}$ = 2.76 TeV, *JHEP 05*, 063 (2012)

[11] J. Adam *et al.* Inclusive, prompt and non-prompt J/ψ production at mid-rapidity in Pb-Pb collisions at $\sqrt{s_{NN}}$ = 2.76 TeV, *JHEP 07*, 051 (2015)

[12] E. G. Ferreiro, Excited charmonium suppression in proton–nucleus collisions as a consequence of comovers, *Phys. Lett.* **B 749** 98–103 (2015)

EPJ Web of Conferences **141**, 05007 (2017) DOI: 10.1051/epjconf/201714105007
ISMD 2016

Bottomia physics at RHIC and LHC

Georg Wolschin[1,a]

[1] *Institut für Theoretische Physik der Universität Heidelberg, Philosophenweg 16, D-69120 Heidelberg, Germany, EU*

Abstract. In UU collisions at RHIC energies and PbPb collisions at LHC energies the suppression of Υ mesons in the hot quark-gluon medium (QGP) versus reduced feed-down is investigated. Our model encompasses screening, collisional damping and gluodissociation in the QGP. For $\Upsilon(1S)$ it is in agreement with both STAR and CMS data provided the relativistic Doppler effect and the reduced feed-down from the $\Upsilon(nS)$ and $\chi_b(nP)$ states are properly considered. At both energies, most of the suppression for the $\Upsilon(1S)$ state is found to be due to reduced feed-down, whereas most of the $\Upsilon(2S)$ suppression is caused by the hot-medium effects. The role of the in-medium effects relative to reduced feed-down increases with energy. The $\Upsilon(1S)$-suppression in PbPb at $\sqrt{s_{NN}} = 5.02$ TeV is predicted.

1 Introduction

The production of heavy mesons and, in particular, of bottomia in initial hard partonic interactions in relativistic heavy-ion collisions at RHIC and LHC energies is of special interest because quarkonia in the fireball can act as a probe to test the properties of the hot medium. The heavier the hadron that is produced in the collision, the shorter its formation time τ_F. Very heavy mesons such as the J/ψ or the Υ meson in their 1S spin-triplet ground states are produced in hard collisions at very short times, typically at $\tau_F \simeq 0.3-0.5$ fm/c. Since the $\Upsilon(1S)$ state is particularly stable, it has a sizeable probability to survive in the hot quark-gluon medium that is produced in the fireball of a heavy-ion collision at LHC energies, even at initial medium temperatures of the order of 400 MeV or above.

There exists meanwhile a considerable literature on the dissociation of quarkonia and in particular of the Υ meson in the hot quark-gluon medium; see [1] and references therein for a review. In [2, 3] we have devised a model that accounts for the gluon-induced dissociation of the various bottomium states in the hot medium (gluodissociation), the damping of the quark-antiquark binding due to the presence of the medium which generates an imaginary part of the temperature-dependent potential, and the screening of the real part of the potential. The latter turns out to be less important for the strongly bound $\Upsilon(1S)$ ground state, but it is relevant for the excited $b\bar{b}$ states, and also for all $c\bar{c}$ bound states.

In this work we utilize our model to quantitatively disentangle the role of bottomia suppression in the quark-gluon plasma (QGP) relative to the role of reduced feed-down for the $\Upsilon(1S)$ ground state as function of energy, comparing with data from both RHIC and LHC. The relativistic Doppler effect is

[a] e-mail: wolschin@uni-hd.de

discussed in detail. We simultaneously consider the $\Upsilon(2S)$ state where the QGP effects are expected to be much more important with respect to reduced feed-down regarding the measured suppression, and verify this expectation in a calculation. We compare with centrality-dependent CMS data [4, 5] for the $\Upsilon(1S)$ and $\Upsilon(2S)$ states. For PbPb we predict the centrality-dependent suppression at the higher LHC energy of $\sqrt{s_{NN}} = 5.02\,\text{TeV}$.

2 Relativistic Doppler effect

The radial Schrödinger equation is solved for the six bottomia states considered in the hot fireball – $\Upsilon(1S)$, $\Upsilon(2S)$, $\Upsilon(3S)$, and $\chi_b(1P)$, $\chi_b(2P)$, $\chi_b(3P)$ – with energies $E_{nl}(T)$ in a complex potential $V_{nl}(r, T)$ and corresponding damping widths $\Gamma_{nl}^{\text{damp}}(T)$ as has been outlined in [3]. We consider the running of the strong coupling in the calculation of the wave function, and the various dissociation processes. Since the coupling constant depends on the solution of the Schrödinger equation for the six bottomium states, we use an iterative method for the solution of the problem [3], together with the one-loop expression for the running of the coupling. In addition to the damping width $\Gamma_{nl}^{\text{damp}}(T)$ we consider the width caused by gluon-induced dissociation $\Gamma_{nl}^{\text{diss}}(T)$ [3]. The total decay width of a given bottomium state is then the incoherent sum $\Gamma_{nl}^{\text{tot}} = \Gamma_{nl}^{\text{damp}} + \Gamma_{nl}^{\text{diss}}$.

Bottomia are too massive to experience a substantial change of their momenta by collisions with the light partons in the medium. Hence, there will be a finite relative velocity between the expanding QGP medium and the $b\bar{b}$ mesons. In the rest frame of the bottomia, the surrounding distribution of massless gluons then appears as a Bose-Einstein distribution with an anisotropic temperature T_D that is determined from the relativistic Doppler effect as

$$T_D(T, |\boldsymbol{u}|, \theta) = T\frac{\sqrt{1-|\boldsymbol{u}|^2}}{1-|\boldsymbol{u}|\cos\theta}. \tag{1}$$

Here, $\boldsymbol{u}$ is the average velocity of the surrounding fluid cell (measured in the bottomium restframe) and θ the angle between $\boldsymbol{u}$ and the direction in which the temperature is evaluated.

In the rest frame of the bottomia the Doppler effect causes a blueshifted temperature for $\theta = 0°$ and a redshifted temperature in the opposite direction $\theta = 180°$. The effects of red- and blueshift get more and more pronounced with increasing relative velocity $|\boldsymbol{u}|$, but the angular range with $T_D < T$ (redshifted region) is growing while the angular range with $T_D > T$ (blueshifted region) is restricted to smaller and smaller angles θ as has been noted in [6]. To account for the effect of the anisotropic temperature T_D on the bottomium dissociation we substitute the total decay width $\Gamma_{nl}^{\text{tot}}(T)$ by an effective value

$$\Gamma_{\text{eff},nl}^{\text{tot}} := \langle\Gamma_{nl}^{\text{tot}}(T_D)\rangle_\Omega = \frac{1}{4\pi}\int d\Omega\,\Gamma_{nl}^{\text{tot}}(T_D). \tag{2}$$

This takes into account the red- or blueshifted temperatures and also correctly describes the non-existence of bound states once the effective temperature in the blueshifted region exceeds the dissociation temperature, see [7] for more details. The p_T-dependence is rather flat and at 2.76 TeV in reasonable agreement with the available CMS data [7]. There is no rapidity dependence in our model, both minimum bias and centrality dependent yields are flat as functions of y.

3 Hydrodynamic expansion

Our treatment of the hydrodynamic flow of the fireball has been outlined in [3] including transverse expansion. The initial time for the evolution is $\tau_{\text{init}} = 0.1\,\text{fm}/c$, the initial conditions in the transverse plane (x^1, x^2) are given in Eqs. (14) – (16) of [3].

We determine the initial central temperature T_0 from a fit of the experimental $R_{AA}(\Upsilon(1S))$ in central collisions, cf. section 5. It is in reasonable agreement with values derived from ideal hydrodynamical calculations of elliptic flow at RHIC and LHC energies, respectively [8].

The same formation time of $\tau_F = 0.4\,\text{fm}/c$ is used for ground and exited states, with theta functions for the decay widths of the six states considered. In the co-moving coordinate system used for the hydrodynamical calculation, time dilation of the formation times is taken into account. We had previously investigated the dependence of R_{AA} on τ_F, cf. Fig. 9 in [3].

To estimate the initial populations $N^{\mathrm{i}}_{AA,nl}$ of the six states, we consider the measured final populations $N^{\mathrm{f}}_{\mathrm{pp},nl}$ of the three $\Upsilon(nS)$-states in pp collisions at the same energy and calculate the decay cascade [9] backwards to obtain the initial populations in pp, $N^{\mathrm{i}}_{\mathrm{pp},nl}$ [7] These are then scaled by the number of binary collisions N_{coll} yielding the initial populations in the heavy-ion case. When the suppression factors are calculated, the number of binary collisions cancels out. The required branching ratios are taken from the Review of Particle Physics [10] or from theory where no experimental values are available (as is the case for $\chi_b(3P)$), see [9] for details and references.

The QGP is approximated as a relativistic perfect fluid consisting of gluons and up-, down- and strange-quarks, described in detail in [3]. We define the QGP-suppression factors $R^{\mathrm{QGP}}_{AA,nl}(c, p_T)$ which quantify the amount of in-medium suppression of bottomia with transverse momentum p_T for PbPb collisions in the centrality bin c, where $b_c \leq b < b_{c+1}$. The QGP-suppression factor is not directly measurable since it accounts only for the amount of suppression inside the fireball due to the three processes of color screening, collisional damping and gluodissociation. It is given by the ratio of the number of bottomia that have survived the fireball to the number of bottomia produced in the collision. The latter scales with the number of binary collisions at a given point in the transverse plane and hence with the nuclear overlap, $N_{b\bar{b}} \propto N_{\mathrm{coll}} \propto T_{AA}$. We calculate $R^{\mathrm{QGP}}_{AA,nl}$ as outlined in [3].

4 In-medium suppression versus feed-down

Once the bottomia states have survived the hot quark-gluon plasma environment, the feed-down cascade from the excited states to the ground state is considered in detail. Due to the rapid melting or depopulation of the excited states caused by the mechanisms in the QGP-phase, the feed-down to the ground state is reduced, resulting in additional $\Upsilon(1S)$-suppression with respect to the situation in pp collisions at the same energy.

The focus of the present investigation is the determination of the relevance of feed-down for a given bottomium state as function of incident energy (RHIC vs. LHC), and of its relative importance for the $\Upsilon(1S)$ and $\Upsilon(2S)$ states, which appears to be a new consideration. As will be shown in the next section, at both RHIC and LHC energies, most of the suppression for the $\Upsilon(1S)$ state is found to be due to reduced feed-down, and even more so at the lower RHIC energy. In contrast, most of the $\Upsilon(2S)$ suppression is caused by the hot-medium effects.

5 Results and comparison to data

The suppression of the spin-triplet ground state at both RHIC and LHC energies is well described by our model for initial central temperatures of $T_0 = 417\,\text{MeV}$ in UU at $\sqrt{s_{\mathrm{NN}}} = 193\,\text{GeV}$ and $T_0 = 480\,\text{MeV}$ in PbPb at 2.76 TeV, and an Υ and χ_b formation time of $\tau_F = 0.4\,\text{fm}/c$; see Fig. 1 for PbPb, and [7] for UU. The parameters for the density distributions of the lead and uranium ions are taken from [11]. Our minimum bias value of the suppression in 2.76 TeV PbPb is $R^{\text{min. bias}}_{AA}(\Upsilon(1S)) = 0.44$.

The relative contributions of in-medium effects and reduced feed-down change as a function of incident energy, see Tab. 1 for detailed minimum-bias results: In 193 GeV UU, only about 20 % of the

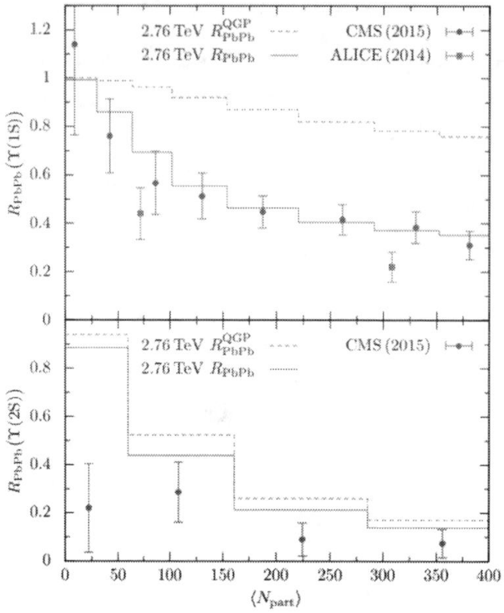

Figure 1. (color online) Top: Calculated suppression factor $R_{AA}(\Upsilon(1S))$ in PbPb collisions at $\sqrt{s_{NN}} = 2.76$ TeV (solid line) together with centrality-dependent data from CMS ($|y| < 2.4$, [5]) and ALICE (squares, $2.5 < y < 4$, [17]) as function of the number of participants N_{part} (averaged over centrality bins). The suppression factor R_{AA}^{QGP} in the QGP-phase without the effect of reduced feed-down is shown as dashed (upper) curve. Bottom: Suppression factor for the first excited state $R_{AA}(\Upsilon(2S))$ in PbPb collisions at $\sqrt{s_{NN}} = 2.76$ TeV (solid line) together with preliminary data from CMS [5]. The suppression factor R_{AA}^{QGP} in the QGP-phase without the effect of reduced feed-down is shown as dashed (upper) curve.

total suppression $(1 - R_{AA})$ is due to the in-medium effects, whereas in 2.76 TeV PbPb the in-medium contribution is already about 30 % and further increases in 5.02 TeV PbPb.

The situation is very different for the first excited state $\Upsilon(2S)$ as shown in Fig. 1 (bottom) for 2.76 TeV PbPb: With the same set of parameters as for the $\Upsilon(1S)$ state, the calculated suppression of the $\Upsilon(2S)$ state is much more pronounced in the QGP-phase than for the $\Upsilon(1S)$ state. Tab. 1 shows that more than 80 % of the total minimum-bias $\Upsilon(2S)$ suppression in UU and more than 90 % in PbPb is due to in-medium effects. Hence the additional contribution of the feed-down cascade on the $\Upsilon(2S)$ suppression is rather marginal and drops below 10 % at LHC energies. Unfortunately, there are no data available that directly quantify the feed-down fractions as functions of centrality. As is obvious from Fig. 1, the comparison with the CMS data [5] leaves room for additional suppression mechanisms in particular in the peripheral region for the $\Upsilon(2S)$ state.

For $\Upsilon(1S)$ we also predict in Fig. 2 the suppression at the current LHC energy of 5.02 TeV PbPb as $R_{AA}^{\text{min. bias}}(\Upsilon(1S)) = 0.40$ using the same formation time, but a scaled initial temperature $T_0 = 513$ MeV. This value is obtained from the proportionality between the initial entropy density, the charged particle multiplicity per unit of rapidity and the cube of the temperature [12–14]. Here the extrapolated $dN_{ch}/d\eta$ for 0 – 5 % centrality PbPb taken from [15] is in agreement with recent data from ALICE [16]. The ensuing enhancement of the suppression is within the experimental error bars of the 2.76 TeV result. It remains to be seen whether this is confirmed by the forthcoming analysis of the mid-rapidity CMS data from LHC Run-II.

DOI: 10.1051/epjconf/201714105007

Table 1. Calculated nuclear suppression factors for the $\Upsilon(1S)$ and $\Upsilon(2S)$ states in minimum-bias 193 GeV UU as well as 2.76 and 5.02 TeV PbPb collisions. The in-medium suppression factor is R_{AA}^{QGP}, the total suppression factor including reduced feed-down is R_{AA}. The last column gives the percentage of the suppression in the medium relative to the total suppression $(1 - R_{AA}^{QGP})/(1 - R_{AA})$.

$\sqrt{s_{NN}}$	System	State	R_{AA}^{QGP}	R_{AA}	$\frac{1-R_{AA}^{QGP}}{1-R_{AA}}$
193 GeV	UU	$\Upsilon(1S)$	0.919	0.568	18.8 %
193 GeV	UU	$\Upsilon(2S)$	0.469	0.399	88.4 %
2.76 TeV	PbPb	$\Upsilon(1S)$	0.834	0.443	29.8 %
2.76 TeV	PbPb	$\Upsilon(2S)$	0.285	0.239	94.0 %
5.02 TeV	PbPb	$\Upsilon(1S)$	0.783	0.396	35.9 %
5.02 TeV	PbPb	$\Upsilon(2S)$	0.226	0.189	95.4 %

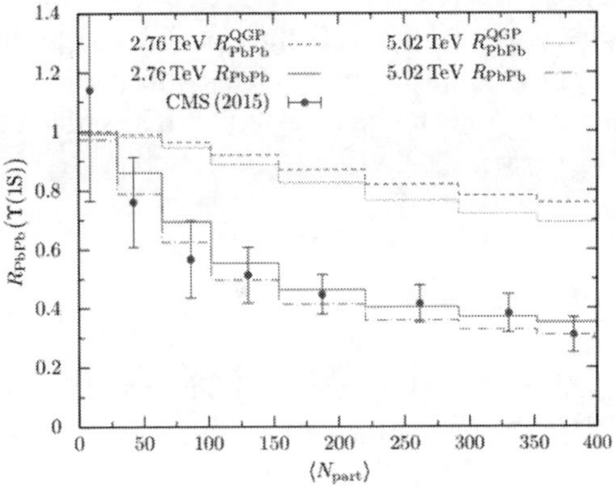

Figure 2. (color online) Calculated suppression factor $R_{AA}(\Upsilon(1S))$ in PbPb collisions at $\sqrt{s_{NN}} = 2.76$ TeV (lower solid line) and prediction at 5.02 TeV (lower dash-dotted line) together with centrality-dependent 2.76 TeV data from CMS ($|y| < 2.4$, [5]) as function of the number of participants N_{part} (averaged over centrality bins). The suppression factors R_{AA}^{QGP} in the QGP-phase without the effect of reduced feed-down are shown as upper curves (dashed and dotted) for both energies, again yielding slightly more suppression at the higher energy.

6 Conclusion

In summary, we have investigated the suppression of the $\Upsilon(nS)$ states in UU and PbPb collisions at RHIC and LHC energies in a model that considers the in-medium processes gluodissociation, screening and damping. The feed-down cascade from the excited bottomia states produces substantial additional ground-state suppression, since the excited states melt through screening, or depopulate through dissociation processes and hence, there is less feed-down to the $\Upsilon(1S)$ ground state as compared to pp at the same energy. In contrast, the suppression of the first excited state $\Upsilon(2S)$ at both RHIC and LHC energies is largely due to the properties of the hot quark-gluon medium.

Our model results for the ground state are in agreement with the centrality-dependent STAR and CMS data [4, 5, 18]. The flat transverse-momentum dependence of the suppression factor is consistent with the preliminary CMS data when the relativistic Doppler effect due to the velocity of the moving bottomia relative to the expanding QGP is properly considered. In minimum-bias

193 GeV UU, only about 20 % of the total ϒ(1S) suppression ($1 - R_{AA}$) is due to the in-medium effects, whereas in 2.76 TeV PbPb the in-medium contribution is already about 30 % and further increases in 5.02 TeV PbPb.

The suppression of the first excited ϒ(2S) state which occurs mostly in the QGP-phase requires additional centrality-dependent dissociation mechanisms in particular for peripheral collisions. For the centrality and transverse momentum dependence of ϒ(1S) we have made predictions at the LHC energy of 5.02 TeV PbPb where results are currently being analyzed. The ALICE Collaboration has meanwhile released preliminary data for the ϒ(1S) centrality dependent yields in 5.02 TeV PbPb [19] at rapidities $2.5 < |y| < 4.0$. Here the suppression is found to be slightly less than at 2.76 TeV, but almost compatible within the experimental error bars. CMS has presented preliminary 5.02 TeV data in the midrapidity region $|y| < 2.4$, but so far only for the double ratio $R_{AA}(\Upsilon(2S))/R_{AA}(\Upsilon(1S))$ [20]. Once their suppressions factors for the individual states are available, the consistency of the ALICE and CMS results, and the agreement with our prediction can be checked.

Acknowledgements

This work has been partially supported by DFG through TRR 33 at the Universities of Bonn, LMU Munich and Heidelberg. I am grateful to my students - in particular, Johannes Hoelck and Felix Nendzig - for their contributions. An extended version of this conference article is found in [7].

References

[1] A. Andronic, F. Arleo, R. Arnaldi, A. Beraudo et al., Eur. Phys. J. C **76**, 107 (2016)
[2] F. Nendzig, G. Wolschin, Phys. Rev. C **87**, 024911 (2013)
[3] F. Nendzig, G. Wolschin, J. Phys. G: Nuclear and Particle Physics **41**, 095003 (2014)
[4] S. Chatrchyan et al. (CMS Collaboration), Phys. Rev. Lett. **109**, 222301 (2012)
[5] M. Jo et al. (CMS Collaboration), Proc. Quark Matter Conf. Kobe 2015, in preparation; CMS-PAS-HIN-15-001 (2015)
[6] M.A. Escobedo, F. Giannuzzi, M. Mannarelli, J. Soto, Phys. Rev. D **87**, 114005 (2013)
[7] J. Hoelck, F. Nendzig, G. Wolschin (2016), 1602.00019
[8] B. Schenke, S. Jeon, C. Gale, Phys. Lett. B **702**, 59 (2011)
[9] F. Vaccaro, F. Nendzig, G. Wolschin, Europhys. Lett. **102**, 42001 (2013)
[10] K.A. Olive et al. (Particle Data Group), Chin. Phys. **C38**, 090001 (2014)
[11] H. de Vries, C.W. de Jager, C. de Vries, At. Data Nucl. Data Tables 36 p. 495 (1987)
[12] J.D. Bjorken, Phys. Rev. D **27**, 140 (1983)
[13] G. Baym, B.L. Friman, J.P. Blaizot, M. Soyeur, W. Czyż, Nucl. Phys. A **407**, 541 (1983)
[14] M. Gyulassy, T. Matsui, Phys. Rev. D **29**, 419 (1984)
[15] G. Wolschin, Phys. Rev. C **91**, 014905 (2015)
[16] J. Adam et al. (ALICE Collaboration), Phys. Rev. Lett. **116**, 222302 (2016)
[17] B. Abelev et al. (ALICE Collaboration), Phys. Lett. B **738**, 361 (2014)
[18] R. Vértesi et al. (STAR Collaboration), Nucl. Part. Phys. Proc. **276**, 269 (2016)
[19] C. Hadjidakis (ALICE Collaboration), Talk at ICNFP2016 Kolymbari, to be published in EPJ-Web of Conf. (2016)
[20] Y. Kim et al. (CMS Collaboration), Proc. Hard Probes, Wuhan (2016)

EPJ Web of Conferences **141**, 05008 (2017) DOI: 10.1051/epjconf/201714105008
ISMD 2016

Measurement of bottomonia states in pp, pPb and PbPb collisions at 2.76 TeV from CMS

Yongsun Kim on behalf of the CMS collaboration[1]

[1] *Korea University, 145 Anam-ro, Seonbuk-gu, Seoul, Republic of Korea*

Abstract. The modification of jets in heavy ion collision at a center of mass energy of $\sqrt{s_{NN}}$ = 2.76 TeV was observed via various analyses carried out with Run I LHC data. In November 2015, LHC delivered the Run II phase lead-lead beams at 5.02 TeV and proton-proton beams at the same energy for reference. The high luminosity PbPb data, more than twice of the previous run, provided an opportunity for elaborate studies based on high-pt probes. This presentation discusses the development of jet finding algorithms and jet observables, including the production rates, fragmentation patterns, and flavor dependence, published by CMS, ALICE and ATLAS collaborations. The results from 2.76 TeV data are briefly reviewed, followed by a presentation of the most recent results from $\sqrt{s_{NN}}$ = 5.02 TeV.

1 Introduction

At extremely high temperature, the quarks and gluons are expected to be deconfined from hadronic states and matter to transit to a new thermal system, a Quark-Gluon Plasma (QGP). A QGP state is predicted to be produced in a laboratory by colliding relativistically accelerated heavy ions. Measurement of high p_T probes is of big interest because it can characterize the QGP in several aspects. The suppression of the high p_T jets is one of the main evidence of QGP formation because an energetic parton would lose its energy as it traverses the medium. As a result, that parton becomes a jet with lower p_T compared to what it is supposed to be in vacuum. Moreover, the jet has been one of the most popular physics object in Heavy Ion community at LHC because of the high production cross section and the advanced detection technologies. Since the first PbPb collisions in 2010, many papers and preliminary notes have been published by LHC heavy ion experiments to quantify the energy loss of high p_T jets. In addition, it was also found that the jets are modified even in pPb collisions, by so-called the cold nuclear matter (CNM) effects. The CNM effects include the modification of the parton distribution function, Cronin effect and nuclear absorption. During the Run I period, the pPb collisions were carried out at $\sqrt{s_{NN}}$ = 5.02 TeV and pp and PbPb are measured at $\sqrt{s_{NN}}$ = 2.76 TeV.

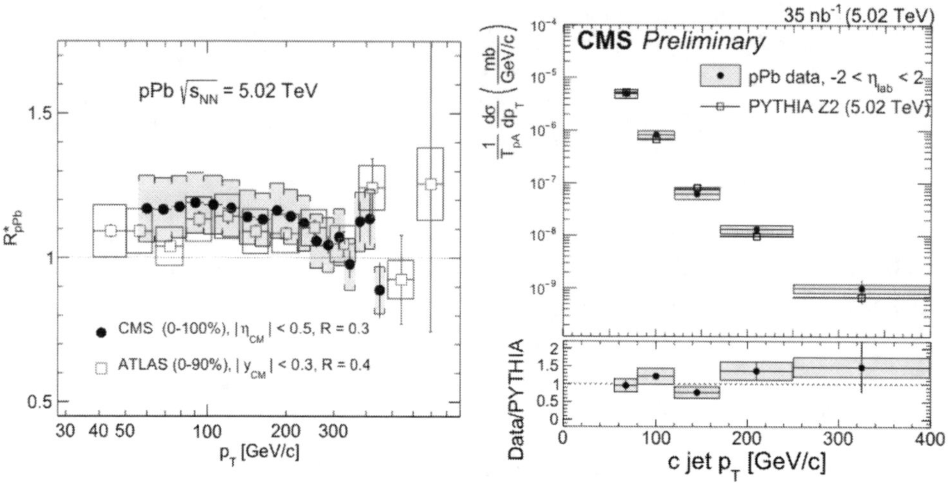

Figure 1. (Left) R_{pA} of inclusive jets measured by ATLAS and CMS (in different jet radius) [3]. (Right) Ratio of cross-sections of b jet in pPb to calculations by PYTHIA generator, after scaling by the nuclear overlap factor from the Glauber model [4].

2 Jets in pPb collisions at $\sqrt{s_{NN}}$ = 5.02 TeV

2.1 R_{pA} of inclusive jets and heavy flavored jets

The spectra of jet in pPb and pp systems were compared, as one of the conventional methods to measure the nuclear matter effect. R_{pA}, which is the ratio of production cross sections in pPb to pp normalized by the binary nucleon-nucleon collisions, was measured by both CMS [1] and ATLAS [2]. The same jet reconstruction algorithm (anti-k_T) were used with different jet cone radii: 0.3 for CMS, and 0.4 for ATLAS. As shown in Fig. 1, the measured R_{pA} is consistent in 1 - 1.5 σ level depending on p_T . This result confirms that the strong suppression of jets discovered in PbPb collisions is due to the hot and dense medium effect. Compared to various theoretical models, the CMS result of R_{pA} favors the NLO calculation with EPS09 nPDF. ATLAS also measured the jet R_{pA} and R_{CP} (Central-to-Peripheral ratio) as a function of the pPb centrality and observed a strong reduction of jet yields pronounced at high p_T above 200 GeV/c in the most central events. CMS also measured the spectra of jets from bottom quarks [5] and charm quarks [6] using the heavy quark tagging technique based on the the secondary vertex reconstruction. As displayed in Fig. 1, the production rate of heavy quark jets is consistent to the Monte Carlo calculation from PYTHIA generator within the measured uncertainties.

2.2 Dijet pair and nPDF

Besides the inclusive jet studies, the dijet events in pPb collision can be good probes for the nPDF because they provide two-dimensional information: the average value of pseudo-rapidities and the mean p_T reflect the x and Q^2 in the PDF respectively. The CMS collaboration made a thorough measurement of di-jet pairs in pPb and pp from the recent Run II data, and provided a strong constraint for nPDF models [7]. Fig. 2 is one example.

[a]e-mail: kimy@cern.ch

 DOI: 10.1051/epjconf/201714105008

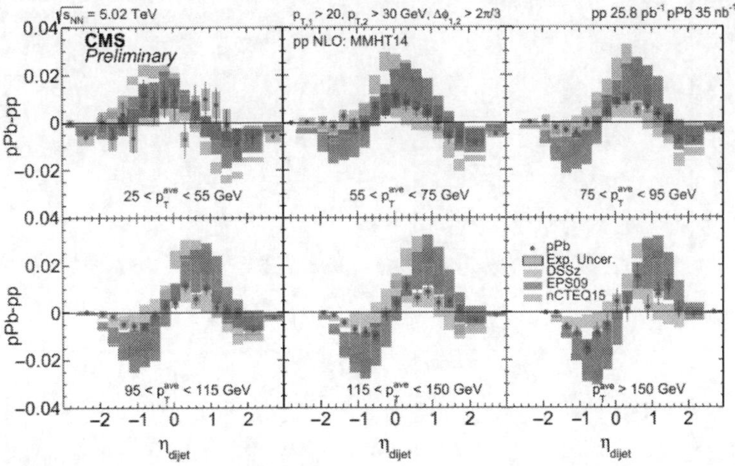

Figure 2. The difference of dijet pseudo-rapidity distributions between pPb and pp in bins of average p_T of dijet pairs [8]. The results are compared to NLO calculations based on nPDF sets DSSZ, EPS09 and nCTEQ15 and proton PDF MMHT14.

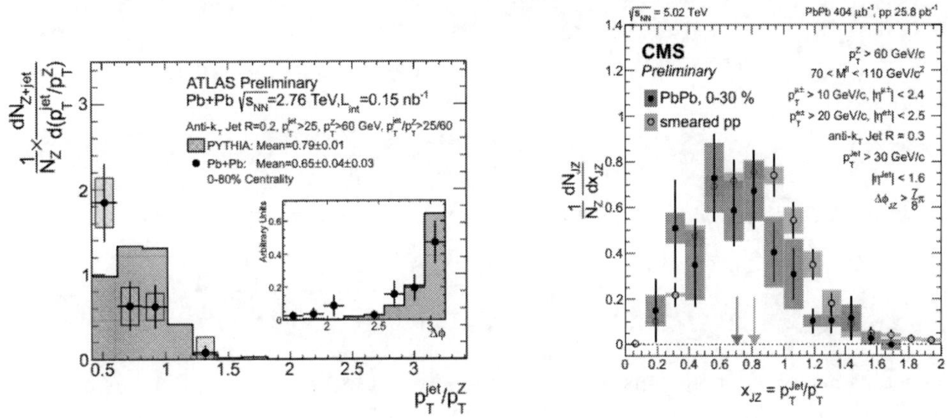

Figure 3. Transverse momentum ratio X_{JZ} between the Z boson and the jets with the azimuthal angle separation $\Delta\phi > 7\pi/8$, measured in ATLAS (Left) [9] and CMS (Right) [10].

3 Jet quenching in PbPb collisions at $\sqrt{s_{NN}}$ = 5.02 TeV

3.1 Photon-jet and Z-jet correlation

The observable of the first jet quenching measurement from CMS Run II data was correlation in Z-jet pairs [11]. The Z boson, like direct photons in γ-jet channel, is color blind. For this reason, the direct comparison of the p_T of a Z boson to the momentum of associated jets provides the information of energy loss of jets. The distribution of p_T ratio between Z and jet was found to be in the similar level of the γ-jet result in the same kinematic range [12] as shown in Fig. 3. Despite its statistical limit, this measurement can be recognized as a meaningful study because it demonstrates the utility of Z

DOI: 10.1051/epjconf/201714105008

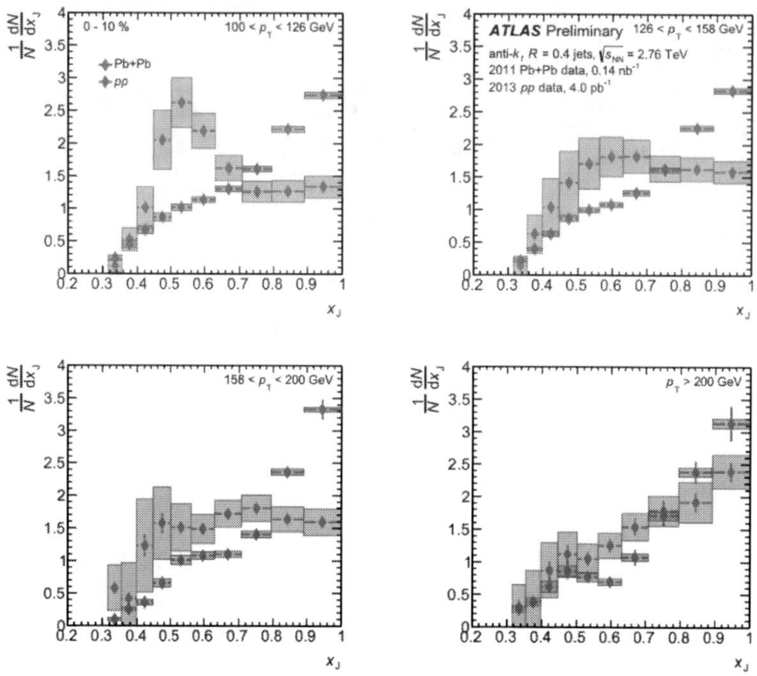

Figure 4. The (1/N)dN/dx distributions for different selections on leading jet p_{T} in 0-10% PbPb events (red) and pp (blue). [13]

bosons for quantitative study of jet quenching. A similar measurement was performed by ATLAS at 2.76 TeV [14].

3.2 Di-jet results

Since the first heavy ion collisions at LHC, several analyses were carried out using the back-to-back dijet pairs in PbPb and pp. Recently ATLAS published the momentum asymmetry results in PbPb, where the jet energy resolution effects are fully unfolded [13]. From the distribution of $x_J = p_{\mathrm{T}}^{subleading jet}/p_{\mathrm{T}}^{leading jet}$ in 0-10% central collisions, a clear peak was observed around $x_J = 0.5$ which provides a constraint in the modeling of jet quenching.

3.3 Jet-track correlation

In order to chase the missing energy dissipated by the jet quenching effect, several analyses of jet-track correlation were carried by the CMS collaboration. In this note, three publications are highlighted. In the first paper [15], the p_{T} distribution of charged particles in dijet events are studied. The p_{T} distribution of tracks projected to the jet axis are measured. The integration of the projected p_{T} values turned out to make up the momentum imbalance of the dijet pair. The redistribution pattern of momentum was also studied for different centrality bins using various selection of radius in jet reconstruction algorithm. In the second analysis [16], the momentum distribution of hadrons are compared for those in peak region (($|\Delta\eta| < 1.5$) and in long-range region ($1.5 < |\Delta\eta| < 2.5$). In Fig 5,

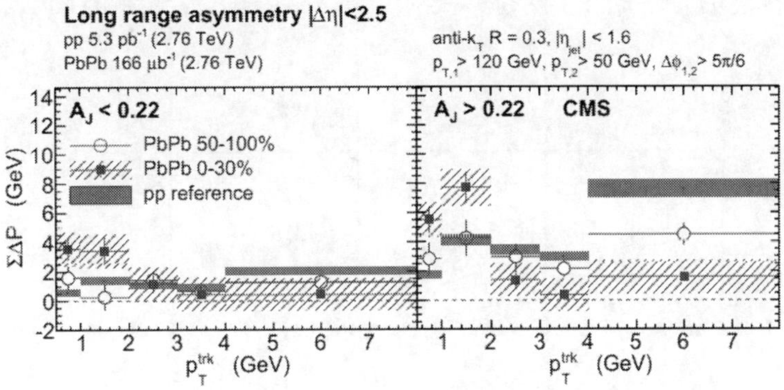

Figure 5. Integrated p_T of tracks in long-range correlated tracks as a function of p_T^{trk} integrated over $|\Delta\phi| < \pi/2$ and $|\Delta\eta| < 2.5$ for pp, peripheral PbPb and central PbPb data balanced dijet events (Left) and unbalanced for dijet events (Right). [16]

the momentum of charged tracks are integrated in the hemisphere of $|\Delta\eta| < 2.5$ and $|\Delta\eta| < 1.5$) for each p_T bins of tracks. For unbalanced dijet events ($A_J > 0.22$),an assymetry was observed for overall range of p_T . For balanced dijet events $A_J < 0.22$, clear modification was observed in p_T range below 2 GeV/c. In the third anlaysis [17], a full census of $\Delta\eta$ and $\Delta\phi$ of track-jet pairs are taken for leading jet and subleading jet for several track and jet p_T bins. A clear modification of distributions is found for PbPb results compared to pp.

3.4 Study of jet splitting

The jet splitting is one of new measurement done by CMS using 5.02 TeV Run II data. Jets reconstructed by anti-kT algorithm and then the internal jet structure is divided into two smaller jets by running the kT algorithm inside. Novel techniques, such as jet Grooming and sub-jet finding algorithms [19], were used for this analysis. The main physics observable is $z_g = \min(p_T^1, p_T^2)/(p_T^1 + p_T^2)$, where p_T^1 and p_T^2 are the transverse momentum of two branches of a jet. As shown in Fig. 6, the distribution of z_g was measured in 4 different centrality bins in PbPb data and in pp data, and their ratio was used to quantify the difference of the splitting patterns. In the central PbPb collisions, the fraction of imbalanced branches were enhanced and balanced cases were suppressed compared to pp result. And, the result in the most peripheral PbPb collision was consistent to pp data within the uncertainties. It was also observed that such a modification is less conspicuous for higher jet p_T selection.

4 Summary

The heavy ion jet results at 2.76 TeV and 5.02 TeV from CMS and ATLAS collaboration were reviewed. The pPb result of nuclear modification factor for inclusive, b-tagged and c-tagged jets confirms that the strong suppression in PbPb collision is attributed to the hot medium effects. The analysis on dijet pair provided a strong constraint for nPDF in anti-shadowing regime. The jet quenching effects at 2.76 TeV are measured in several hard probe channels including dijet, photon-jet and Z-jet. Plus, the precision measurement to change the missing energy was done by studying the jet-track correlation. In addition, many results using the new data at 5.02 TeV from LHC Run II have been released. The Z-jet result proved the role of Z bosons for elaborate jet quenching analyses in near

 DOI: 10.1051/epjconf/201714105008

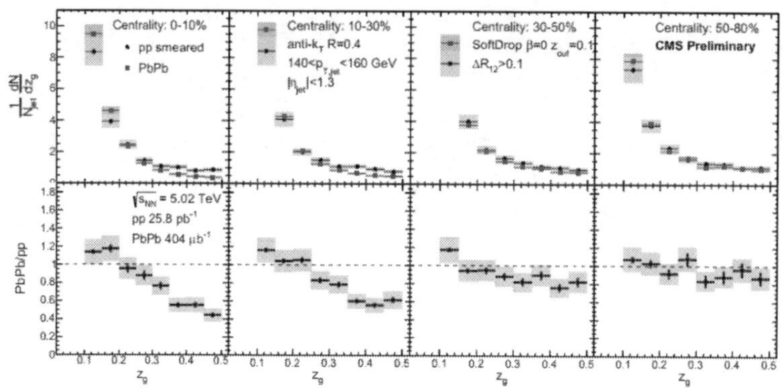

Figure 6. Splitting function of jets in PbPb for $140 < p_T < 160$ GeV/c in several centrality ranges compared to pp data. [18]

future. And the jet splitting phenomena measured using the sub-jet finding technique shed light on the modification of hadronization pattern of jets at early stage.

References

[1] CMS Collaboration, EPJC 76 (2016) 372.
[2] ATLAS Collaboration, EPLB 748 (2015) 392.
[3] CMS Collaboration and ATLAS Collaboration, Retrieved from http://cms-results.web.cern.ch/cms-results/public-results/publications/HIN-14-001/.
[4] CMS Collaboration, Retrieved from http://cms-results.web.cern.ch/cms-results/public-results/publications/HIN-14-007/.
[5] CMS Collaboration, PLB 754 (2016) 59.
[6] CMS Collaboration, CMS-HIN-15-012 (2015).
[7] CMS Collaboration, CMS-PAS-HIN-16-003 (2016).
[8] CMS Collaboration, Retrieved from https://cms-results.web.cern.ch/cms-results/public-results/preliminary-results/HIN-16-003/index.html.
[9] ATLAS Collaboration, Retrieved from https://atlas.web.cern.ch/Atlas/GROUPS/PHYSICS/CONFNOTES/ATLAS-CONF-2012-119/.
[10] CMS Collaboration, Retrieved from https://cms-results.web.cern.ch/cms-results/public-results/preliminary-results/HIN-16-002/index.html.
[11] CMS Collaboration, CMS-PAS-HIN-15-013 (2016).
[12] CMS Collaboration, CMS-PAS-HIN-13-006 (2016).
[13] ATLAS Collaboration, ATLAS-CONF-2015-052 (2015).
[14] ATLAS Collaboration, ATLAS-CONF-2012-119 (2012).
[15] CMS Collaboration, JHEP 01 (2016) 006.
[16] CMS Collaboration, CMS-HIN-15-011 (2016).
[17] CMS Collaboration, JHEP 1602 (2016) 156.
[18] CMS Collaboration, CMS-PAS-HIN-16-006 (2016).
[19] CMS Collaboration, Retrieved from https://cms-results.web.cern.ch/cms-results/public-results/preliminary-results/HIN-16-006/index.html.

On wounded constituents in nuclear collisions*

Piotr Bożek[1,a], Wojciech Broniowski[2,3,b], and Maciej Rybczyński[3,c]

[1] *AGH University of Science and Technology, Faculty of Physics and Applied Computer Science, 30-059 Cracow, Poland*
[2] *The H. Niewodniczański Institute of Nuclear Physics, Polish Academy of Sciences, 31-342 Cracow, Poland*
[3] *Institute of Physics, Jan Kochanowski University, 25-406 Kielce, Poland*

Abstract. In this talk we summarize the main results of our recent paper [1], where we explore predictions of the wounded quark model for particle production and the properties of the initial state formed in ultra-relativistic collisions of atomic nuclei.

Particle-particle interactions at high energies usually lead to copious production of new particles whose number rises as a function of the collision energy. The charged particle multiplicity ($N_{\rm ch}$) and its pseudorapidity density ($dN_{\rm ch}/d\eta$) are fundamental measurable quantities which serve as important characteristics of the global properties of the system. The data on production of particles in relativistic collisions of atomic nuclei collected during the operation of RHIC [2, 3] and the LHC [4] accelerators indicate that the standard wounded nucleon model [5] (according to the definition, a wounded nucleon is the one which underwent at least one inelastic collision) does not describe the observed centrality dependence of particle multiplicities in the A+A collisions, unless a component proportional to the binary collisions [6] is added.

We are interested in the impact of subnucleonic components of matter on the dynamics of the of the early stage of the collisions, and explore th wounded quark model [10–13]. One of the interesting results of [1] is that taking into account subnucleonic degrees of freedom in the description of the early phase of the collision one obtains the linear dependence of production of particles as a function of the number of wounded constituents, namely

$$dN_{\rm ch}/d\eta \propto Q_{\rm W}, \qquad (1)$$

where $Q_{\rm W}$ is the number of wounded constituents involved in the collision. Such scaling has been suggested in [14] for the RHIC data. The wounded quark scaling also works for the SPS energy range [15]. More recent studies were reported in [16–19].

Our results for particle production at midrapidity are shown in Fig. 1. We note an approximately flat centrality dependence of $dN_{\rm ch}/d\eta/Q_{\rm W}$ in qualitative agreement with the earlier studies [8, 14, 15, 20]. In contrast, the flatness is not the case in the wounded nucleon model, where, as is well known, the ratio $dN_{\rm ch}/d\eta/N_{\rm W}$ substantially increases with the number of wounded nucleons, $N_{\rm W}$. The value of $dN_{\rm ch}/d\eta/Q_{\rm W}$ (i.e., the average number of charged hadrons per unit of rapidity coming from a single

*Presented by M. Rybczyński
[a] e-mail: Piotr.Bozek@fis.agh.edu.pl
[b] e-mail: Wojciech.Broniowski@ifj.edu.pl
[c] e-mail: Maciej.Rybczynski@ujk.edu.pl

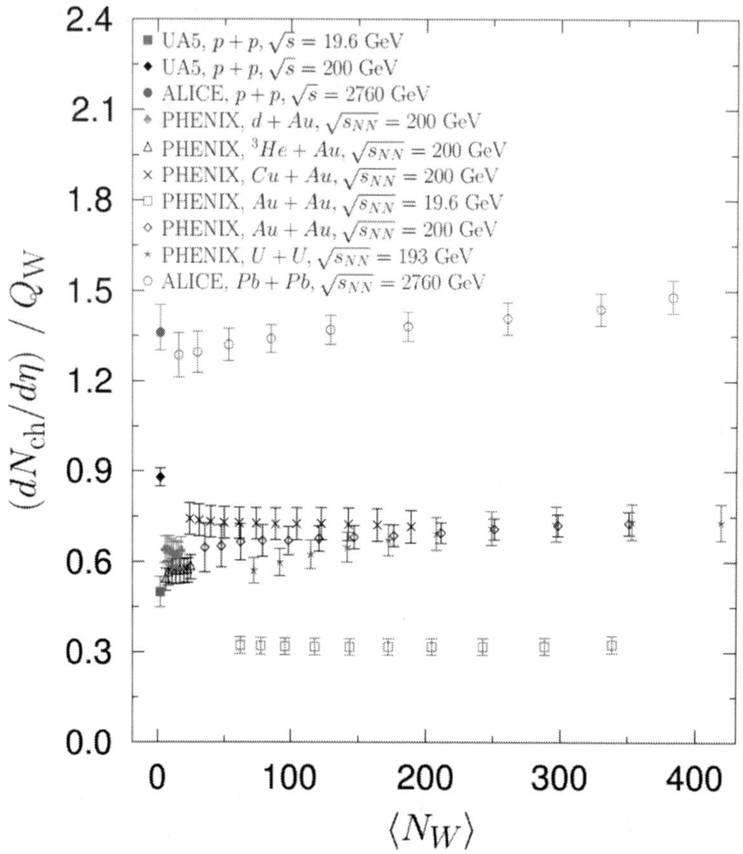

Figure 1. Experimental multiplicity of charged hadrons per unit of pseudorapidity (at mid-rapidity) divided by the number of wounded constituents, $dN_{ch}/d\eta/Q_W$, plotted as function of centrality expressed via the number of wounded nucleons. We also show the results for the p+p collisions (filled symbols at $\langle N_W \rangle = 2$). The data are from Refs. [4, 7–9].

wounded quark), is at a given energy roughly similar for various considered reactions. For the LHC energies it is also consistent with the production in elementary p+p collisions. However, at RHIC energies the p+p point is noticeably higher (about 30%) than the band for the A+A collisions.

The basic effect of the subnucleonic degrees of freedom in particle production is a stronger combinatorics when compared to wounded nucleon model, which accomplishes the approximately linear scaling of production with the number of constituents. An intermediate combinatorics is realized in the quark-diquark model [21], which led to proper description of the RHIC data as well as the proton-proton scattering amplitude (including the differential elastic cross section) at the CERN ISR energies.

Modeling with use of subnucleonic components allows for examination of p+p collisions with the methods typically used for larger systems. Since on physical grounds the sources of particle production must have a certain transverse size, we use Gaussian smearing for the entropy density

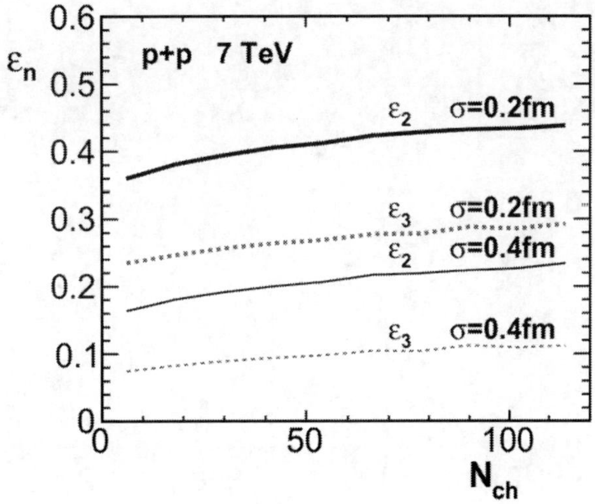

Figure 2. Ellipticity ε_2 (solid lines) and triangularity ε_3 (dotted lines) of the fireball in p+p collisions at $\sqrt{s} = 7$ TeV. The eccentricities are plotted as functions of the mean charged multiplicity at $|\eta| < 2.4$.

from a single source centered at a transverse point (x_0, y_0), namely

$$g(x,y) = \frac{1}{2\pi\sigma^2} \exp\left(-\frac{(x-x_0)^2 + (y-y_0)^2}{2\sigma^2}\right). \tag{2}$$

It should be noted that in small systems the fireball eccentricities depend largely on the value of the smearing width σ. In Fig. 2 we show fireball eccentricities resulting from the wounded constituents in p+p collisions at the LHC. Both the ellipticity and triangularity are large, hence may lead to substantial harmonic flow, in accordance to the collectivity mechanism expected for collisions of small systems [22–25] at sufficiently high multiplicity. The eccentricities do not depend strongly on the multiplicity of produced particles (centrality), which shows that they originate from fluctuations and not the geometry of the collision.

Next, we present our new results for centrality dependence of pseudorapidity density in Au+Au collision at RHIC. The BRAHMS [26] and PHENIX [8] Collaborations data on $dN_{ch}/d\eta$ have been obtained in very different kinematic regions of particle production: from mid-rapidity, $\eta = 0$, up to $\eta = 4.5$. The number of wounded constituents has been simulated through the use of accordingly modified GLISSANDO code [27, 28]. The obtained linear dependence of $dN_{ch}/d\eta/Q_W$ on centrality, shown in Fig. 3, indicates that the wounded constituent scaling in A+A collisions is universal, i.e., independent of the phase-space region in pseudorapidity.

Our conclusions are as follows:

- The production of particles at midrapidity follows the wounded constituent scaling, with the quantity $dN_{ch}/d\eta/Q_W$ displaying approximately a flat behavior with centrality. At the LHC the optimal number of constituents is three.
- $dN_{ch}/d\eta/Q_W$ scales with the number of wounded constituents at different kinematic regions of phase-space.

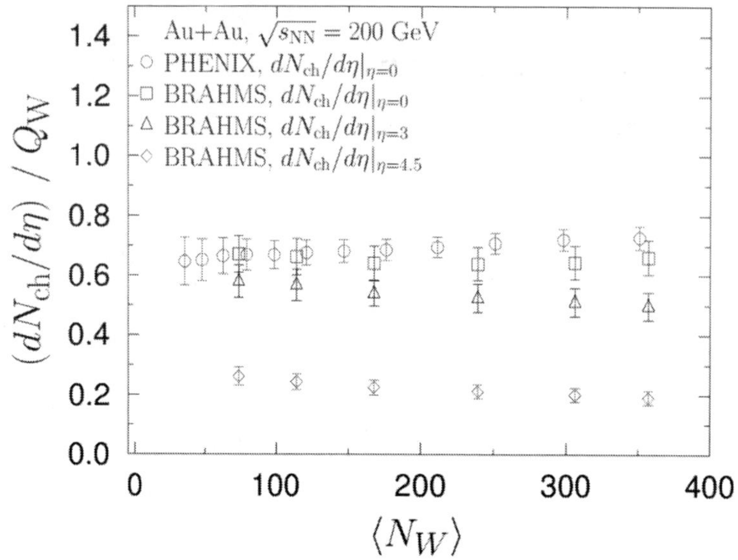

Figure 3. Experimental multiplicity of charged hadrons per unit of pseudorapidity obtained at different phase-space kinematic regions ($0 < \eta < 4.5$) divided by the number of wounded constituents, $dN_{ch}/d\eta/Q_W$, plotted as function of centrality expressed via the number of wounded nucleons. The data are from Refs. [8, 26].

- For p+p collisions, the elliptic and triangular eccentricities are large, hence it is expected that the corresponding flow coefficients will be large. We also expect that the resulting triangular flow should be significantly smaller from the elliptic flow.

Acknowledgements: Research supported by the Polish National Science Centre grants 2015/17/B/ST2/00101, 2015/19/B/ST2/00937, and 2015/18/M/ST2/00125.

References

[1] P. Bożek, W. Broniowski, M. Rybczyński, Phys. Rev. **C94**, 014902 (2016), `1604.07697`
[2] B.B. Back et al. (PHOBOS), Phys. Rev. **C65**, 031901 (2002), `nucl-ex/0105011`
[3] B.B. Back et al. (PHOBOS), Phys. Rev. **C70**, 021902 (2004), `nucl-ex/0405027`
[4] K. Aamodt et al. (ALICE Collaboration), Phys. Rev. Lett. **106**, 032301 (2011), `1012.1657`
[5] A. Białas, M. Błeszyński, W. Czyż, Nucl. Phys. **B111**, 461 (1976)
[6] D. Kharzeev, M. Nardi, Phys. Lett. **B507**, 121 (2001), `nucl-th/0012025`
[7] G.J. Alner et al. (UA5), Z. Phys. **C33**, 1 (1986)
[8] A. Adare et al. (PHENIX), Phys. Rev. **C93**, 024901 (2016), `1509.06727`
[9] J. Adam et al. (ALICE) (2015), `1509.07541`
[10] A. Białas, W. Czyż, W. Furmański, Acta Phys. Polon. **B8**, 585 (1977)
[11] A. Białas, K. Fiałkowski, W. Słomiński, M. Zieliński, Acta Phys. Polon. **B8**, 855 (1977)
[12] A. Białas, W. Czyż, Acta Phys. Polon. **B10**, 831 (1979)
[13] V.V. Anisovich, Yu.M. Shabelski, V.M. Shekhter, Nucl. Phys. **B133**, 477 (1978)

[14] S. Eremin, S. Voloshin, Phys. Rev. **C67**, 064905 (2003), nucl-th/0302071
[15] P. Kumar Netrakanti, B. Mohanty, Phys. Rev. **C70**, 027901 (2004), nucl-ex/0401036
[16] L. Zheng, Z. Yin, Eur. Phys. J. **A52**, 45 (2016), 1603.02515
[17] R.A. Lacey, P. Liu, N. Magdy, M. Csanád, B. Schweid, N.N. Ajitanand, J. Alexander, R. Pak (2016), 1601.06001
[18] J.T. Mitchell, D.V. Perepelitsa, M.J. Tannenbaum, P.W. Stankus (2016), 1603.08836
[19] C. Loizides (2016), 1603.07375
[20] S.S. Adler et al. (PHENIX), Phys. Rev. **C89**, 044905 (2014), 1312.6676
[21] A. Białas, A. Bzdak, Acta Phys. Polon. **B38**, 159 (2007), hep-ph/0612038
[22] P. Bożek, Phys. Rev. **C85**, 014911 (2012), 1112.0915
[23] P. Bożek, W. Broniowski, Phys. Lett. **B718**, 1557 (2013), 1211.0845
[24] P. Bożek, W. Broniowski, Phys. Rev. **C88**, 014903 (2013), 1304.3044
[25] P. Bożek, W. Broniowski, G. Torrieri, Phys. Rev. Lett. **111**, 172303 (2013), 1307.5060
[26] I.G. Bearden et al. (BRAHMS), Phys. Rev. Lett. **88**, 202301 (2002), nucl-ex/0112001
[27] W. Broniowski, M. Rybczyński, P. Bożek, Comput. Phys. Commun. **180**, 69 (2009), 0710.5731
[28] M. Rybczyński, G. Stefanek, W. Broniowski, P. Bożek, Comput. Phys. Commun. **185**, 1759 (2014), 1310.5475

Evolution equations for the double parton distributions

Initial conditions and transverse momentum dependence

Anna M. Staśto[1,a] and Krzysztof Golec-Biernat[2,3,b]

[1] *Department of Physics, The Pennsylvania State University, University Park, PA 16802, U.S.A.*
[2] *H. Niewodniczański Institute of Nuclear Physics, Polish Academy of Sciences, 31-342 Kraków, Poland*
[3] *Faculty of Mathematics and Natural Sciences, University of Rzeszów, 35-959 Rzeszów, Poland*

Abstract. In the first part of this contribution we discuss the problem of the initial conditions for the evolution of double parton distribution functions (PDFs). We show that one can construct a framework based on the expansion in terms of the Dirichlet functions in which both single and double PDFs satisfy momentum sum rules. In the second part, we propose how to include the transverse momentum dependence for the double parton distribution functions using the extension of the Kimber-Martin-Ryskin framework previously applied to the single PDFs.

1 Introduction

Hard processes in hadron collisions are usually described by the collinear factorization. This factorization can be schematically written as follows

$$d\sigma = D_1^f(x_1, Q^2) \otimes \hat{\sigma}^{ff'}(\hat{s}, Q^2) \otimes D_1^{f'}(x_1', Q^2) + O(1/Q^2)\,, \tag{1}$$

where $D_1^f(x, Q^2)$ is the standard non-perturbative collinear PDF for the partons of type f (and simiularly for f') and $\hat{\sigma}^{ff'}(\hat{s}, Q^2)$ is the partonic cross section which can be evaluated perturbatively. The hard scale Q^2 could be a heavy quark mass, transverse energy of a jet or invariant mass of the Drell-Yan pair. The above factorization formula is appropriate for the description of inclusive cross sections for single hard processes. In the new regime of high energies, such as those probed at Tevatron and more recently at the Large Hadron Collider, multiple scatterings occur more frequently. This is related to the fact that at these high energies the parton luminosities increase rapidly with the decreasing fractions of the longitudinal momenta. For the description of the processes in which two pairs of partons scatter simultaneously in one hadronic encounter, the following formula is usually utilized

$$d\sigma = D_2^{f_1f_2}(x_1, x_2; Q_1^2, Q_2^2; q_T) \otimes \hat{\sigma}^{f_1f_1'}(\hat{s}_1, Q_1^2) \otimes \hat{\sigma}^{f_2f_2'}(\hat{s}_2, Q_2^2) \otimes D_2^{f_1'f_2'}(x_1', x_2'; Q_1^2, Q_2^2; q_T)\,, \tag{2}$$

where $D_2^{f_1f_2}(x_1, x_2; Q_1^2, Q_2^2; q_T)$ is the double collinear PDF. The formula in Eq. (2) is meant to represent the scattering of two pairs of partons, with the hard scattering described by the two hard parton

[a]e-mail: ams52@psu.edu
[b]e-mail: golec@ifj.edu.pl

cross sections at hard scales Q_1 and Q_2. The additional scale q_T is the relative transverse momentum of the two partons. The above formula is widely used in the phenomenology of the double parton scattering, though strictly speaking it has not yet been rigorously proven. Important steps towards the proof for the double Drell-Yan process have been provided in [1].

The single PDFs are non-perturbative quantities, but their evolution with the hard scale can be described by the DGLAP evolution equations, with the splitting functions which can be calculated perturbatively. The double PDFs also obey the evolution equations which have in the leading logarithmic approximation the following DGLAP-like form, see e.g. [2] for more details,

$$\partial_t D_{f_1 f_2}(x_1, x_2; t) = \sum_{f'} \int_0^{1-x_2} du\, \mathcal{K}_{f_1 f'}(x_1, u, t)\, D_{f' f_2}(u, x_2; t)$$

$$+ \sum_{f'} \int_0^{1-x_1} du\, \mathcal{K}_{f_2 f'}(x_2, u, t)\, D_{f_1 f'}(x_1, u; t) + \sum_{f'} \mathcal{K}^R_{f' \to f_1 f_2}(x_1, x_2, t) D_{f'}(x_1 + x_2; t)\,. \quad (3)$$

Here, we have used the notation in which $D_{f_1 f_2}(x_1, x_2; t)$ denotes the double PDF which depends on two equal hard scales $Q_1 = Q_2 = Q$ and the evolution variable is defined as $t = \ln Q^2/Q_0^2$. Also, the above evolution equations are defined for the case of zero q_T momentum. The kernels $\mathcal{K}$ are the Altarelli-Parisi splitting functions, with $\mathcal{K}^R$ being the real part of the splitting function only. The physical interpretation of these evolution equations is as follows: the first two terms describe the situation where one of the two partons undergoes the collinear splitting, and the third term describes the situation when one parton collinearly splits into two. The first two terms are homogeneous contribution, whereas the third, splitting term, is the non-homogeneous contribution since it depends on the single parton distribution function. The above equations were first discussed in the context of the jet structure, and later on derived for the the parton distribution functions.

2 Momentum sum rules

The evolution equations (3) need to be solved simultaneously with the DGLAP equations for the single parton distirbution functions. In addition, there are sum rules which must be obeyed by the solutions to the evolution equations for both the double and single PDFs. There are known sum rules for the single parton distribution functions like the momentum sum rule and the quark number sum rule. The corresponding momentum sum rule for the double PDFs reads

$$\sum_{f_1} \int_0^{1-x_2} dx_1\, x_1 \frac{D_{f_1 f_2}(x_1, x_2; t)}{D_{f_2}(x_2; t)} = 1 - x_2\,. \quad (4)$$

The integrand in this sum rule is an expression of the conditional probability to find the parton f_1 with the momentum fraction x_1 while keeping the second parton f_2 with momentum x_2 fixed. The total momentum fraction carried by the partons f_1 under this condition is equal to $1 - x_2$. In the similar manner, one can construct the quark number sum rule for the double PDFs [2].

The important feature of the evolution equations is that they conserve the sum rules. If the PDFs satisfy the sum rules at certain scale t_0, then they will also satisfy these sum rules at higher scale t_1 after the evolution. The presence of the splitting term in the evolution equation for the double PDFs is essential to guarantee this property.

2.1 Initial conditions

Thus for the practical purposes, in phenomenological applications one needs to impose suitable initial conditions both for the single and double PDFs simultaneously. At the same time, the momentum sum rules should be also satisfied. First attempts to resolve this problem were given in [2]. In reference [3] another solution was proposed which, however, did not have the symmetry with respect to the exchange of the partons. Another proposal was given in [4] where the construction was based on the Dirichlet distributions. The obtained solution was positive definite and symmetric with respect to the exchange of the two partons.

The framework of reference [4] rests on the observation that the natural functions which obey the sum rules for the single PDFs and double PDFs are the Dirichlet distributions. Let us consider the following form for the single and double PDFs,

$$D(x) = N_1 x^{-\alpha}(1-x)^{\beta}\,, \qquad D(x_1,x_2) = N_2 x_1^{-\tilde{\alpha}} x_2^{-\tilde{\alpha}}(1-x_1-x_2)^{\tilde{\beta}}\,. \tag{5}$$

One can rewrite them using the Mellin transform,

$$\tilde{D}(n) = \int_0^1 dx\, x^{n-1}\, D(x)\,, \qquad \tilde{D}(n_1,n_2) = \int_0^1 dx_1\, x_1^{n_1-1} \int_0^{1-x_1} dx_2\, x_2^{n_2-1}\, D(x_1,x_2)\,, \tag{6}$$

to obtain

$$\tilde{D}(n) = N_1 \frac{\Gamma(n-\alpha)\Gamma(\beta+1)}{\Gamma(n-\alpha+\beta+1)}, \qquad \tilde{D}(n_1,n_2) = N_2 \frac{\Gamma(n_1-\tilde{\alpha})\Gamma(n_2-\tilde{\alpha})\Gamma(\tilde{\beta}+1)}{\Gamma(n_1+n_2-2\tilde{\alpha}+\tilde{\beta}+1)}\,. \tag{7}$$

The momentum sum rule in the Mellin space can be rewritten as:

$$\tilde{D}(n_1,2) = \tilde{D}(n_1) - \tilde{D}(n_1+1)\,, \tag{8}$$

and similarily for the second parton. It can be easily seen, using some basic properties of the Beta function that the form for the single and double PDFs satisfies the above momentum sum rule. In particular the powers of the two distributions need to satisfy

$$\alpha = \tilde{\alpha}, \qquad \tilde{\beta} = \beta + \alpha - 1\,, \tag{9}$$

with the normalizations also being uniquely defined. We see therefore that the small x powers for the single and double PDFs are the same. On the other hand, the large x power of the correlating factor of the double PDF is related to the sum of the small and large x powers of the single PDF.

In practice, the parametrizations for the PDFs need to be more complicated than the single Dirichlet distribution, nevertheless the framework can be generalized to the sum of arbitrary number of such terms. In such a case, for each term in the series there would be a condition similar to Eq. (9). Thus the presented framework allows for the unique construction of the functional form of the initial conditions for the double parton distributions, using the knowledge of the functional form of the single parton distribution functions if the latter ones can be expressed in terms of the series of the Beta distributions. The powers of all the terms in the expansion for the double PDFs are uniquely fixed by the powers of the single PDFs. In the case of the single channel, the normalization for the double PDFs is also fixed in this framework.

As an example of the application we have used the MSTW2008 LO parametrization for the gluon distribution to construct the double gluon distribution D_{gg}. To illustrate the correlations in the double

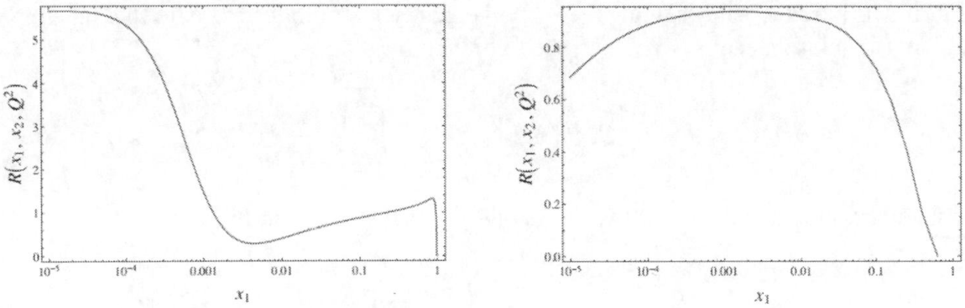

Figure 1. The ratio of the double PDF to the product of single PDFs, constructed in the framework using Dirichlet distributions from MSTW2008 LO parametrization. Left: the ratio at the initial scale $Q^2 = 1$ GeV^2 for $x_2 = 0.01$. Right: the ratio evolved to $Q^2 = 25$ GeV^2 and $x_2 = 0.03$.

distributions it is useful to present the results in the form of the ratio of the double PDF and the product of the single PDFs

$$R_{gg}(x_1, x_2; t) = \frac{D_{gg}(x_1, x_2; t)}{D_g(x_1; t) D_g(x_2; t)} . \tag{10}$$

In the left plot in figure 1 the ratio is showed for the initial scale $Q^2 = 1$ GeV^2. We see that for this very low scale, the ratio is significantly different than unity for wide range of x, which implies that the correlations are very large. On the right hand plot, we show the ratio after the single and double PDFs have been evolved to a higher scale using the evolution equations. We have checked explicitly that the evolution does preserve the momentum sum rule which has been imposed onto the initial conditions within the presented construction. The evolved ratio indicates that most of the correlations are washed out by the LO evolution, except at the highest value of the longitudinal momentum

3 Unintegrated double PDFs

An important aspect for the description of the processes in hadronic collisions is the transverse momentum dependence of the parton distribution functions. The collinear description has certain limitations in that it does not correctly describe the kinematics of the reaction, particularly when one considers more exclusive processes. In that context, the transverse momentum dependent (or unintegrated parton distributions) gained a lot of attention due to the fact that they encode more correct kinematics of the process. The question thus can be posed as to how to construct the unintegrated double parton distribution functions. In [5] a practical approach has been suggested which is based on the formulation originally proposed in [6], called the KMR (Kimber-Martin-Ryskin) approach for the single unintegrated PDFs. The idea behind this approach, which is accurate to the leading power, is that the unintegrated PDFs can be obtained using the integrated PDFs and the Sudakov form factors. The benefits are that one can utilize the integrated PDFs obtained from the standard DGLAP equations to introduce the transverse momentum dependence.

The construction of the KMR framework of the unintegrated PDFs starts from the DGLAP evolution equation written in the following form

$$\frac{\partial D_a(x,\mu)}{\partial \ln \mu^2} = \sum_{a'} \int_x^{1-\Delta} \frac{dz}{z} P_{aa'}(z,\mu) D_{a'}\Big(\frac{x}{z},\mu\Big) - D_a(x,\mu) \sum_{a'} \int_0^{1-\Delta} dz z P_{a'a}(z,\mu) , \tag{11}$$

where the first term comes from the real emissions and the second term is the virtual contribution. The cutoff Δ is introduced to regulate the soft divergencies in both terms. After the integration of the virtual part the solution can be written in the following form

$$D_a(x,Q) = T_a(Q,Q_0,)D_a(x,Q_0) + \int_{Q_0^2}^{Q^2} \frac{dk_\perp^2}{k_\perp^2} f_a(x,k_\perp,Q)\,, \tag{12}$$

where the unintegrated parton distribution function (UPDF) are given by

$$f_a(x,k_\perp,Q) \equiv T_a(Q,k_\perp) \sum_{a'} \int_x^{1-\Delta} \frac{dz}{z} P_{aa'}(z,k_\perp) D_{a'}\Big(\frac{x}{z},k_\perp\Big)\,, \tag{13}$$

with $T_a(Q,k_\perp)$ being the Sudakov form factor. The transverse momentum dependent density is thus built from the standard integrated density in the last step of the evolution through the inclusion of the Sudakov form factor. The cutoff parameter needs to be specified in order to fully fix the unintegrated PDF. In [6] $\Delta = k_\perp/Q$ was chosen in the spirit of the DGLAP strongly ordered emissions. In the following works, for example [7], angular ordering was considered which results in $\Delta = k_\perp/(k_\perp + Q)$ and allows for the smooth transition into the region $k_\perp > Q$.

In order to extend the KMR construction to the double PDFs one can use the parton-to-parton evolution functions, which evolve the parton densities from initial scale to final scale, i.e.

$$\tilde{D}_a(n,\mu) = \sum_b \tilde{E}_{ab}(n,\mu,\mu_0)\, \tilde{D}_b(n,\mu_0)\,, \tag{14}$$

where Mellin space representation was used, with variable n, instead of the direct x space (see Eq. (6)). These parton-to-parton evolution functions $\tilde{E}$ obey the DGLAP equations, and therefore the virtual part can also be integrated out.

Then, one can represent the solution to the evolution equations for the double PDFs (3) by means of these evolution functions [8],

$$\begin{aligned}\tilde{D}_{a_1a_2}(n_1,n_2,Q_1,Q_2) = \sum_{a',a''} \Big\{ &\tilde{E}_{a_1a'}(n_1,Q_1,Q_0)\, \tilde{E}_{a_2a''}(n_2,Q_2,Q_0)\, \tilde{D}_{a'a''}(n_1,n_2,Q_0,Q_0) \\ &+ \int_{Q_0^2}^{Q^2_{\rm min}} \frac{dQ_s^2}{Q_s^2} \tilde{E}_{a_1a'}(n_1,Q_1,Q_s)\, \tilde{E}_{a_2a''}(n_2,Q_2,Q_s)\, \tilde{D}^{(sp)}_{a'a''}(n_1,n_2,Q_s) \Big\}\,, \end{aligned} \tag{15}$$

where $Q^2_{\rm min} = \min\{Q_1^2, Q_2^2\}$ and $\tilde{D}^{(sp)}_{a'a''}$ is the non-homogeneous contribution originating from the single parton density. Using the solutions for the parton-to-parton evolution functions one can perform the same substitution as was done for the case of the single PDFs and find the corresponding unintegrated double parton distribution function. It turns out however, that in that case there are three distinct regions of momenta, where different forms of the unintegrated double PDFs are obtained:

For the case when $k_{1\perp} \leq Q_0$ and $k_{2\perp} > Q_0$, we have

$$\tilde{f}^{(h)}_{a_1a_2}(n_1,n_2,k_{2\perp},Q_1,Q_2) = T_{a_1}(Q_1,Q_0)\, T_{a_2}(Q_2,k_{2\perp}) \sum_b \tilde{P}_{a_2b}(n_2,k_{2\perp})\, \tilde{D}^{(h)}_{a_1b}(n_1,n_2,Q_0,k_{2\perp})\,. \tag{16}$$

The dependence of the transverse momentum $k_{1\perp}$ is integrated over up to Q_0 in such a case and $k_{1\perp}$ is not present among the arguments of the defined function. The effect of such an integration is hidden in the integrated double PDFs on the r.h.s. taken at the scale Q_0 for the first parton.

Similarly, for $k_{1\perp} > Q_0$ and $k_{2\perp} \leq Q_0$, we have

$$\tilde{f}^{(h)}_{a_1a_2}(n_1, n_2, k_{1\perp}, Q_1, Q_2) = T_{a_1}(Q_1, k_{1\perp}) T_{a_2}(Q_2, Q_0) \sum_b \tilde{P}_{a_1b}(n_1, k_{1\perp}) \tilde{D}^{(h)}_{ba_2}(n_1, n_2, k_{1\perp}, Q_0) . \quad (17)$$

Now the momentum $k_{2\perp}$ is integrated up to the scale Q_0 and only $k_{1\perp}$ dependence is present. Finally, for $k_{1\perp}, k_{2\perp} > Q_0$ the unintegrated double PDF has the form

$$\begin{aligned} \tilde{f}^{(h)}_{a_1a_2}(n_1, n_2, k_{1\perp}, k_{2\perp}, Q_1, Q_2) &= T_{a_1}(Q_1, k_{1\perp}) T_{a_2}(Q_2, k_{2\perp}) \\ &\times \sum_{b,c} \tilde{P}_{a_1b}(n_1, k_{1\perp}) \tilde{P}_{a_2c}(n_2, k_{2\perp}) \tilde{D}^{(h)}_{bc}(n_1, n_2, k_{1\perp}, k_{2\perp}) . \quad (18) \end{aligned}$$

The treatment of the non-homogeneous term is much more complicated because there are two potential contributions to the transverse momentum dependence. One can have the transverse momentum either from the evolution of one or both partons after the splitting or from the splitting vertex itself. In the first case, it is sufficient to consider the inhomogeneous term from formula (15) and use the solution for the parton evolution functions as we have done before. For the second contribution, one needs to compute the exact form of the vertex including the transverse momentum dependence, and therefore treat the splitting beyond the collinear approximation.

4 Summary

In this presentation we have discussed two issues related with the double parton distribution functions. In the first part, we have proposed the framework for the construction of the initial conditions for the double PDFs using the information from the single PDFs and the constraints from the sum rules. It turns out that (in the single channel case) there is a unique solutions, which allows for the construction of such initial conditions. In the second part, we have proposed a generalization of the KMR framework for the unintegrated PDFs to the double PDFs. The treatment of the homogeneous term is relatively straightforward and one arrives at three different perturbative contributions for the unintegrated double PDFs depending on the relevant hierarchy of hard scales. The treatment of the non-homogeneous term is more complicated as it involves two sources of the transverse momenta either from the evolution or from the splitting vertex itself, and therefore in principle it goes beyond the collinear approximation. The presented framework for unintegrated DPDFs relies on the use of the integrated double PDFs and can be implemented numerically.

This work was supported by the National Science Center, Poland, Grant No. 2015/17/B/ST2/01838, by the Department of Energy Grant No. DE-SC-0002145 and by the Center for Innovation and Transfer of Natural Sciences and Engineering Knowledge in Rzeszów.

References

[1] M. Diehl, J. R. Gaunt, D. Ostermeier, P. Ploessl and A. Schaefer, JHEP **1601**, 076 (2016).
[2] J. R. Gaunt and W. J. Stirling, JHEP **1106**, 048 (2011), [1103.1888].
[3] K. Golec-Biernat and E. Lewandowska, Phys. Rev. D **90**, no. 1, 014032 (2014).
[4] K. Golec-Biernat, E. Lewandowska, M. Serino, Z. Snyder and A. M. Stasto, Phys. Lett. B **750**, 559 (2015).
[5] K. Golec-Biernat and A. M. Stasto, arXiv:1611.02033 [hep-ph].
[6] M. A. Kimber, A. D. Martin and M. G. Ryskin, Eur. Phys. J. C **12**, 655 (2000).
[7] M. A. Kimber, J. Kwiecinski, A. D. Martin and A. M. Stasto, Phys. Rev. D **62**, 094006 (2000).
[8] F. A. Ceccopieri, Phys. Lett. B **697**, 482 (2011).

EPJ Web of Conferences **141**, 06002 (2017) DOI: 10.1051/epjconf/201714106002
ISMD 2016

Hadron tomography studies by generalized parton distributions and distribution amplitudes

S. Kumano[1,2]

[1] *KEK Theory Center, Institute of Particle and Nuclear Studies, KEK and Department of Particle and Nuclear Physics, Graduate University for Advanced Studies, 1-1, Ooho, Tsukuba, Ibaraki, 305-0801, Japan*
[2] *J-PARC Branch, KEK Theory Center, Institute of Particle and Nuclear Studies, KEK and Theory Group, Particle and Nuclear Physics Division, J-PARC Center, 203-1, Shirakata, Tokai, Ibaraki, 319-1106, Japan*

Abstract. We discuss hadron-tomography studies for the nucleon and exotic hadrons by high-energy hadron reactions. First, the constituent-counting rule is explained for determining internal quark configurations of exotic-hadron candidates by scaling properties of high-energy exclusive cross sections. Next, possibilities are discussed for investigating the generalized parton distributions (GPDs) of the nucleon and exotic hadrons at J-PARC. In particular, we study hadronic $2 \to 3$ process $p + p \to N + \pi + B$, exclusive Drell-Yan process, and exotic-hadron GPDs. For determining three-dimensional structure of unstable exotic hadrons, we consider s-t crossed quantities of the GPDs called generalized distribution amplitudes (GDAs), which can be investigated at KEKB. We explain possible studies of the GDAs by two-photon processes.

1 Introduction

Internal structure of the nucleon has been investigated for a long time by lepton and hadron scattering processes at various energies, and major properties are now understood. However, the origin of the nucleon spin has not been clarified yet although it is one of fundamental physics quantities. For understanding the whole picture of the nucleon spin in terms of quarks and gluons, it is essential to understand partonic orbital-angular-momentum contributions. Therefore, it became necessary to establish three-dimensional (3D) picture of the nucleon, and this field is called hadron tomography. It is now under active investigations through the 3D structure functions of generalized parton distributions (GPDs) and transverse-momentum-dependent parton distributions (TMDs). There are also generalized distribution amplitudes (GDAs) which are s-t crossed quantities to the GPDs.

These 3D structure functions are key quantities in solving the nucleon-spin puzzle. However, it is interesting to use them for exotic-hadron studies because it is not easy to find internal quark-gluon configurations, namely undoubted evidence of exotic nature, solely by low-energy observables such as masses and decay widths. In general, high-energy reactions should be useful for finding internal structure of exotic hadrons by using, for example, fragmentation functions [1], constituent-counting rule [2, 3], and the 3D structure functions [4] because the appropriate degrees of freedom are quark and gluons.

DOI: 10.1051/epjconf/201714106002

In this article, possible exotic-hadron studies are proposed by using high-energy reactions, together with the usual nucleon GPDs, at facilities such as J-PARC and KEKB. First, we explain the constituent-counting rule, predicted by perturbative QCD, as a useful guidance to probe the internal quark-gluon configurations of exotic-hadron candidates [2, 3] in Sec. 2. Then, the GPDs for the nucleon and exotic-hadron candidates are discussed in Sec. 3 including possible experimental studies at J-PARC [4–6]. Since there is no stable exotic hadron, the direct GPD studies do not seem to be possible for exotic hadrons, except for transition GPDs such as the ones from the nucleon to an exotic hadron. In this respect, the GDAs could be appropriate 3D structure functions for studying exotic hadrons [4] because an exotic-hadron pair could be produced in the final state even for an unstable hadron. We explain such GDAs in Sec. 4, and it should be a possible project as two-photon processes at KEKB [4, 7]. We summarize our studies in Sec. 5.

2 Constituent-counting rule for exotic hadrons

The constituent-counting rule was predicted in hard exclusive reactions by perturbative QCD. As shown typically in Fig. 1, the two-body hadron reaction $a + b \to c + d$ with large p_T should occur due to hard gluon exchanges between quarks in the hadrons. By considering the hard gluon and quark propagators together with other kinematical factors, we obtain the scaling relation for the cross section as [2]

$$\frac{d\sigma_{ab\to cd}}{dt} = \frac{1}{s^{n-2}} f_{ab\to cd}(t/s), \qquad (1)$$

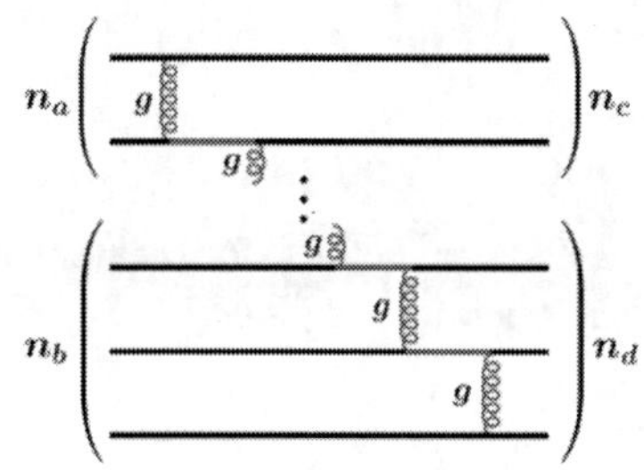

Figure 1. A hard-gluon exchange process [2].

where s and t are Mandelstam variables, $f(t/s)$ is a function depends on the scattering angle, and n is the total number of constituents given by $n = n_a + n_b + n_c + n_d$. The high-energy cross section is proportional to $1/s^{n-2}$ with the number of constituents (n), and it is called the constituent-counting rule. It has been confirmed experimentally by various hard processes in lepton-hadron and hadron-hadron exclusive reactions, even for compound systems of hadrons such as the deuteron, ^{3}H, and ^{3}He. Since it is not easy to judge whether or not discovered hadrons are actually exotic only by low-energy observables, we think that it is appropriate to use high-energy reaction processes to find internal constituents of exotic hadrons. Especially, the appropriate degrees of freedom are quarks and gluons at high energies, so that such approaches could be promising as future exotic-hadron studies.

For example, we investigate an exotic-hadron candidate $\Lambda(1405)$, which is considered to be $\bar{K}N$ molecule or pentaquark state [8], by the constituent counting rule. First, we look at the ground Λ

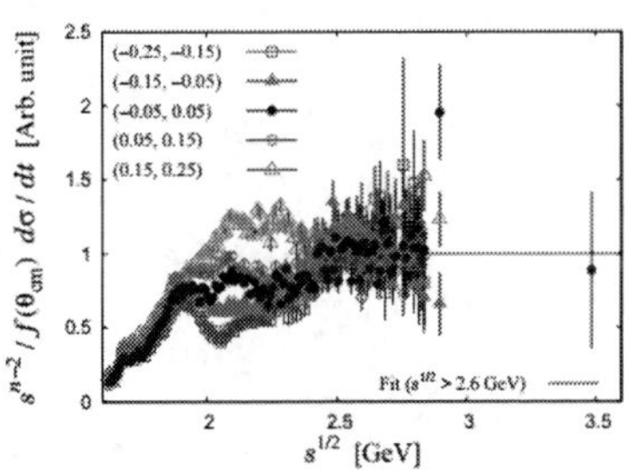

Figure 2. Scaling in $\gamma + p \to K^+ + \Lambda$ [3].

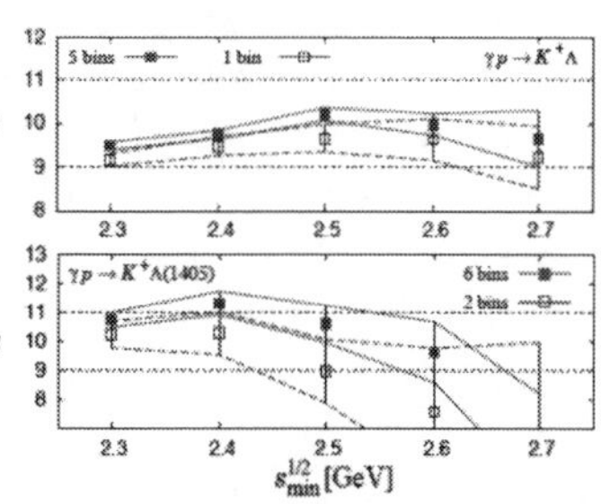

Figure 3. Energy dependence of factor n [3].

production $\gamma + p \rightarrow K^+ + \Lambda$, and the cross section scales as shown in Fig. 2. The scaling factor n determined from the data depends the minimum energy $\sqrt{s_{min}}$ for the scaling region as shown in Fig. 3. The factor $n = 9$ $(= 1 + 3 + 2 + 3)$ means that the ground Λ is consistent with the qqq picture.

There are also data for the $\Lambda(1405)$ production $\gamma + p \rightarrow K^+ + \Lambda(1405)$. We analyzed its data and obtained the scaling factor n as shown in Fig. 3. The figure implies an interesting tendency that $\Lambda(1405)$ looks like a five-quark state ($\bar{K}N$ molecule or pentaquark state) at relatively-low energies but it becomes three-quark one at high energies. However, the errors of n are too large to draw a solid conclusion at this stage. Future experimental studies including by a 12-GeV JLab experiment are needed. Similar studies should be possible by hadron accelerator facilities such as J-PARC. As an example, we estimated the cross sections for $\pi + p \rightarrow K^0 + \Lambda(1405)$ by taking $n_{\Lambda(1405)} = 3$ or 5. There are large differences between the curves, so that the internal structure of $\Lambda(1405)$ can be investigated through the counting rule at J-PARC.

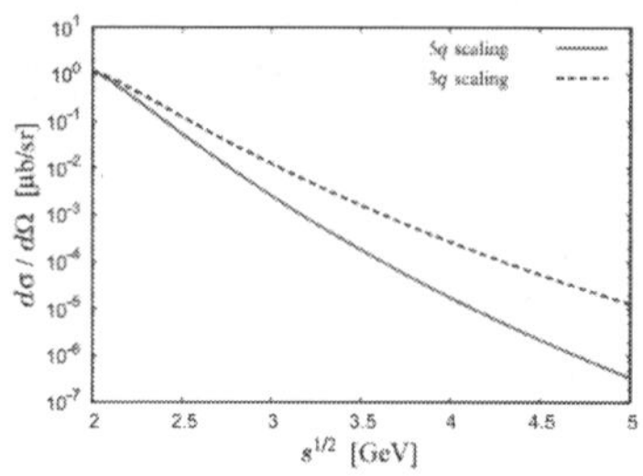

Figure 4. $\pi p \rightarrow K^0\Lambda(1405)$ cross sections [2].

3 Generalized parton distributions and J-PARC project

The GPDs are defined for the nucleon by off-forward nucleon matrix elements of quark (and gluon) bilocal operators as [4–6]

$$\int \frac{dy^-}{4\pi} e^{ixP^+y^-} \langle p' \,|\bar{q}(-y/2)\gamma^+ q(y/2)|\, p\rangle \Big|_{y^+=\vec{y}_\perp=0} = \frac{1}{2P^+}\,\bar{u}(p')\left[H^q(x,\xi,t)\gamma^+ + E^q(x,\xi,t)\frac{i\sigma^{+\alpha}\Delta_\alpha}{2m_N}\right]u(p),$$

$$\int \frac{dy^-}{4\pi} e^{ixP^+y^-} \langle p' \,|\bar{q}(-y/2)\gamma^+\gamma_5 q(y/2)|\, p\rangle \Big|_{y^+=\vec{y}_\perp=0} = \frac{1}{2P^+}\,\bar{u}(p')\left[\tilde{H}^q(x,\xi,t)\gamma^+\gamma_5 + \tilde{E}^q(x,\xi,t)\frac{\gamma_5\Delta^+}{2m_N}\right]u(p), \tag{2}$$

where $\sigma^{\alpha\beta}$ is $\sigma^{\alpha\beta} = (i/2)[\gamma^\alpha, \gamma^\beta]$, gauge-link operators for the gauge invariance are not explicitly written, $H^q(x,\xi,t)$ and $E^q(x,\xi,t)$ are the unpolarized quark GPDs, and $\tilde{H}^q(x,\xi,t)$ and $\tilde{E}^q(x,\xi,t)$ are the polarized ones. By the momentum assignments in Fig. 5, the variables x and ξ are defined by

$$x = \frac{((k-\Delta/2)+(k+\Delta/2))^+}{(p+p')^+} = \frac{k^+}{P^+},$$

$$\xi = \frac{(p-p')^+}{(p+p')^+} = \frac{-\Delta^+}{2P^+} = \frac{((k-\Delta/2)-(k+\Delta/2))^+}{(p+p')^+}, \tag{3}$$

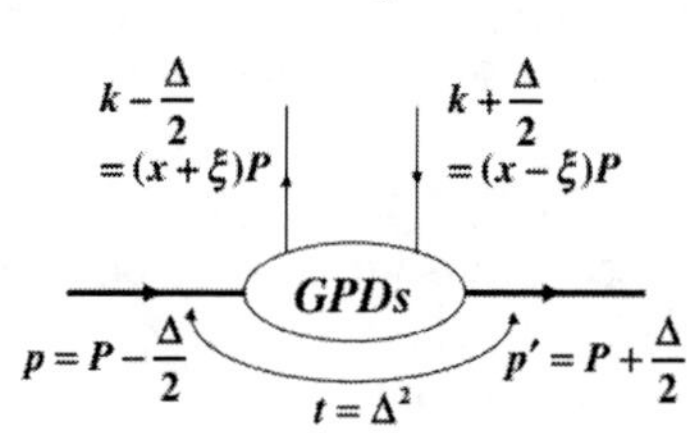

Figure 5. Kinematics of GPDs [6].

so that x and 2ξ are the light-cone momentum fractions of the average momentum and momentum transfer for the relevant quarks to the average momentum of the parent nucleon. The range of x is from -1 to 1, whereas ξ is from 0 to 1. The GPDs have necessary information to find 3D structure of the nucleon, and they are key quantities in the studies of hadron tomography. There are three important features in the GPDs. (I) In the forward limit ($\Delta \rightarrow 0$, $\xi \rightarrow 0$ and $t \rightarrow 0$), they become unpolarized and longitudinally-polarized parton distribution functions (PDFs) for the nucleon: $H^q(x,0,0) = q(x)$ and $\tilde{H}^q(x,0,0) = \Delta q(x)$. (II) The first moments of the unpolarized GPDs are Dirac and Pauli form factors of the nucleon, and those of the polarized GPDs are axial and pseudoscalar form factors:

$$\int_{-1}^{1} dx\, H^q(x,\xi,t) = F_1^q(t),\ \int_{-1}^{1} dx\, E^q(x,\xi,t) = F_2^q(t),\ \int_{-1}^{1} dx\, \tilde{H}^q(x,\xi,t) = g_A^q(t),\ \int_{-1}^{1} dx\, \tilde{E}^q(x,\xi,t) = g_P^q(t). \tag{4}$$

(III) The second moment of a GPD combination becomes the quark total angular momentum: $J^q = \int_{-1}^{1} dx\, x\, [H^q(x, \xi, t = 0) + E^q(x, \xi, t = 0)]/2$, which makes it possible to find the the quark orbital-angular-momentum contribution to the nucleon spin by $J^q = \Delta q/2 + L^q$. We now explain how the GPDs can be investigated at hadron facilities such as J-PARC in addition to the usual deeply virtual Compton scattering (DVCS) in the following subsections.

3.1 Generalized parton distributions in $p + p \to N + \pi + B$

The GPDs have been experimentally investigated mainly in the DVCS. However, it is possible to study them in hadron reactions. Hadron-reaction rates are generally larger than the electromagnetic Compton scattering, so that extended kinematical regions could be probed. Furthermore, it is a unique opportunity that the interesting Efremov-Radyushkin-Brodsky-Lepage (ERBL) region can be studied in the reaction $N + N \to N + \pi + B$ of Fig. 6 [5]. There are three kinematical regions in the GPDs as shown in Fig. 7. The regions (*a*) and (*c*) are the DGLAP (Dokshitzer-Gribov-Lipatov-Altarelli-Parisi) regions, and their GPDs corresponds to the antiquark and quark distributions, respectively. The (*b*) is the ERBL region, and the GPDs of this region indicate quark-antiquark distributions.

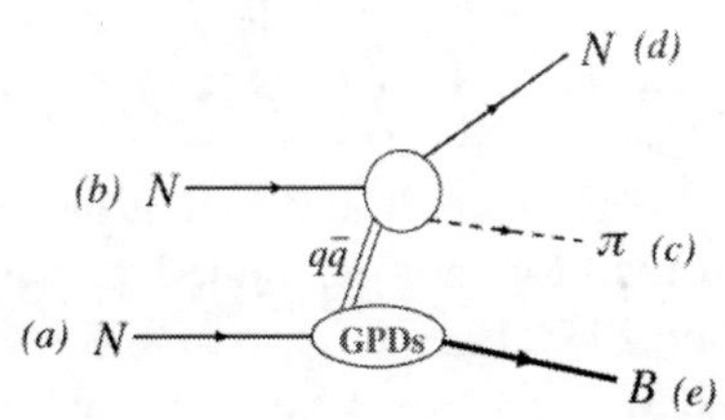

Figure 6. $NN \to N\pi B$ process for GPDs.

We define the Mandelstam variables as $s' = (p_c + p_d)^2$, $t' = (p_b - p_d)^2$, and $u' = (p_b - p_c)^2$. For investigating the GPDs, we consider the hard region $s', |t'|, |u'| \gg m_N^2$. In the reaction $N + N \to N + \pi + B$ of Fig. 6, it is possible to probe the ERBL region. We estimated its cross sections for J-PARC kinematics of 30 GeV and 50 GeV proton-beam energies in Fig. 8 by taking $t = (p_a - p_e)^2 = -0.3$ GeV^2 and $t' = -5.0$ GeV^2 as an example [5]. Here, the neutron or Δ^0 are assumed for the final-state baryon B, and the pion- and ρ-pole terms are used for the ERBL GPDs. These cross-section estimates are intended for proposing a future J-PARC experiment. The GPDs, H, E, $\tilde{H}$, and $\tilde{E}$, for the nucleon and the $N \to \Delta$ transition are involved in the cross section, and they are theoretically estimated by the meson-pole contributions. The interactions for $\pi N \to \pi N$ and $\rho N \to \pi N$, which correspond to the upper part of Fig. 6, are determined from the measurements of the BNL-E838 experiment [5]. The hadron facility measurements at J-PARC and possibly at GSI-FAIR should provide new information on the GPDs in the unique ERBL region. In order to determine the GPDs for the studies of the nucleon spin and 3D structure, all the kinematical regions should be investigated, so that the J-PARC measurement is valuable and complementary to the usual DVCS experiments.

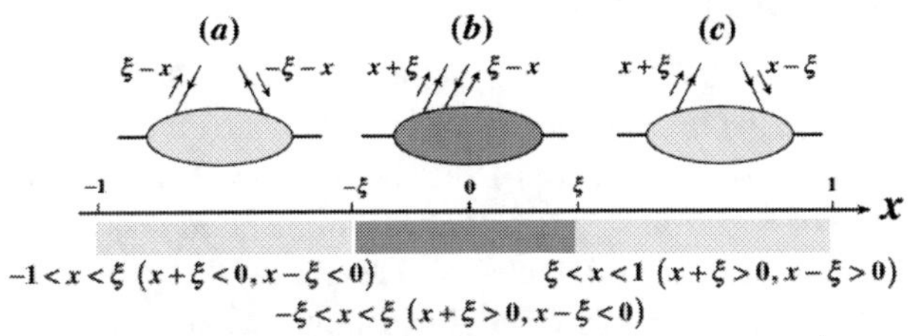

Figure 7. Three kinematical regions of GPDs: (*a*) emission and reabsorption of an antiquark, (*b*) emission and a quark and an antiquark, (*c*) emission and reabsorption of a quark.

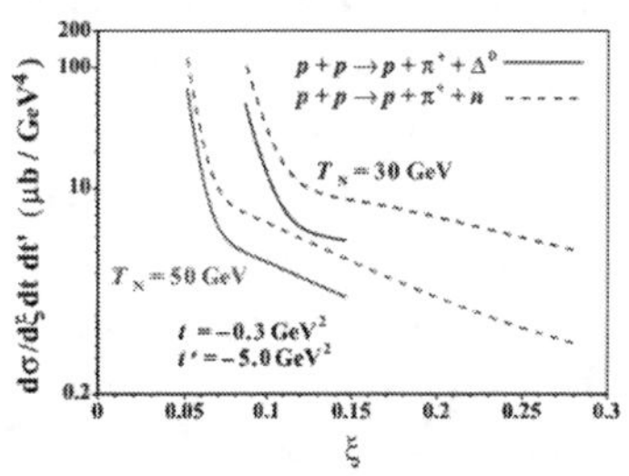

Figure 8. Cross sections as the function of ξ for J-PARC kinematics [5].

3.2 Generalized parton distributions in exclusive Drell-Yan

There is another possibility to investigate the GPDs at J-PARC by using an exclusive Drell-Yan process $\pi^- + p \to \mu^+\mu^- + n$ [6] because high-momentum unseparated-hadron (essentially pion) beamline will become available soon. A typical GPD process for the exclusive dimuon production is shown in Fig. 9, where the process is described by the GPDs combined with the pion distribution amplitude (DA). However, the distribution amplitude should be constrained by other measurements such as on the $\gamma \to \pi$ transition form factor of the Belle and BaBar experiments, although there are still some ambiguities whether it is the asymptotic form, the Chernyak-Zhitnitsky one, or something in between.

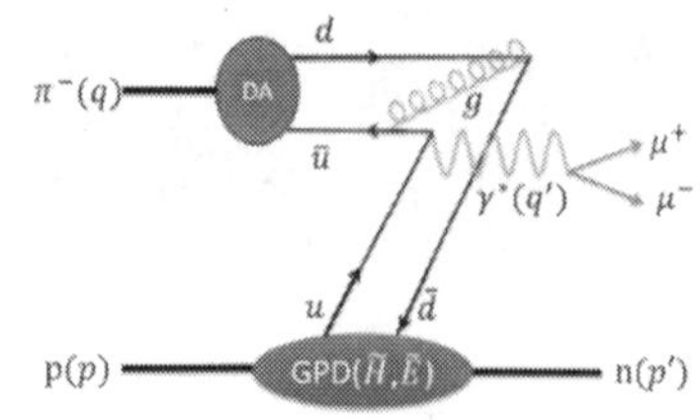

Figure 9. GPD contribution to exclusive Drell-Yan [6].

The GPDs $\tilde{H}$ and $\tilde{E}$ contribute to the cross section, and a typical GPD parametrization GK2013 is used for showing numerical results in Fig. 10, where three possible pion momenta, $p_\pi = 10$, 15, and 20 GeV, are taken and t_0 is the limiting value of the Mandelstam variable t at the zero scattering angle. This estimate is intended for a J-PARC proposal, and actual GPDs should be determined from future measurements at J-PARC together with other data including DVCS ones. Next, Monte Carlo simulations of missing-mass (M_X) spectra of $\mu^+\mu^-$ events are shown in Fig. 11. There are contributions from various sources: exclusive Drell-Yan (red), inclusive Drell-Yan (blue), J/ψ (green), and random background (purple). It is obvious that the exclusive Drell-Yan dominates the events at $M_X \simeq 1$ GeV, so that it is a feasible experiment at J-PARC. At JLab, there is a proposed experiment to measure the opposite process $\gamma^* N \to \pi N$; however, it is a timelike photon process at J-PARC and the probed x region is smaller than the JLab case [6], so that the J-PARC experiment is a valuable complementary contribution to the GPD studies.

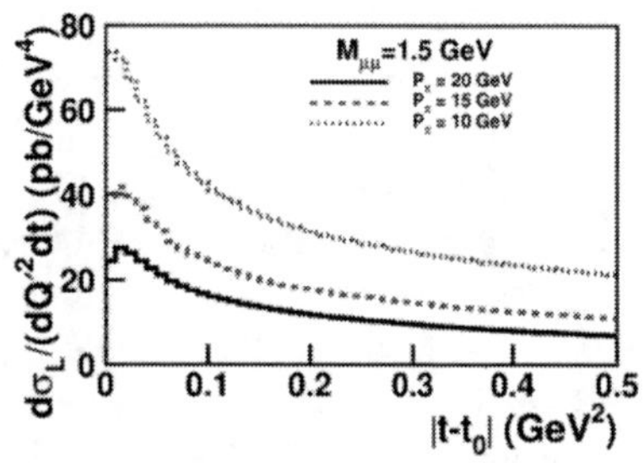

Figure 10. t dependence of $\pi^- + p \to \mu^+\mu^- + n$ cross sections for J-PARC kinematics [6].

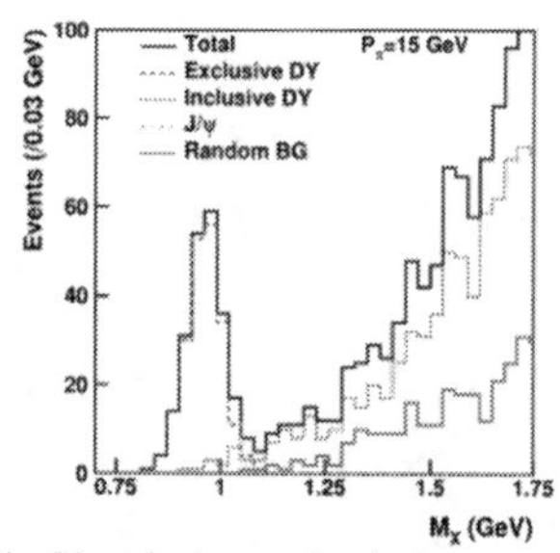

Figure 11. Simulations of missing-mass (M_X) spectra of $\mu^+\mu^-$ events [6].

3.3 Generalized parton distributions for exotic hadrons

In addition to the constituent counting rule in Sec. 2, the GPDs, in general the 3D structure functions, can be used for exotic-hadron studies [4] because the relevant degrees of freedom are quarks and gluons at high energies and because exotic signatures should appear in the 3D structure. The GPDs have two types of information on internal structure of a hadron: (1) longitudinal-momentum distributions, namely PDFs ($f_n(x)$), (2) transverse form factors ($F_n^h(t, x)$) which also depend on the longitudinal-momentum fraction x. Therefore, the GPDs could be expressed by their multiplication as a simple form:

$$H_q^h(x, \xi = 0, t) = f_n(x)\, F_n^h(t, x), \tag{5}$$

where n indicates the number of valence quarks.

A simple x-dependent form of the longitudinal PDFs is $f_n(x) = C_n x^{\alpha_n} (1-x)^{\beta_n}$. We consider valence-quark distributions in this form. Then, if β_n is known, the parameters C_n and α_n are determined by the valence-quark number $\int_0^1 dx\, f_n(x) = n$, and the quark momentum $\int_0^1 dx\, x\, f_n(x) = \langle x \rangle_q$. The parameter β_n is associated with the functional behavior in the elastic limit $x \to 1$, and it is determined by the constituent counting rule as $\beta_n = 2n - 3 + 2\Delta S_z$ with the spin factor $\Delta S_z = |S_z^q - S_z^h|$. From these three conditions, we theoretically obtain the PDFs not only for the ordinary hadrons, such as pion and nucleon, but also for exotic tetraquark and pentaquark hadrons, and the results are shown by the solid curves in Fig. 12. In comparison, typical PDFs of the pion and the proton extracted from experimental measurements are shown by the dotted curves at $Q^2 = 2$ GeV^2. Our theoretical curves and the PDF parametrizations agree with each other. Therefore, the predications for the new PDFs of the tetraquark and pentaquark hadrons should be interesting if they are measured by experiments.

Next, we explain that the transverse form factor $F_n^h(t,x)$ also has information on exotic nature [4]. A simple form of this function is an exponential form $F_n^h(t,x) = e^{(1-x)t/(x\Lambda^2)}$, where Λ is cutoff parameter for the transverse momentum and it is related to the the root-mean-square radius by $\langle r_\perp^2 \rangle = 4(1-x)/(x\Lambda^2)$. Since the pion and nucleon form factors are roughly given by the monopole and dipole forms $(1/(|t| + \Lambda^2)$, $1/(|t| + \Lambda^2)^2)$, the Gaussian form could be too steep as a function of $q_\perp^2$ and it is realized only in light nuclei. We do not step into such details in this article. The results are shown in Fig. 13 for $x = 0.2$ and 0.4. Depending on compact $q\bar{q}/qqq$ hadrons or diffuse molecular exotic hadrons, the transverse form factors are different, so it should be an appropriate quantity to find the internal structure of exotic-hadron candidates.

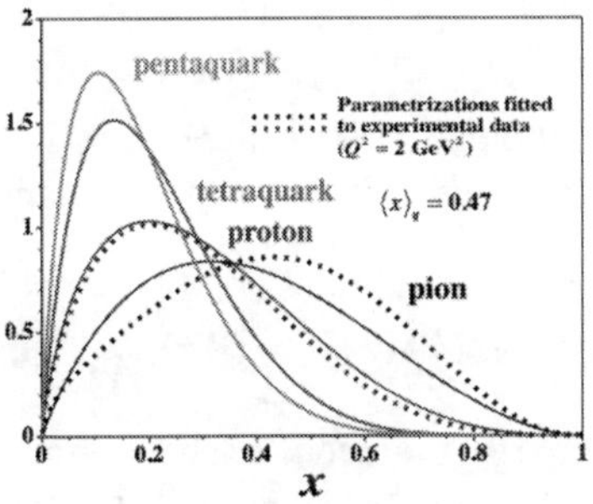

Figure 12. Valence-quark distributions for pion, proton, tetraquark, and pentaquark [4].

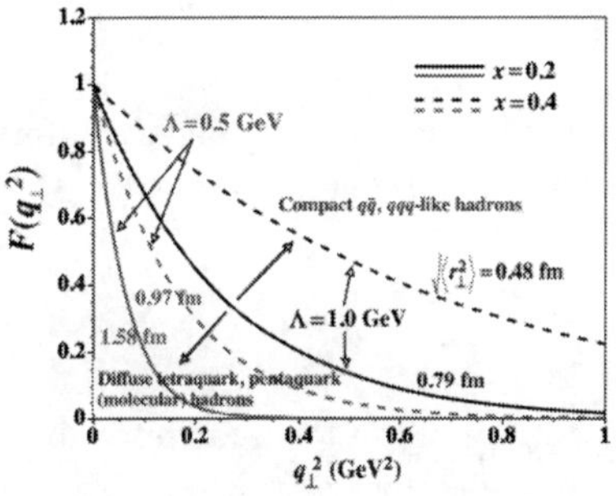

Figure 13. Transverse form factors at $x = 0.2, 0.4$ [4].

The GPDs of the exotic hadrons could not be observed directly at least at this stage because the exotic hadrons cannot exist as stable fixed targets. Of course, transition GPDs such as for proton $\to$ an exotic hadron can be investigated in future. For example, by using high-momentum kaon beam which will become available at J-PARC, it is possible to investigate the transition GPDs of $p \to \Lambda(1405)$ as shown in Fig. 14, and they should reflect the exotic nature of $\Lambda(1405)$. Since the transition GPDs, especially on exotic hadrons, are not well studied, theoretical and experimental efforts are needed to push this kind of experimental projects in future.

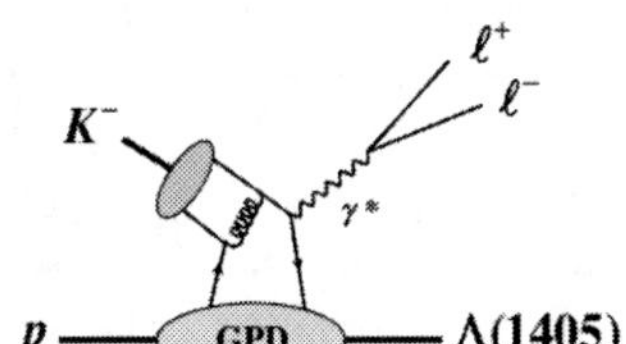

Figure 14. Transition GPDs at J-PARC.

4 Generalized distribution amplitudes and KEKB project

It is an interesting idea to use the transition GPDs for exotic-hadron studies; however, another promising way of studying 3D structure functions for exotic hadrons is to use general distribution amplitudes (GDAs) because exotic hadrons can be produced in s-channel processes [4]. A two-photon process to produce an exotic-hadron pair $\gamma\gamma^* \to h\bar{h}$ is shown in Fig. 15. If the kinematical condition, $Q^2\,(= -q^2) \gg W^2\,(= (p+p')^2),\ \Lambda^2$, is satisfied, we have the factorization of the process in terms of the hard photon interaction and the GDAs. This two photon process is the s-t crossed one to the DVCS ($\gamma^* h \to \gamma h$) where the GPDs are investigated, so that the s-t crossed 3D structure functions of the GPDs are the GDAs.

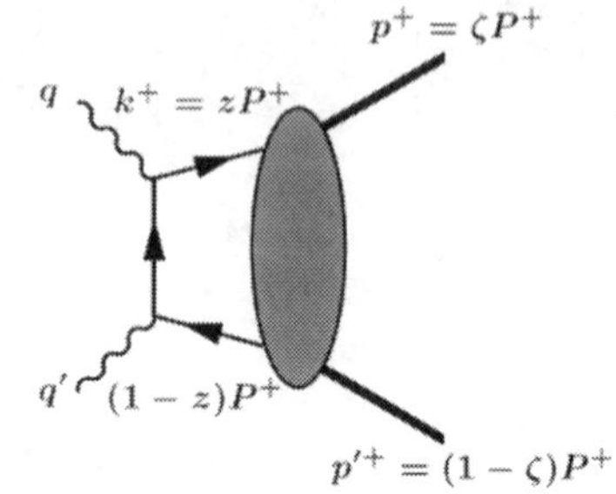

Figure 15. $\gamma\gamma \to h\bar{h}$ and GDAs [4].

The quark GDAs are defined by the same lightcone operator to the GPDs between the vacuum and the hadron pair $h\bar{h}$:

$$\Phi_q^{h\bar{h}}(z,\zeta,s) = \int \frac{dy^-}{2\pi}\, e^{i(2z-1)P^+y^-} \langle\, h(p)\,\bar{h}(p')\,|\,\bar{\psi}(-y/2)\gamma^+\psi(y/2)\,|\,0\rangle\Big|_{y^+=\vec{y}_\perp=0}, \tag{6}$$

and the gluon GDA is defined in the similar way. The s-t crossing relates the GDAs to GPDs, and the relation is given by

$$\Phi_q^{h\bar{h}}(z,\zeta,W^2) \longleftrightarrow H_q^h\left(x = \frac{1-2z}{1-2\zeta}, \xi = \frac{1}{1-2\zeta}, t = W^2\right). \tag{7}$$

It means that the GDAs contain the information of GPDs not only in the usual physical region but also in the unphysical regions: $0 \le |x| < \infty$, $0 \le |\xi| < \infty$, $|x| \le |\xi|$, and $t \ge 0$. A simple functional form is given by

$$\Phi_q^{h\bar{h}(I=0)}(z,\zeta,W^2) = N_{h(q)}\, z^\alpha (1-z)^\beta (2z-1)\,\zeta(1-\zeta)\, F_{h(q)}(W^2), \tag{8}$$

where $F_{h(q)}(W^2)$ is a form factor of the quark part of the energy-momentum tensor. According to the constituent counting rule, it is expressed as $F_{h(q)}(W^2) = 1/[1+(W^2-4m_h^2)/\Lambda^2]^{n-1}$, where the factor n is $n=2$ for ordinary $q\bar{q}$ mesons and $n=4$ for tetraquark hadrons.

The $f_0(980)$ and $a_0(980)$ mesons have been controversial hadrons which seem to be tetraquark states or $K\bar{K}$ molecules [9]. Here, we show W^2 dependence of the cross section $e\gamma \to e'h\bar{h}$ in Fig. 16 for $h = f_0(980)$ or $a_0(980)$ by using the simple GDAs in Eq. (8) with the form factor suggested by the constituent counting rule. There are large differences in the cross sections depending on the constituent number n, $n=2$ (ordinary meson, $q\bar{q}$) or $n=4$ (tetraquark type, $qq\bar{q}\bar{q}$). Even for the $n=4$ hadron, it is possible distinguish a compact $qq\bar{q}\bar{q}$ state from a diffuse molecular state from the measurement. Therefore, the GDAs should provide valuable information on internal structure of exotic hadrons, and such two-photon physics is certainly a possible project at KEKB [7].

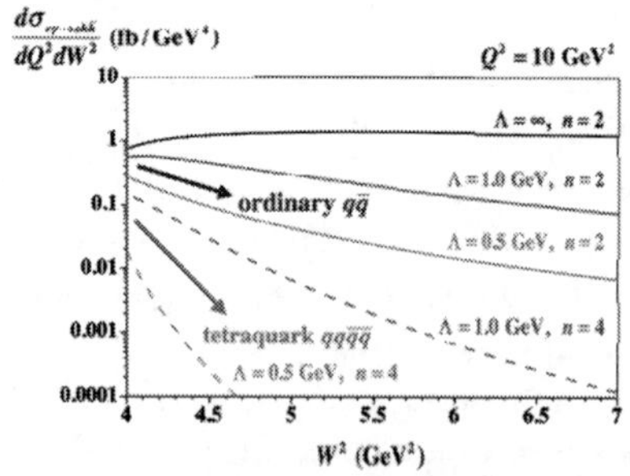

Figure 16. Cross section for $h\bar{h}$-pair production [4].

5 Summary

It became crucial to investigate the 3D structure functions for finding the original of nucleon spin including partonic orbital-angular-momentum contributions. One of such 3D structure functions is the GPDs which have been investigated by the DVCS so far. However, it is also possible to investigate them at hadron facilities such as J-PARC, for example, by using the $2 \to 3$ process $p + p \to N + \pi + B$ and the exclusive Drell-Yan process $\pi^- + p \to \mu^+\mu^- + n$. In addition, the 3D structure functions, the GPDs and GDAs, can be used for finding internal structure of exotic hadrons. In particular, the GDAs should be appropriate 3D quantities for the exotic-hadron studies because the exotic hadrons can be created in the final state, whereas it is difficult to measure the exotic-hadron GPDs directly because unstable exotic hadrons cannot be used as stable fixed targets. The GDAs can be investigated by the two-photon processes, for example, at KEKB and it is a new promising direction of exotic-hadron studies.

Acknowledgements

This work was supported by Japan Society for the Promotion of Science (JSPS) Grants-in-Aid for Scientific Research (KAKENHI) Grant No. JP25105010.

References

[1] M. Hirai, S. Kumano, M. Oka, and K. Sudoh: Phys. Rev. D **77**, 017504 (2008).
[2] H. Kawamura, S. Kumano, and T. Sekihara, Phys. Rev. D **88**, 034010 (2013).
[3] W.-C. Chang, S. Kumano, and T. Sekihara, Phys. Rev. D **93**, 034006 (2016).
[4] H. Kawamura and S. Kumano, Phys. Rev. D **89**, 054007 (2014).
[5] S. Kumano, M. Strikman, and K. Sudoh, Phys. Rev. D **80**, 074003 (2009).
[6] T. Sawada, W.-C. Chang, S. Kumano, J.-C. Peng, S. Sawada, and K. Tanaka, Phys. Rev. D **93**, 114034 (2016).
[7] S. Kumano and Q.-T. Song, research in progress.
[8] T. Sekihara and S. Kumano, Phys. Rev. C **89**, 025202 (2014).
[9] S. Kumano and V. R. Pandharipande, Phys. Rev. D **38**, 146 (1988); F. E. Close, N. Isgur, and S. Kumano, Nucl. Phys. B **389**, 513 (1993); T. Sekihara and S. Kumano, Phys. Rev. D **92**, 034010 (2015).

EPJ Web of Conferences **141**, 07001 (2017) DOI: 10.1051/epjconf/201714107001
ISMD 2016

Interactions of Cosmic Rays around the Universe

Models for UHECR data interpretation

Denise Boncioli[1,a]

[1] *DESY, Platanenalle 6, D-15738 Zeuthen, Germany*

Abstract. Ultra high energy cosmic rays (UHECRs) are expected to be accelerated in astrophysical sources and to travel through extragalactic space before hitting the Earth atmosphere. They interact both with the environment in the source and with the intergalactic photon fields they encounter, causing different processes at various scales depending on the photon energy in the nucleus rest frame. UHECR interactions are sensitive to uncertainties in the extragalactic background spectrum and in the photo-disintegration models.

1 Introduction

The origin and nature of ultra high energy cosmic rays (UHECRs) are unknown: where these particles, as energetic as $\approx 10^{20}$ eV, come from, their chemical composition and which mechanisms are involved in their acceleration are still open issues. Extragalactic sources are considered as candidate for cosmic ray acceleration at the highest energies because of their characteristic size and magnetic fields [1].

In order to predict the expected flux of UHECRs at Earth, interactions of particles with the photon fields both in the source and in the extragalactic space have to be taken into account. The suppression of the flux [2, 3] measured by UHECR experiments as the Pierre Auger Observatory [4] and the Telescope Array [5] can be attributed to the energy losses suffered in these interactions and/or to the maximal energy reached at the source. If due to interactions, the suppression can be generated by different phenomena depending on the mass of the cosmic ray nucleus.

Measurements of observables sensitive to UHECR mass composition are not conclusive. The Telescope Array experiment claims to find a consistency of their X_{max} distributions with light composition above 10^{18} eV [6]. On the contrary, the Auger collaboration finds that the primary mass is decreasing reaching the minimal values at around $10^{18.3}$ eV and starts to increase for the higher energies; at the same time the spread of the mass is almost constant till $10^{18.3}$ eV and then starts to decrease, indicating that the relative fraction of protons becomes smaller for higher energies [7]. By comparing the X_{max} of the two experiments taking into account different resolutions, acceptances and analysis strategies, the two results are found to be in good agreement within systematic uncertainties [8].

At the highest energies, the spectrum features are related to characteristics of propagation of cosmic rays in the extragalactic background fields. If UHECRs above 10^{18} eV are mainly extragalactic protons, they interact with the cosmic microwave background (CMB) and the infrared, optical and

[a] e-mail: denise.boncioli@desy.de

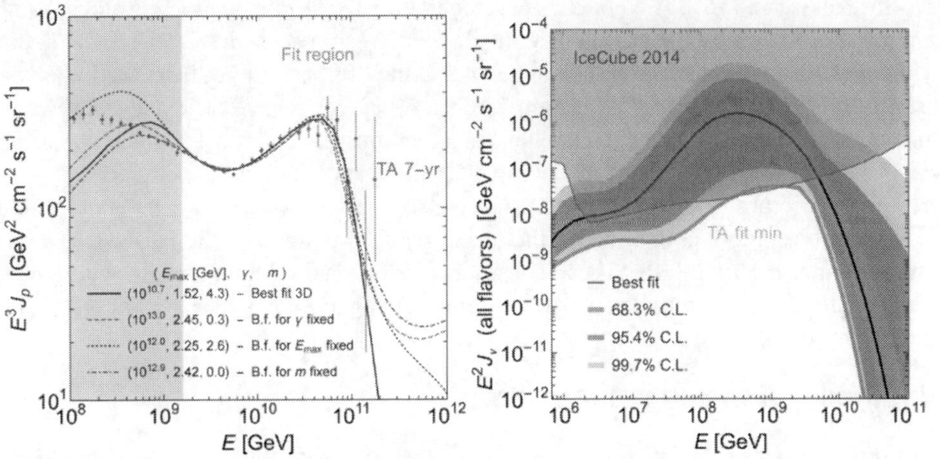

Figure 1. Right: Best-fit spectrum (black solid line) superimposed to the TA energy spectrum [3], as taken from [15]. Left: All-flavor flux of cosmogenic neutrinos predicted by the best fit to the TA energy spectrum [15].

ultraviolet background (named as extragalactic background light, EBL). The ankle, *i.e.* the hardening at about 5×10^{18} eV, in the proton dip model, is expected to be due to energy losses due to electron-positron pair production mainly on CMB [9]. In the same context the suppression of the flux can be justified with the energy losses suffered by protons in photo-meson reactions, as it was predicted in [10, 11] (the GZK effect). The astrophysical scenario is complicated by the fact that source properties, as for example their distribution and their acceleration power, are unknown: the suppression of the flux, even in the pure-proton composition case, cannot be addressed to GZK cutoff in absence of knowledge about sources [12]. Since the threshold energy for photo-meson production (roughly) increases $\propto A$, the suppression of the flux in case of a mixed composition is expected to be due, at least partly, to the photo-disintegration of nuclei, occurring after the excitation of the giant dipole resonance (GDR, ≈8 MeV in the NRF) that follows the absorption of a CMB or EBL photon [13, 14]. From the point of view of the models, a self-consistent interpretation of the spectrum and composition results of the Auger Observatory and the Telescope Array experiment can be used to limit the astrophysical scenarios compatible with data.

2 Hypothesis of pure-proton composition

The fundamental assumption in this case is that of an extragalactic pure proton UHECR composition, as in the proton dip model. This assumption can be tested in light of the UHECR energy spectrum measurements from the TA collaboration [3] and the upper bound on the flux of cosmogenic neutrinos from the IceCube collaboration [16], as done in [15]. The distribution of (identical) sources is taken as homogeneous, and the source evolution is parametrized relative to the star formation rate (SFR). A fit of the TA spectrum is performed (and it is shown in Figure 1, left panel): the fit slightly prefers low maximal energies, hard spectra, and strong source evolution. The high value of m implies that the contribution of distant sources is enhanced with respect to SFR evolution.
Cosmogenic neutrinos are produced by the decay of pions, muons, kaons, and neutrons produced in photo-hadronic interactions during cosmic-ray propagation. Unlike protons or nuclei, neutrinos created at high redshifts reach Earth because they rarely interact and only undergo adiabatic energy

losses and flavor mixing. Figure 1 (right panel) shows the flux of cosmogenic neutrinos associated to the best-fit obtained in [15]. The bottom curve–marked "TA fit min"–is the envelope of all possible neutrino fluxes allowed by the cosmic-ray fit. It is in tension with the TA UHECR data at 99.7% C.L. Moreover, the event rate can be calculated using the fluxes resulting from the fit, as for example explained in [17]: in particular, "TA fit min" yields 4.9 events. Since only one event was observed, this flux can be excluded at 95% C.L. We have then demonstrated that the limits on neutrino flux can be used as a test of the proton dip model completely independent of composition data.
The presence of protons in UHECRs can also be constrained by considering the photons that are produced during propagation, in the light of the recent results obtained by the Fermi-LAT collaboration about the origin of the EBL [18]. These constraints, additional or complementary to the ones given by the expectations from neutrino production during propagation, are discussed in [19–22].

3 Hypothesis of mixed composition

Several analytical and Monte Carlo codes have been developed to simulate the propagation of UHECRs. Here we mainly discuss the accuracy of our knowledge of relevant quantities for modeling nuclei interactions, making use of a publicly available Monte Carlo code, *SimProp* [23]. The expansion of the Universe, the spectrum and evolution of the CMB and the cross section for pair production are known with good accuracy and enter in calculating interactions of both protons and nuclei. For the case of mixed composition, photo-disintegration processes are possible both with CMB and EBL photons. The quantities here involved are poorly known.
The choice of the EBL model does not significantly affect the UHECR spectrum at Earth if the composition is assumed as pure protons. It is well known that it is important for neutrino production (however, this uncertainty is smaller than the ones in the cosmological evolution of sources [24]).

In the case of nuclei, due to GDR, at energies $\approx 10^{19}$ eV photo-disintegration can occur via interaction with photons with energies $\approx 10 - 100$ meV in the laboratory frame: the EBL plays a fundamental role in the propagation of nuclei. As shown in Figure 2, the discrepancies among different models are larger for larger wavelengths, which are the most relevant ones for UHECR propagation. The effect of these uncertainites is shown in Figure 4 (left panel): the stronger far-IR peak in the EBL spectrum in Dominguez EBL [26] with respect to the Gilmore EBL model [27] causes more interactions, resulting in a higher production of light particles at Earth and a softer spectrum.

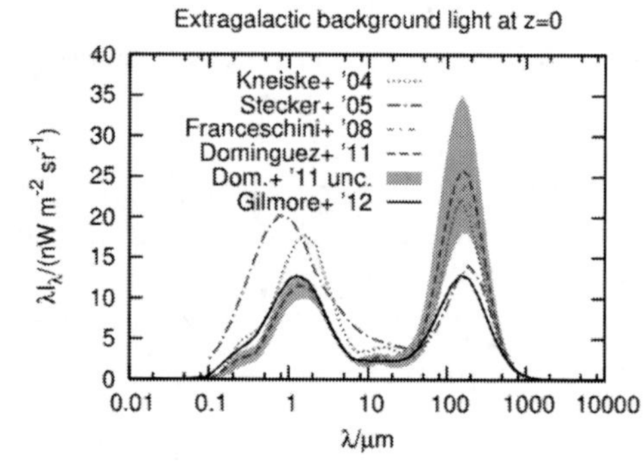

Figure 2. Intensity of EBL at $z = 0$. See [25] and references therein.

The other main source of uncertainties in calculations of cosmic rays interactions comes from nuclear physics. Data on photo-nuclear cross sections are scarse and the existent models use photo-neutron cross sections where available, or empirical parametrizations [28]. In Figure 3 the experimental situation on photo-nuclear cross sections is shown, based on data from EXFOR database [29] in the GDR and quasi-deuteron energy band, for nuclear isotopes interesting for cosmic-ray astrophysics. The availability of nuclear models is also shown; moreover, it can be seen that there are models, like TALYS [30], which predict the population of isotopes off the main diagonal, while more simplified models as PSB [31, 32] have only one isotope for each A. Moreover, the TALYS model allows the ejection of small fragments and not only nucleons as in the PSB model. The effect of the use of these different models in UHECR propagation is shown in Figure 4 (right panel): the spectrum of residual primary nuclei at Earth is expected to be more eroded at the highest energies in the TALYS model than in the PSB one, because of the more efficient photo-disintegration.

For the same reason, the flux of small fragments is enhanced.

A fit of the Auger spectrum [2] and $X_{\max}$ data [7] with a simplified model of UHECR sources has been performed [33, 34], with the assumptions of identical sources homogeneously distributed in a comoving volume. The injection consists only of ^{1}H, ^{4}He, ^{14}N and ^{56}Fe. In this study two different simulation codes for propagation have been used, *SimProp* and CRPropa [35] with different settings for the EBL and cross section models, in order to quantify the effect of such differences on the fit. In particular, if the propagation is simulated with *SimProp* v2r3 [36] using the Gilmore EBL [27] and the PSB photo-disintegration model [31, 32] and air showers are simulated with CONEX using the EPOS-LHC hadronic interaction model, the best fit is found with relatively hard spectral index, relatively low cutoff rigidity and a very metal-rich composition. The corresponding simulated spectrum and average and standard deviation of $X_{\max}$ distributions are shown and compared to Auger data in Figure 5. The fit has been repeated by using different Monte-Carlo codes, EBL models, photo-disintegration models, air interaction models and with data shifted by their systematical uncertainties, finding that the best-fit parameter values are strongly dependent on the models used, though certain features of the best fit are shared by almost all scenarios.

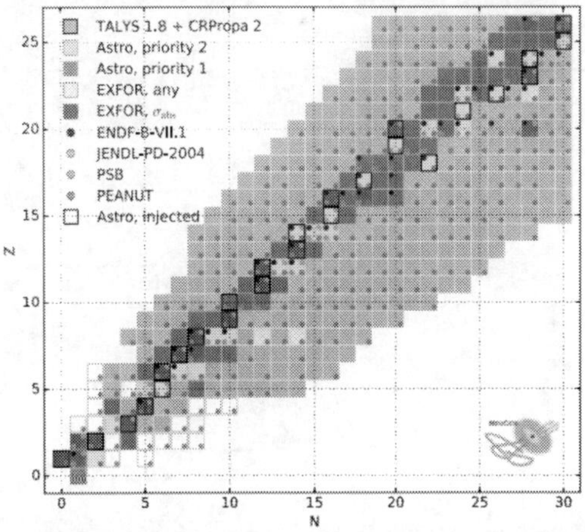

Figure 3. Experimental situation versus astrophysical requirements for nuclear isotopes interesting for cosmic ray astrophysics (gray boxes, from TALYS and CRPropa). Experimental measurements (from EXFOR database [29]) are marked by red and yellow boxes for the total absorption and any inclusive cross sections, respectively. Models are marked by dots (see [28], from which the figure is taken, and references therein).

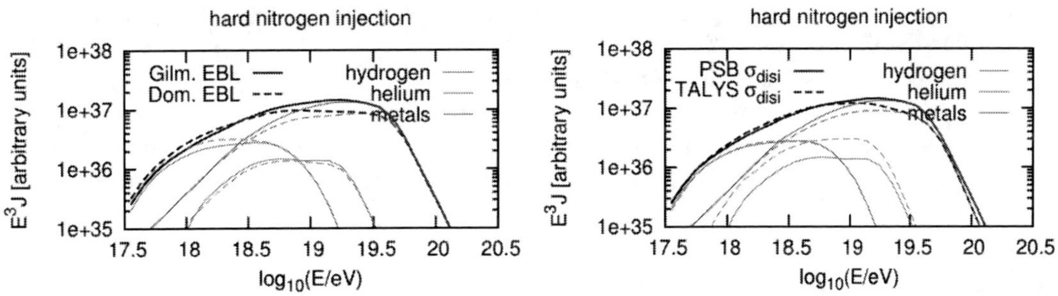

Figure 4. Effect of different EBL models (left panel) and of different photo-disintegration models (right panel) on propagated UHECR fluxes [34].

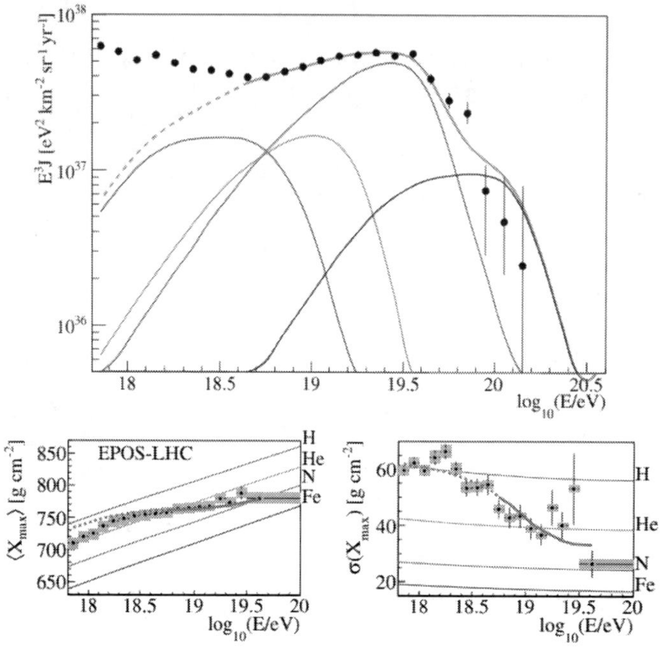

Figure 5. Simulated UHECR fluxes, $\langle X_{\max}\rangle$ and $\sigma(X_{\max})$ in the best-fit scenario found in [33] (thick brown: total; red: $A = 1$; grey: $2 \leq A \leq 4$; green: $5 \leq A \leq 26$; blue: $A \geq 27$) compared to Auger data [2, 7].

4 Conclusions

At the present time, the proton dip, ankle and mixed-composition models all remain valid. The predicted flux of cosmogenic neutrinos can be used as a tool to test the proton dip model, as done in [15], independently from composition data. This method has in principle the power to rule out candidates to UHECR acceleration that have very high source evolution. However, predictions of neutrinos from interactions inside the source and their connections with the cosmic-ray escape mechanism can also be used as a complementary information to the cosmogenic neutrino studies [37].
One possible alternative to the proton dip model is the mixed-composition scenario, that has been tested for example in [33] by fitting the Auger spectrum and composition results. In a mixed-composition hypothesis several poorly-known quantities enter in the modeling of extragalactic propagation, such as the EBL models and the photo-disintegration cross sections. The fit of experimental data is affected by uncertainties in the models.
Uncertainties in details of photon field and photo-disintegration affect not only the extragalactic propagation but also the modeling of interactions in the source ambient. The missing information from the nuclear physics side is described in [28] and it is valid for cosmic-ray astrophysics in general.

Acknowledgments

This work has received funding from the European Research Council (ERC) under the European Union's Horizon 2020 research and innovation programme (Grant No. 646623).

References

[1] A.M. Hillas, Ann. Rev. Astron. Astrophys. **22**, 425 (1984)
[2] I. Valiño (for The Pierre Auger Collaboration) (PoS(ICRC2015)271, 2015)
[3] D. Ivanov (for Telescope Array Collaboration) (PoS(ICRC2015)349, 2015)
[4] A. Aab et al. (Pierre Auger Collaboration), Nucl. Instrum. Meth. **A798**, 172 (2015), 1502.01323
[5] T. Abu-Zayyad et al. (Telescope Array Collaboration), Nucl. Instrum. Meth. **A689**, 87 (2013), 1201.4964
[6] J. Belz (for Telescope Array Collaboration) (PoS(ICRC2015)351, 2015)
[7] A. Porcelli (for The Pierre Auger Collaboration) (PoS(ICRC2015)420, 2015)
[8] M. Unger (for The Pierre Auger and Telescope Array Collaboration) (PoS(ICRC2015)307, 2015)
[9] V.S. Berezinsky et al., Phys. Rev. **D74**, 043005 (2006), hep-ph/0204357
[10] K. Greisen, Phys. Rev. Lett. **16**, 748 (1966)
[11] G.T. Zatsepin, V.A. Kuzmin, JETP Lett. **4**, 78 (1966), [Pisma Zh. Eksp. Teor. Fiz.4,114(1966)]
[12] R. Aloisio, D. Boncioli, Astropart. Phys. **35**, 152 (2011), 1002.4134
[13] R. Aloisio, V. Berezinsky, S. Grigorieva, Astropart. Phys. **41**, 73 (2013), 0802.4452
[14] R. Aloisio, V. Berezinsky, S. Grigorieva, Astropart. Phys. **41**, 94 (2013), 1006.2484
[15] J. Heinze, D. Boncioli, M. Bustamante, W. Winter, Astrophys. J. **825**, 122 (2016), 1512.05988
[16] M.G. Aartsen et al. (IceCube Collaboration) (2016), 1607.05886
[17] P. Baerwald, M. Bustamante, W. Winter, Astropart. Phys. **62**, 66 (2015), 1401.1820
[18] M. Ackermann et al. (Fermi-LAT Collaboration), Astrophys. J. **799**, 86 (2015), 1410.3696
[19] V. Berezinsky, A. Gazizov, O. Kalashev, Astropart. Phys. **84**, 52 (2016), 1606.09293
[20] A. Supanitsky, Phys. Rev. **D94**, 063002 (2016), 1607.00290
[21] A. van Vliet, in *ARENA 2016* (2016), 1609.03336
[22] O. Kalashev, in *Quarks 2016* (2016), 1608.07530
[23] R. Aloisio et al., JCAP **1210**, 007 (2012), 1204.2970
[24] R. Aloisio et al., JCAP **1510**, 006 (2015), 1505.04020
[25] R. Alves Batista et al., JCAP **1510**, 063 (2015), 1508.01824
[26] A. Dominguez et al., Mon. Not. Roy. Astron. Soc. **410**, 2556 (2011), 1007.1459
[27] R.C. Gilmore et al., Mon. Not. Roy. Astron. Soc. **422**, 3189 (2012), 1104.0671
[28] D. Boncioli, A. Fedynitch, W. Winter (2016), 1607.07989
[29] N. Otuka et al., Nuclear Data Sheets **120**, 272 (2014)
[30] A.J. Koning et al., in *Int. Conf. on Nuclear Data for Science and Tecnology* (2007)
[31] J.L. Puget, F.W. Stecker, J.H. Bredekamp, Astrophys. J. **205**, 638 (1976)
[32] F.W. Stecker, M.H. Salamon, Astrophys. J. **512**, 521 (1999), astro-ph/9808110
[33] A. di Matteo (for The Pierre Auger Collaboration) (PoS(ICRC2015)249, 2015)
[34] D. Boncioli, A. di Matteo, A.F. Grillo (for the Pierre Auger Collaboration), in *CRIS 2015* (2015), 1512.02314
[35] R. Alves Batista et al., JCAP **1605**, 038 (2016), 1603.07142
[36] R. Aloisio, D. Boncioli, A. di Matteo, A. Grillo, S. Petrera, F. Salamida (2016), 1602.01239
[37] P. Baerwald, M. Bustamante, W. Winter, Astrophys. J. **768**, 186 (2013), 1301.6163

Prompt atmospheric neutrino flux from the various QCD models

Yu Seon Jeong[1,a], Atri Bhattacharya[2], Rikard Enberg[3], C. S. Kim[4], Mary Hall Reno[5], Ina Sarcevic[6], and Anna Stasto[7]

[1] *National Institute of Supercomputing and Networking, KISTI, Daejeon 34141, Korea*
[2] *Space sciences, Technologies and Astrophysics Research (STAR) Institute, Université de Liège, Bât. B5a, 4000 Liège, Belgium*
[3] *Department of Physics and Astronomy, Uppsala University, Uppsala, Sweden*
[4] *Department of Physics and IPAP, Yonsei University, Seoul 03722, Korea*
[5] *Department of Physics and Astronomy, University of Iowa, Iowa City, Iowa 52242*
[6] *Department of Astronomy, University of Arizona, 933 N. Cherry Ave., Tucson, AZ 85721*
[7] *Department of Physics, The Pennsylvania State University, University Park, PA 16802*

Abstract. We evaluate the prompt atmospheric neutrino flux using the different QCD models for heavy quark production including the b quark contribution. We include the nuclear correction and find it reduces the fluxes by 10% – 50% according to the models. Our heavy quark results are compared with experimental data from RHIC, LHC and LHCb.

1 Introduction

The IceCube collaboration reported that they have observed 54 high energy events for the neutrino candidate events and rejected a pure atmospheric origin scenario with about 7σ [1]. For astrophysical neutrinos, the main background is the atmospheric neutrinos produced from the interaction of cosmic rays with air nuclei. In estimating the atmospheric neutrino flux, the most important factor is the cross sections for production of quarks fragmented into hadrons, which subsequently decay and generate neutrinos. Above $\sim$1 PeV of energies, atmospheric neutrinos from the decay of heavy hadrons (called prompt neutrinos) are more important than those from the π or K decay (conventional neutrinos).

In this work, we investigate the effect of the QCD models for heavy quark production to the prompt atmospheric neutrino flux. We evaluated the prompt flux using three different models, next-to-leading order (NLO) perturbative QCD, the dipole model and k_T factorization, and we include B hadron contributions. To investigate the uncertainties, we vary the QCD scales so that the total cross sections for $q\bar{q}$ pair production are consistent with the data from RHIC and LHC experiments. We find that the scale ranges used here for the cross sections are acceptable by comparing the rapidity distributions of the differential cross sections with the corresponding LHCb data. For other uncertainty factors, we include the nuclear corrections and used the modern cosmic ray fluxes parameterized by their composition and source component. In this contribution, we present the prompt neutrino fluxes evaluated from the different models for only one cosmic ray spectrum. Our comprehensive results can be found in [2].

[a] e-mail: ysjeong@kisti.re.kr

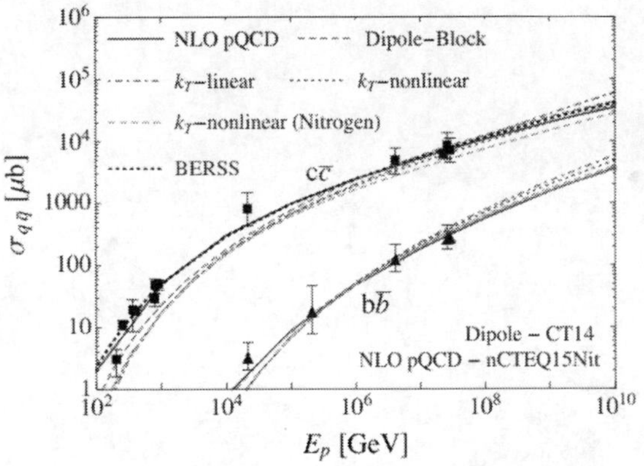

Figure 1. Total cross section for $c\bar{c}$ and $b\bar{b}$ production as a function of the incident proton energy. The results are presented for the different models described in section 2. For comparison, the previous evaluation at NLO in pQCD approach is presented as BERSS [14].

2 QCD models for heavy quark production

2.1 Perturbative QCD

The standard method to evaluate heavy quark production is to use the parton model in perturbative QCD (pQCD) with a collinear approximation. In the pQCD framework, the differential cross section for $q\bar{q}$ pair production is given by

$$\frac{d\sigma(pp \to q\bar{q}X)}{dx_F} = \int \frac{dM^2_{q\bar{q}}}{(x_1 + x_2)s} \sigma_{f_a f_b \to q\bar{q}}(\hat{s}) f_a(x_1, M_F^2) f_b(x_2, M_F^2) \quad (1)$$

with the parton distribution function (PDF) $f_{a(b)}$, their momentum fraction $x_{1(2)}$ and the Feynman variable x_F. Momentum fractions can be expressed as $x_{1,2} = ((x_F^2 + 4M^2_{q\bar{q}}/s)^{1/2} \pm x_F)/2$. For the forward scattering that dominates the prompt flux, the partons from the incoming cosmic ray hadrons have the large momentum fraction x_1 while those from the target nucleus side have the small x_2.

We compute the heavy quark production cross section at next-to-leading order (NLO) to evaluate the prompt neutrino flux including the nuclear corrections. Nuclear corrections are included through the recent nuclear PDF, nCTEQ15 [3]. For the factorization (M_F) and renormalization (M_R) scales, we use scales proportional to the transverse mass m_T, $(2.1, 1.6)m_T$ for the central values, as experimentally constrained by the RHIC and LHC data [4].

At the high energies where the prompt flux dominates, gluon fusion to $q\bar{q}$ is the main process, hence the essential factor is the poorly constrained gluon distribution at small-x. Therefore we investigate other possible approaches to the extrapolation with the resummation of large logs of x.

2.2 Dipole model

One of the alternative approaches is a color dipole model, which can include the parton saturation effect through the so-called dipole cross section. In the dipole model, the heavy quarks are produced through two sequential processes: a fluctuation of incoming gluon into $q\bar{q}$ pair (color dipole), and the interaction of the color dipoles with target nuclei. The dipole cross section is given by

$$\sigma^{gp\to q\bar{q}X}(x, M_R, Q^2) = \int dz\, d^2\vec{r}\, |\Psi_g^q(z, \vec{r}, M_R, Q^2)|^2 \sigma_d(x, \vec{r}) \quad (2)$$

with the wave function squared ($|\Psi_g^q|^2$) for a gluon fluctuation to a color dipole and the dipole cross section (σ_d) for the interaction between the dipole and the target. The x_F differential cross section for

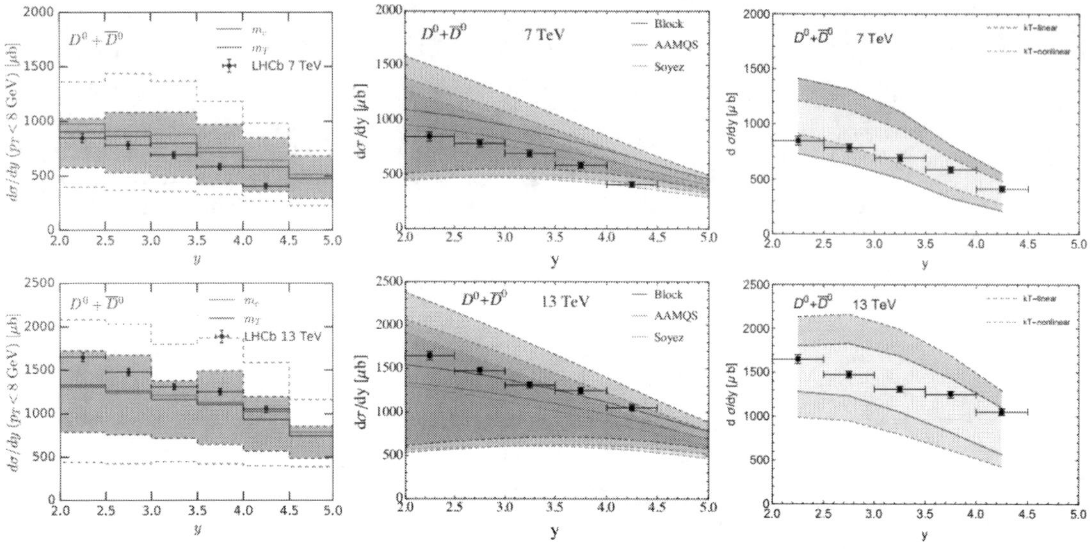

Figure 2. Rapidity distribution for $pp \to D^0/\bar{D}^0\,X$ at $\sqrt{s} = 7$ TeV (upper) and at 13 TeV (lower) from NLO perturbative QCD with nCTEQ15-01 PDFs, three dipole models with the CT14 LO gluon PDF and k_T factorization approach. Experimental data are from the LHCb measurement [16].

heavy quark pair production in pp collision can be written as

$$\frac{d\sigma(pp \to q\bar{q}X)}{dx_F} \simeq \frac{x_1}{x_1 + x_2} g(x_1, M_F)\sigma^{gp\to q\bar{q}X}(x_2, M_R, Q^2 = 0)\,. \tag{3}$$

In this work, three different dipole cross sections are used to evaluate the prompt atmospheric neutrino flux, referred to as Soyez-, AAMQS- and Block-dipole, respectively. The Soyez dipole cross section [5] is parameterized with the approximate form by [6] for the solution of the BK equation, which includes saturation [7]. The parameters are determined by fitting to the HERA data, and this dipole result is used in the earlier evaluation in Ref. [8]. The AAMQS dipole presented in ref. [9] is improved by including the running coupling corrections to the BK equation (rcBK). The third dipole referred to as Block-dipole [10] is found approximately from the parameterization of the structure function $F_2(x, Q^2)$ [11].

For the nuclear correction in the dipole model, we use the Glauber-Gribov method, formulated as

$$\sigma_d^A(x,r) = \int d^2\vec{b}\, 2\left[1 - \exp\left(-\frac{1}{2}AT_A(b)\sigma_d^p(x,r)\right)\right], \tag{4}$$

with the impact parameter b. The nuclear profile function $T_A(b) = \int dz\rho_A(z,\vec{b})$ is normalized to unity, $\int d^2\vec{b}\,T_A(b) = 1$. Here, we use a Gaussian distribution for nuclear density, $\rho_A(z,\vec{b})$.

2.3 k_T factorization

Another possible approach is the so-called k_T factorization scheme, which incorporates the transverse momentum of the partons. Therefore, the cross section in this scheme is expressed in terms of the k_T

dependent partonic cross section and PDFs. The cross section for this case is formulated as

$$\frac{d\sigma}{dx_F}(s, m_Q^2) = \int \frac{dx_1}{x_1}\frac{dx_2}{x_2} dz\, \delta(zx_1 - x_F)\, x_1 g(x_1, M_F) \int \frac{dk_T^2}{k_T^2} \hat{\sigma}^{\rm off}(z, \hat{s}, k_T)\, f(x_2, k_T^2)\,, \quad (5)$$

with collinear approximation and the integrated distributions for incoming partons and the unintegrated PDFs, which incorporates the small-x resummation for partons in the target. This is known as the *hybrid* formalism [12]. In eq. (5), $g(x_1, M_F)$ is the integrated PDF and $f(x_2, k_T^2)$ is the unintegrated PDF. Parton saturation effect can be included through nonlinear evolution of the unintegrated PDF. We investigate both cases with (k_T-nonlinear) or without (k_T-linear) saturation effects. Nuclear effects in k_T factorization can also be included through the unintegrated parton density, more specifically the nonlinear term in its evolution equation. The corresponding equation is presented in ref. [13] with detailed descriptions.

2.4 Total and differential cross section

Fig. 1 shows the total cross sections for the charm and bottom production from the different models described in the previous sections, perturbative QCD, dipole model and k_T factorization. In order to incorporate the nuclear effect, we take nitrogen for the averaged air nuclei. For comparison, the earlier calculation in the perturbative QCD approach in ref. [14] (BERSS) is also shown as well as the experimental data from RHIC, LHC and fixed target experiments [15, 16]. At low energies, only perturbative QCD results agree well with the data. However, at the high energies focused in this work, all of our evaluations are consistent and have a good agreement with the data.

We also evaluate the differential cross sections in rapidity to determine the acceptable range of the scales comparing with the LHCb data [16]. For the NLO perturbative QCD case, we take the scale variation as $(1.25, 1.48)m_T$ and $(4.65, 1.71)m_T$ for the lower and upper limits as guided by ref. [4]. In the left panels of fig. 2, we present the rapidity distributions for $D^0 + \bar{D}^0$ production in pp collisions for $2.0 < y < 4.5$ and $p_T \leq 8$ GeV with the corresponding data from the LHCb measurements [16]. The blue banded area presents the results evaluated with the m_T dependent scales while the magenta dashed lines show the results with the $m_c = 1.27$ GeV dependent scales. The experimental data are in the narrower error bands from the scales proportional to m_T, hence we use the m_T dependent scales for the uncertainties of the flux evaluations.

Fig. 2 also presents the predictions from the dipole models (center) and k_T factorization (right). We have found that the differential cross sections in rapidity for $D^0 + \bar{D}^0$ production from the three different dipole models are consistent with experimental data at the factorization scale, $1m_c \leq M_F \leq 4m_c$. For the central values shown in the plot, we take $M_F = 2m_c$. For the dipole model evaluation, since there is no explicit p_T dependence we do not apply the p_T cut, which does not have the large effect in the results where the p_T cut can be made.

For the k_T factorization results, the magenta and blue bands represent the results calculated with and without a saturation effect, respectively. The band width depends on the upper limit of the integration over k_T in eq. (5), for which is used $2.5m_T$ and k_{max} here. The evaluations from the k_T-linear with $2.5m_T$ and k_T-nonlinear with k_{max} gives the gray bands in the right panels of fig. 2. We use this range for the uncertainties by the scales in the prompt neutrino fluxes.

3 Prompt atmospheric neutrino fluxes

3.1 Cosmic ray fluxes

In calculating the atmospheric neutrino fluxes, the incident cosmic ray flux is one of the essential and most important factors. In many earlier calculations, the broken power law (BPL) spectrum

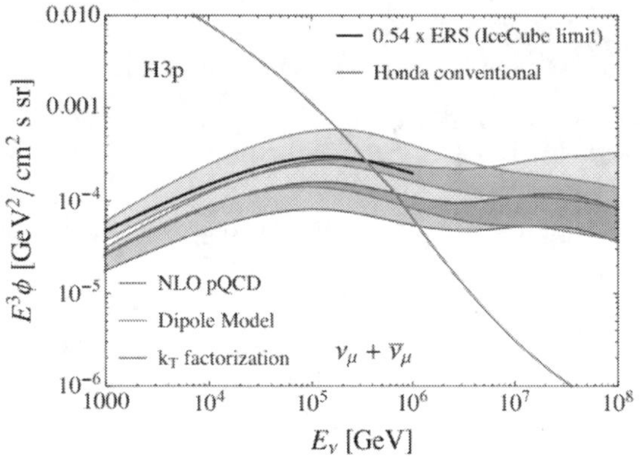

Figure 3. The prompt atmospheric neutrino fluxes from the different QCD models for the H3p cosmic ray spectrum. The upper limit from the IceCube collaboration and the conventional neutrino flux are also presented for comparison.

has been used with the assumption that all cosmic rays are protons. The BPL is parameterized as $\phi_N(E) = 1.7\ (E/\mathrm{GeV})^{-2.7}$ for $E < 5 \cdot 10^6$ GeV and $174\ (E/\mathrm{GeV})^{-3}$ for $E > 5 \cdot 10^6$ GeV. However, the recent study on the cosmic ray spectrum has provided several new parameterizations based on the models for the composition and sources of the cosmic rays. In our comprehensive work [2], we used four spectra: the BPL to compare with other predictions, two parameterizations from the model for three source components [17] and one spectrum for four source components (called GST) [18]. The two spectra in [17] is distinguished by the composition of cosmic rays from the extragalactic sources: one has a mixed composition (H3a), and the other has only protons (H3p).

In this contribution, our focus is on the effect of the different models to the prompt fluxes, therefore it would be sufficient to compare the results with only one kind of spectrum, H3p as shown in fig. 3.

3.2 Prompt muon neutrino flux

The atmospheric neutrino flux can be evaluated from the cascade equations which describe the propagation of high energy particles in the atmosphere. The coupled cascade equations for protons, hadrons and neutrinos can be solved using the approximate Z-moment method. The Z moments are the rescaled generation functions with only energy dependence, and they are associated with the production of hadrons from the interactions and their decays to neutrinos. The solutions of the coupled cascade equations can be written in terms of the Z moments for production and decay in the high energy and low energy limits. The flux of neutrinos are eventually obtained by interpolating the two approximate solutions and doing the sum over hadrons. More details for the Z-moment method are described in ref. [19]. In evaluating the muon neutrino fluxes, we take D^0, D^+, D_s, Λ_c (B^0, B^+, B_s, Λ_b) into account for the charmed (bottom) hadrons.

Fig. 3 shows the resulting neutrino fluxes with the H3p spectrum and all the approaches introduced in section 2. We have found that the nuclear correction has different effect according to the models: about 20% – 30% for the NLO pQCD, 10% – 20% for the dipole models, and 30% – 50% for the k_T factorization at $10^5 - 10^8$ GeV. The predictions using the dipole model are larger than the fluxes from the NLO pQCD calculation all over the energy range. For the k_T factorization case, the flux with the linear evolution is the largest above $E \sim 10^7$ GeV, while that from nonlinear evolution with nuclear correction is close to the lower bound of the NLO pQCD evaluation. The upper limit from the IceCube data based on 3 year observation are also presented for comparisons [1]. This IceCube limit excludes most of the dipole model results and quite constrains the upper values of the k_T factorization results in this work. Only new predictions in the NLO pQCD and k_T factorization with nuclear corrections are placed safely below the IceCube limit.

4 Summary

We have investigated the prompt atmospheric neutrino flux in the different frameworks for heavy quark production, NLO pQCD, three kinds of the dipole models and k_T factorization. We set the allowed QCD scales by comparisons of the total cross sections for heavy quark production with the experimental data from RHIC and LHC [4]. We checked that the differential cross sections for $D^0/\bar{D}^0$ production with these scales are consistent with the LHCb data. The overall prompt neutrino flux is lower from NLO pQCD calculation than from the other two evaluations. The nuclear corrections are most significant in the k_T factorization case based on the nonlinear evolution of the gluon density. In this work, we find that the nuclear corrected NLO pQCD prediction and the nuclear corrected k_T factorization with saturation (nonlinear evolution) can safely survive from the IceCube limit based on the 3 year data. However, a new limit was recently released based on the 6 year data by IceCube after our study [20], and we note that our prediction from all approaches are below this new limit.

Acknowledgments

This research was supported by the US DOE, the NRF of Korea, the NSC of Poland, the SRC of Sweden and the FNRS of Belgium.

References

[1] M. G. Aartsen et al. [IceCube Collaboration], arXiv:1510.05223
[2] A. Bhattacharya et al., arXiv:1607.00193
[3] K. Kovaric K et al., Phys. Rev. **D93**, 085037 (2016)
[4] R. E. Nelson, R. Vogt and A. D. Frawley, Phys. Rev. **C87**, 014908 (2013)
[5] G. Soyez, Phys. Lett. **B655**, 32-38 (2007)
[6] E. Iancu, K. Itakura and S. Munier, Phys. Lett. **B590**, 199-208 (2004)
[7] I. Balitsky, Nucl. Phys. **B463**, 99-160 (1996); Y. V. Kovchegov, Phys, ReV. **D60**, 034008 (1999)
[8] R. Enberg, M. H. Reno and I. Sarcevic, Phys. Rev. **D78**, 043005 (2008)
[9] J. L. Albacete et al., Eur. Phys. J. **C71**, 1705 (2011)
[10] C. Ewerz et al., JHEP **03**, 062 (2011); Y. S. Jeong et al., JHEP **11**, 025 (2014); C. Arguelles et al., Phys. Rev. **D92**, 074040 (2015)
[11] M. M. Block, L. Durand and P. Ha, Phys. Rev. **D89**, 094027 (2014)
[12] T. Altinoluk, N. Armesto, G. Beuf, A. Kovner and M. Lublinsky, Phys. Rev. **D93**, 054049 (2016)
[13] K. Kutak and S. Sapeta, Phys. Rev. **D86**, 094043 (2012)
[14] A. Bhattacharya, R. Enberg, M. H. Reno, I. Sarcevic and A. Stasto JHEP **06**, 110 (2015)
[15] UA1 Collaboration, Phys. Lett. **B256** (1991) 121; LHCb Collaboration, Eur. Phys. J. **C71**, 1645 (2011); ALICE Collaboration, JHEP **07**, 191 (2012); JHEP **11**, 065 (2012); Phys. Lett. **B738**, 97 (2014); ATLAS Collaboration, Nucl. Phys **B907**, 717-763 (2016); PHENIX Collaboration, Phys. Rev. Lett. **97**, 252002 (2006); Phys. Rev. **C91**, 014907 (2015); STAR Collaboration, Phys. Rev. **D86**, 072013 (2012); C. Lourenco and H. K. Wohri, Phys. Rept. **433**, 127 (2006); HERA-B Collaboration, Eur. Phys. J. **C52**, 531 (2007);
[16] LHCb Collaboration, R. Aaij et al., Nucl. Phys. **B871**, 1-10 (2013); JHEP **03**, 159 (2016)
[17] T. K. Gaisser, Astropart. Phys. **35**, 801 (2012)
[18] T. K. Gaisser, T. Stanev and S. Tilav, Front. Phys. (Beijing) **8**, 748-758 (2013)
[19] P. Lipari, Astropart. Phys. **1**, 195-227 (1993)
[20] M. G. Aartsen et al. [IceCube Collaboration], arXiv:1607.08006.

EPJ Web of Conferences **141**, 08001 (2017) DOI: 10.1051/epjconf/201714108001
ISMD 2016

Ultra-peripheral heavy-ion collisions with the CMS experiment

The CMS Collaboration

Abstract. Ultraperipheral collisions (UPCs) of heavy ions involve long range electromagnetic interactions at impact parameters larger than sum of their radii where hadronic interaction is largely suppressed and the exchanged photon materializes into qq(bar) bound state after interacting with the gluonic field of the target proton or ion. Photoproduction of heavy vector mesons (J/psi, Upsilon) thus provide direct information on the gluon distribution functions in the nucleon/nucleus at very low values of Bjorken-x. The CMS experiment has excellent capabilities for the measurement of the heavy vector mesons in the dimuon decay channel using the tracker and the muon chambers. The measured coherent J/psi photoproduction cross section in ultraperipheral Pb-Pb collisions using 2011 PbPb data and Upsilon photoproduction in ultraperipheral pPb collisions during 2013, will be presented. The prospects for future measurements using the data collected in the 2015 PbPb run will also be described.

1 Introduction

1.1 Coherent J/ψ production in PbPb UPCs

Ultraperipheral collisions (UPCs) of protons and ions involve electromagnetic interactions at impact parameters larger than the sum of the radii of colliding nuclei. These photon-nucleus and photon-photon collisions are produced at the LHC in abundance [1][2]. The cross section for photon-nucleus collisions is proportional to Z^2, where Z is the charge of the nucleus. The cross section for producing vector mesons such as J/ψ, $\psi(2S)$ and Υ is proportional to the square of the nuclear gluon density [3][4]. the gluon distribution function within nuclei is poorly constrained, especially at low Bijorken-x. This is currently a subject of strong interest since it dominates the initial state of heavy ion collisions. Photoproduction of vector mesons has been studied with lepton beams [5] and later at the electron-proton collider HERA [6][7]. The STAR and PHENIX collaborations at RHIC have studied ρ^0 and J/ψ production in ultraperipheral AuAu collisions at $\sqrt{s_{NN}}$ = 200 GeV [8][9]. These studies were not able to considerably constrain the nuclear gluon distribution function due to the small sample size. The production rate for UPC physics processes is much higher at the LHC. By measuring coherent J/ψ photoproduction in ultraperipheral PbPb collisions at $\sqrt{s_{NN}}$ = 2.76 TeV, the ALICE collaboration has found evidence that the nuclear gluon density is below that expected for a simple superposition of protons and neutrons in the nucleus[10][11].

1.2 Exclusive Υ production in pPb UPCs

In UPCs, hadrons (ions) interact electromagnetically where an exchanged (quasi)real photon materializes into a $q\bar{q}$ bound state after interacting with the gluon field of the target proton (ion) [2]. Since the

DOI: 10.1051/epjconf/201714108001

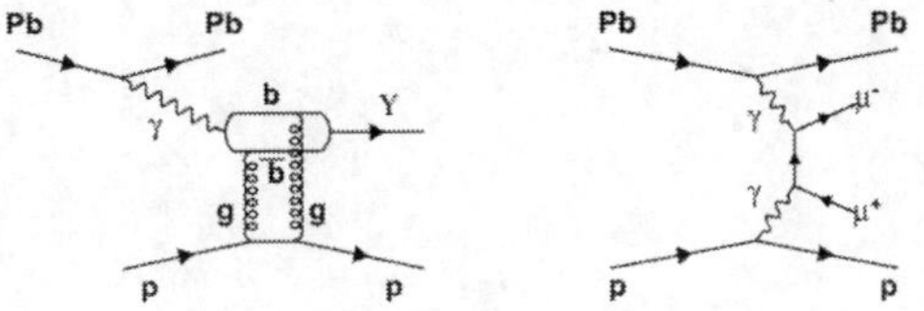

Figure 1. Diagrams representing exclusive Υ photoproduction (left), and exclusive dimuon QED continuum (right) in pPb collisions.

incoming hadrons remain intact after the interaction and no other particles are produced in the event, the process is called "exclusive". Given that the photon flux scales with the square of the emitting charge (Z^2), the emission of quarireal photons from the Pb-ion is strongly enhanced compared to that from the proton. Figure 1 (left) shows the diagram for exclusive Υ photoproduction in pPb collisions, dominated by the process pPb → (γp) → pΥ(nS)Pb. Similarly, the exchanged photon can interact with another equivalently radiated photon from the other hadron leading to two-photon collisions followed by exclusive dilepton production as shown in Fig. 1 (right) [2][12]. Since we are interested in studying exclusive Υ production via its dilepton decay, such photon-photon continuum constitutes a non-irreducible background for our signal.

2 Results

2.1 Coherent J/ψ photoproduction in ultra-peripheral PbPb collisions at $\sqrt{s_{NN}} = 2.76$ TeV

The coherent J/ψ photoproduction cross section in ultraperipheral PbPb collisions at $\sqrt{s_{NN}} = 2.76$ TeV, in conjunction with at least one neutron on the side of the interaction point and no neutron activity on the other side, is measured to be $d\sigma^{coh}_{X_nOn}/dy(J/\psi) = 0.36 \pm 0.04(\text{stat}) \pm 0.04(\text{syst})$ mb in the rapidity interval $1.8 < |y| < 2.3$. This measurement is extrapolated to the total coherent J/ψ cross section, resulting in $d\sigma^{coh}/dy(J/\psi) = 1.82 \pm 0.22(\text{stat}) \pm 0.20(\text{syst}) \pm 0.19(\text{theo})$ mb in the measured rapidity interval. These results complement recent measurements on coherent J/ψ photoproduction in ultraperipheral PbPb collisions at $\sqrt{s_{NN}} = 2.76$ TeV by the ALICE collaboration. An impulse approximation model prediction is strongly disfavored, indicating that nuclear effects expected to be present at low x and Q^2 values are needed to describe the data. The prediction given by the leading twist approximation, which includes nuclear gluon shadowing, is consistent with the data.

2.2 Exclusive Υ photoproduction in ultra-peripheral pPb collisions at $\sqrt{s_{NN}} = 5.02$ TeV

We reported the first measurement of the exclusive photoproduction of Υ(1S, 2S, 3S) mesons in the $\mu^+\mu^-$ decay modes in ultraperipheral pPb collisions at $\sqrt{s_{NN}} = 5.02$ TeV, using data collected with the CMS detector in 2013, corresponding to an integrated luminosity of 32.6 inverse nb. The exclusive photoproduction cross sections have been measured as a function of the photon-proton center of mass energy in the range $91 < W_{\gamma p} < 826$ GeV, probing the region of parton fractional momenta in the proton $x \approx 10^{-4} - 10^{-2}$, bridging a previously unexplored region between HERA and LHCb measurements. Our data are compatible with a power law dependence of $\sigma(W_\gamma)$, disfavouring faster rising extrapolations of HERA results (STARLIGHT) and LO pQCD predictions. The differential cross section $d\sigma/d|t|$, has been also measured in the range $|t| < 1.0\ GeV^2$, and the exponential spectral slope $b = 4.5 \pm 1.7(\text{stat}) \pm 0.6(\text{syst})\ GeV^{-2}$ has been extracted, in agreement with earlier measurements and consistent with predictions based on pQCD models. The data presented here provide new constraints for the evolution of the gluon density at low values of fractional momenta x in the proton.

 DOI: 10.1051/epjconf/201714108001

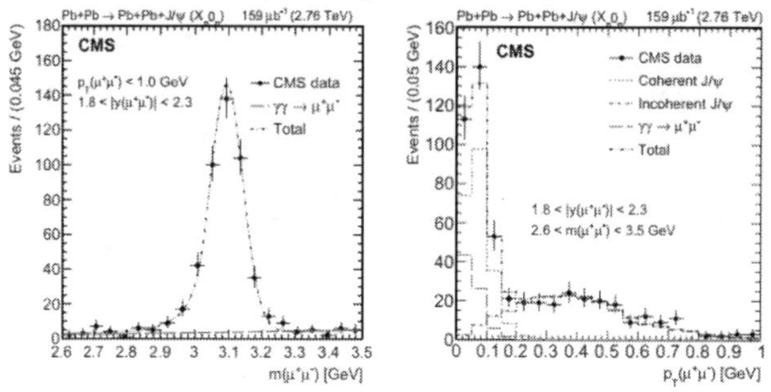

Figure 2. Results from the simultaneous fit to dimuon invariant mass (left) and p_T (right) distributions from opposite-sign muon pairs with $p_T < 10$ GeV, $1.8 < |y| < 2.3$ and $2.6 < m(\mu^+\mu^-) < 3.5$ GeV for the X_nO_n break-up mode, after all selections are applied. In the left panel the green curve represents the $\gamma + \gamma$ component (second-order polynomial) and the black curve the sum of the $\gamma + \gamma$, coherent J/ψ, and incoherent J/ψ components. In the right panel the green, red, and blue curves represent $\gamma + \gamma$, coherent J/ψ and incoherent J/ψ conponents. Only statistical uncertaintied are shown. The data are not corrected by acceptance and efficiencies, and the MC templates are folded with the detector response simulation.

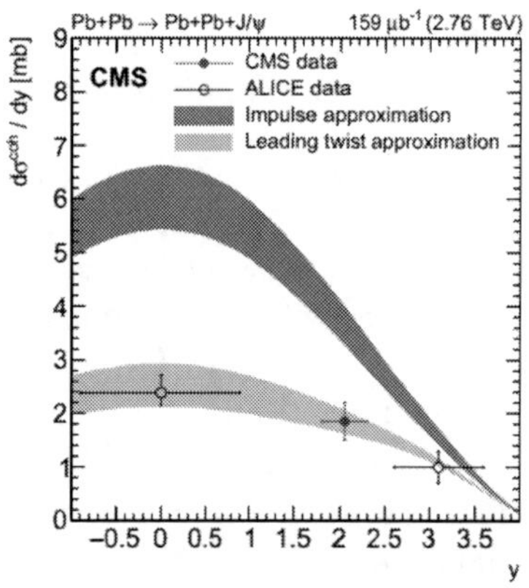

Figure 3. Differential cross section versus rapidity for coherent J/ψ production in ultraperipheral PbPb collisions at $\sqrt{s_{NN}}$ = 2.76 TeV, measured by ALICE[2] and CMS. the vertical error bars include the statistical and systematic uncertainties added in quadrature, and the horizontal bars represent the range of the measurements in y. Also the impulse approximation and the leading twist approximation calculations are shown.

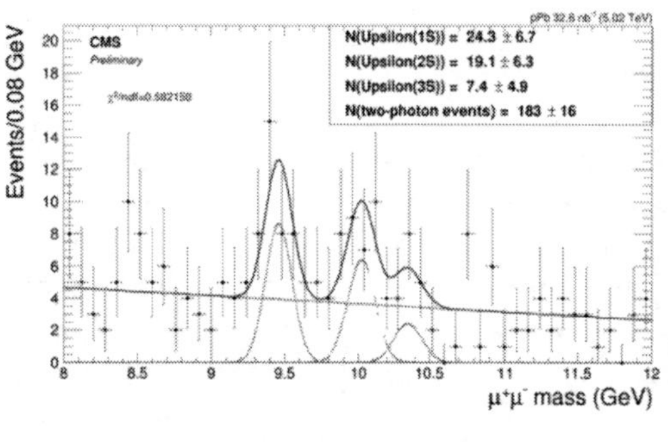

Figure 4. Invariant mass distribution for the $\mu^+\mu^-$ pairs in the range $8 < M_{\mu^+\mu^-} < 12$ GeV. The three peaks correspond to the $\Upsilon(1S)$, $\Upsilon(2S)$, and $\Upsilon(3S)$ mesons. The fit to the data is performed with ROOFIT. The blue line corresponds to the polynomial fit to the two-photon QED continuum, the red dashed line to the Gaussian fit of the resonances. The corresponding number of events are indicated in the legend.

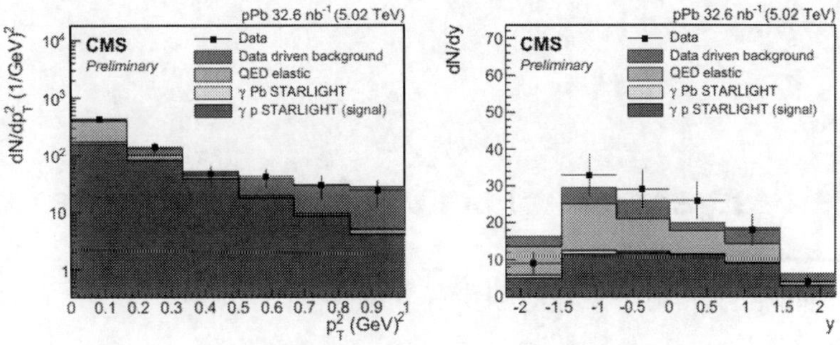

Figure 5. Distributions of the transverse momentum squared p_T^2 and rapidity y of the muon pairs with invariant mass $9.12 < m_{\mu^+\mu^-} < 10.64$ GeV in the combined pPb and Pbp samples.

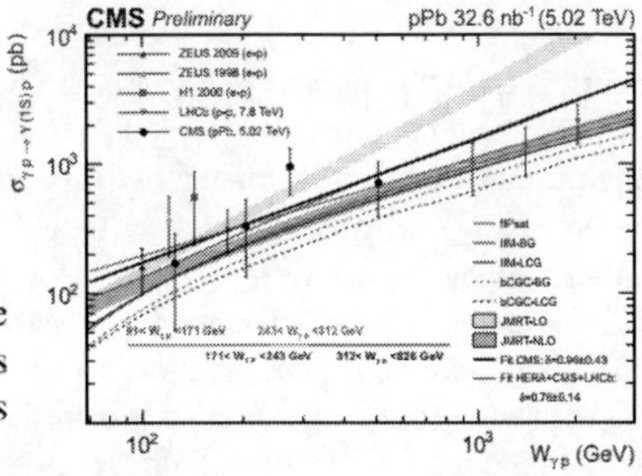

Figure 6. Cross section for exclusive Υ(1S) photoproduction, $\gamma p \to \Upsilon(1S)p$ as a function of photon-proton center-of-mass energy.

References

[1] C. A. Bertulani, S. R. Klein, and J. Nystrand, "Physics of ultra-peripheral nuclear collisions" **Annual Review of Nuclear and Particle science 55, no. 1**, 271-310 (2005)

[2] A. Baltz et al., "The physics of ultraperipheral collisions at the LHC" **Physics Reports 458, no. 13**, 1-171 (2008)

[3] M. Ryskin, "Diffractive J/ψ electroproduction in LLA QCD" **Z. Phys. C57**, 89-92 (1993)

[4] S. J. Brodsky et al., "Diffractive leptoproduction of vector mesons in QCD" **Phys. Rev. D50**, 3134-3144 (1994)

[5] L. J. Lanzerotti et al., "High-Energy Photoproduction of Neutral Rho Mesons" **Phys. Rev. Lett. 15**, 210-213 (1965)

[6] ZEUS Collaboration, "Exclusive photoproduction of J/ψ mesons at HERA" **Eur.Phys.J. C24**, 345-360 (2002)

[7] ZEUS Collaboration, "Exclusive electroproduction of J/ψ mesons at HERA" **Nucl.Phys. B695**, 3-37 (2004)

[8] STAR Collaboration, "ρ^0 photoproduction in ultraperipheral relativistic heavy ion collisions at $\sqrt{s_{NN}}$ = 200 GeV" **Phys. Rev. C 77**, 034910 (2008)

[9] S. Afanasiev et al., "Photoproduction of and of high mass in ultra-peripheral Au + Au collisions at $\sqrt{s_{NN}}$ = 200 GeV" **Phys. Lett. B 679, no. 4**, 321-329 (2009)

[10] ALICE Collaboration, "Coherent J/ψ photoproduction in ultra-peripheral Pb-Pb collisions at $\sqrt{s_{NN}}$ = 2.76 TeV" **Phys. Lett. B718**, 1273-1283 (2013)

[11] V. Guzey, E. Kryshen, M. Strikman, and M. Zhalov, "Evidence for nuclear gluon shadowing from the ALICE measurements of PbPb ultraperipheral exclusive J/ψ production" **Phys.Lett. B726**, 290-295 (2013)

[12] V. Budnev et al., "The Two photon particle production mechanism. Physical problems, Applications, Equivalent photon approximation" **Phys. Rep. 15**. 181 (1975)

Measurements of beauty-decay electrons in ALICE at the LHC

Minjung Kim for the ALICE Collaboration[1,a]

[1] *Inha University, Incheon, Republic of Korea*

Abstract. Heavy quarks are essential probes of the evolution of the Quark-Gluon Plasma (QGP) created in ultra-relativistic heavy-ion collisions. In particular, the dependencies of the medium-induced parton energy loss on the parton mass and colour charge can be investigated via measurements of open heavy-flavour hadrons. To quantify medium effects in AA collisions, one needs to study p-A collisions in order to disentangle possible cold nuclear matter effects. Thanks to the excellent vertex and impact parameter resolution of ITS and electron-identification capability provided by TPC and TOF in the ALICE experimental setup, measurements of beauty production in pp, p-Pb and Pb-Pb collisions could be done via electrons from semi-leptonic decays of beauty hadrons. The recent measurements of beauty-decay electrons in p-Pb collisions at $\sqrt{s_{\rm NN}}$ = 5.02 TeV and in Pb-Pb collisions at $\sqrt{s_{\rm NN}}$ = 2.76 TeV in ALICE are presented.

1 Introduction

In ultra-relativistic heavy-ion collisions, heavy quarks (charm and beauty) are predominantly created in the initial hard scattering processes due to their large masses [1]. Since their production is expected to be earlier than the formation of the Quark-Gluon Plasma (QGP), heavy quarks can traverse the hot and dense medium created in heavy-ion collisions, interacting with its constituents and experience the full system evolution. The medium-induced parton energy loss of quarks is smaller than that of gluons due to their smaller colour coupling to the medium with respect to gluons. Furthermore, the medium-induced parton energy loss of heavy partons is expected to be suppressed compared to light partons in both radiative energy loss process [2–4] and collisional energy loss process [5]. The energy loss hierarchy among different parton species can be studied via measurements of open heavy-flavour hadrons nuclear modification factor ($R_{\rm AA}$) and comparison with those of light-flavour hadrons. $R_{\rm AA}$ is defined as :

$$R_{\rm AA} = \frac{1}{< T_{\rm AA} >} \frac{{\rm d}N_{\rm AA}/{\rm d}p_{\rm T}}{{\rm d}\sigma_{\rm pp}/{\rm d}p_{\rm T}}, \quad (1)$$

where ${\rm d}N_{\rm AA}/{\rm d}p_{\rm T}$ is the $p_{\rm T}$-differential yield in AA collisions, ${\rm d}\sigma_{\rm pp}/{\rm d}p_{\rm T}$ is the $p_{\rm T}$-differential cross section in pp collisions, and $< T_{\rm AA} >$ is the average nuclear overlap function for a given centrality class in AA collisions. Especially, separate measurements of beauty hadrons from charm hadrons can test the mass dependence of medium-induced parton energy loss even at high $p_{\rm T}$ [6].
To quantify medium effects in AA collisions through the $R_{\rm AA}$, one needs to disentangle the possible cold nuclear matter (CNM) effects, such as the modification of parton distribution functions in the

[a]e-mail: minjung.kim@cern.ch

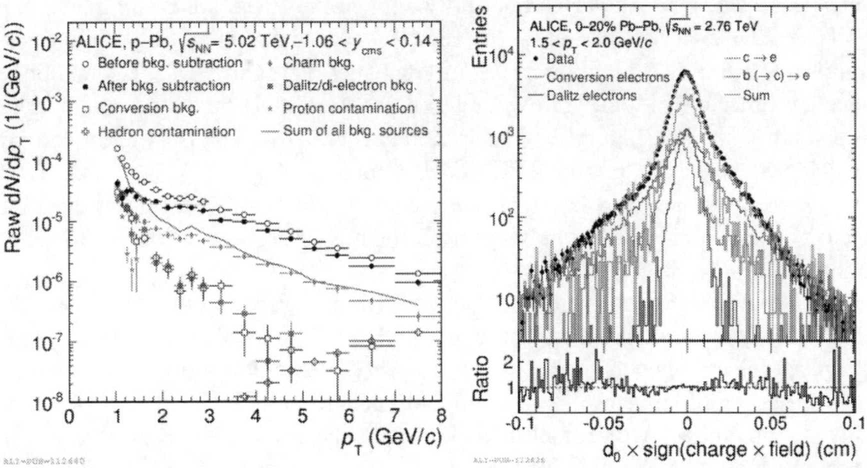

Figure 1. Raw p_T distributions of electrons after applying eID and minimum d_0 cut compared with background components in p-Pb collisions (left). d_0 distributions of electrons in the 20% most central Pb-Pb collisions (right). The individual distributions for each sources were obtained from Monte Carlo simulations with data-drive corrections [12]

nucleus [7] and the transverse momentum broadening from multiple scattering [8], by measuring the nuclear modification factor in p-A collisions (R_{pA}).
ALICE (A Large Ion Collider Experiment) is dedicated to studying QGP properties [9]. The ALICE detector provides electron identification capability down to very low p_T and excellent momentum and track impact parameter resolution at mid-rapidity region [10, 11]. Beauty-hadron production at mid-rapidity have been measured via electrons from semi-leptonic decays of beauty hadrons.

2 Analysis

In the mid-rapidity region, charged particles were reconstructed with Inner Tracking System (ITS), which consists of the 6-layered silicon detector and Time Projection Chamber (TPC) embedded in a magnetic field of 0.5 T, providing p_T resolution better than 2.5% for $p_T \leq 30$ GeV/c and transverse impact parameter (d_0) resolution better than 75 μm for $p_T > 1$ GeV/c [13]. d_0 is defined as the track distance of closest approach to the primary vertex in the plane transverse to the beam axis. Electron identification (eID) was performed by requiring the dE/dx in the TPC to be within -0.5σ and +3σ with respect to the expected dE/dx of electrons in both p-Pb and Pb-Pb analyses. The remaining hadron contamination was determined after requiring the measured time-of-flight in Time Of Flight (TOF) detector to be within ±3σ of the expected value for electrons for whole measured p_T range in the Pb-Pb analysis and for $p_T \geq 2.5$ GeV/c in the p-Pb analysis for further hadron rejection [12].
The electron candidates fulfilling above selection criteria are composed of signal, i.e. beauty-decay electrons, and background electrons coming from various sources [14]. At low p_T, the dominant sources of background electrons are Dalitz decays of neutral light mesons, di-electron decays of light-vector mesons and photon conversions in the beam pipe and inner most layer of ITS. At high p_T, semi-leptonic decays of charm hadrons are the main background source. Long lifetime of beauty hadrons leads to larger d_0 of electrons from beauty-hadron decays than those of background electrons. Based on impact parameter information, two different analysis strategies were used.

In p-Pb analysis, requirement of minimum d_0 for electron candidates was used for extracting the signal from electron candidates, which was successfully applied in previous measurements in pp collisions at $\sqrt{s}$ = 7 TeV and $\sqrt{s}$ = 2.76 TeV [14, 15]. Electron candidates satisfying minimum d_0 are shown as open circle spectrum as a function of p_T in left panel of Figure 1. The remaining background electrons were estimated by Monte Carlo simulations re-weighted to match the p_T distributions of the background sources in data using ALICE measurements [16–18]. The fraction of hadron contamination was calculated by parameterizing TPC dE/dx at high p_T ($p_T \geq 4\,\mathrm{GeV}/c$) and ITS dE/dx at low p_T ($1\,\mathrm{GeV}/c < p_T < 1.2\,\mathrm{GeV}/c$). By subtracting background components from electron candidates, the raw yield of beauty-decay electrons shown in left panel of Figure 1 as black-solid circle spectrum was obtained.
In Pb-Pb collisions, the signal extraction was done by fitting d_0 distribution of electron candidates with signal (beauty-decay electrons) and background templates obtained from Monte Carlo simulations as shown in right panel of Figure 1. The discrepancy of d_0 distributions between data and Monte Carlo simulations caused by difference of resolution and background-source spectra was corrected based on data-driven method and ALICE measured D meson spectrum [19], respectively. The remaining imperfection of Monte Carlo simulations on describing data was propagated to final systematic uncertainty [12]. Since both the hadron contamination and Dalitz electrons originate from or close to interaction point, their d_0 distributions are similar. Thus, the contribution of hadron contamination was absorbed into Dalitz electrons.
After correction for geometrical acceptance as well as efficiency of the track reconstruction, matching and track selection criteria, the p_T-differential spectrum of electrons from beauty-hadron decays could be obtained in p-Pb collisions and in Pb-Pb collisions. In order to calculate R_{pPb} as well as R_{PbPb}, corresponding pp reference spectra at $\sqrt{s}$ = 5.02 TeV and $\sqrt{s}$ = 2.76 TeV, respectively, was obtained by a pQCD-driven $\sqrt{s}$-scaling of the cross section of electrons from beauty-hadron decays measured in pp collisions at $\sqrt{s}$=7 TeV [14].

3 Results

The nuclear modification factor of electrons from beauty-hadron decays in p-Pb collisions (R_{pPb}) and in the 20% most central Pb-Pb collisions (R_{PbPb}), measured as a function of p_T, is shown in Figure 2. The R_{pPb} is compatible with unity within uncertainties for measured p_T range. The result is well-described by theoretical model calculations including CNM effects and blast-wave model calculations considering possible hydrodynamic effects [20]. The measured R_{PbPb} is described by theoretical models of beauty-quark energy loss in the QGP within uncertainties. Thus, the suppression of electrons from beauty-hadron decays in the 20% most central Pb-Pb collisions for $p_T > 3\,\mathrm{GeV}/c$ can be understood as an effect of the hot and dense medium created in Pb-Pb collisions on beauty quarks.

Acknowledgements

This work was supported by National Research Foundation of Korea (NRF), Basic Science Research Program through the National Research Foundation of Korea funded by the Ministry of Education, Science and Technology (Grant number: 2014R1A1A1008246) and by Inha University Research Grant (INHA-47297).

References

[1] A. Andronic et al., Eur. Phys. J. **C76**, 107 (2016), arXiv:1506.03981
[2] Y.L. Dokshitzer, D.E. Kharzeev, Phys. Lett. **B519**, 199 (2001), hep-ph/0106202

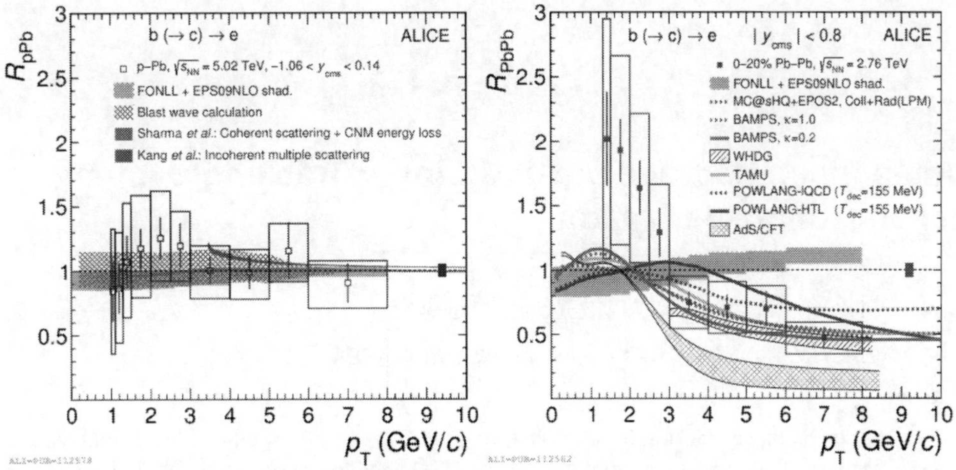

Figure 2. Nuclear modification factor of electrons from beauty-hadron decays at mid-rapidity in p-Pb collisions at $\sqrt{s_{NN}}$ = 5.02 TeV (left) and in Pb-Pb collisions at $\sqrt{s_{NN}}$ = 2.76 TeV in the 20% most central collisions (right) with several theoretical model predictions [12]

[3] M. Djordjevic, M. Gyulassy, Nucl. Phys. **A733**, 265 (2004), nucl-th/0310076
[4] N. Armesto, C.A. Salgado, U.A. Wiedemann, Phys. Rev. **D69**, 114003 (2004), hep-ph/0312106
[5] H. van Hees, V. Greco, R. Rapp, Phys. Rev. **C73**, 034913 (2006), 0508055
[6] N. Armesto, A. Dainese, C.A. Salgado, U.A. Wiedemann, Phys. Rev. D **71**, 054027 (2005)
[7] K.J. Eskola, H. Paukkunen, C.A. Salgado, JHEP **04**, 065 (2009), arXiv:0902.4154
[8] I. Vitev, Phys. Rev. C **75**, 064906 (2007)
[9] K. Aamodt et al., (ALICE Collaboration), JINST **3**, S08002 (2008)
[10] F. Carminati et al., (ALICE Collaboration), J. Phys. G: Nucl. Part. Phys. **30**, 1517 (2004)
[11] B. Alessandro et al., (ALICE Collaboration), J. Phys. G: Nucl. Part. Phys. **32**, 1295 (2004)
[12] J. Adam et al., (ALICE Collaboration) (2016), arXiv:1609.03898
[13] B. Abelev et al., (ALICE Collaboration), Int. J. Mod. Phys. **A29**, 1430044 (2014), arXiv:1402.4476
[14] B. Abelev et al., (ALICE Collaboration), Phys. Lett. **B721**, 13 (2013), arXiv:1208.1902
[15] B. Abelev et al., (ALICE Collaboration), Phys. Lett. **B738**, 97 (2014), arXiv:1405.4144
[16] B. Abelev et al., (ALICE Collaboration), Phys. Lett. **B728**, 25 (2014), arXiv:1307.6796
[17] J. Adam et al., (ALICE Collaboration) (2016), arXiv:1601.03658
[18] B. Abelev et al., (ALICE Collaboration), Phys. Rev. Lett. **113**, 232301 (2014), arXiv:1405.3452
[19] B. Abelev et al., (ALICE Collaboration), JHEP **09**, 112 (2012), arXiv:1203.2160
[20] A.M. Sickles, Phys. Lett. **B731**, 51 (2014), arXiv:1309.6924

Studies for an upgrade of ALICE Inner Tracking System: Pixel chip characterization

Jonghan PARK on behalf of the ALICE collaboration[1,a]

[1] *INHA UNIVERSITY, 100 Inharo, Nam-gu Incheon 22212, KOREA*

Abstract. Inner Tracking System (ITS) of ALICE is used for vertex determination and tracking. Future heavy-ion program at the LHC aims to run with high luminosity. To address this challenge, upgrade program of ITS is underway, which aims at better position resolution (factor of 3), high detection efficiency (>99%), high-rate readout capabilities (100 kHz for Pb-Pb) and moderate radiation hardness (> 700 krad). The new ITS will be composed with 7 layers of silicon pixel chip based on Monolithic Active Pixel Sensor (MAPS) technology. The characterization test of various version of prototype chips at different phases of development has been performed. This contribution will provide the main characterization results obtained from the measurements performed at laboratories and using test beam for finalizing the pixel chip specification.

1 Introduction

ALICE experiment is a dedicated heavy-ion experiment at the LHC. The main goal of ALICE experiment is to study properties of hot and dense matter (regarded as Quark-Gluon Plasma) expected to be produced heavy-ion collisions [1]. Up to now, ALICE has confirmed the creation of hot hadronic matter at high density and temperature by measurement of light and heavy flavours, flow, transverse momentum, jet, etc. However, the current experimental setup is not fully optimized to measure the rare probes over a broad range of p_T. Further progress on the characterization of QGP properties requires precision measurements of rare probes over a large kinematic range (from low p_T to high p_T) and as a function of multi-differential observables. For significant improvement of vertexing and tracking capabilities at low p_T, ALICE ITS upgrade is underway, which aims at better position resolution, high detection efficiency and high-rate readout capabilities [2]. The ALICE collaboration confirmed the chip design after successfully performing chip characterization test for different prototypes in various phases. The results of the chip characterization test will be presented in the following sections.

2 Upgrade plan of Inner Tracking System (ITS)

The ITS is located in a region closest to the beam pipe and used for vertex determination and tracking. The current ITS is constructed by 6 layers of three different technologies: 2 layers of silicon pixel

[a] e-mail: jonghan@cern.ch

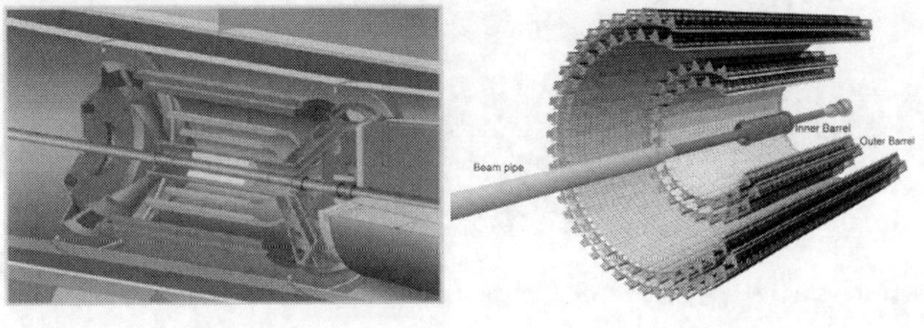

Figure 1. Current inner tracking system(left) and new inner tracking system(right) schematics [4]. The current ITS is constructed by three different types of detectors (silicon pixel, silicon drift, silicon strip). The new ITS will be composed of only single type of detector, i.e. monolithic active pixel sensor having seven layers: three inner barrels and four outer barrels

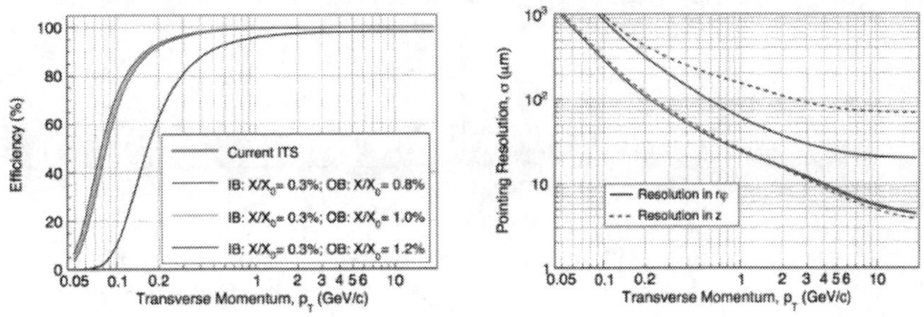

Figure 2. Simulation results of tracking efficiency(left) and pointing resolution(right) for pion [5]. The results are compared between current ITS and new ITS with different material budget for new ITS outer barrel.

detector(Inner barrel), 2 layers of silicon drift detector and 2 layers of silicon strip detector (outer barrel) as shown in Figure 1. The silicon pixel detector of inner barrel is based on hybrid silicon pixel technology. During the second long shutdown (LS2) of the LHC in 2019-20, the current ITS will be replaced by a new ITS, which will be constructed by 7 layers (3 inner layers and 4 outer layers) of silicon pixel detector based on monolithic active pixel sensor technology. A schematics of the current and new ITS are shown in Figure 1.

The new ITS will achieve better position resolution by a factor of 3 and tracking efficiency higher than 99%. To improve position resolution, the innermost layer is moved close to the beam pipe from 39mm to 22mm, reducing material budget from 1.14% to 0.3% for inner barrel and reducing pixel size to 28μm×28μm (currently, 50μm×425μm) [3, 4]. The tracking efficiency and the p_T resolution will be increased thanks to the increase of the radial extension and granularity of the detector. The simulation results for the tracking efficiency and pointing resolution of the new ITS compared to current ITS are shown in Figure 2. Furthermore, the new ITS will allow fast readout (Pb-Pb interaction rate of 100 kHz) and easy maintenance (fast insertion and removal of non-functioning detector modules) during yearly shutdown.

The new ITS silicon pixel chip is based on Monolithic Active Pixel Sensors (MAPS) technology with 0.18μm CMOS imaging sensor by Tower Jazz. The CMOS circuitry is implemented on high-resistivity (>1kΩ cm) p-type epitaxial layer on p-type substrate shown in Figure 3. The electron-hole pairs are generated in expitaixal layer by charged particles and collected at NWELL collection diode.

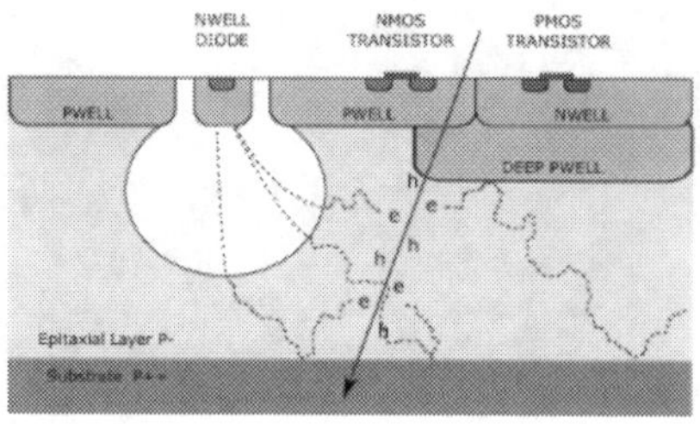

Figure 3. Cross-sectional diagram of the MAPS showing collection diode and transistors [4].

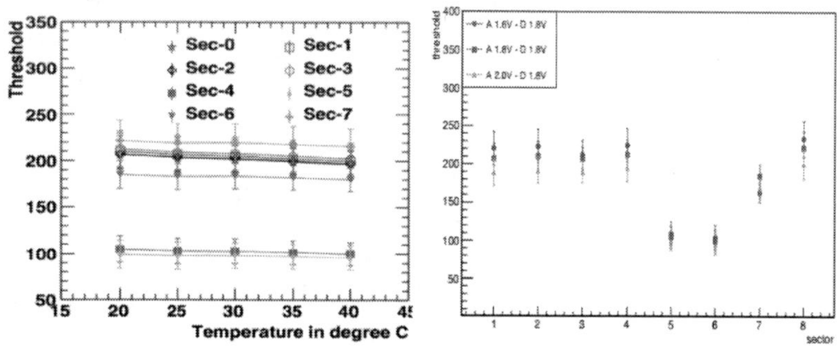

Figure 4. Temperature dependence (left) and analog supply voltage dependence (right) results. pALPIDE-3b chip is used with the back-bias voltage of -3V. The results show that the chip response is not changed in each conditions.

The DEEP PWELL shields NWELL containing PMOS transistor from collecting signal charge. It allows of full CMOS circuitry within active area [6].

3 Chip characterization test

The ALPIDE prototypes have 15 mm $\times$ 30 mm dimensions and contain 1024 $\times$ 512 pixels. The ALPIDE prototype is UBS-based test system used both for laboratory measurements and test beam. It provides all readout and control functionality as well as ADCs for current measurements and has flexibility, which could be easily adapted for the tests of the other chips.

The laboratory test verified the dependence of the chip environment on the chip responses. For temperature dependence study, the chip temperature is changed between 20 and 40 degree by 5 degree steps and the threshold values are compared with different temperature shown in Figure 4. Threshold values remain same with increasing temperature for the back-bias voltage of -3V. For supply voltage dependence study, the analog supply voltage is changed from 1.6V to 2.2V by 0.2V steps. When the analog supply voltage is changed, the chip responses such as the threshold and the noise are also changed because real voltage and current on in-pixel circuitry are changed. However, change in analog supply voltage is compensated by tuning DAC values as shown in Figure 4. At the condition of the tuned DAC values, the threshold and noise values remain almost constant for all back-bias voltage condition. For the both studies, the cluster multiplicities were measured. The cluster multiplicities are found to be independent of chip temperature and of the analogue supply voltage.

DOI: 10.1051/epjconf/201714108003

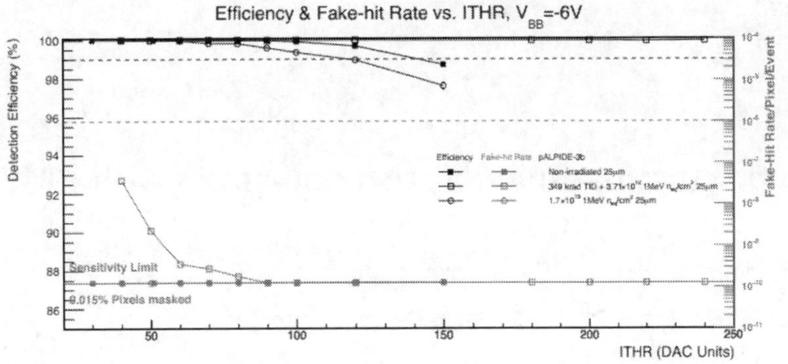

Figure 5. Detection efficiency and fake-hit rates for pALPIDE-3 with the back-bias voltage of -6V before and after irradiation.

The characterization of chip by test beam has been progressing at following sites: CREN, Pohang (Korea), SLRI (Thailand), Frascati (Italy). The ALPIDE prototypes are tested before and after irradiation up to the expected maximum levels of the radiation. The ALPIDE maintains the detection efficiency of $\geq 99\%$ over a wide threshold value range and achieved very low fake-hit rates before and after irradiation as shown in Figure 5.

4 Summary

The ALICE ITS upgrade program is underway for significant improvement of vertexing and tracking capabilities at low p_T and high readout rate. The chip which will be used in new ITS is developed after characterization of various prototype chips at various level to optimize its engineering design. The design goal of the chip is achieved from lab test and test beam. The ALPIDE prototype has been confirmed high detection efficiency($\geq$ 99%), low fake hit rates($<10^{-6}$), temperature and radiation hardness, etc. In future heavy-ion experiment, the new ITS will allow to measure rare probes and will improve the accuracy of the measurements.

References

[1] T.A. Collaboration, K. Aamodt et al., Journal of Instrumentation **3**, S08002 (2008)
[2] B.A. et al (ALICE), Journal of Physics G: Nuclear and Particle Physics **41**, 087001 (2014)
[3] G. Contin, Journal of Instrumentation **7**, C06007 (2012)
[4] B. Abelev et al. (ALICE Collaboration), Tech. Rep. CERN-LHCC-2013-024. ALICE-TDR-017 (2013), `http://cds.cern.ch/record/1625842`
[5] P. Riedler, Nuclear Physics A **956**, 866 (2016), the {XXV} International Conference on Ultrarelativistic Nucleus-Nucleus Collisions: Quark Matter 2015The {XXV} International Conference on Ultrarelativistic Nucleus-Nucleus Collisions: Quark Matter 2015
[6] M. Mager, Nuclear Instruments and Methods in Physics Research Section A: Accelerators, Spectrometers, Detectors and Associated Equipment **824**, 434 (2016), frontier Detectors for Frontier Physics: Proceedings of the 13th Pisa Meeting on Advanced Detectors

EPJ Web of Conferences **141**, 08004 (2017) DOI: 10.1051/epjconf/201714108004
ISMD 2016

D meson production in heavy ion collisions with CMS

Hyunchul Kim[1,a] on behalf of the CMS Collaboration

[1] *Chonnam National University, 61186, Gwangju, Republic of Korea*

Abstract. The nuclear modification factor, R_{AA}, of the D^0 meson production has been measured in pp and PbPb collisions at $\sqrt{s_{NN}}$ = 2.76 TeV and 5.02 TeV with CMS detector. This measurement is important in investigating the properties of the high-density QCD (quantum chromodynamics) matter. The dependence on the centrality and transverse momentum up to 100 GeV were presented in a poster at this conference.

1 Introduction

The production of heavy quarks is an attractive probe of understanding the mechanism of heavy quark interaction with the medium created in heavy ion collisions. Usually the heavy quarks (charm and beauty) are produced mainly in the initial stage of heavy ion collisions, before the formation of the hot and dense QCD matter. When they propagate through the matter, their properties are modified, therefore, the measurement of heavy-flavor mesons (a quark antiquark light-heavy quarks) probes the medium-induced energy loss of partons created in heavy ion collisions. Theoretical models expect a flavor dependent energy loss in heavy ion collisions, which is limited by dead cone effect (suppression of gluon radiation from massive partons at small angles [1]) for heavier beauty and charm quarks, compared to gluons and up, down or strange quarks.

At the LHC (Large Hadron Collider) energies, the charm cross-section, and hence that of charm mesons, like D meson, is high. The CMS (Compact Muon Solenoid) Collaboration measured the production of the prompt D^0 mesons in PbPb collisions at $\sqrt{s_{NN}}$ = 2.76 TeV [2] with Run 1 data (using an extrapolated pp reference), and at $\sqrt{s_{NN}}$ = 5.02 TeV [3] with both pp and PbPb data recorded in the LHC Run 2. In this presentation, there were presented the analysis procedure and the results of the nuclear modification factor R_{AA} (the ratio of the corrected yield of D meson in PbPb and pp collisions) at the two collision energy.

2 Experimental environment and analysis procedure

The D^0 meson and its charge conjugate are reconstructed in the mid-rapidity region ($|y| < 1$) of the CMS detector via the hadronic decay channel $D^0 \rightarrow K\pi$. A detailed description of the CMS detector can be found in Ref. [4]. Out of whole CMS detector system, the inner silicon tracker is the main detector used for this analysis. The silicon tracker, located in the 3.8 T magnetic field generated by the superconducting solenoid, has an excellent track impact parameter resolution: 200 μm at 1 GeV/c, and 10 μm at 20 GeV/c.

[a] e-mail: hyunchul.kim@cern.ch

These studies are performed using 2011 PbPb data collected at $\sqrt{s_{NN}}$ = 2.76 TeV, and 2015 pp and PbPb at $\sqrt{s_{NN}}$ = 5.02 TeV. The 2011 and 2015 PbPb samples correspond to an integrated luminosity of 166 μb^{-1} and 404 μb^{-1}, respectively, while the 2015 pp sample corresponds to an integrated luminosity of 25.8 pb^{-1}. Both 2011 and 2015 data analyses use large minimum bias sample to measure D^0 candidates in the p_T ranges of 2.5 – 40 GeV/c (2011) and 2 – 16 GeV/c (2015). In order to enhance the amount of data at high p_T, dedicated high-level trigger (HLT) were used for both the 2015 pp and PbPb data taking in the higher p_T region (p_T > 16 GeV/c).

D^0 candidates are reconstructed by combining pairs of oppositely charged tracks and requiring an invariant mass within 0.2 GeV/c^2 of the nominal D^0 mass provided by the particle data group [5]. Kaon and pion candidates from D^0 are reconstructed from the inner trackers in the mid-rapidity region ($|\eta| < 1$) and p_T larger than 1 GeV/c by assuming one of the particles has the mass of the pion while the other has the mass of kaon, and vice-versa. In order to reduce background and enhance signal, several topological cuts are applied such as two-dimensional decay length normalized by its uncertainty, pointing angle which is the angle between total momentum of D^0 meson and the vector from primary vertex to secondary vertex in transverse plane and chi-square fit probability of D^0 vertex.

For the reference, pp data and FONLL expectation [6] are used. For the 2.76 TeV analysis, the ALICE measurement in pp collisions at 7 TeV is used, rescaled to 2.76 TeV using FONLL calculations in the lower p_T range (p_T < 16 GeV/c), while in the high p_T range (p_T > 16 GeV/c) FONLL calculation itself is used. In the 5.02 TeV analysis, 25.8 pb^{-1} pp collision events were collected by CMS, and the cross-section of prompt D^0 meson is calculated with the same procedure as for the PbPb analysis. The results from pp data is confirmed by FONLL expectation at the same energy scale.

3 Results

Figure 1 shows two example of invariant mass distributions for prompt D^0 meson candidates in PbPb collisions at 2.76 TeV (left) and 5.02 TeV (right). The signal shape is modeled by two Gaussian functions with same mean but different widths. The combinatorial background from random combination of pairs of tracks not produced from a D^0 meson decay is modeled by an exponential function in the 2.76 TeV analysis, and by a third order polynomial function for the 5.02 TeV data analysis. In addition, the background from mis-identified D^0 by swapped kaon-pion mass is shaped by a single Gaussian function.

The measured R_{AA}, defined in Eq. 1, and theoretical calculations for two centrality categories, 0 - 10% and 0 - 100%, are compared in Fig. 2. R_{AA} is decreasing up to p_T of 10 GeV/c, and increasing up to 100 GeV/c. The results are compared to various theoretical models in Fig. 2 [9–17].

$$R_{AA}(p_T) = \frac{1}{T_{AA}} \frac{dN^{D^0}_{PbPb}}{dp_T} / \frac{d\sigma^{D^0}_{pp}}{dp_T} \tag{1}$$

where T_{AA} is the nuclear overlap function (T_{AA} = 5.58 mb^{-1} for inclusive PbPb collisions and T_{AA} = 23.2 mb^{-1} for events in the centrality interval 0 - 10%) [7]. Results have similar p_T dependence for both 2.76 and 5.02 TeV. Also no significant centrality difference in the shape is observed. Lastly, Fig. 3 shows a comparison of R_{AA} for prompt D^0 meson, charged hadrons and non-prompt J/ψ at $\sqrt{s_{NN}}$ = 2.76 TeV. The open charm is more suppressed than the open beauty in the most central events.

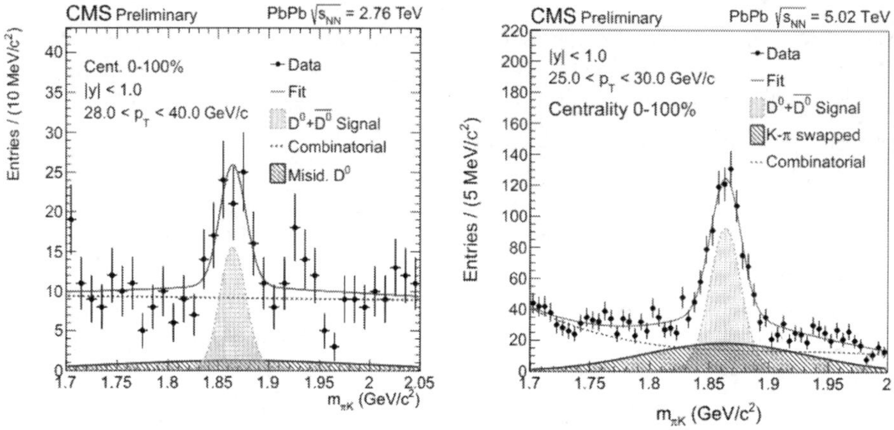

Figure 1. Invariant mass distributions of prompt D^0 meson candidates in an arbitrary p_T region in PbPb collisions at 2.76 TeV (left) [2] and 5.02 TeV (right) [3]. The orange areas represent the fitted prompt D^0 meson signal distributions, and the green-hatched areas show the background from kaon-pion mass swapping. The blue-dotted lines represent the combinatorial backgrounds.

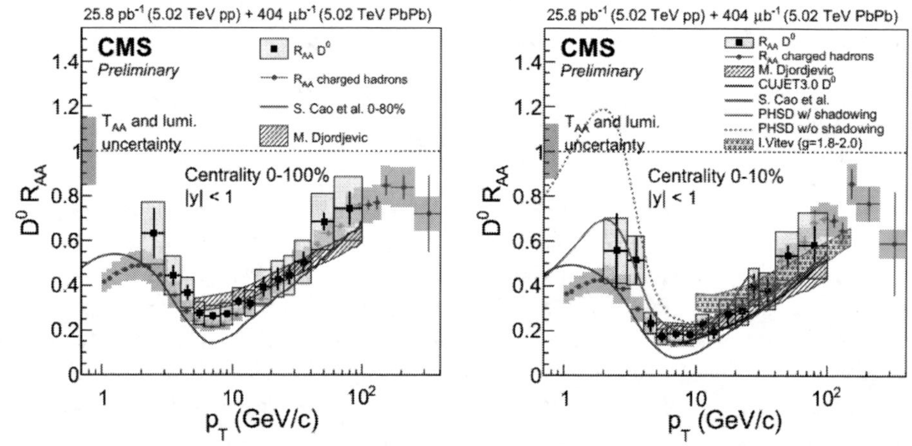

Figure 2. The R_{AA} of prompt D^0 meson as function of transverse momentum, for centrality inclusive range $(0 - 100\%)$ (left) and most central range $(0 - 10\%)$ (right), at 5.02 TeV [3]. The R_{AA} of inclusive charged particles are superimposed for equivalent event selections [8]. Experimental results are compared to various theoretical calculations [9–17].

4 Summary

The measurements of prompt D^0 meson production in 2.76 TeV and 5.02 TeV PbPb collisions from the CMS Collaboration were presented. Similar suppression trends between prompt D^0 meson and charged hadrons are observed in the kinematic regions analyzed.

 DOI: 10.1051/epjconf/201714108004

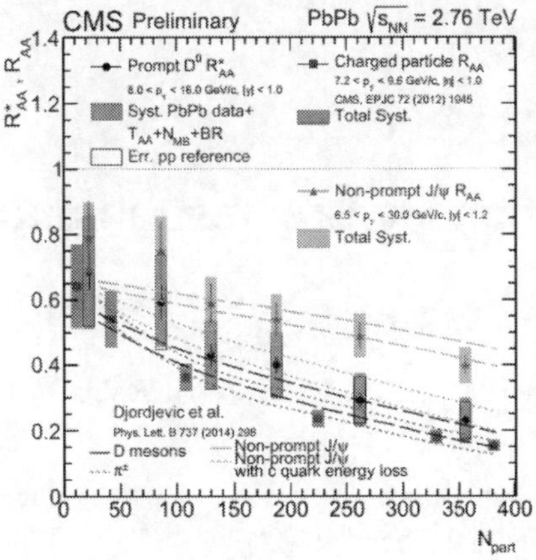

Figure 3. The R_{AA} of prompt D^0 meson (black circles) [2] compared to those of charged particle (blue squares) [18], and non-prompt J/ψ (green triangles) [19] from b quark, at 2.76 TeV, as function of number of participants. The systematic errors of prompt D^0 meson, charged particle and non-prompt J/ψ R_{AA} are drawn as gray, blue and green boxes respectively.

References

[1] Armesto N., Salgado C. A. and Wiedemann U. A., Phys. Rev. D **69**, 114003 (2004)

[2] CMS Collaboration, CMS Physics Analysis Summary CMS-PAS-HIN-15-005 (2015)

[3] CMS Collaboration, CMS Physics Analysis Summary CMS-PAS-HIN-16-001 (2016)

[4] CMS Collaboration, JINST **3**, S08004 (2008)

[5] C. Patrignani et al. (Particle Data Group), Chin. Phys. C **40**, 100001 (2016)

[6] M. Cacciari, M. Greco and P. Nason, JHEP **05**, 007 (1998) 007, M. Cacciari, S. Frixione and P. Nason, JHEP **03**, 006 (2001)

[7] Michael L. Miller, Klaus Reygers, Stephen J. Sanders, and Peter Steinberg, Annu. Rev. Nucl. Part. Sci. **57**, 205–243 (2007)

[8] CMS Collaboration, CMS Physics Analysis Summary CMS-PAS-HIN-15-015 (2016)

[9] J. Xu, J. Liao, and M. Gyulassy, JHEP **02**, 169 (2016)

[10] J. Xu, A. Buzzatti, and M. Gyulassy, JHEP **08**, 063 (2014)

[11] X. Jiechen, L. Jinfeng, and M. Gyulassy, Chin. Phys. Lett. **32**, no. 9, 092501 (2015)

[12] M. Djordjevic and M. Djordjevic, Phys. Rev. C **92**, 024918, (2015)

[13] G.-Y. Q. X.-N. W. Shanshan Cao, Tan Luo, Phys. Rev. C**94**, 014909 (2016)

[14] T. Song et al., Phys. Rev. C **92**, 014910 (2015)

[15] T. Song et al., Phys. Rev. C **93**, 034906 (2016)

[16] Z.-B. Kang et al., Phys. Rev. Lett. **114**, 092002 (2015)

[17] Y.-T. Chien et al., Phys. Rev. D **93**, 074030 (2016)

[18] CMS Collaboration, CMS Physics Analysis Summary CMS-PAS-HIN-12-014 (2012)

[19] CMS Collaboration, Eur. Phys. J. C **72**, 1945 (2012)

EPJ Web of Conferences **141**, 08005 (2017) DOI: 10.1051/epjconf/201714108005
ISMD 2016

Two-particle correlation via Bremsstrahlung

Soyeon Cho[1,a], Kayoung Park[1], and Jin-Hee Yoon[1,b]

[1] *Department of Physics, Inha University, Incheon, Republic of Korea*

Abstract. Ridge is the well-known structure in two-particle angular correlations at high-energy heavy-ion collisions. This structure is physically understood through elliptic and higher-order flows at nucleus-nucleus collisions. This behavior is also found in small systems, such as proton-proton collisions, recently. However, Ridge structure in small system is hard to be understood using hydrodynamics, since small systems are not dense enough to produce the Quark-Gluon plasma. Thus, we try to describe this phenomena through kinematic interaction between jets and medium partons. In high-energy heavy-ion collision, the energetic particles called jets go out in specific direction and lose their energy while passing through the medium. During such process, photons/gluons are emitted from interaction between jets and medium partons. We concentrate on energy loss via photon radiations, known as Bremsstrahlung. Recently, two symmetric double scattering processes between jet particle and medium parton are reported to be able to produce certain constructive interference, which gives collective motion and medium partons are aligned along incoming jet particles. We conjecture that similar behavior might happen in Bremsstrahlung processes, and therefore we consider the two symmetric diagrams of photon emission and medium parton scattering. We expect these two amplitudes to give constructive interference leading to the collective motion of medium. We check the correlation between emitted photon and final jet, and those between medium parton and final jet for high-energy jet. To describe parton momentum distribution in medium, we use the Maxwell-Boltzmann distribution. We discover collective motion in both angular correlations. We also check the tendency of the angular correlation for two particles according to the incident angle of jet particle, energy of emitted photon and temperature of systems, respectively. We can conclude that collective motion is able to be understood through kinematic description.

1 Introduction

Ridge structure which is pseudo-rapidity independent shape is found in two-particle angular correlation for high-energy heavy-ion collisions. This structure can be explained by hydrodynamics, especially elliptic and higer-order flows. Recently, Ridge structure is also discovered in small systems such as proton-proton collisions with high energies. Such case is also understood through hydrodynamics but many physicist do not expect that hot and dense medium (Quark-Gluon plasma state) is produced at small systems, hence it should be understood carefully. Thus we try to describe Ridge structure through kinematic interaction between jets and medium, such as energy-loss mechanism.

[a]e-mail: soyeon.cho@cern.ch
[b]e-mail: jinyoon@inha.ac.kr

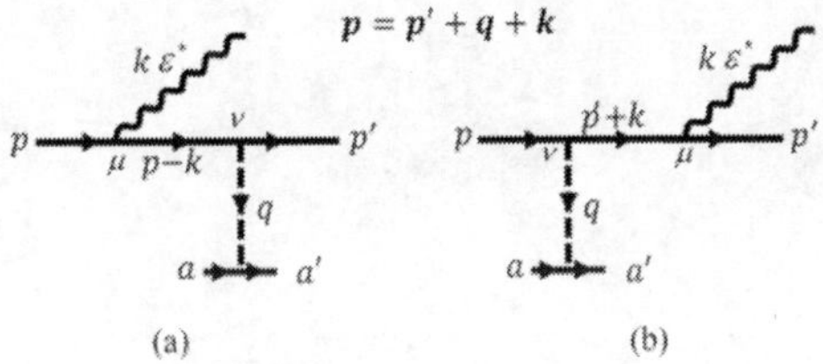

Figure 1. Two symmetric Feynman diagrams for Bremsstrahlung process.

In high-energy heavy-ion collisions, high-energy particles which come out in a particular direction are called as jet. When jets pass through the medium, they lose their partial energy by interacting medium partons. The main interaction is gluon or photon radiation process, which gluons or photons are emitted by scattering between jet particles and medium partons. In our study, we attempt to produce Ridge structure from energy-loss mechanism of jets by emitting gluons or phtons. First of all, we consider only phton radiation process, known as Bremsstrahlung.

2 Formalism

Wong reported that certain constructive interference appears in two symmetric double scattering processes between jet particle and medium parton. [4] This interference produces collective motion and medium partons are aligned along incident jet particles.

In our study, we conjecture that such constructive interference also arises in the Bremsstrahlung process. Thus, we consider two symmetric diagrams of γ emission and medium parton scattering as shown as Figure 1. We expect these two amplitudes give constructive interference leading to collective motion for medium, since the cross section is the absolute-square of the sum of amplitudes, as in equation (1).

$$d\sigma \sim |M_{(a)} + M_{(b)}|^2 = |M_{(a)}|^2 + |M_{(b)}|^2 + (interference) \tag{1}$$

Here each amplitudes are given as

$$\begin{aligned} M_{(a)} &= -i\bar{u}\,(p')\,(-ig\gamma^{\nu})\,\frac{1}{q^2}\bar{u}\,(a')\,(-ig\gamma_{\nu})\,u\,(a)\,\frac{i\left(\not{p} - \not{k} + m\right)}{(p-k)^2 - m^2 + i\epsilon}\,(-ig\gamma^{\mu})\,\epsilon^*_{\mu}u\,(p) \\ M_{(b)} &= -i\bar{u}\,(p')\,\epsilon^*_{\mu}\,(-ig\gamma^{\mu})\,\frac{i\,(\not{p}' + \not{k} + m)}{(p'+k)^2 - m^2 + i\epsilon}\,(-ig\gamma^{\nu})\,\frac{1}{q^2}\bar{u}\,(a')\,(-ig\gamma_{\nu})\,u\,(a)\,u\,(p)\,. \end{aligned} \tag{2}$$

3 Results

We calculate the angular correlation for final jet particle and emitted photon and that for final jet particle and final medium parton. We check the tendency of the angular correlation for two particles according to the incident angle of jet particle, energy of emitted photon and temperature of systems, separately. We set the outgoing direction of jet particle along z-axis and incoming and outgoing jets are on x-z plane. We set the initial and final energy of incoming jet particle to be 10 GeV and 9 GeV. To describe momentum distribution of initial medium partons, we apply Maxwell-Boltzmann distribution function. We also confirmed that results are not changed significantly when we use Fermi-Dirac distribution function instead of Maxwell-Boltzmann distribution function.

The Figure 2 shows angular correlation between final jet particle and emitted photon for various incident angles of jet particle θ_p, energies of emitted photon E_k and temperatures of system k_BT. We

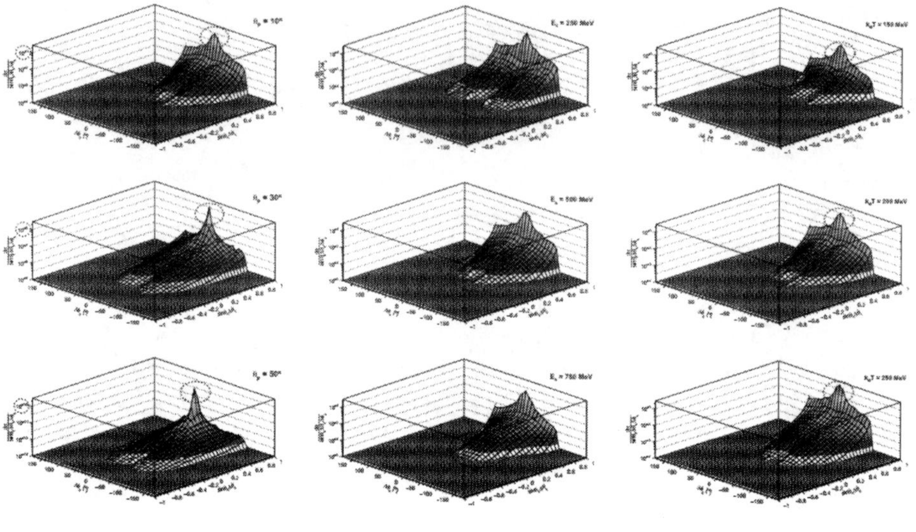

Figure 2. Angular correlation between final jet particle and emitted phton.

change θ_p from 10° to 50° by 20° with fixed values of E_k = 500 MeV and k_BT = 200 MeV, which are shown in the first column of Figure 2. As θ_p increases, the position of peak is shifted along incoming direction of jet particle rather than outgoing direction of jet particle. And the amplitude of peak is reduced by $\sim 10^{-5}$, roughly when θ_p increases from 10° to 50°. We can conclude that peak aligns to the incoming jet particle. In the second column of Figure 2, we vary E_k from 250 MeV to 750 MeV by 250 MeV with fixed values of θ_p = 10° and k_BT = 200 MeV. When E_k increases, tails of $d\sigma$ are combined at $\Delta\phi \approx 0$ and possible region of $d\sigma \neq 0$ is getting shrunk. It means that photon with higher energy emits more along the outgoing jet particle. Also we change k_BT from 150 MeV to 250 MeV by 50 MeV with fixed values of θ_p = 10° and E_k = 500 MeV, whose results are displayed in the third column of Figure 2. As k_BT increases, the amplitude of peak increase by $\sim$ 10 and the number of collisions between jet particle and medium parton increases due to the active thermal motion of medium partons.

The Figure 3 shows angular correlation between final jet particle and final medium parton for various incident angles of jet particle θ_p and temperatures of system k_BT. In this correlation, we find that Ridge structure appears at peak and away side. When θ_p increases from 10° to 50° by 20° with fixed k_BT = 200 MeV, as shown in the first column of Figure 3, the peak-side and away-side Ridge structure moves in the opposite direction to each other. The peak for the medium parton appears almost perpendicular to the pear for the photon because of four-momentum conservation. As k_BT increases from 150 MeV to 250 MeV by 50 MeV, we can get same results as that between final jet particle and emitted photon.

4 Conclusion & Outlooks

We calculate the angular correlation for final jet particle and emitted photon and that for final jet particle and final medium parton. We check the dependency of the angular correlation for two particles on the incident angle of jet particle, energy of emitted photon and temperature of systems, separately. We set the initial and final energy of incoming jet particle to be 10 GeV and 9 GeV. We discover

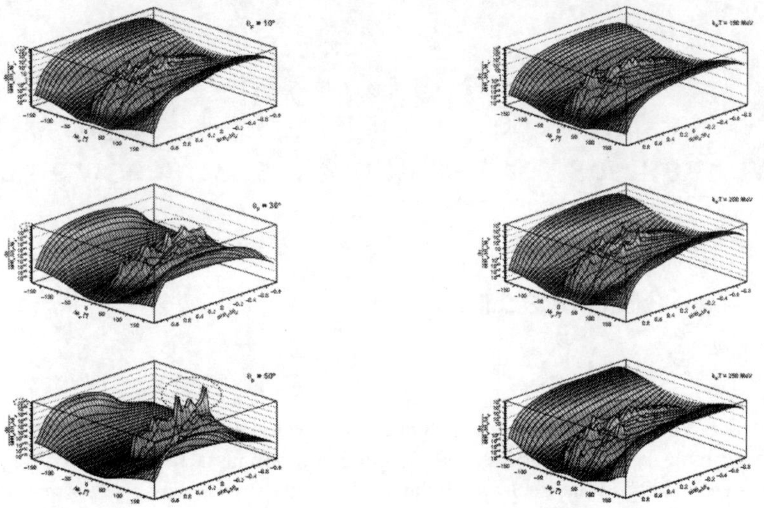

Figure 3. Angular correlation between final jet particle and final medium parton.

collective motion in both angular correlations. The peak aligns to incoming jet particle in correlation between final jet particle and emitted photon. And Ridge structure appears in correlation between final jet particle and final medium parton. From these results, we can conclude that collective motion such as Ridge structure is able to be understood through kinematic interpretation. However, we have to study further effect to fully describe the experimental result. For example, we need to consider boost and flow effects on the momentum distribution for initial medium partons. Also multiple scattering process should be counted. Eventually, we need to include gluon radiation process, which has more important contribution in high-energy heavy-ion collisions than Bremsstrahlung.

Acknowledgements

This research is funded by Inha University.

References

[1] C. Y. Wong, *Introduction to high-energy heavy-ion collision* (World scientific, Singapore, 1994)
[2] C. Itzykson and J-. B. Zuber, *Quantum field theory* (McGraw-Hill, New York, 1980)
[3] M. E. Peskin and D. V. Schroeder, *An introduction to quantum field theory* (Westview, USA, 1995)
[4] C. Y. Wong, Physical Review C **85**, 064909 (2012)
[5] ATLAS collaboration, Physical Review Letters **116**,172301 (2016)
[6] CMS collaboration, Physics Letters B **724**, 213240 (2013)

EPJ Web of Conferences 141, 08006 (2017) DOI: 10.1051/epjconf/201714108006
ISMD 2016

Alternative methods for top quark mass measurements at the CMS

Ji Hyun Kim (on behalf of the CMS collaborations)[1,a]

[1] *University of Seoul*

Abstract. The top quark mass is a fundamental parameter of the standard model and together with the W boson mass and the Higgs boson mass. It provides a strong self-consistency check of the electroweak theory. Recently several new measurements of the top quark mass using alternative observables and reconstruction methods are performed by the CMS collaborations at the CERN LHC. Alternative methods can give a insight by providing different systematic sensitivities while standard ones are currently limited by jet energy uncertainties. We introduce various results from new methods including the one using a charmed meson, which are found to be consistent with what is obtained in standard measurements.

1 Introduction

The top quark is the heaviest elementary particle observed so far, and thus is expected to have the highest of the Yukawa couplings with the Higgs boson. Thus, its mass, m_t, is an important input to global fits of electroweak parameters along with measurements of the W boson and Higgs boson masses, and gives significant cross-check of the consistency of the electroweak theory. In addition, by comparing precision electroweak measurements and theoretical predictions, an exact measurement of m_t can place strong constraints on contributions from physics beyond the standard model (SM). The top quark mass has been measured with ever-increasing precision, using a full kinematic reconstruction of the $t\bar{t}$ final states since it was discovered at the Tevatron [1, 2]. Most precise measurements reconstruct top quarks decayed in hadronic jets and calibrate the energy of hadronic jets in situ, using constraints from the reconstructed W boson mass [3–5]. Currently, the world's best measurement of the top quark mass at 7 and 8 TeV by the CMS experiment [6] is value of 172.44 ± 0.48 GeV, reaching a precision of 0.28% [7] and is in good agreement with the 2014 world average combined from four experiments (ATLAS, CDF, CMS, and D0) [8]. Experimental uncertainties stem from the calibration of reconstructed jet energies and their resolution are one of the biggest limits of the most precise top quark mass measurements. The other important uncertainties are concerning the modeling of the bottom quark fragmentation and hadronization. Alternative measurements can give us a hand to improve the precision of the top quark mass measurement and give a insight by providing different systematic sensitivities. In this note, alternative measurements of the top quark mass performed at CMS without using jets or by extracting it from the theoretically calculable observables or by using single-top event topologies are described.

[a] e-mail: jkim@cern.ch

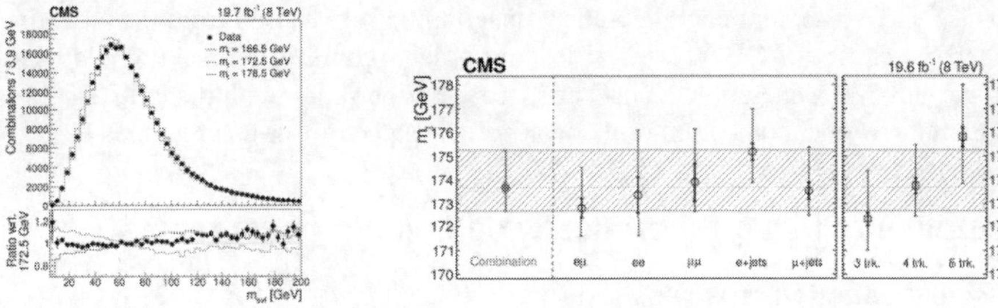

Figure 1. Lepton-SV invariant mass distribution for a combination of all five channels, compared to simulated distributions at three different generated top quark mass values (left) and results of the m_t measurement in the individual channels and their combination (right).

2 Measurements without jets

2.1 Using secondary vertices and leptons

The mass of the top quark is estimated by measuring the invariant mass of a charged lepton from the W boson decay and the tracks used in the reconstruction of a secondary vertex (SV) resulting from the b hadrons. This secondary vertex-lepton invariant mass system (m_{svl}) is highly sensitive to the top quark mass and lessens experimental uncertainties from the calibration of jets, but brings about uncertainties in the modeling of top quark decays and b hadronization. In the recent results of extracting the top mass by using m_{svl} [9], m_t is obtained as 173.68 ± 0.2(stat) $^{+1.58}_{-0.97}$(syst) GeV. Figure 1 shows the observed m_{svl} distribution for a combination of all five channels($e\mu$, ee, $\mu\mu$, $e + jets$ and $\mu + jets$), compared to simulated distributions at three different generated top quark mass values (left) and results of the m_t measurement in the individual channels and their combination (right). The leading source of systematic uncertainty is the modeling of b-quark fragmentation in the simulation. Experimental systematic uncertainty and the overall modeling of hadronization do not impact on the analysis strongly, hence it can be complementary to the standard methods.

2.2 Using charm mesons and leptons

The idea on estimation of the top quark mass using J/ψ was suggested in [10] and implemented by the CMS collaboration using LHC 8 TeV data for the first time [11]. Correlation with the invariant mass of J/ψ from the B hadron decays and an isolated lepton from the W boson decay and top mass is used to extract the top quark mass. The measured top quark mass is 173.5 GeV ± 3.0 GeV (stat.) ± 0.9 GeV (syst.) GeV. As the top quark mass is reconstructed by using only leptons, several systematic uncertainties related to QCD, jet reconstruction and b-tagging techniques, are considerably reduced, while the b quark fragmentation modeling is to be dominant. With the statistics available in the Run I dataset this is one of the few top mass measurements which is not yet dominated by systematics. As such it is one of the more promising methods to follow up in Run II of the LHC.

2.3 Using lepton kinematics

Another alternative approach [12] is to use observables that rely only on the kinematics of lepton originating from the decay of the top quark. Amongst several kinematic observables, $p_t(l^+l^-)$ is found to be the most sensitive to m_t and the more robust against the uncertainties in the modelling of the dynamics of top production. m_t obtained by the CMS is 171.7 ± 1.1 (stat.) ± 0.5 (exp.) $^{+2.5}_{-3.1}$ (th.) $^{+0.8}_{-0.0}$ (p_t(t)) GeV [13]. The dominant theoretical uncertainties are caused from the modeling of the top quark p_T and the choice of QCD scales of leading order signal MC, hence it could be significantly reduced in the next MC generation at NLO at 13 TeV. Furthermore, with the LHC larger datasets and a better calibration of the momentum of the leptons, we expect lower uncertainties.

3 Measurements using theoretically calculable observables

3.1 Inclusive production cross section

The top quark pole mass can be inferred by comparing the measured cross section to a precise theoretical prediction of the dependence of the inclusive $t\bar{t}$ cross-section on m_t. At center-of-mass energy of 7 or 8 TeV, precision of the top mass measured by CMS at NNLO with NNLL resummation reached below 2 GeV [14]. Recently, CMS extracted the top pole mass m_t = 172.3 $^{+2.7}_{-2.3}$ GeV at 13 TeV [15]. Figure 2 (left) shows the measured production cross section as a function of the assumed top-quark mass, compared with the mass-dependent prediction at NNLO+NNLL, for 7 and 8 TeV in LHC run I. The top quark mass is extracted by maximizing the product of the likelihoods corresponding to the measured cross-section at those energies and uncertainties are assigned to the cross-section to account for the LHC luminosity uncertainties.

3.2 $t\bar{t}$ + jet invariant mass

A method using the normalized differential cross section as a function of the invariant mass of the $t\bar{t}$ system and the leading extra jet which is not coming from top quark decay is proposed in [16] and experimentally tested by the CMS experiments [17]. The measured differential cross section is compared to the predicted cross sections for each bin of the ρ_s ($=2m_0/m(t\bar{t}, jet)$) distribution using different top quark masses and the most probable top quark mass is extracted. The precision is mostly limited by the systematic uncertainties arising from modeling sources and the theoretical uncertainties from POWHEG $t\bar{t}$+jet simulation.

4 Measurements in alternative topologies

While most measurements of the top quark mass to date are obtained from $t\bar{t}$ events, measuring the top quark mass in single-top quark production has performed because of sufficient statistics of those events and systematic uncertainties partially uncorrelated from those considered in $t\bar{t}$ production. In this analysis, top quark candidates are reconstructed from the muon, missing transverse energy, and b-jet and constraint is that the muon and the neutrino come from W boson decay. The measured value is m_t = 172.60 ± 0.77 (stat.) $^{+0.97}_{-0.93}$ (syst) GeV [18] and shows very good agreement with the current world average, based on measurements with $t\bar{t}$ events. The potential differences in sensitivity to the hard scattering and the modeling of underlying events and color reconnection are so far not observed and the measurements is limited by jet energy scale related uncertainties.

5 Summary and prospects

Figure 2 (right) [19] summarizes the most recent top mass measurements performed by the CMS Collaboration employing alternative methods. The precision of the standard measurements has approached ≈500 MeV, remains still unchallenged by the precision of the alternative methods explored

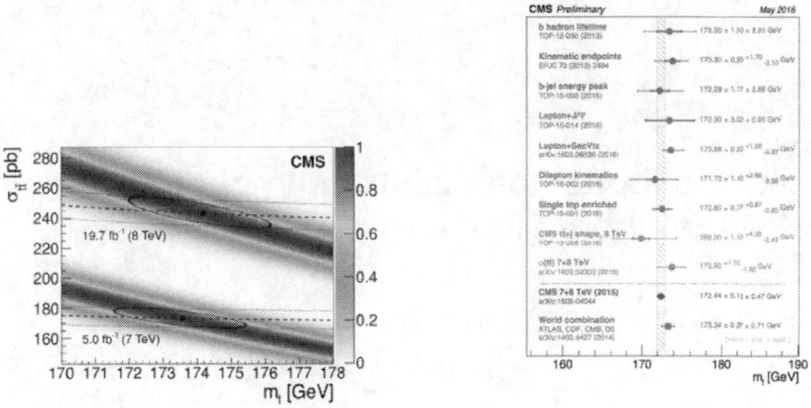

Figure 2. Predicted dependence of the $t\bar{t}$ production cross section on the top quark pole mass at NNLO+NNLL for 7 and 8 TeV (left) and recent alternative top-quark mass measurements from the CMS collaboration compared to the combination of standard measurements from CMS and to the 2014 world average (right).

so far. Nevertheless, alternative mass measurements can provide insights in the modeling of top quark events and bottom quark hadronization and produce important cross checks by using different mass definitions. With the expected higher integrated luminosity and $t\bar{t}$ cross sections at the Run II of the LHC, many of these methods will be more important and might play a crucial role in improving the overall precision on the top quark mass.

References

[1] F. Abe et al. (CDF Collaboration), Phys. Rev. Lett. **74**, 2626 (1995).
[2] S. Abachi et al. (D0) Collaboration, Phys. Rev. Lett. **74**, 2632 (1995).
[3] S. Chatrchyan et al. (CMS Collaboration), J. High Energy Phys. **12**, 105 (2012).
[4] CMS Collaboration, Eur. Phys. J. **C 74**, 2758 (2014).
[5] ATLAS Collaboration, Eur. Phys. J. **C 75**, 330 (2015).
[6] S. Chatrchyan et al. (CMS Collaboration) JINST **3**, S08004 (2008).
[7] CMS Collaboration, Phys. Rev. **D 93**, 072004 (2016).
[8] ATLAS, CDF, CMS, and D0 Collaborations, arXiv:1403.4427.
[9] V. Khachatryan et al. (CMS Collaboration), Phys. Rev. **D 93**, 092006 (2016).
[10] A. Kharchilava, Phys. Lett. **B 476**, 73-78 (2000).
[11] CMS collaboration, CMS PAS TOP-15-014.
[12] S. Frixione and A. Mitov, JHEP **09**, 012 (2014).
[13] CMS collaboration, CMS PAS TOP-16-002.
[14] CMS collaboration, CMS PAS TOP-13-004, arXiv:1603.02303.
[15] CMS collaboration, CMS PAS TOP-16-006.
[16] S. Alioli, et al., EPJC **73**, 2438 (2013).
[17] CMS collaboration, CMS PAS TOP-13-006.
[18] CMS collaboration, CMS PAS TOP-15-001.
[19] CMS collaboration, https://twiki.cern.ch/twiki/bin/view/CMSPublic/PhysicsResultsTOPSummaryFigures.

Measurement of bottomonium states in pp, pPb and PbPb collisions at 2.76 TeV from CMS

JaeBeom Park on behalf of the CMS collaboration[1,a]

[1] *Korea University, 145 Anam-ro, Seonbuk-gu, Seoul, Republic of Korea*

Abstract. The bottomonium states are one of the cleanest probes to study the characteristics of the quark-gluon plasma (QGP) as they are produced through hard scattering at the early stage of the collision. Their relative production cross section in PbPb collisions compared to pp collisions, the nuclear modification factor R_{AA}, has been studied and strong suppression was found in PbPb collisions. The R_{AA} has been obtained in centrality intervals for Υ(1S) & Υ(2S) and also reported as a function of p_T and rapidity. Also the yield ratios of the excited states Υ(nS) compared to the ground state Υ(1S) in both pp and PbPb collisions are combined to form the double ratio, which is equivalent to the ratio of the nuclear modification factor R_{AA}. This quantity has been studied for the two excited states Υ(2S) and Υ(3S) in pPb and PbPb collisions.

1 Introduction

A medium consist of deconfined quarks and gluons at high density and high temperature, the Quark-Gluon Plasma (QGP) is predicted to be formed at the early stage of high energy heavy ion collisions. Measurement of quarkonium states is a promising way to study the characteristics of the QGP. Suppression of those quarkonium states is one of the most important signatures in understanding the property of the medium, which was expected to be in the order of the binding energy due to the debye screening of the neighboring color charges [1]. The suppression of the various states is therefore expected to be denoted in a sequential pattern and can provide information of the dissociation temperature in the medium which is dependent on the binding energy of each quarkonia. Since the binding energy of the bottomonium states is stronger for the ground state than the other excited states, the ground state (Υ(1S)) has a larger dissociation temperature compared to the others (Υ(2S) & Υ(3S)) and that would lead the three states to follow the expected sequential suppression. The modification of the quarkonium production is not only affected by the hot nuclear matter but also by the effect of the nucleus itself which is so-called the cold nuclear matter (CNM) effect. The CNM effect includes the modification of the parton distribution function, cronin effect and nuclear absorption [2]. The bottomonium production was measured with the data collected at $\sqrt{s_{NN}}$ = 5.02 TeV in pPb collisions and $\sqrt{s_{NN}}$ = 2.76 TeV in pp and PbPb collisions. In this conference proceeding, the latest results of the Υ measurement will be described for the three different collision system.

[a] e-mail: jaebeom.park@cern.ch

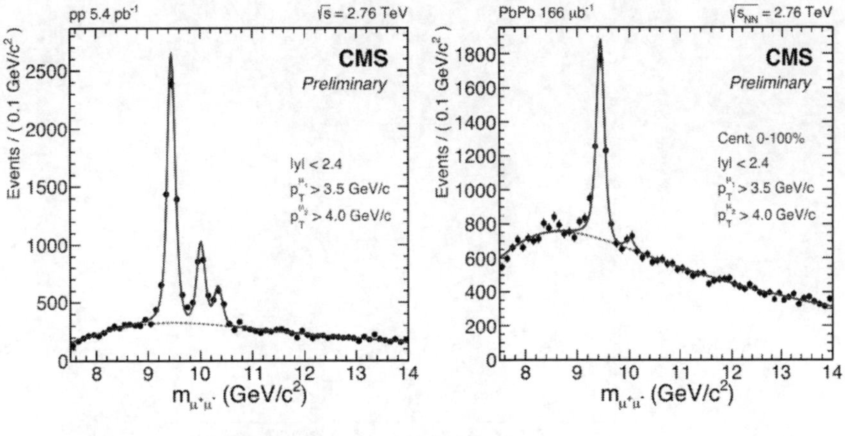

Figure 1. Dimuon mass distribution in pp and PbPb collisions at 2.76 TeV [4].

2 Signal Extraction of Υ states

In this analysis, the Υ states were measured via the di-muon decay channel which is easy to detect from the CMS detector. The detailed description of the CMS detector can be found in Ref. [3]. The single muons were reconstructed within 7.5-14 GeV/c^2 di-muon mass window applied with the single muon p_T cut ($p_T^{\mu_1}$ > 3.5 GeV/c & $p_T^{\mu_2}$ > 4.0 GeV/c). The mass spectrum in pp and PbPb collisions at $\sqrt{s_{NN}}$ = 2.76 TeV is shown in figure 1. An unbinned maximum likelihood fit was performed to describe the $m_{\mu^+\mu^-}$ mass spectrum by a signal PDF and a background PDF in each collision system. The sum of two Crystal Ball (CB) functions was used for the signal PDF while the parameters were fixed to the obtained values from the Monte Carlo (MC) simulation. Those fixed parameters were then released one by one in order to compute the systematic uncertainty of the signal function variation. The background PDF consists of an error function multiplied by an exponential function in order to fit the low-mass turn on. Additional function to the nominal background PDF such as first and second order Chebychev polynomial was used to estimate the systematics of the background PDF variation. The extracted yields were corrected by the acceptance and efficiency which was obtained from MC simulation and also corrected by a data-drived technique called tag-and probe [5].

3 Results

3.1 Υ(nS) in double ratios

The double ratio of the excited states to the ground states is calculated by dividing the single ratios in pp & PbPb (pPb) collisions as in Eq 1. It quantifies how much more the excited states are suppressed compared to the ground state and has an advantage in the cancellation of the acceptance and efficiency while it also eliminates the fraction of the systematic uncertainties. The double ratio in PbPb collisions from Ref. [6] and pPb collisions is shown on Fig 2 over the integrated centrality and event activity. Strong suppression was found for both excited states in PbPb collisions and the large amount of suppression shows that the effect of the hot nuclear matter (HNM) is dominant in PbPb collisions. Although the suppression is much larger in PbPb collisions, the double ratio results in pPb collisions shows a significant amount of suppression which can be interpreted as the cold nuclear matter (CNM) effect.

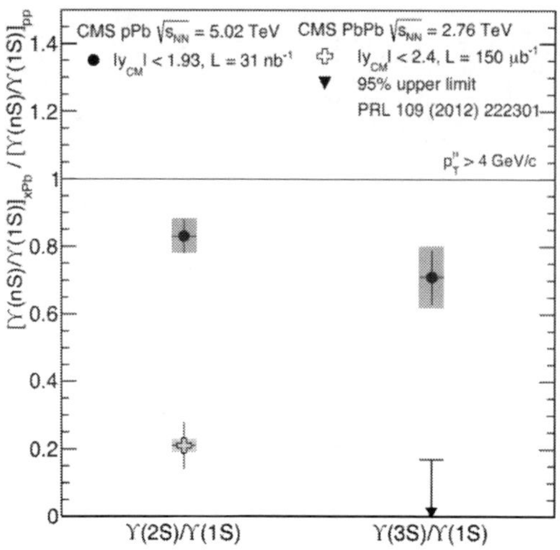

Figure 2. Double ratios of the Υ(nS) to the Υ(1S) in pPb collisions (circles) and the corresponding ratios in PbPb collisions (cross) [7].

$$\frac{\Upsilon(2S)/\Upsilon(1S)|_{PbPb(pPb)}}{\Upsilon(2S)/\Upsilon(1S)|_{pp}} = \frac{R_{AA(pPb)}(\Upsilon(2S))}{R_{AA(pPb)}(\Upsilon(1S))} \tag{1}$$

3.2 R_{AA} of Υ states

The normalized yield ratio of the bottomonium states in PbPb collisions compared to pp collisions, quantified as the nuclear modification factor R_{AA}, is one of the direct signal of the quarkonium suppression due to the color screening. The R_{AA} can be obtained by dividing the yields of Υ in PbPb by the pp cross sections and nuclear overlap function T_{AA} as in Eq 2. Fig 3 shows the R_{AA} results as a function of centrality, p_T and rapidity. The centrality dependent left plot in Fig 3 shows gradual decrease of Υ(1S) and strong suppressions of Υ(2S) in all centrality bins. Hence no clear peak of Υ(3S) was observed within the given statistics in PbPb collisions, an upper limit of the R_{AA} was calculated as less than 0.14 at 95% C.L. over the whole centrality and kinematic range. Υ(1S) and Υ(2S) in the integrated 0-100 % bin shows suppression by a factor of 2 and 10, respectively in PbPb collisions compared to pp collisions. The sequential suppression of the three bottomonium states indicates the pattern of the sequential binding energy that was expected from the screening effect of the heavy quark potential. The middle and right plot shows the p_T and rapidity dependent R_{AA} in the kinematic range of $|y| < 2.4$ & $p_T < 20$ GeV/c and no significant dependence was observed.

$$R_{AA} = \frac{1}{T_{AA}} \frac{d^2N^{AA}/dp_T d\eta}{d^2\sigma^{NN}/dp_T d\eta} \tag{2}$$

4 Summary

The three bottomonium states Υ(1S), Υ(2S) and Υ(3S) were measured in pp, pPb and PbPb collisions from the data collected during RUN I of the CMS experiment. The double ratio results in pPb &

DOI: 10.1051/epjconf/201714108007

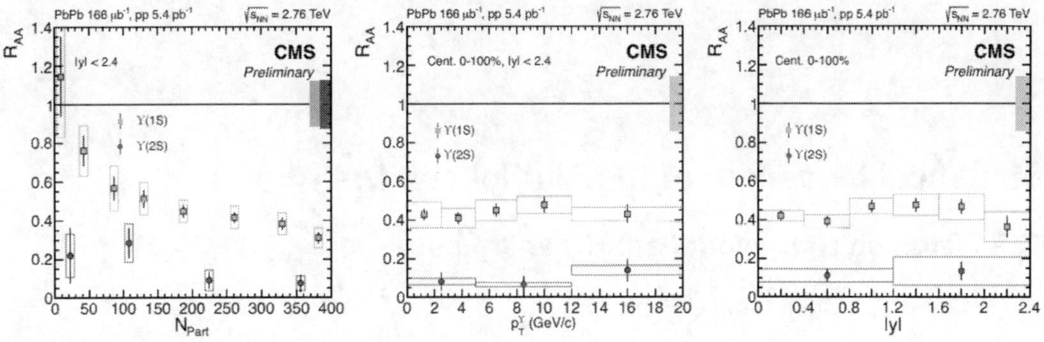

Figure 3. $R_{\mathrm{AA}}$for Υ(1S)and Υ(2S)as a function of centrality (left), p_{T} (middle) and y (right) in pp and PbPb collisions [4].

PbPb collisions has been reported and the results of the nuclear modification factor R_{AA} has been shown in PbPb collisions. The cold nuclear matter (CNM) effect was probed in the double ratio with a distinct amount of suppression, while much more suppression appeared in PbPb collisions due to the hot dense matter. R_{AA} has been presented as a function of N_{part}, p_{T} and $|y|$ and strong suppression of Υ(1S) and Υ(2S) with sequential order was observed with no kinematic dependence. Gradual decrease in the centrality dependent R_{AA} was observed for Υ(1S) while Υ(2S) was strongly suppressed in all centrality intervals.

References

[1] T.MatsuiandH.Satz, Phys. Lett. B **178**, 416 (1986).
[2] A. Andronic, et al., Eur.Phys.J. C76 (2016) 107.
[3] The CMS Collaboration, JINST 3 (2008) S08004.
[4] The CMS Collaboration, HIN-15-001.
[5] The CMS Collaboration, JHEP 1101 (2011) 080.
[6] The CMS Collaboration, Phys. Rev. Lett. 109 (2012) 222301.
[7] The CMS Collaboration, JHEP 04 (2014) 103.

Statistical Modelling of the Multiplicity Distribution

The Weighted GMD model at LHC energies

W.Y.Wang[1,a], A.H.Chan[1], and C.H.Oh[1]

[1] *Physics Department, National University of Singapore, Singapore*

Abstract. The multiplicity distribution is described in terms of models incorporating the stochastic branching of quarks and gluons, along with fluctuations in their production. Results show a reasonable description of charged-particle multiplicity distributions measured in pp collisions by the CMS collaboration at centre-of-mass collision energies of 0.9, 2.36, and 7 TeV for $|\eta| < 0.5$, 1.0, 1.5, 2.0, and 2.4. A prediction for 13 TeV is also given.

1 The Weighted GMD Model

Charged-particle multiplicities are one of the most inclusive variables that can be measured in high energy collisions. Their inclusiveness make them one of the most straightforward variables to define and measure but difficult to understand. However it is clear that the characteristics of various particle production processes drive its evolution across different centre-of-mass energies, and its distribution in phase space. A Poisson distributed (PD) multiplicity for example, indicates the independent emission of single particles, while particle cascades that develop from energetic collisions lead to a broader distribution like the Negative Binomial Distribution (NBD) [1].

Charged-particle multiplicity distributions can be described by the Generalised Multiplicity Distribution (GMD) [2–4] which is a general solution of the stochastic branching equation [5] and is reducible to the the PD, NBD, and Furry-Yule distribution [6]. Given m and k' initial number of quarks and gluons, the stochastic branching equation yields the GMD:

$$P_{GMD}(n;p,k,k') = \frac{\Gamma(n+k)}{\Gamma(n-k'+1)\Gamma(k'+k)}(1-p)^{n-k'}(p)^{k'+k}, \tag{1}$$

where n is particle multiplicity, k and p are related the initial quark/gluon numbers and their branching probabilities [2–4]. The average multiplicity of final state hadrons is given by

$$\bar{n} = \frac{k'+k}{p} - k. \tag{2}$$

Within this framework [5], the QCD evolution of quark and gluon numbers during fragmentation are modeled as Markov chains with the following processes: quark bremsstrahlung ($q \to q+g$) and gluon fission ($g \to g+g$). This intuitive picture has been shown to be equivalent to an algorithm

[a] e-mail: w.y.wang@u.nus.edu

calculating the multi-parton distributions within a QCD jet in a leading logarithmic approximation [7].

The GMD however is modified when there are fluctuations in the initial gluon numbers, distributed in a way that is determined by the dynamics of the collision. In pp collisions, various dynamical causes like Parton Distribution Functions (PDFs), low momentum transfer collisions, or interactions in the non-perturbative QCD regime in general play a role. These are not well understood from first principles and a statistical approach may provide some insights.

The modification of the GMD due to initial gluon number fluctuations is in general described by the weighted GMD (WGMD) model [8]. Suppose each independent collision produces k' number of gluons distributed according to a distribution $P(k'; x_1, ..., x_r)$ with r parameters, then the WGMD is given by

$$P_{WGMD}(n; x_1, ..., x_r, p, k) = \sum_{k'=0}^{n} P(k'; x_1, ..., x_r) \times P_{GMD}(n; p, k, k'). \tag{3}$$

The observed final state multiplicity distribution is thus a weighted sum of the GMD over the probabilities of the initial states.

The WGMD describes an ensemble of events with varying gluon numbers described by the weight factor. This model opens up possibilities to probe the average gluon number as a function of the collision centre-of-mass energy and pseudorapidity acceptance. Given a different weight distribution (e.g. NBD, Bose-Einstein), correlations in gluon production can also be studied.

In the simplest case, gluons emitted independently by identical sources give rise to a Poisson distribution, i.e. $P(k'; \overline{k'}) = \overline{k'}^{k'} \exp\left(-\overline{k'}\right)/k'!$, where $\overline{k'}$ is the average number of gluons. In this case, the fluctuation can be quantified by the variance which is equal to the mean of the Poisson distribution. This results in a final particle multiplicity with mean $\overline{n} = \frac{\overline{k'}+k}{p} - k$.

2 Results

The Poisson weighted GMD (PGMD) is applied to describe charged-particle multiplicity distributions measured by the CMS collaboration [9]. Figure 1 shows the distributions at centre-of-mass energies of (top left) 0.9, (top right) 2.36, and (bottom left) 7 TeV. Comparison is made with the best fit PGMD and GMD. The interior point algorithm [10, 11] is used to find the parameters that give the minimum χ^2.

In general the evolution of multiplicities across all pseudorapidity windows and centre-of-mass energies is well described by both GMD and PGMD. While the PGMD describes the tail ends of the multiplicity distributions better than the GMD, it does not do as well at low multiplicities. It is however expected that both models would not be able to accommodate both the low multiplicity peak as well as the broadened "shoulder" at higher multiplicities, since diffractive events which have lower multiplicities are not factored in both models.

The advantage that PGMD has over the GMD can be seen in the lower multiplicity bins, especially the $n = 0$ spike, which is is evident in both CMS data and PGMD. This is not possible with the GMD since a Markov branching model with fixed initial number of gluons k' will not be able to describe multiplicities with $n < k'$. This is reflected in the GMD where the best description is reduced to the NBD i.e. $k' = 0$. This constraint limits the applicability of the GMD in the low multiplicity regime.

At 0.9, 2.36, and 7 TeV, the PGMD gives an average number of 1.07, 0.741, and 0.865 gluons respectively in the widest pseudorapidity window of 2.4. For increasing centre-of-mass energies, the parameter p decreases monotonically from 0.135 to 0.0614, reflecting an increasing initial parton

DOI: 10.1051/epjconf/201714108008

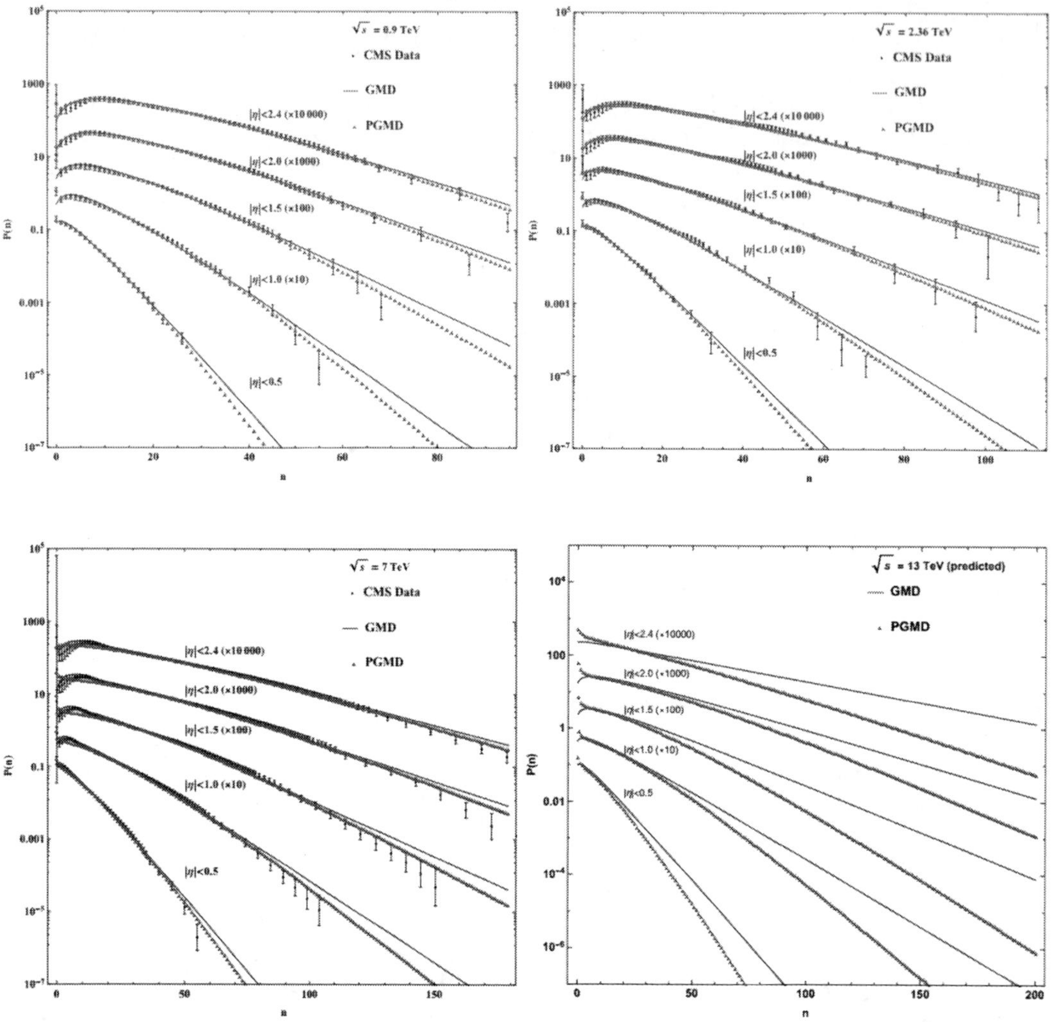

Figure 1. Charged-particle multiplicity distributions at (top left) $\sqrt{s} = 0.9$, (top right) 2.36, and (bottom left) 7 TeV for $|\eta| < 0.5$, 1.0, 1.5, 2.0, and 2.4 as measured by the CMS collaboration (circles). Comparison is made between the best fit PGMD (triangle) and GMD (line). The vertical lines in the data points represent the statistical errors and systematic uncertainties added in quadrature. The multiplicity distributions at (bottom right) $\sqrt{s} = 13$ TeV is predicted by the models.

invariant mass Q as a function of the centre-of-mass energy, given a fixed branching probability [8]. The corresponding mean hadron numbers 18.1, 23.3, and 30.4 are compatible with the experimental values measured by the CMS collaboration [9].

Figure 1 (bottom right) shows predictions of the multiplicity distributions for all pseudorapidity windows at a centre-of-mass energy of 13 TeV. The centre-of-mass energy dependence of the parameter p, which is related to the mean multiplicity $\overline{n}$, follows a power law.

3 Conclusion

The weighted Generalised Multiplicity Distribution model is obtained from a weighted sum of the Generalised Multiplicity Distribution. The model incorporates event-by-event fluctuations in the initial gluon numbers into the Generalised Multiplicity Distribution, giving a more realistic description of pp collisions as well as providing a method of obtaining insights into the collision dynamics.

In the case where the initial gluons are produced independently, the gluon numbers are distributed according the the Poisson distribution. The Poisson weighted Generalised Multiplicity Distribution is applied to charged-particle multiplicity distributions measured by the CMS collaboration at centre-of-mass energies of 0.9, 2.36, and 7 TeV. Without the incorporation of statistics describing diffractive events which mainly factor into the low multiplicity bins, both the Generalised Multiplicity Distribution and the Poisson weighted Generalised Multiplicity Distribution fail to describe low multiplicity peak and broad "shoulder" together. However, the reasonable description of the high multiplicity tail and the lowest $n = 0$ spike by the Poisson weighted Multiplicity Distribution suggests that with the addition of diffractive processes, both the peak and the "shoulder" may be described.

Prediction of charged-particle multiplicity distributions for $|\eta| < 0.5$, 1.0, 1.5, 2.0, and 2.4 at the centre-of-mass energy of 13 TeV is given. The main features of the multiplicity distribution are reproduced.

The application of Poisson weights to describe the gluon number production implies that the sources produce independent and uncorrelated single gluons. The use of other distributions as weights, for example the Negative Binomial Distribution or Bose-Einstein distribution may be more physically motivated and would suggest a different picture where the gluons are produced in a correlated manner. Insight on gluon production may be obtained by the comparison of such distributions.

References

[1] A. Giovannini, L. Van Hove, A. Phys. Pol. B **19**, 495 (1988)
[2] C.K. Chew, D. Kiang, H. Zhou, Phys. Lett. B **186**, 411 (1987)
[3] A.H. Chan, C.K. Chew, Phys. Rev. D **41**, 851 (1990)
[4] A.H. Chan, C.K. Chew, Z. Phys. C **55**, 503 (1992)
[5] A. Giovannini, Nucl. Phys. B **161**, 429 (1979)
[6] R.C. Hwa, C.S. Lam, Phys. Lett. B **173**, 346 (1986)
[7] K. Konishi, A. Ukawa, G. Veneziano, Nucl. Phys. B **157**, 45 (1979)
[8] W.Y. Wang, S. Seah, J. Setianegara, A.H. Chan, C.H. Oh, EPJ Web of Conferences **120**, 03004 (2016)
[9] CMS Collaboration, JHEP **01**, 79 (2011)
[10] Wolfram Research, Inc., *Mathematica*, version 9.0 edn. (Wolfram Research, Inc., Champaign, Illinois, 2012)
[11] S. Mehrotra, SIAM J. Optim. **2**, 575 (1992)

EPJ Web of Conferences 141, 08009 (2017) DOI: 10.1051/epjconf/201714108009
ISMD 2016

Photon reconstruction and identification in heavy ion collisions with the CMS detector

Yeonju Go on behalf of the CMS Collaboration[1,a]

[1] *Korea University, 145 Anam-ro, Seonbuk-gu, Seoul, Republic of Korea*

Abstract. Measurement of isolated photon gives a possibility to constrain the nuclear parton distribution functions (nPDF). The isolated photon is also the clear tag in the photon-jet channel for studying in-medium modifications of fragmentation functions of parton initiated jets. The performance of isolated photon reconstruction and identification has been studied in PbPb collisions at $\sqrt{s_{NN}}$ = 5.02 TeV. Trigger, reconstruction and isolation efficiencies are measured in PYTHIA+HYDJET Monte-Carlo simulation. We describe signal template method to determine the purity of isolated photons.

1 Introduction

Prompt photons coming from the primary vertex of the high energy hadron collisions are unaffected by the strongly interacting medium produced in the reaction. Prompt photons are the 'clean' probes of the initial state of the collision and play key roles in better understanding the Quark-Gluon Plasma (QGP). Isolated photons include prompt photons and collinear fragmentation photons of a parton producing large transverse momentum. The performance of reconstruction and identification of isolated photons has been studied with data collected in 2015 PbPb collisions at $\sqrt{s_{NN}}$ = 5.02 TeV and PYTHIA+HYDJET Monte Carlo simulation. In this analysis, isolated photon performances have been studied with respect to its transverse momentum and centrality, which is defined by the impact parameter b. Experimentally, centrality is expressed as a percentage of the total nuclear interaction cross-section.

2 Efficiency

2.1 Trigger Efficiency

Figure 1 shows trigger efficiency for different photon threshold (left) and reconstructed photon spectra coming from all the photon trigger combination without isolation condition (right). All trigger efficiencies are over 99 % and have sharp turn-on feature for each threshold.

[a]e-mail: ygo@cern.ch

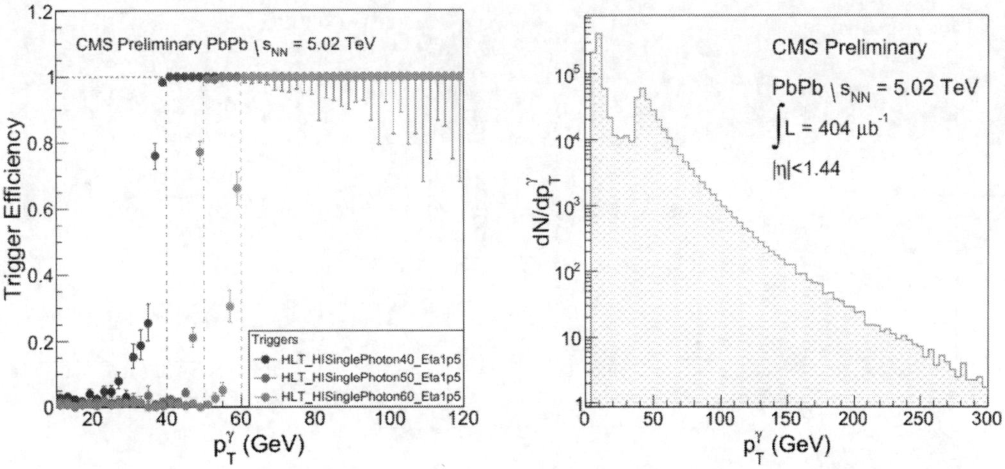

Figure 1. Photon trigger efficiency in PbPb data as a function of the reconstructed E_T (left), computed using a minimum bias data sample. It shows that for photons with an E_T > 40 GeV/c the trigger was nearly 100% efficient. Reconstructed photon spectra as a function of p_T (right) have been measured without isolation condition.

2.2 Reconstruction Efficiency

Reconstruction efficiency is the fraction of reconstructed photons in generated photons without isolation condition as shown in Figure 2 left panel. There is no centrality dependence of the reconstruction efficiency for all p_T region. Reconstruction efficiency is over 98 % in all centrality and p_T region.

2.3 Isolation Efficiency

Isolation Efficiency is defined as reconstructed photons passing isolation selection in all reconstructed photons. Table 1 summarizes the isolated photon selections.

Table 1. Isolated photon selection

Isolated photon selection
H/E < 0.1
SumIso < 1 GeV
$\sigma_{i\eta i\eta}$ < 0.01

The H/E ratio is defined as the fraction of hadronic energy to the electromagnetic energy in an angular region of cone radius $R = \sqrt{(\Delta\eta)^2 + (\Delta\phi)^2} = 0.15$ which is computed from the energy deposition in ECAL and HCAL. This variable is used to reject candidates originating from jets, which carry not only electromagnetic energy but also hadronic energy. In this analysis, the H/E ratio of the photon candidates is required to be smaller than 0.1 in order to reject candidates originating from jets.

Isolation variable 'SumIso' is defined as energy sum in a $R = \sqrt{(\Delta\eta)^2 + (\Delta\phi)^2} = 0.4$ cone around photon candidates, which is tracker E_T + ecal E_T + hcal E_T. SumIso condition selects isolated photons.

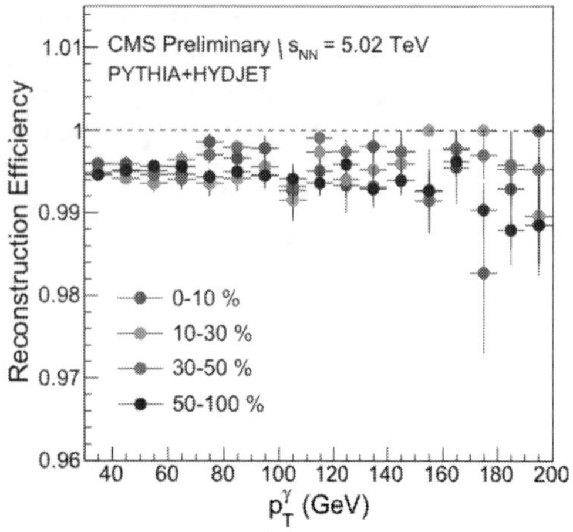

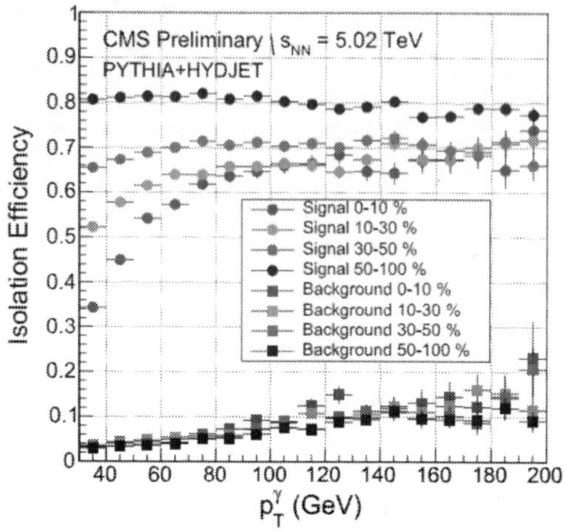

Figure 2. Reconstruction efficiency (left) and isolation efficiency (right) as a function of reconstructed photon p_T based on PYTHIA+HYDJET samples. Different colors represent different centrality range.

The $\sigma_{i\eta i\eta}$ characterizes the shower shape in the η direction which is defined as:

$$\sigma^2_{i\eta i\eta} = \frac{\sum_i^{5\times5} w_i(\eta_i - \eta_{5\times5})^2}{\sum_i^{5\times5} w_i}, w_i = \max(0, c + \ln\frac{E_i}{E_{5\times5}}), \quad (1)$$

where E_i and η_i are the energy and pseudorapidity of the ith crystal within the 5×5 electromagnetic cluster. $\sigma_{i\eta i\eta}$ is calculated with logarithmic weights w_i. $E_{5\times5}$ and $\eta_{5\times5}$ are the energy and averaged η of the 5×5 crystals. The value of c is a constant which was set to 4.7 in the pp studies and is effectively a cutoff on the crystal energy. The isolated photon candidates tend to have smaller $\sigma_{i\eta i\eta}$ while hadrons and decay photon from π^0s tend to have larger $\sigma_{i\eta i\eta}$.[1]

Isolation efficiencies for the signal and the background are shown in Figure 2 right panel for different centrality bins. Most of the background photons are rejected by isolation condition according to photon p_T. These background photons are mainly decay photons from neutral meson such as π^0 and η and also non-isolated prompt and fragmentation photons such as jets. Isolation efficiency for the signal depends on centrality. It decreases with increasing centrality and decreasing p_T. But, isolation efficiency for the background has no strong dependence on the centrality.

3 Purity

The selection criteria described above yield a relatively pure sample of isolated photons. Yet, there are still non-prompt photons, such as those from an isolated π^0 carrying a large fraction of the parent fragmenting parton energy, which can pass the isolation cuts. The fraction of those remaining backgrounds is statistically calculated using a two-component fit of the shape of the electromagnetic shower in ECAL ($\sigma_{i\eta i\eta}$), called the Template method.

A template is defined as the probability distribution function of $\sigma_{i\eta i\eta}$ of either pure photons (signal template) or pure background (e.g. π^0, η) (background template). The signal template was obtained from MC photon samples (PYTHIA+HYDJET for PbPb or PYTHIA for pp), and the background template

 DOI: 10.1051/epjconf/201714108009

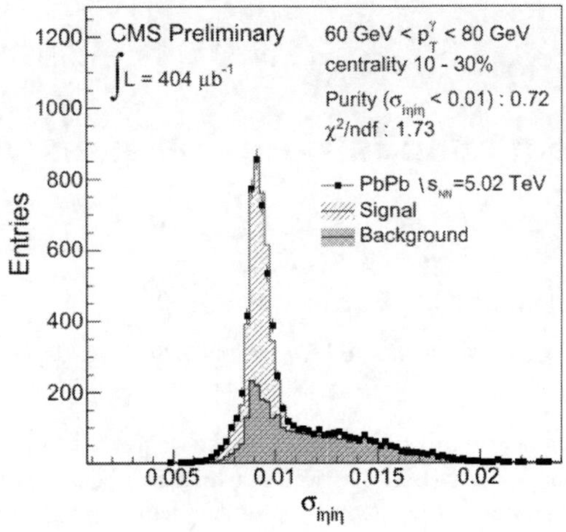

Figure 3. The shower shape variable ($\sigma_{i\eta i\eta}$) of photons from mid-central (10-30%) PbPb collisions. The black points are PbPb data, the red histogram is the signal template from PYTHIA+HYDJET, and the green histogram is the background template from a non-isolated data sideband. The purity is defined as the fraction of isolated photons in the signal region $\sigma_{i\eta i\eta}$< 0.01.

was obtained from data in a sumIso sideband (10 GeV< SumIso < 20 GeV) where the background is enriched. The yield of signal photons is estimated with a binned maximum likelihood fit to the $\sigma_{i\eta i\eta}$ distribution of data with the signal and background templates. The purity is defined as the fraction of photons with $\sigma_{i\eta i\eta}$ < 0.01 (chosen to maximize the purity value) which are accounted for by the signal template.[1]

An example of the template fitting method is shown in Figure 3.

4 Summary

Reconstruction and identification of isolated photons have been developed in heavy ion collisions from the data collected during run 2 CMS experiment and PYTHIA+HYDJET Monte-carlo simulation. Trigger, reconstruction and isolation efficiencies are presented and purity value using data-driven method is observed. An isolation condition is optimized to maximize the yield of prompt photon and reject the background photons from neutral meson decay. Isolated photon measurement will give a better understanding of the strongly interacting medium in heavy ion collisions.

References

[1] CMS collaboration, "Photon-Jet correlations in PbPb and pp collisions at center of mass energy 5.02 TeV", CMS Physics Analysis Summary CMS-PAS-HIN-16-002, CERN (2016)

Neutral meson production in p-Pb collisions with ALICE at the LHC

Tsubasa Okubo[1,a] on behalf of the ALICE collaboration

[1]*Hiroshima University, 1-3-1 Kagamiyama, Higashi-Hiroshima 739-8526, Japan*

Abstract. Strong suppression of high p_T π^0 has been observed in heavy-ion collisions at the Large Hadron Collider (LHC) energies, which can be interpreted by involving various transport properties of the quantum chromodynamics (QCD) medium and initial state effects. Comparing particle production in pp, p-A, A-A reactions has frequently been used to separate initial state effects of colliding nuclei from final state effects in quark matter created by the collisions.

We have measured π^0 and η meson emitted in p-Pb collisions at $\sqrt{s_{NN}}$ = 5.02 TeV via complementary methods, using the ALICE electromagnetic calorimeters, PHOS and EMCal, and Photon Conversion Method (PCM), which identifies photons converted to e^+e^- pairs in the material of the ALICE inner detectors, TPC and ITS. In this paper, π^0 and η meson spectra in p-Pb collisions as well as the nuclear modification factor for π^0 ($R^{\pi^0}_{\text{p-Pb}}$) will be shown.

1 Introduction

The π^0 and η meson measurement in pp collisions is useful on one hand to understand particle production and provides a test of perturbative quantum chromodynamic (pQCD) predictions. On the other hand, in heavy-ion collisions it is needed to study parton energy loss and particle suppression by the hot and dense medium called the Quark-Gluon Plasma (QGP). Strong suppression of high p_T π^0 has been observed in central Pb-Pb collisions at $\sqrt{s_{NN}}$ = 2.76 TeV [1]. This suppression is stronger than the one measured at RHIC or at SPS [2–4]. The suppression mechanism can be explained by various processes involving transport properties of the QCD medium and initial state effects. Therefore studies in p-Pb collisions which are intermediate between pp collisions and heavy-ion collisions in terms of system size and number of produced particles, are needed to disentangle suppression originating from initial state effects in the colliding nuclei or from final state effects in the Pb-Pb collisions.

2 Neutral meson reconstruction in ALICE

The ALICE experiment [5, 6] measured π^0 and η mesons in p-Pb collisions at $\sqrt{s_{NN}}$ = 5.02 TeV with different methods, using the electromagnetic calorimeters PHOS [7] and EMCal [8], and the Photon Conversion Method (PCM). This report is based on minimum bias triggered data with about

[a]e-mail: tsubasa.okubo@cern.ch

10^8 events which corresponds to an integrated luminosity of 50 μb^{-1} collected in the beginning of 2013. The ALICE minimum bias trigger requires a coincidence signal in both scintillator arrays of V0A and V0C. They cover $2.8 < \eta < 5.1$ and $-3.7 < \eta < -1.7$, respectively. The π^0 and η mesons are measured via their photon decay channels $\pi^0 \rightarrow 2\gamma$ and $\eta \rightarrow 2\gamma$. PHOS and EMCal measure energy and hit coordinates of photons and electrons. The PHOS detector consists of $PbWO_4$ crystals and APD readout at 4.6 m radius from the interaction point. It has high energy resolution ($\sigma_{E(\mathrm{GeV})}/E = 0.01/E \oplus 0.04/\sqrt{E} \oplus 0.01$) and high granularity. Its coverage is $|\eta| < 0.13$ and $260° < \varphi < 320°$. The EMCal detector is a lead-scintillator sandwich calorimeter with 77 alternating layers of 1.4 mm lead and 1.7 mm scintillator. It is located at 4.4 m from the interaction point and has an energy resolution of $\sigma_{E(\mathrm{GeV})}/E = 0.05/E \oplus 0.11/\sqrt{E} \oplus 0.02$. Its coverage is $|\eta| < 0.7$ and $80° < \varphi < 180°$. PHOS and EMCal cover the intermediate to high p_T regions (1 GeV/c < p_T < 20 GeV/c). PCM reconstructs and identifies photons converted into e^+e^- pairs in the material of the inner detectors, ITS [9] and TPC [10], via 2γ decay channel and γ-Dalitz decay channel ($\pi^0 \rightarrow \gamma^*\gamma \rightarrow e^+e^-\gamma$, denoted as PCM-Dalitz). PCM has a wide acceptance with $|\eta| < 0.9$ and full azimuthal but the photon conversion probability is small (~8.5 %). PCM covers the low to intermediate p_T regions (0.3 GeV/c < p_T < 14 GeV/c).

The π^0 and η meson yields are counted from invariant mass distribution of photon candidates pairs with a numerical integral of the number of entries around the reconstructed meson mass. The mass position is determined via a fit after background subtraction. The background is estimated with an event mixing technique. Corrections including acceptance and reconstruction efficiencies are applied to the π^0 and η meson raw yields to obtain the invariant yields.

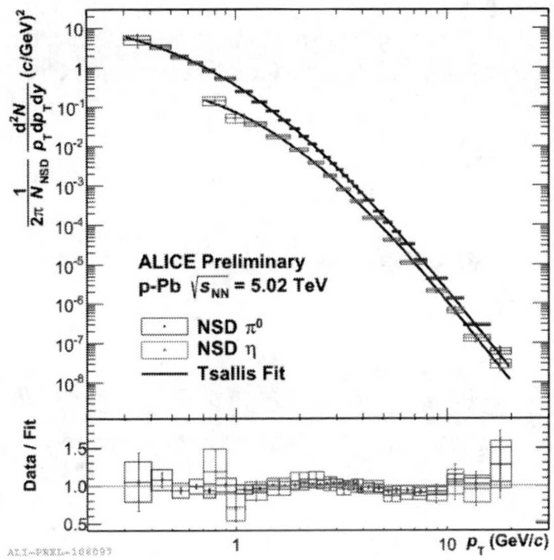

Figure 1. π^0 and η invariant yields in p-Pb collisions with Tsallis fit.

3 π^0 and η invariant yields in p-Pb collisions at $\sqrt{s_{NN}}$ = 5.02 TeV

ALICE has measured π^0 and η meson invariant differential yields in p-Pb collisions at $\sqrt{s_{NN}}$ = 5.02 TeV for the first time (Fig. 1). The π^0 yields are measured by PHOS, EMCal, PCM and PCM-Dalitz within the p_T range 0.3-20 GeV/c and the η meson yields are measured by EMCal and PCM within the p_T range 0.7-20 GeV/c. There is not enough statistics and small acceptance for η meson measurement in PHOS and PCM-Dalitz. Each yield is measured individually with the different methods and they are weighted according to their uncertainties in the combination. The invariant yields are fitted by a Tsallis function and the ratio of measured yield to the corresponding fit is shown in the bottom panel of the Fig. 1.

4 η/π^0 ratio in p-Pb collisions

The η/π^0 ratio in p-Pb collisions has been measured and it is shown in Fig. 2. The ratio is calculated individually with PCM and EMCal and then combined because some systematic uncertainties of the yield measurement cancel this way. The ratio in p-Pb collisions increases in the low p_T region below 4 GeV/c and reaches a plateau of 0.5 in the high p_T region from 4 GeV/c. It is compared to the ALICE

DOI: 10.1051/epjconf/201714108010

pp measurement at $\sqrt{s}$ = 7 TeV [11]. The measurement in both systems indicate similar magnitude and behavior within uncertainties.

5 Nuclear modification factor

ALICE has measured the nuclear modification factor for π^0 in p-Pb collisions, $R^{\pi^0}_{\text{p-Pb}}$ defined as

$$R_{\text{p-Pb}}(p_{\text{T}}) = \frac{\mathrm{d}^2N/\mathrm{d}p_{\text{T}}\mathrm{d}y|_{\text{p-Pb}}}{\langle T_{\text{p-Pb}}\rangle \times \mathrm{d}^2\sigma/\mathrm{d}p_{\text{T}}\mathrm{d}y|_{\text{pp}}} \quad (1)$$

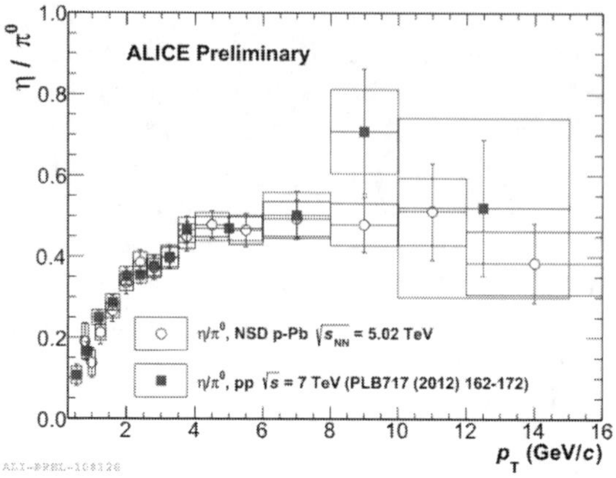

Figure 2. The η/π^0 ratio in p-Pb collisions compared to that in pp collisions at $\sqrt{s}$ = 7 TeV.

and it is shown in Fig. 3. Here, the nuclear overlap function $\langle T_{\text{p-Pb}}\rangle$ is related to the average number of inelastic proton-nucleon collisions, calculated by a Glauber model [12, 13], which gives $\langle T_{\text{p-Pb}}\rangle = \langle N_{\text{coll}}\rangle/\sigma_{\text{NN}} = 0.0983 \pm 0.0035$ mb^{-1}, with $\langle N_{\text{coll}}\rangle = 6.9 \pm 0.7$ and σ_{NN} = 70 ± 5 mb. The referenced spectrum is estimated with the published π^0 spectrum in pp collisions at $\sqrt{s}$ = 2.76 TeV and 7 TeV for interpolation with a power law fit. $R^{\pi^0}_{\text{p-Pb}}$ is calculated individually for each method due to cancellation of some systematic uncertainties and then combined. It increases at p_{T} < 2 GeV/c and agrees with unity at p_{T} > 2 GeV/c. It is compared to next-to-leading order pQCD (NLO pQCD) calculations which use three different fragmentation functions (FF), KKP, AKK, and fDSS [14], and Color Glass Condensate (CGC) [15]. The model predictions reproduce the data reasonably well.

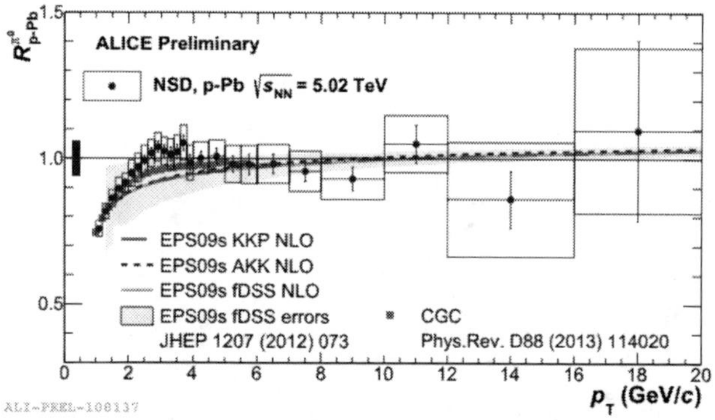

Figure 3. $R^{\pi^0}_{\text{p-Pb}}$ at $\sqrt{s_{\text{NN}}}$ = 5.02 TeV compared to model predictions.

6 Summary

ALICE has measured π^0 and η meson with different methods in p-Pb collisions at $\sqrt{s_{\text{NN}}}$ = 5.02 TeV for the first time. The π^0 invariant yields are measured by PHOS, EMCal, PCM and PCM-Dalitz and the η meson invariant yields are measured by EMCal and PCM. The η/π^0 ratio increases at p_{T} < 4 GeV/c and arrives a constant value at p_{T} > 4 GeV/c. The ratio in p-Pb collisions is consistent with that in pp collisions. $R^{\pi^0}_{\text{p-Pb}}$ is calculated with interpolated pp spectrum as reference. It agrees with unity at p_{T} > 2 GeV/c and is well described by NLO pQCD calculations.

References

[1] B. Abelev, et al. [ALICE Collaboration], Eur. Phys. J.C **74** 3108 (2014)
[2] A. Adare, et al. [PHENIX Collaboration], Phys. Rev. Lett. **109** 152301 (2012)

[3] A. Adare, et al. [PHENIX Collaboration], Phys. Rev. Lett. **101** 232301 (2008)
[4] M. M. Aggarwal et al. [WA98 Collaboration], Phys. Rev. Lett. **100** 242301 (2008)
[5] K. Aamodt, et al. [ALICE Collaboration], J. Instrum. **3** S08002 (2008)
[6] B. Abelev, et al. [ALICE Collaboration], Int. J. Mod. Phys. A **29** 1430044 (2014)
[7] ALICE Collaboration, CERN-LHCC-99-004 (1999)
[8] ALICE Collaboration, CERN-LHCC-2008-014 (2008)
[9] K. Aamodt, et al. [ALICE Collaboration], J. Instrum. **5** P03003 (2010)
[10] J. Alme, et al., Nucl. Instrum. Meth. A622 316-367 (2010)
[11] B. Abelev et al. [ALICE Collaboration], Phys. Lett. B **717** 162-172 (2012)
[12] B. Alver, M. Backer, C. Loinzides, and P. Steinberg, arXiv:0805.4411 [nucl-ex] (2008)
[13] M. L. Miller, K. Reygers, S. J. Sanders, P. Steinberg, Ann. Rev. Nucl. Part. Sci. **57** 205-243 (2007)
[14] I. Helenius et al., JHEP **1207** 073 (2012)
[15] T. Lappi, et al., Phys. Rev. D **88** 114020 (2013)

Measurement of Normalized Differential Cross Section for the $t\bar{t}$ Production in the Dilepton Channel in pp Collisions at a center-of-mass energy of 13 TeV

Dajeong Jeon[1,a] on behalf of the CMS Collaboration

[1] *University of Seoul, South Korea*

Abstract. Differential cross sections of top-quark pair production as function of the kinematic variables of leptons, b-jets and top-quarks at particle level are measured in the dilepton decay channel with proton-proton collisions at a center-of-mass energy of 13 TeV. The measurements are performed with Run II data using the CMS detector at the Large Hadron Collider.

1 Introduction

The precise measurement of the $t\bar{t}$ differential cross section can lead to a better understanding of background contributions [1] and provides a good test of perturbative QCD calculations. The normalized $t\bar{t}$ differential cross section is measured in the dilepton channel as a function of the $t\bar{t}$ system, top quark, daughter lepton and b-jet kinematic properties such as transverse momentum, rapidity, invariant mass and azimuthal decorrelation [2]. We present the measurement at particle level with final state objects defined in a theoretically safe and unambiguous way. The particle level measurements are expected to reduce the dependence on the theoretical model, so variables are mainly corrected for detector effects.

2 Differential cross section measurement

2.1 Data and simulated samples

We use $\sqrt{s} = 13$ TeV proton-proton (pp) collision data, corresponding to an integrated luminosity of $2.2 fb^{-1}$collected in 2015 with the CMS detector at the CERN LHC [3]. At least two leptons (electrons or muons) are required at trigger level. Monte Carlo (MC) simulated event samples are used to model the $t\bar{t}$ signal and the background processes. The detector response to the final state particles is simulated using `GEANT4` [4], and the events are reconstructed and analyzed with the same software used to process the data.

[a] e-mail: dajeong.jeon@cern.ch

2.2 Object and event selection

The dilepton final state of the $t\bar{t}$ decay consists of two leptons (electrons or muons), at least two jets, and missing transverse momentum from two neutrinos.

We require events with exactly two opposite-charge leptons with $M(l^{+}l^{-}) > 20$ GeV, and at least two jets, of which at least one jet identified as a b-quark jet. In addition, for the same-flavor lepton channels (ee and $\mu\mu$), additional selection criteria are applied which are designed to reject events from Drell-Yan production: $E_T^{miss} > 40$ GeV and $|M(l^{+}l^{-}) - M_Z| > 15$ GeV, where $M_Z = 91$ GeV.

2.3 Signal definition at particle level

We define the particle level top quark following the generator level prescriptions used in Ref. [5]. This approach avoids differences in the top quark definition due to possible differences in the decay history of different generators, and leads to results that are expected to be independent of the generator implementation and tuning.

The top quarks are reconstructed starting from the final state particles in $t\bar{t}$ events at generator level as summarized in Tab.1. To avoid the ambiguity from additional leptons at generator level, a distinction is made between prompt particles and particles from hadron decays.

Table 1. Summary of object definitions at the particle level

Object	Definition	Selection criteria
Prompt neutrino	neutrinos not from hadron decay	none
Dressed lepton	anti-k_T jet with a distance parameter of 0.1 using electrons, muons and photons not from hadron decay	$p_T > 20$ GeV, $\|\eta\| < 2.4$
b quark jet	anti-k_T jet with a distance parameter of 0.4 using all particles and ghost B hadrons not including any neutrinos nor particles used in dressed leptons	$p_T > 30$ GeV, $\|\eta\| < 2.4$ with ghost B hadrons

Pseudo-W bosons are reconstructed by combining a dressed lepton and a prompt neutrino. A pair of pseudo-W bosons is chosen among the possible combinations to minimize the scalar sum of invariant mass differences with respect to the W boson mass of 80.4 GeV. Similarly, the pseudo-top quarks are defined with an invariant mass requirement of 172.5 GeV. The visible phase space is defined to have a pair of pseudo-top quarks, constructed from prompt neutrinos, dressed leptons and b-quark jets. Events that are not in the visible phase space are considered as background.

2.4 Systematic uncertainties

We consider the following experimental and theoretical systematic uncertainties: Pileup contribution, lepton identification and isolation, lepton trigger efficiency scale factors, jet energy scale and jet energy resolution, b jet tagging scale factor, MC modeling, PDF, factorization and Renormalization scales, hadronization, top quark mass and background. The dominant contributions to the systematic uncertainty are due to the modeling (12%), hadronization (6%) and background estimation (6%). MC modeling uncertainties are measured using different generator such as `MG5_aMC@NLO` and `POWHEG` sample. To estimate the hadronisation uncertainties, we compare `POWHEG` +`HERWIG++` from `POWHEG` +`PYTHIA` 8 samples.

DOI: 10.1051/epjconf/201714108011

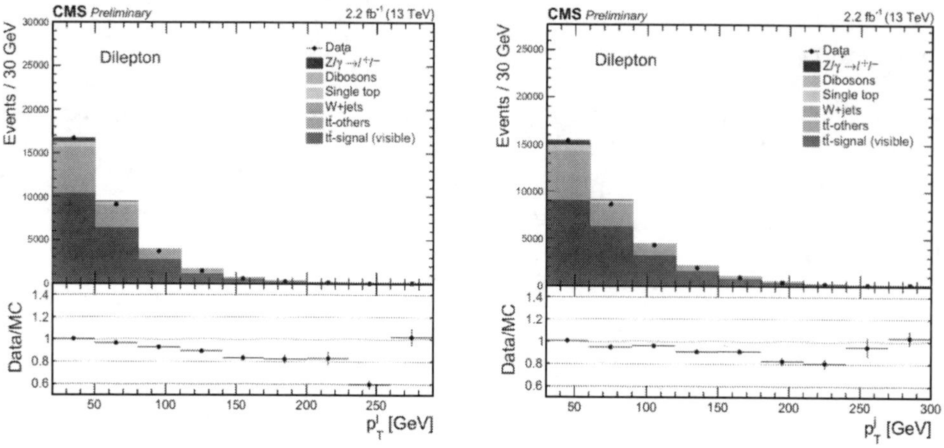

Figure 1. Reconstructed p_T^l, p_T^j distributions. All corrections described in the text are applied to the simulation.

3 Measurements of the normalized differential cross section

We measure the normalized differential $t\bar{t}$ cross section $(1/\sigma)(\mathrm{d}\sigma/\mathrm{d}X)$ as a function of the transverse momenta of leptons (p_T^l), jets (p_T^j), top quarks (p_T^t), and the $t\bar{t}$ system ($p_T^{t\bar{t}}$), the rapidities of the top quarks (y^t) and the $t\bar{t}$ system ($y^{t\bar{t}}$), and the invariant mass of the $t\bar{t}$ system ($M^{t\bar{t}}$). In the $t\bar{t}$ dilepton channel, the reconstruction of the neutrino and anti-neutrino is crucial in determining the top quark kinematics. Using an analytical approach [6, 7], the six unknown neutrino degrees of freedom are constrained by the two measured components of the missing transverse momentum and the assumed invariant masses of both W boson and top quark systems.

To allow for the finite resolution of the measured objects the $t\bar{t}$ system is reconstructed for 100 random variations within their simulated resolution functions, and the W boson mass is also allowed to vary over its Breit-Wigner distribution. In each trial there are up to eight solutions, and we select the solution with the minimum invariant mass of the $t\bar{t}$ system. For each trial, a weight is calculated using the expected invariant mass distribution of lepton and b quark jet pairs (M_{lb}) at particle level. The reconstructed neutrino momentum is taken from the weighted average over the trials. The final solution of the $t\bar{t}$ system is the solution with the maximum weight. The efficiency of the kinematic reconstruction is approximately 90%. Figure 1 shows the distributions of p_T^l and p_T^j.

Detector resolution and reconstruction efficiency effects are corrected by using an unfolding procedure. The method relies on a response matrix which maps the expected relation between the true and reconstructed variables. D'Agostini's method is employed to perform the unfolding [8–10].

Figures 2 show the normalized differential $t\bar{t}$ cross section as a function of $p_T^{t\bar{t}}$, $y^{t\bar{t}}$ and $M^{t\bar{t}}$ at particle level in the visible phase space. The measurements are generally found to be in agreement with the standard model predictions.

References

[1] A. Denner, S. Dittmaier, S. Kallweit, S. Pozzorini, PoS **LL2012**, 015 (2012), arXiv:1208.4053
[2] S. Choi, H.S. Lee, Phys. Rev. D **87**, 034012 (2013), arXiv:1207.1484
[3] Journal of Instrumentation **3**, S08004 (2008)

 DOI: 10.1051/epjconf/201714108011

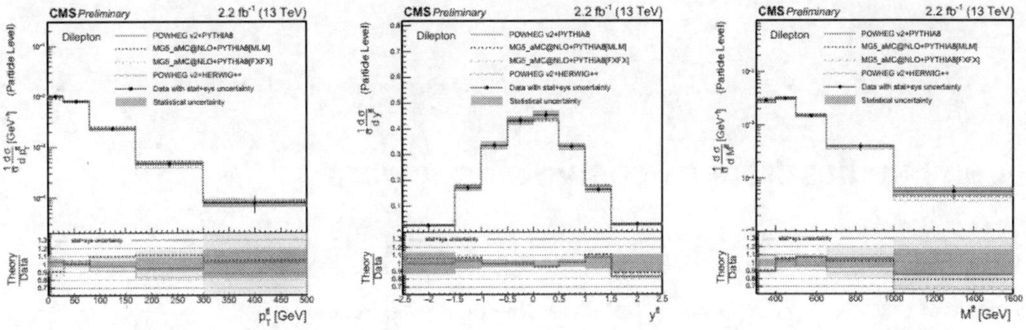

Figure 2. Normalized differential cross section as a function of $p_T^{t\bar{t}}$ (left), $y^{t\bar{t}}$ (center) and $M^{t\bar{t}}$ (right) for the top quarks or anti-quarks, measured at particle level in the visible phase space. The error bars on the data points indicate the total (combined statistical and systematic) uncertainties, while the dark shaded band shows the statistical uncertainty.

[4] J. Allison et al., IEEE Trans. Nucl. Sci. **53**, 270 (2006)

[5] CMS Physics Analysis Summary CMS-PAS-TOP-15-011 (2015), `https://cds.cern.ch/record/2110635`

[6] L. Sonnenschein, Phys. Rev. D **73**, 054015 (2006), [Erratum: Phys. Rev.D78,079902(2008)], `hep-ph/0603011`

[7] R.H. Dalitz, G.R. Goldstein, Phys. Lett. B **287**, 225 (1992)

[8] T. Adye, Tech. Rep. arXiv:1105.1160 (2011), `https://cds.cern.ch/record/1349242`

[9] G.D'Agostini., Nuclear Instruments and Methods in Physics Research A **362:487-498** (February 1995)

[10] G. D'Agostini, ArXiv (2010), `arXiv:1010.0632`

Muon bundles from extensive air showers

Maciej Rybczyński[1,a], Zbigniew Włodarczyk[1,b], and Grzegorz Wilk[2,c]

[1] *Institute of Physics, Jan Kochanowski University, 25-406 Kielce, Poland*
[2] *National Centre for Nuclear Research, Department of Fundamental Research, 00-681 Warsaw, Poland*

Abstract. In this talk we present a possible explanation of the presence of high muon multiplicity events registered recenty by CERN ALICE experiment in its dedicated cosmic ray run.

1 Introduction

A very important aspect of understanding of primary cosmic ray flux and its composition is a proper description of recent measurement done by the ALICE experiment at CERN LHC, in its dedicated cosmic ray run [1]. ALICE Collaboration registered the presence of large groups of muons produced in EAS by cosmic ray interactions in the upper atmosphere [1]. A special emphasis has been given to the study of high multiplicity events containing more than 100 reconstructed muons. Similar events have been studied in previous underground experiments such as ALEPH [2], DELPHI [3] and L3 [4] at CERN Large Electron-Positron Collider. While these experiments were able to reproduce the measured muon multiplicity distribution with Monte Carlo simulations at low and intermediate multiplicities, their simulations failed to describe the frequency of the highest multiplicity events. The muon multiplicity distribution measured by ALICE when compared with the fits obtained from CORSIKA simulations with proton or iron primary cosmic rays indicates that the expected rate of higher multiplicity muon events is sensitive to assumptions made about the dominant hadronic production mechanisms in air shower development. Although, assuming the presence of only the Fe nuclei in the primary flux it is roughly possible to describe the presence of the low and intermediate muon bundles, the same assumption used in the conventional models of EAS development completely fails to describe observed muon groups at high multiplicities. Surprisingly, the ALICE Collaboration shows in [1] only fits for muon multiplicities up to about 70, neglecting many events with measured muon multiplicities up to 270.

In this talk we discuss the hypothesis that muon bundles of extremely high multiplicity observed recently by ALICE detector can originate from small lumps of Strange Quark Matter (SQM) colliding with the atmosphere. We demonstrate that extremely large groups of muons can be very well described by a relatively minute (of the order of 10^{-5} of total primary flux) admixture of SQM of the same total energy. Our estimate of SQM flux do not contradict results obtained recently by the SLIM Collaboration [5].

[a] e-mail: maciej.rybczynski@ujk.edu.pl
[b] e-mail: zbigniew.wlodarczyk@ujk.edu.pl
[c] e-mail: Grzegorz.Wilk@ncbj.gov.pl

2 Strange Quark Matter in cosmic rays

Following [6, 7] it is fully sensible to search for SQM in cosmic ray experiments because the specific features of strangelets [8] allow them to penetrate deep into atmosphere [9–12]. The point is that there exists some critical size of the strangelet given by the critical value of its mass number, $A = A_{crit} \sim 300 - 400$, such that for $A > A_{crit}$ strangelets are absolutely stable against neutron emission. Below this limit strangelets decay rapidly evaporating neutrons.

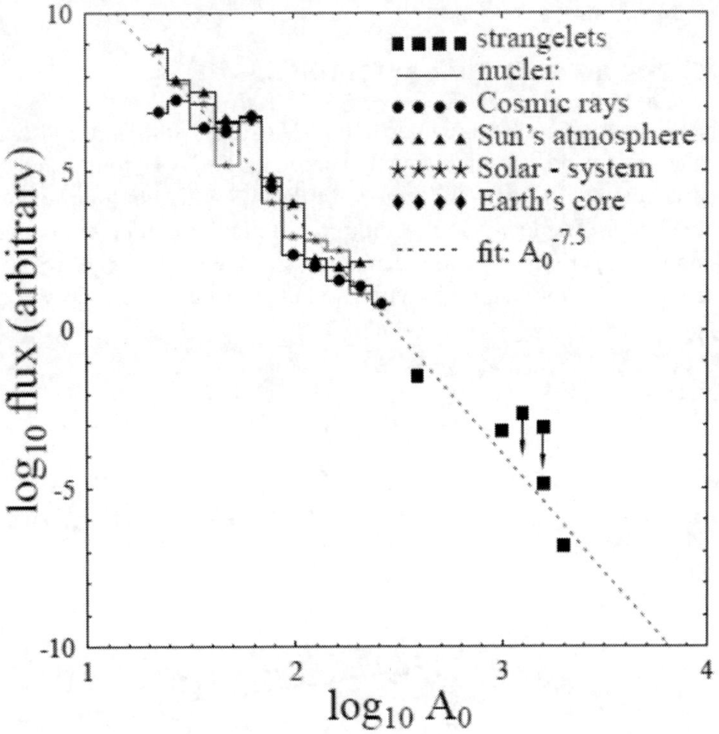

Figure 1. Comparison of the estimated mass spectrum $N(A_0)$ for strangelets with the known abundance of elements in the Universe [17].

The geometrical radii of strangelets turn out to be comparable to the radii of ordinary nuclei [6], i.e., their geometrical cross sections are similar to the normal nuclear ones. To account for their strong penetrability one has to accept that strangelets reaching deeply into atmosphere are formed in many successive interactions with air nuclei by the initially very heavy lumps of SQM entering the atmosphere and decreasing due to the collisions with air nuclei (until their A reaches the critical value A_{crit} [6]). Such scenario is fully consistent with all present experiments [6]. In this scenario interaction of strangelet with target nucleus involves all quarks of the target located in the geometrical intersection of the colliding air nucleus and strangelet.

There are several reports suggesting existence of direct candidates for SQM [13] (characterized mainly by their very small ratios of Z/A). All of them have mass numbers A near or slightly exceeding A_{crit}. Analysis of these candidates for SQM shows [6] that the abundance of strangelets in the primary cosmic ray flux is $F_S\ (A = A_{crit})\ /F_{tot} \sim 2.4 \cdot 10^{-5}$ at the same energy per particle. For normal flux of primary cosmic rays [14] the expected flux of strangelets is then equal to $F_S = 7 \cdot 10^{-6}$ m^{-2}h^{-1}sr^{-1} for the energy above 10 GeV per initial strangelet.

The experimental data mentioned before lead to the flux of strangelets which is consistent with the astrophysical limits and with the upper limits given experimentally [15]. It follows the $A^{-7.5}$ behaviour, which coincides with the behaviour of abundance of normal nuclei in the Universe, see Fig. 1 for comparison.

Chart of nuclides presented in Ref. [16] shows all known forms of stable matter. Between the heaviest atomic elements and neutron stars, which are giant nuclei, lies a vast, unpopulated nuclear desert. This void may actually be filled with strange quark matter.

3 High multiplicity muon bundles from SQM

In this study we have used the suitably modified SHOWERSIM [18] modular software. We performed Monte Carlo simulations for primary nuclei composed with 50% of protons and 50% of iron nuclei and for primary strangelets with mass A taken from the $A^{-7.5}$ distribution. In Fig. 2 the results of our simulations are shown. The lower and medium multiplicities can be reproduced by the ordinary nuclei. The extremely large groups of muons can be very well described by a relatively minute (of the order of 10^{-5} of total primary flux) admixture of SQM of the same total energy.

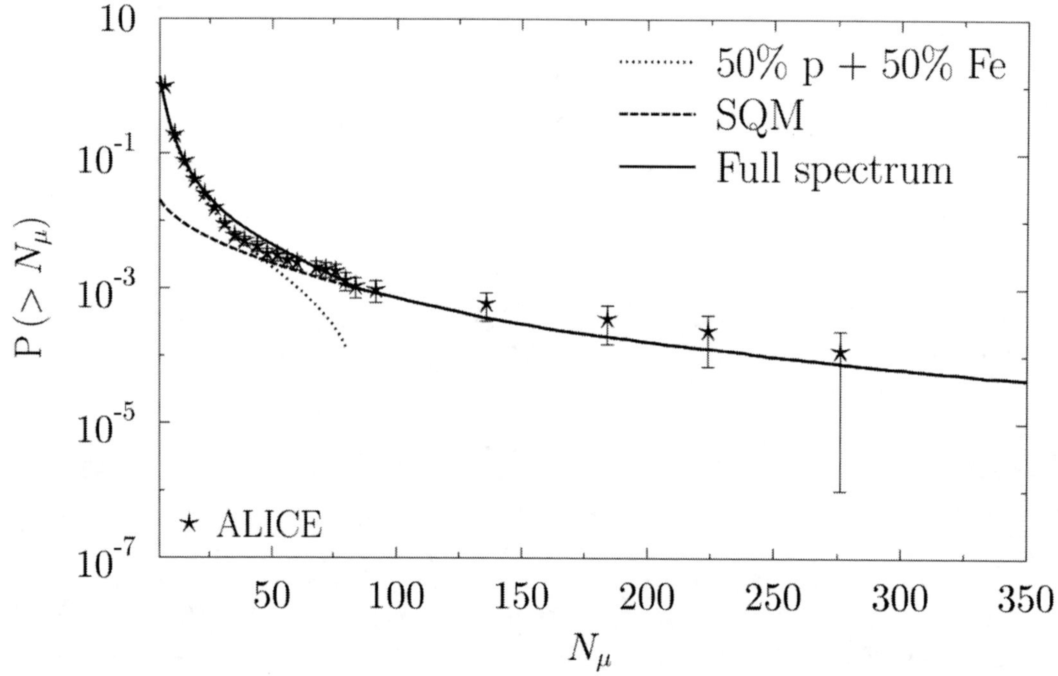

Figure 2. Integral multiplicity distribution of muons for the ALICE data [1] (stars). Monte Carlo simulations for primary nuclei composed with 50% of protons and 50% of iron nuclei (dotted line) and for primary strangelets with mass A taken from the $A^{-7.5}$ distribution (broken line). Full line shows the summary (calculated) distribution. Simulations were done using suitable modifications of the SHOWERSIM [18] modular software.

4 Conclusions

Our conclusions are as follows:

- Accelerator apparata can be suitable for cosmic-ray physics: LEP experiments were the pioneers on this topic. The LHC ALICE have interesting results, apart from the global physics studies used in model tuning of hadronic interactions.
- The low multiplicities of muon groups measured by the CERN ALICE experiment favor light nuclei as primaries, whereas medium multiplicities show behavior specific for heavier primaries.
- At high multiplicities of muon groups the common interaction models fail to describe muon bundles.
- A relatively small (of the order of 10^{-5} of total primary flux) admixture of SQM of the same total energy allows to reproduce the high muon multiplicity groups.
- Our estimate of SQM flux do not contradict the results obtained recently by the SLIM Collaboration [5].

Acknowledgements: Research supported by the Polish National Science Centre grant UMO-2015/18/M/ST2/00125

References

[1] J. Adam *et al.* [ALICE Collaboration], JCAP **1601**, no. 01, 032 (2016) [arXiv:1507.07577 [astro-ph.HE]].
[2] V. Avati *et al.*, Astropart. Phys. **19**, 513 (2003).
[3] J. Abdallah *et al.* [DELPHI Collaboration], Astropart. Phys. **28**, 273 (2007) [arXiv:0706.2561 [astro-ph].
[4] P. Achard *et al.* [L3 Collaboration], Phys. Lett. B **598**, 15 (2004) [hep-ex/0408114].
[5] Z. Sahnoun *et al.*, Radiat. Meas. **44**, 894 (2009) [arXiv:0812.3248 [hep-ex]].
[6] G. Wilk and Z. Wlodarczyk, Heavy Ion Phys. **4**, 395 (1996) [[Acta Phys. Hung. **4**, 395 (1996)].
[7] G. Wilk and Z. Wlodarczyk, J. Phys. G **23**, 2057 (1997).
[8] C. Alcock and E. Farhi, Phys. Rev. D **32**, 1273 (1985).
[9] G. Wilk and Z. Włodarczyk, Nucl. Phys. Proc. Suppl. **52B**, 215 (1997).
[10] E. Gładysz-Dziaduś and Z. Włodarczyk, J. Phys. G **23**, 2057 (1997).
[11] M. Rybczyński, Z. Włodarczyk and G. Wilk, Acta Phys. Polon. B **33**, 277 (2002).
[12] M. Rybczyński, Z. Włodarczyk and G. Wilk, Nucl. Phys. B (Proc. Suppl.) **151**, 341 (2006).
[13] T. Saito, Y. Hatano, Y. Fukada and H. Oda, Phys. Rev. Lett. **65**, 2094 (1990).
[14] T. Shibata, Nuovo Cim. C **19**, 713 (1996).
[15] P. B. Price, Phys. Rev. D **38**, 3813 (1988).
[16] H. J. Crawford and C. H. Greiner, Sci. Am. **270**, 58 (1994).
[17] M. Rybczyński, Z. Wlodarczyk and G. Wilk, Int. J. Mod. Phys. A **20**, 6724 (2005) [hep-ph/0410065].
[18] A. Wrotniak, SHOWERSIM/84, University of Maryland Preprint, 85 (1984).

J/ψ production in pPb collisions from CMS

Songkyo Lee on behalf of the CMS collaboration[1,a]

[1] *Korea University, 145 Anam-ro, Seongbuk-gu, Seoul, Republic of Korea*

Abstract. We report on the prompt and nonprompt J/ψ meson production from CMS, using 34.6 nb^{-1} of proton-lead (pPb) data recorded in 2013 at 5.02 TeV. The differential production cross sections are analyzed in the transverse momentum range $2 < p_T < 30$ GeV/c, and the center-of-mass (CM) rapidity region $-2.87 < y_{CM} < 1.93$. The ratio of J/ψ yields at forward (proton-going direction) and backward (Pb-going direction) is studied in several ranges of p_T, rapidity, and the event-activity regions.

1 Introduction

Charmonia represent an important tool for studying the properties of the de-confined medium created in ultra-relativistic heavy-ion collisions, the quark-gluon plasma (QGP). Due to the Debye screening of the heavy-quark potential in QGP, charmonium states are predicted to dissociate sequentially reflecting their binding energies [1]. Given the abundance of charm quarks produced at LHC energies, charmonia can also be produced at the hadronization stage via the recombination of initially uncorrelated charm and anticharm quarks [2]. On the other hand, the measurement of charmonium production in proton-nucleus collisions enables to investigate the so-called "cold nuclear matter" (CNM) effects, which provide new information to examine quantitatively the genuine hot-medium effects in nucleus-nucleus collisions. These effects include modifications of the nuclear parton distribution functions (nPDFs), initial or final state parton energy loss, and nuclear absorption [3]. For example, the depletion of gluon distributions in nuclei from those in a free proton can lead to the suppression of the charmonium yields at small values of x, the momentum fraction carried by a gluon. Such a process is known as "shadowing". The values of x explored in this analysis correspond to $10^{-4} < x < 10^{-2}$. Similar nuclear effects are also expected to affect the open-heavy-flavor production, which can be indirectly examined by the measurement of nonprompt J/ψ mesons coming from b hadron decays.

2 The CMS detector

A detailed description of the CMS detector can be found in Ref. [4]. Its central feature is a superconducting solenoid, providing a magnetic field of 3.8 T. Within the field are the silicon tracker, the lead-tungstate crystal electromagnetic calorimeter, and the brass/scintillator hadronic calorimeter. The

[a]e-mail: songkyo.lee@cern.ch

tracker consists of 66 million pixel and 10 million strip sensor elements to measure charged-particle trajectories in $|\eta| < 2.5$. Muons are measured within $|\eta| < 2.4$ in gas ionization detectors embedded in the steel return yoke, using the detection planes based on three technologies: drift Tubes, cathode strip chambers, and resistive plate chambers. The CMS apparatus also has extensive forward calorimetry, composed of two steel/quartz-fiber Cherenkov hadron forward calorimeters (HF), which cover $2.9 < |\eta| < 5.2$. These detectors are utilized for online event selection and the impact parameter characterization of the events in pPb collisions, where the impact parameter refers to the transverse distance between the two centers of the incoming projectiles.

3 Analysis procedure

The J/ψ mesons were measured via their dimuon decay channel. The invariant mass spectrum of $\mu^+\mu^-$ pairs was modelled by the sum of a Crystal Ball (CB) function and a Gaussian function for the J/ψ signal, and by an exponential function for the underlying continuum background. The prompt and nonprompt J/ψ mesons were separated using the pseudo-proper decay length, $l_{J/\psi} = L_{xy}\, m_{J/\psi}/p_T$, where L_{xy} is the transverse distance between the primary and secondary vertices in the laboratory frame. The invariant-mass spectrum and the $l_{J/\psi}$ distribution of $\mu^+\mu^-$ pairs were fitted sequentially in an extended unbinned maximum-likelihood fit [5].
Prompt and nonprompt J/ψ yields extracted from the fit were corrected for acceptances and efficiencies, which were evaluated from PYTHIA Monte Carlo (MC) simulation. Additional correction factors were applied to minimize the discrepancies between MC simulations and real detector responses, based on a data-driven technique, tag-and-probe [6].

4 Results

4.1 Prompt J/ψ

The rapidity dependence of nuclear effect is quantified using the forward-to-backward production ratio, $R_{FB}(p_T, y > 0) = [d^2\sigma(p_T, +y)/dp_T dy]/[d^2\sigma(p_T, -y)/dp_T dy]$, where the $d^2\sigma/dp_T dy$ is the J/ψ production cross section decayed into muon pairs in a given p_T and y bin, and forward regions ($+y$) are defined by the proton-going direction.

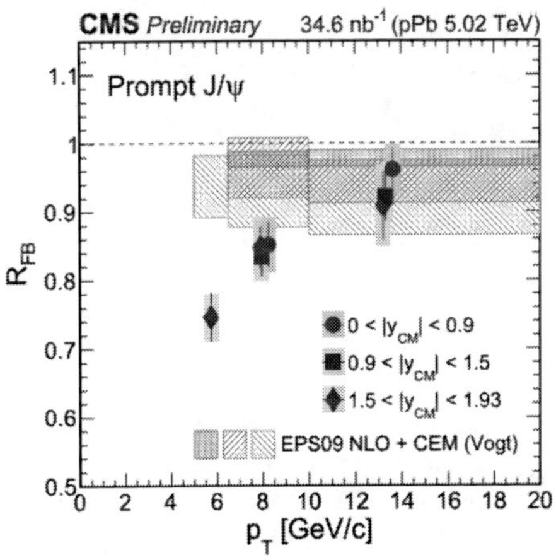

Figure 1. p_T dependence of prompt J/ψ R_{FB} in three rapidity ranges [5]. The shadowing prediction based on the EPS09 (NLO) parameterization is also shown [7]. The error bars show the statistical uncertainties and the shaded boxes represent the systematic uncertainties.

Figure 1 shows the p_T dependence of R_{FB} for prompt J/ψ mesons in three different rapidity ranges. The data points are plotted at the bin-averaged values. The R_{FB} is closer to unity at high p_T, and decreases at low p_T. The EPS09 nuclear parton distribution function, with next-to-leading order Color Evaporation Model (CEM) for the J/ψ production, is also shown by bands [7]. Data indicates lower R_{FB} values than the model prediction at $5 < p_T < 6.5$ GeV/c, suggesting hints for other nuclear effects beyond the single nPDF parameterization.

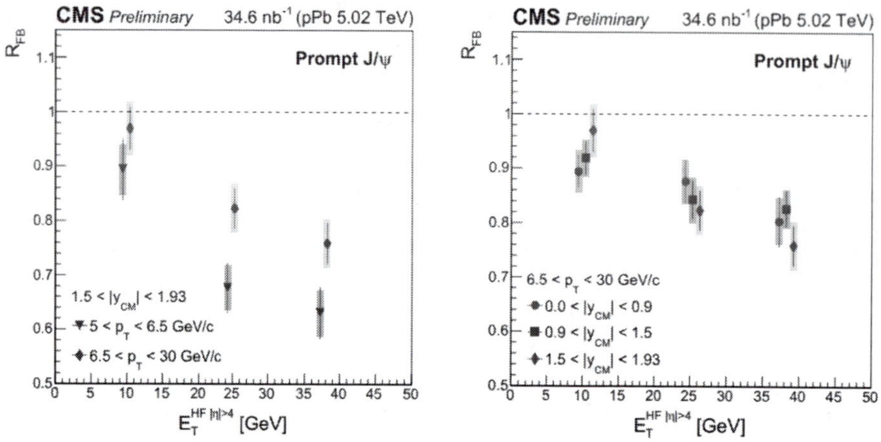

Figure 2. $E_T^{HF|\eta|>4}$ dependence of prompt J/ψ R_{FB} with two different p_T ranges in the most forward region (left), and with the same p_T range in three different rapidity intervals (right) [5]. The error bars show the statistical uncertainties and the shaded boxes represent the quadratic sum of systematic uncertainties.

The R_{FB} is further analyzed in terms of an event activity variable $E_T^{HF|\eta|>4}$, the transverse energy deposited in the hadron forward calorimeter in $4 < |\eta| < 5.2$, as shown in Fig. 2. The clear decreasing tendency of the R_{FB} with increasing $E_T^{HF|\eta|>4}$ has been observed over the whole kinematic interval studied.

4.2 Nonprompt J/ψ

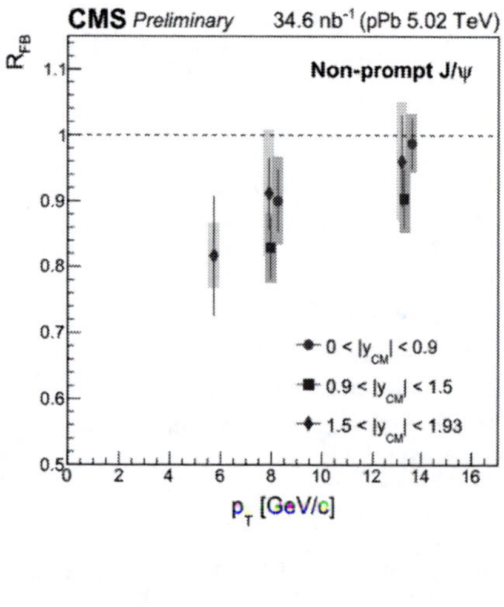

Figure 3. p_T dependence of nonprompt J/ψ R_{FB} in three rapidity ranges [5]. The error bars show the statistical uncertainties and the shaded boxes represent the systematic uncertainties.

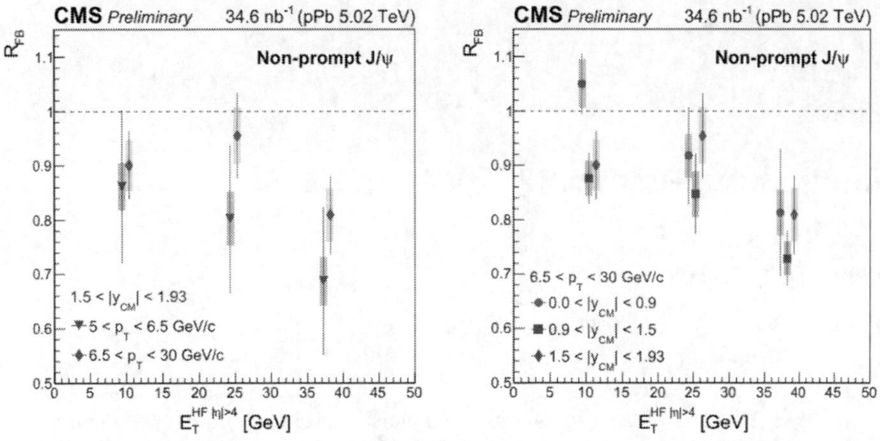

Figure 4. $E_T^{HF|\eta|>4}$ dependence of nonprompt J/ψ $R_{\rm FB}$ with two different $p_{\rm T}$ ranges in the most forward region (left), and with the same $p_{\rm T}$ range in three rapidity intervals (right) [5]. The error bars show the statistical uncertainties and the shaded boxes represent the systematic uncertainties.

The $R_{\rm FB}$ as functions of $p_{\rm T}$ and $E_{\rm T}^{\rm HF|\eta|>4}$ for nonprompt J/ψ mesons are also investigated and shown in Figs. 3–4. The nonprompt J/ψ result exhibits similar features as that of prompt J/ψ mesons, while the significance is less within larger uncertainties.

5 Summary

The measurements of prompt and nonprompt J/ψ production in pPb collisions at $\sqrt{s_{\rm NN}} = 5.02$ TeV are presented. The forward-to-backward production ratio $R_{\rm FB}$ is observed to be smaller than unity at $p_{\rm T} \lesssim 10{\rm GeV}/c$, and compatible with unity for higher $p_{\rm T}$. The $R_{\rm FB}$ has been studied as a function of the event activity, and found to be more asymmetric with increasing $E_{\rm T}^{\rm HF|\eta|>4}$ both for prompt and nonprompt J/ψ mesons. The results hint to nuclear effects present in pPb collisions.

References

[1] T. Matsui, H. Satz, Phys. Lett. B **178**, 416 (1986)
[2] X. Zhao, R. Rapp, Nucl. Phys. A **859**, 114 (2011), 1102.2194
[3] A. Andronic et al. (2015), 1506.03981
[4] CMS Collaboration, JINST **3**, S08004 (2008)
[5] CMS Collaboration, CMS Physics Analysis Summary CMS-PAS-HIN-14-009, CERN (2015), http://cds.cern.ch/record/2044699
[6] CMS Collaboration, JHEP **1101**, 080 (2011), 1012.2466
[7] R. Vogt, Phys. Rev. C **92**, 034909 (2015), 1507.04418

 DOI: 10.1051/epjconf/201714108014

B production in pPb at 5.02 TeV from CMS

Kisoo Lee (for the CMS collaboration)[1]

[1] *Korea University, Seoul 136-701, Republic of Korea*

Abstract. Hadrons with heavy quarks are promising probes to investigate the detailed properties of hot and dense medium generated by heavy-ion collisions at collider energies. Since heavy quarks are sensitive to the transport properties of the medium, the energy-loss pattern of them is expected to be quite different from that of light quarks in a strongly-interacting matter. On the other hand, in order to elicit the actual effects caused by the hot and dense medium, it is necessary to understand the cold nuclear matter effect in pA collisions. For example, the pPb data is expected to provide a baseline for the study of the b-quark energy loss in medium produced by PbPb collisions. Therefore, the CMS Collaboration at the Large Hadron Collider (LHC) has analyzed the production cross sections of B^+, B^0, B_s^0 mesons in pPb collisions as a function of rapidity and the transverse momentum at the nucleon-nucleon center-of-mass energy of 5.02 TeV. In addition, the nuclear modification factors of the B mesons have been constructed using the theoretical pp reference spectra estimated by the perturbative Quantum ChromoDynamics (pQCD) model. Finally CMS collected pp and PbPb collisions at the center-of-mass energy of 5.02 TeV at the LHC in the end of 2015. Analysis of B production in pp and PbPb is ongoing.

1 Introduction

In the heavy-ion collisions, heavy quarks are useful probe to study hot and dense QCD medium because they are produced in the early stage of collision and interact with medium. Besides, there may be effects due to nuclear modifications to the parton distributions, known as the cold nuclear matter effect, such as nuclear modifications to the parton distributions. The pPb collisions are considered as a useful system to study the cold nuclear matter effect because they are not expected to form an extended hot medium. With large cross section of b quark production at the LHC, the CMS collaboration has measured production cross section of b quarks. The CMS collaboration studied the b-hadron production in PbPb collisions at $\sqrt{s_{NN}} = 2.76$ TeV by using inclusive non-prompt J/ψ in the 2010 and 2011 data sets [1]. For the first time, CMS collaboration reconstructed exclusive B mesons via the combination of charged particle tracks to the reconstructed J/ψ in the pPb collisions at $\sqrt{s_{NN}}$ = 5.02 TeV recorded in 2013.

The B meson species analyzed for this study are B^+, B^0, B_s^0 that consist of u (or ū), d (or d̄), and s (or s̄) quark respectively, in addition to the b (or b̄) quark. From many decay modes of each B meson, the specific channel decays to J/ψ and a strange meson is chosen for this analysis.

$$
\begin{aligned}
B^{+} &\rightarrow J/\psi + K^{\pm} \rightarrow \mu^{+} + \mu^{-} + K^{\pm} \\
B^{0} &\rightarrow J/\psi + K_{0}^{*}(892) \rightarrow \mu^{+} + \mu^{-} + K^{+} + \pi^{-} \\
B_{s}^{0} &\rightarrow J/\psi + \phi \rightarrow \mu^{+} + \mu^{-} + K^{+} + K^{-}
\end{aligned} \tag{1}
$$

As this analysis does not separate corresponding antiparticle, B^{+} and B^{-}, B^{0} and $\bar{B^{0}}$, B_{s}^{0} and $\bar{B_{s}^{0}}$ are denoted by B^{+}, B^{0}, B_{s}^{0}. For the final cross section values, the combined results are divided by two to obtain an average.

The CMS detector has highly efficient muon detection system and the high-resolution silicon tracker which is crucial to reconstruct B meson decays. The muon detector system, consisting of Cathode Strip Chambers (CSC), Drift Tubes (DT) and Resistive Plate chambers (RPC), is dedicated to identify and track the muons. The silicon tracker is located in the 3.8 T magnetic field generated by the superconducting solenoid, and consists of 1440 silicon pixel and 15148 silicon strip detector modules with a typical momentum resolution of 1.5 % in p_T for nonisolated particles of $1 < p_T < 10$ GeV/c and $|\eta| < 1.4$, where η is the pseudo-rapidity in the laboratory frame. The detailed description of the CMS detector elements are given in Ref. [2].

For the pPb run lead nuclei of 1.58 TeV per nucleon collide with protons of 4 TeV. The center of mass energy is $\sqrt{s_{NN}} = 5.02$ TeV. Because of the asymmetric beam energies, all produced particles from the collisions are detected with rapidity shift of 0.465 towards the proton-going side in the laboratory frame. The data sample used in this analysis corresponds to an integrated recorded luminosity of (34.6 ± 1.2) nb^{-1}.

2 Analysis

The events used in this analysis are collected with a singe-muon trigger that requires at least one muon with $p_T > 3$ GeV/c. The electromagnetic and beam-gas collision events are triggered by methods described in Ref.[3] The B reconstruction starts with the muons that are reconstructed from the inner tracker system and the outer muon system. The muons must satisfy the muon selection criteria including the kinematic limits imposed by the geometry of the detectors: $p_T > 3.3$ GeV/c for $|\eta| < 1.3$, $p > 2.9$ GeV/c for $1.3 < |\eta| < 2.2$ and $p_T > 0.8$ GeV/c for $2.2 < |\eta| < 2.4$. The opposite-signed muon pairs used for this analysis are within ±300 MeV/c^2 from its nominal J/ψ mass found in the Particle Data Group[4]. The charged hadrons are reconstructed by the inner tracker system within $|\eta| < 2.4$ in the laboratory frame. To reconstruct B^{+}, the kaon mass is assigned to each charged particle and, then, the invariant-mass distribution of J/ψ plus kaon candidates are produced. The B^{0} and B_{s}^{0} analyses are more complicated than the charged B's as they decay into the intermediate states which can be reconstructed by the two charged hadrons. Therefore, for the reconstructions of B^{0} and B_{s}^{0}, the pion and kaon masses are assigned to any pair of the charged particles.

In order to maximize the statistical significance of the B meson signals, selection is optimized for each meson species using a multivariate technique. The χ^2 of the secondary vertex fit, the distance between the primary and the secondary vertices (normalized by its uncertainty) and the cosine of the angle between the momentum vector of B and the position vector of the secondary vertex measured from the primary vertex in the transverse plane, etc. For the B^{0} and B_{s}^{0} reconstructions, additional constraints on the respective proper mass window of the invariant mass for the intermediate meson states are re-imposed. In the case of multiple B mesons being reconstructed in one event, the candidate with the best χ^2 of the secondary vertex is chosen to prevent the possible double counting.(especially, for the B^{0} case.)

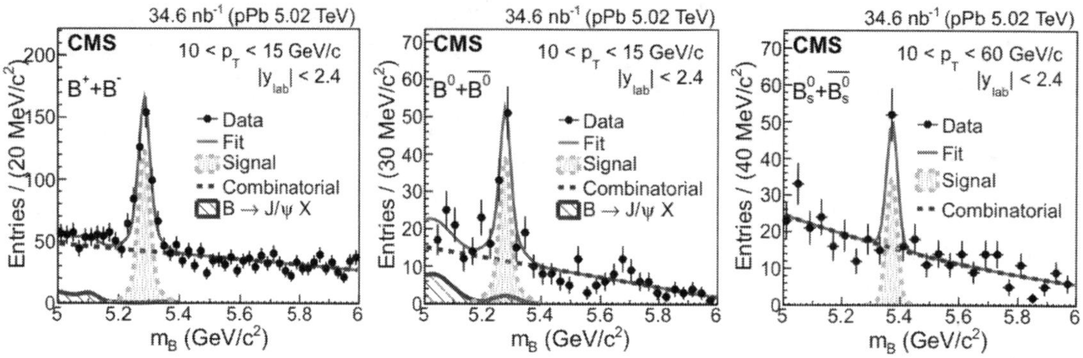

Figure 1. Invariant-mass distributions of B^++B^-(left), B^0+$\bar{B}^0$(center), and B_s^0+$\bar{B}_s^0$(right) candidates in the transverse momentum ranges of 10-15, 10-15 and 10-60 GeV/c respectively. The orange-dotted area represent the fitted B signal distributions and the green-hatched areas represent the peaking backgrounds from other B meson species. The blue-dotted lines represent the combinatorial backgrounds.

3 Results

All B candidates are reconstructed in $10 < p_T < 60$ GeV/c and $|y_{CM}| < 1.93$ where y_{CM} is the rapidity in center-of-mass frame. Fig. 1 shows the invariant-mass distribution of B^+, B^0 and B_s^0 in their respective lowest p_T bin. The raw yields of B^+, B^0 and B_s^0 are extracted using a binned maximum likelihood fit to the B meson invariant-mass distributions in the mass range $5 < m_B < 6$ GeV/c^2. The combinatorial backgrounds, fitted by first or second order polynomial functions, are dominated by inclusive J/ψ's and charged hadrons associated with different mother particles. The so-called peaking background, shown by the green hatched areas in Fig. 1, is caused by the mis-reconstructed B mesons. In order to understand the signal and background distribution, Monte Carlo simulations have been performed. The particles in the Monte Carlo samples are generated by the PYTHIA event generator [5] and decayed by the EvtGen particle decay simulation package [6]. The initial and final state radiations are controlled by the PHOTOS package [7]. Furthermore, the signal particles are embedded in the HIJING minimum-bias events for pPb collisions to mimic more realistic environment [8]. After the proper background subtraction, several functions are tried to fit the signal peak shapes for each channel. Finally, sum of two Gaussian functions with same mean is chosen as the signal function.

The differential cross section for each B species has been obtained by

$$\left.\frac{d\sigma^B}{dp_T}\right|_{|y_{CM}|<1.93} = \frac{1}{2}\frac{1}{\Delta p_T}\frac{N^B\big|_{|y_{CM}|<1.93}}{(\alpha\times\epsilon)\cdot BR\cdot\mathcal{L}} \tag{2}$$

where σ^B is cross-section of B, Δp_T is the width of the p_T bin, N^B is number of B,α is the geometrical acceptance, ϵ is the efficiency estimated by the Monte Carlo simulation, BR is the branching ratio of the decay mode, and $\mathcal{L}$ is the integrated luminosity. Note that the additional factor 1/2 is multiplied to take into account the charge conjugate of each B meson species such as B^- for B^+. For this analysis, the Fixed Order plus Next-to-Leading Logarithm(FONLL) calculations [9, 10] are used as the pp reference because the experimental pp data are not available at 5.02 TeV in the 2013 dataset. Note that FONLL has successfully reproduced the CDF data at 1.96 TeV and the CMS/ATLAS data at 7 TeV within the experimental and theoretical uncertainties [11, 12]. Using the analyzed differential

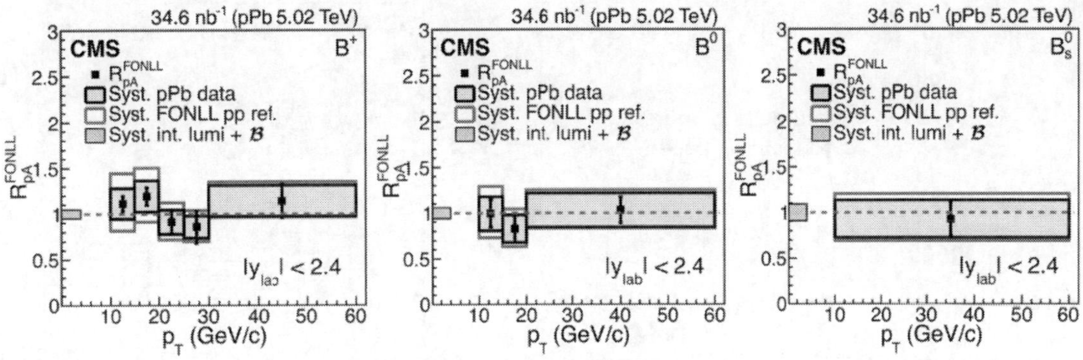

Figure 2. The nuclear modification factors $R^{FONLL}_{pA}(p_T)$ of B^+(left), B^0(center), B^0_s(right) measured in pPb collisions at $\sqrt{s_{NN}}$ = 5.02 TeV. The statistical and systematic uncertainties on the pPb data are shown as bars and yellow boxes around the data points respectively. The systematic uncertainties from the FONLL predictions are plotted separately as open blue boxes. The global systematic uncertainties are shown as full grey boxes at unity, and are not included in the data points.

cross section data and the FONLL calculations, the nuclear modification factors are calculated by

$$R^{FONLL}_{pA}(p_T) = \frac{(d\sigma/dp_T)_{pPb}}{A \times (d\sigma/dp_T)_{pp}} \tag{3}$$

where A is the number of nucleons in Pb. The results on R^{FONLL}_{pA} for B^+, B^0 andB^0_s are shown in Fig. 2. Within the systematic and statistical uncertainties, R^{FONLL}_{pA} of all three particle species are consistent with unity and, thus, significant cold nuclear matter effects are not observed. The nuclear modification factor as a function of rapidity in the center-of-mass frame is calculated only for the B^+ which has the best statistical precision among the three analyzed particle species. In the Fig. 3 the left panel is differential cross section and right panel is nuclear modification factor of the B^+. Although the most backward data point toward the Pb-going side is higher than the others, the experimental uncertainties are large and no rapidity dependence can be claimed.

4 Conclusions

For the first time, the B^+, B^0, and B^0_s mesons are exclusively reconstructed for $10 < p_T < 60$ GeV/c and $|y_{CM}|$ <1.93 in pPb collisions at $\sqrt{s_{NN}}$ = 5.02 TeV by the CMS Collaboration at the LHC. The range of transverse momentum covered is 10 to 60 GeV/c, for the rapidity range (evaluated in the center-of-mass frame) of $|y_{CM}| < 1.93$. Statistically significant peaks are observed for all three B-meson species in the invariant-mass distributions. The nuclear modification factors for all three B-meson species are estimated using FONLL calculations as pp references. The nuclear modification factors are consistent with unity within the uncertainties. Presently, the CMS Collaboration is analyzing the data recorded in 2015 with the 5.02 TeV PbPb collision. The present results will be baseline for the B-meson R_{AA}.

References

[1] CMS Collaboration, J. High Energy Phys. 05 (2012) 063.

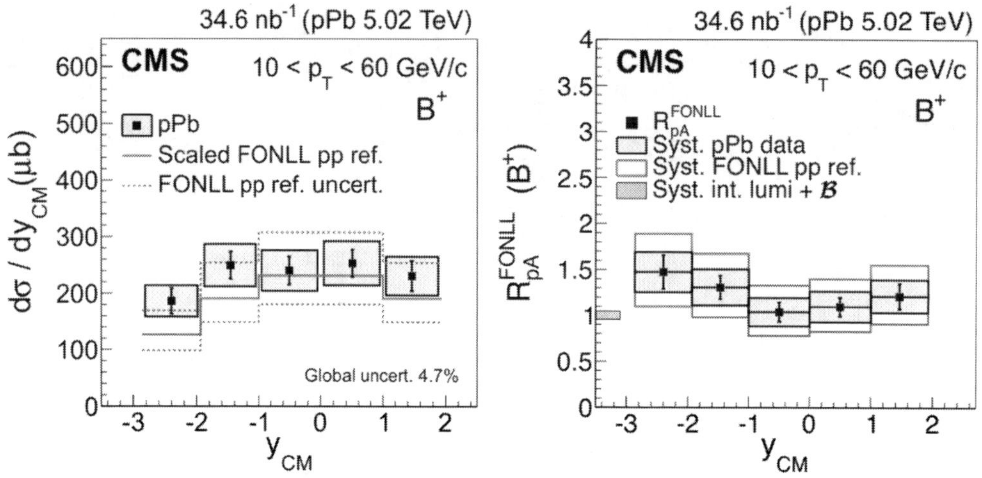

Figure 3. (left) The y_{CM}-differential production cross section of B^+ measured in pPb collisions at $\sqrt{s_{NN}} = 5.02$ TeV. Vertical bars (the boxes) correspond to statistical (systematic) uncertainties The listed global systematic uncertainty is not included in the data points. The result is not included in the data points. The result is compared to a FONLL calculation represented by a continuous histogram. The dashed histograms represent the theoretical uncertainties for the FONLL reference. (right) The nuclear modification factor $R_{pA}^{FONLL}(y_{CM})$ of B^+ as a function of y_{CM}. The statistical and systematic uncertainties on the pPb data are shown as bars and yellow boxes around the data points, respectively. The systematic uncertainty from the FONLL reference is plotted separately as open blue boxes. The global systematic uncertainty is shown as a full grey box at unity, and is not included in the data points.

[2] CMS Collaboration, J Instruments. 3 (2008) S08004.
[3] CMS Collaboration, Eur. Phys. J. C 74 (2014) 2951.
[4] Particle Data Group Collaboration, Phys. Rev. D 86 (2012) 010001.
[5] T. Sjöstrand, S. Mrenna, P. Skands, J. High Energy Phys. 05 (2006) 026.
[6] D.J. Lange, Nucl. Instrum. Methods Phys. Res., Sect. A, Accel. Spectrom. Detect. Assoc. Equip. 462 (2001) 152.
[7] E. Barberio, B. van Eijk, Z. Was, Comput. Phys. Commun. 66 (1991) 115.
[8] X.-N. Wang, M. Gyulassy, Phys. Rev. D 44 (1991) 3501.
[9] M. Cacciari, S. Frixione, P. Nason, J. High Energy Phys. 0103 (2001) 006.
[10] M. Cacciari, M. Greco, P. Nason, J. High Energy Phys. 9805 (1998) 007.
[11] CDF Collaboration, Phys. Rev. D 75 (2007) 012010.
[12] ATLAS Collaboration, J. High Energy Phys. 10 (2013) 042.

EPJ Web of Conferences **141**, 08015 (2017) DOI: 10.1051/epjconf/201714108015
ISMD 2016

Searching for charged Higgs bosons with boosted top and boosted bottom jets

Zack Sullivan[1,a] and Keith Pedersen[1]

[1] *Department of Physics, Illinois Institute of Technology, Chicago, Illinois 60616-3793, USA*

Abstract. At moderate values of $\tan\beta$, a supersymmetric charged Higgs boson $H^{\pm}$ is difficult to find due its small cross section and large backgrounds. Using realistic boosted tagging rates, we present preliminary predictions for the reach for TeV-scale charged Higgs bosons at 14 TeV and 100 TeV colliders in top-Higgs associated production. We conclude that moderate values of $\tan\beta$ will be possible to probe at a 100 TeV collider.

1 Introduction

After the discovery of a 126 GeV Higgs-like boson at the Large Hadron Collider at CERN, the emphasis has shifted to whether this is *the* Higgs boson of the Standard Model, or if it is merely one degree of freedom in a larger model like supersymmetry (SUSY). SUSY, like other two Higgs-doublet models, contains additional neutral (H^0, A^0) and charged ($H^{\pm}$) Higgs states. Experiment already constrains SUSY models to be in the "alignment limit," where the masses of these new Higgs states are nearly degenerate. Hence, the $H^{\pm}$ bosons couple almost exclusively to third generation quarks.

Given the strength of the $tbH^{\pm}$ Yukawa coupling, the dominant production mode for charged Higgs bosons at the LHC is in association with a top quark, where the Higgs decays to a boosted top jet and boosted bottom jet final state. Last winter, excitement was generated by a claim [1] that the "wedge region" in $\tan\beta$ ($\tan\beta \sim 6$ where the h^0 shares equal coupling to top and bottom at leading order) could be explored up to 2 TeV in $H^{\pm}$ mass at a 14 TeV LHC through the channel $tbH^{\pm} \to tb(tb)$. On the other hand, a previous paper [2] found that even a mass of 500 GeV could not be probed at the LHC. Here we utilize our new boosted bottom jet tag [3, 4] to examine whether the $\sim$ 2 TeV limit can be reached in the SUSY wedge region at the LHC or at a future 100 TeV pp collider.

2 Analysis

Charged Higgs boson-top quark associated production has two distinguishable final states (ttb and $ttbb$ as depicted in Fig. 1). The dominant measurable cross section is ttb, where the top and bottom quark from the Higgs boson hadronize into boosted jets, and the associated top quark decays to a lepton plus missing energy and a low energy b jet. Technically a higher-order QCD correction, the $ttbb$ final state adds an additional low energy b jet, and 65% to the measurable cross section. Since the measurement is signal constrained, we consider the $ttb + X$ inclusive measurement.

[a] e-mail: Zack.Sullivan@IIT.edu

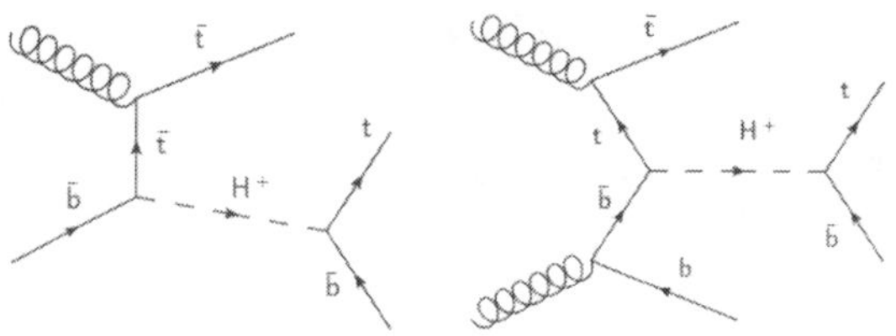

Figure 1. Feynman diagrams for $H^+\bar{t}$-associated production. (Left) The dominant cross section $g\bar{b} \rightarrow \bar{t}H^+ \rightarrow \bar{t}(t\bar{b})$ has a boosted-top jet, a boosted-bottom jet, and a low energy top quark. (Right) Part of the higher-order correction $gg \rightarrow \bar{t}bH^+ \rightarrow \bar{t}b(t\bar{b})$ adds a taggable low energy b jet.

We model the signal and backgrounds using MadEvent 5 [5] and CTEQ 14 PDFs [6], perform showering with PYTHIA 8 [7, 8], and use the DELPHES 3 [9] detector simulation. The dominant backgrounds to this process are from boosted jets faking top or bottom jets in Standard Model $t\bar{t}j + X$ production. Direct $t\bar{t}b\bar{b}/t\bar{t}c\bar{c}$ increase the background by 20%. In order to correctly model the fake rates we use the top tagging (and fake) rates provided by the CMS Collaboration [10], and the μ_x boosted b algorithm [3, 4] with code provided by Ref. [11].

The search strategy for a charged Higgs boson consists of reconstructing the boosted top and boosted bottom jet invariant mass $M_{H^\pm}$, and looking for a peak above background in a window $[0.9, 1.15] \times M_{H^\pm}$. We first reconstruct the boosted top with a Cambridge-Aachen algorithm and $R = 0.8$, and all other jet candidates use anti-k_T jets with $R = 0.4$ in FastJet [12]. We require both boosted jets to have $p_T > 350$ GeV, and other jets to have $p_{Tj} > 20(40)$ GeV and $|\eta| < 2.1(3)$ at a 14(100) TeV pp collider. We suppress backgrounds by identifying exactly one isolated lepton with $p_{Tl} > 15(25)$ GeV, and at least one low-energy b tag that satisfies 70 GeV $< m_{bl} <$ 180 GeV; we cannot fully reconstruct the low energy top quark because the missing energy is poorly constrained in this system.

After all cuts, the signal to background is 1:2(~1:20) at 14(100) TeV. The 95% confidence level (C.L.) exclusion reach at leading order in QCD is shown for both collider energies in Fig. 2. Unfortunately, the production cross section a moderate $\tan\beta$ (the "wedge region") is simply too small to observe at 14 TeV for masses of the charged Higgs boson above 1 TeV. This result looks like a continuation of Fig. 20 of Ref. [2] (which stops at 1 TeV), and hence a charged Higgs boson is very unlikely to be observable at the LHC. In contrast, at a 100 TeV collider it should be possible to exclude charged Higgs bosons up to 2 TeV for all values of $\tan\beta$, and have significant reach up to 6 TeV.

3 Conclusions

At ISMD this year we presented the results of an analysis to find the reach for TeV-scale charged Higgs bosons at the LHC and at a future 100 TeV collider. Given current measurements of the lightest neutral Higgs boson, a charged Higgs boson would be produced in association with a low energy top quark, and the charged Higgs boson will decay to a boosted top and boosted bottom quark. Despite promising hints [1] that t–$H^\pm$ associated production could be identified at the LHC for moderate values of $\tan\beta = 1$–10, our more complete analysis suggests that the LHC will have very little sensitivity to charged Higgs bosons with mass above 1 TeV. However, we find strong motivation for a 100 TeV collider in that charged Higgs bosons up to 6 TeV can probed for nearly all values of $\tan\beta$.

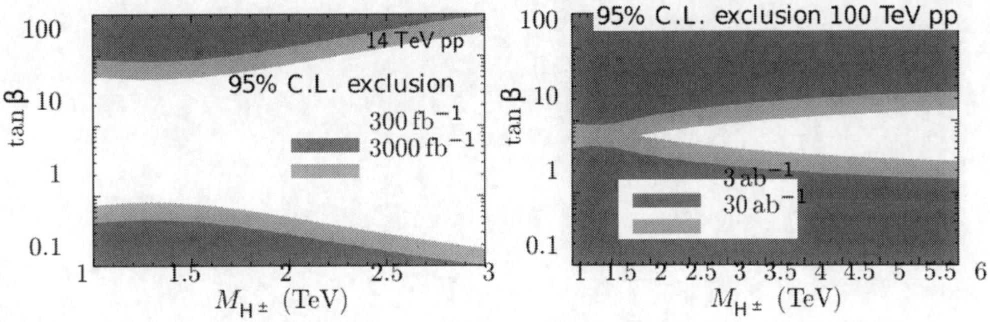

Figure 2. 95% confidence level exclusion limits obtainable at a (left) 14 TeV LHC or (right) 100 TeV pp collider, for $\tan\beta$ vs. charged Higgs boson $H^{\pm}$ mass.

Acknowledgments

Z.S. would like to thank the ISMD organizers for an exceptional symposium and environment for lively discussions. This work was supported by the U.S. Department of Energy under award No. DE-SC0008347.

References

[1] J. Hajer, Y. Y. Li, T. Liu, and J. F. H. Shiu, JHEP **1511**, 124 (2015) [arXiv:1504.07617 [hep-ph]].
[2] N. Craig, F. D'Eramo, P. Draper, S. Thomas, and H. Zhang, JHEP **1506**, 137 (2015) [arXiv:1504.04630 [hep-ph]].
[3] Zack Sullivan and Keith Pedersen, "Flavor tagging TeV jets for physics beyond the Standard Model," in *XLVth International Symposium on Multiparticle Dynamics (ISMD 2015)*, edited by T. Barillari, S. Bethke, S. Kluth, and S. Menke, EPJ Web of Conferences 120, 03001 (2016).
[4] K. Pedersen and Z. Sullivan, Phys. Rev. D **93**, no. 1, 014014 (2016) [arXiv:1511.05990 [hep-ph]].
[5] J. Alwall *et al.*, J. High Energy Phys. **07**, 079 (2014) [arXiv:1405.0301 [hep-ph]].
[6] S. Dulat *et al.*, arXiv:1506.07443 [hep-ph].
[7] T. Sjostrand, S. Mrenna, and P.Z. Skands, J. High Energy Phys. **05**, 026 (2006) [hep-ph/0603175].
[8] T. Sjostrand, S. Mrenna, and P.Z. Skands, Comput. Phys. Commun. **178**, 852 (2008) [arXiv:0710.3820 [hep-ph]].
[9] J. de Favereau *et al.* (DELPHES 3 Collaboration), J. High Energy Phys. **02**, 057 (2014) [arXiv:1307.6346 [hep-ex]].
[10] CMS Collaboration, CMS-PAS-JME-09-001; CMS-PAS-JME-13-007.
[11] K. Pedersen, https://github.com/keith-pedersen/delphes/tree/MuXboostedBTagging.
[12] M. Cacciari, G. P. Salam, and G. Soyez, Eur. Phys. J. C **72**, 1896 (2012) [arXiv:1111.6097 [hep-ph]].

AUTHOR INDEX